United States Nuclear Regulatory Commission
Protecting People and the Environment

NUREG-1947

Final Supplemental Environmental Impact Statement for Combined Licenses (COLs) for Vogtle Electric Generating Plant Units 3 and 4

Final Report

Manuscript Completed: March 2011
Date Published: March 2011

Office of New Reactors

Abstract

This supplemental environmental impact statement (SEIS) documents the U.S. Nuclear Regulatory Commission (NRC) staff's analysis and conclusions regarding the environmental impacts of constructing and operating two new nuclear units (Units 3 and 4) at the Vogtle Electric Generating Plant (VEGP) site near Waynesboro, Georgia, and the mitigation measures available for reducing or avoiding adverse environmental impacts.

On August 26, 2009, NRC issued Early Site Permit (ESP)-004 to Southern Nuclear Operating Company, Inc. (Southern) and several co-applicants (i.e., Georgia Power Company, Oglethorpe Power Corporation, Municipal Electric Authority of Georgia, and the City of Dalton, Georgia) for the VEGP ESP site (the site of the proposed Units 3 and 4). An ESP is an NRC approval of a site as suitable for construction and operation of one or more new nuclear units. As requested in the ESP application, the VEGP ESP also included a Limited Work Authorization (LWA) that authorized certain limited construction activities at the site in accordance with Title 10 of the Code of Federal Regulations (CFR) Subparts 50.10 and 52.24(c). In response to subsequent license amendment applications from Southern relating to the activities authorized by the ESP LWA, the NRC issued three amendments to the ESP in May, June, and July 2010, respectively. These amendments authorized Southern to use Category-1 and Category-2 backfill material from additional onsite sources and to use engineered granular backfill over the side slopes of the Units 3 and 4 excavations.

On March 31, 2008, Southern (on behalf of itself and its four co-applicants) submitted an application for combined licenses (COLs) for two new units at the VEGP site, referencing the VEGP ESP. A COL is a Commission approval for the construction and operation of one or more nuclear power facilities. Southern subsequently updated its COL application to reference the issued ESP-004.

For a COL application that references an ESP, the NRC staff, pursuant to 10 CFR 51.75(c), prepares a supplement to the ESP EIS in accordance with 10 CFR 51.92(e). NRC regulations related to the environmental review of COL applications are in 10 CFR Part 51 and 10 CFR Part 52, Subpart C. Pursuant to NRC regulations in 10 CFR 51.50(c)(1), a COL applicant referencing an ESP need not submit information or analyses regarding environmental issues that were resolved in the ESP EIS, except to the extent the COL applicant has identified new and significant information regarding such issues. In addition, pursuant to 10 CFR 52.39, matters resolved in the ESP proceedings are considered to be resolved in any subsequent proceedings, absent identification of new and significant information.

In October 2009, Southern supplemented its COL application to include a second request for an LWA. The second LWA, in accordance with 10 CFR 50.10 (d), would authorize installation of reinforcing steel, sumps, drain lines, and other embedded items along with placement of concrete for the nuclear island foundation base slab.

After considering the environmental aspects of the proposed action, the NRC staff's recommendation to the Commission is that the COLs and LWA be issued. This recommendation is based on (1) the application, including the environmental report and responses to staff requests for additional information, submitted by Southern; (2) the staff's review conducted for the ESP application and documented in the ESP EIS; (3) the staff's review conducted for the ESP license amendments as documented in the staff's Environmental Assessments; (4) consultation with Federal, State, Tribal, and local agencies; (5) the staff's own independent review of potential new and significant information available since preparation and publication of the ESP EIS; and (6) the assessments summarized in this SEIS, including the potential mitigation measures identified and consideration of public comments received on the draft SEIS. The staff's evaluation of the safety and security aspects of the proposed action will be addressed in the staff's Safety Evaluation Report.

Paperwork Reduction Act Statement

This NUREG references information collection requirements that are subject to the Paperwork Reduction Act of 1995 (44 U.S.C. 3501 et seq.). These information collections were approved by the Office of Management and Budget, approval numbers 3150-0014; 3150-0011; 3150-0021; 3150-0151; and 3150-0093.

Public Protection Notification

The NRC may not conduct or sponsor, and a person is not required to respond to, a request for information or an information collection requirement unless the requesting documents displays a currently valid OMB control number.

Contents

Abstract iii
Executive Summary xi
Abbreviations/Acronyms xv
1.0 Introduction 1-1
    1.1 Background 1-1
        1.1.1 COL Application and Review 1-2
        1.1.2 Concurrent Reviews 1-4
    1.2 The Proposed Federal Action 1-4
    1.3 The Purpose and Need for the Proposed Action 1-5
    1.4 Alternatives to the Proposed Action 1-6
    1.5 Compliance and Consultations 1-6
    1.6 New and Significant Information Review 1-6
        1.6.1 Applicant's Process 1-7
        1.6.2 Staff Evaluation 1-8
        1.6.3 Conclusion 1-9
    1.7 Report Contents 1-9
    1.8 References 1-9
2.0 Affected Environment 2-1
    2.1 Site Location 2-1
    2.2 Land 2-1
    2.3 Meteorology and Air Quality 2-2
    2.4 Geology 2-2
    2.5 Radiological Environment 2-3
    2.6 Water 2-3
    2.7 Ecology 2-3
        2.7.1 Terrestrial Ecology 2-4
        2.7.2 Aquatic Ecology 2-7
    2.8 Socioeconomics 2-7
        2.8.1 Demographics 2-8
        2.8.2 Community Characteristics 2-8

	2.9	Historic and Cultural Resources...2-9
	2.10	Environmental Justice..2-9
	2.11	Related Federal Projects and Consultations 2-10
	2.12	References..2-10
3.0	Site Layout and Plant Description...3-1
	3.1	External Appearance and Plant Layout...3-1
	3.2	Plant Description 3-1
		3.2.1 Plant Water Use 3-4
		3.2.2 Cooling System..3-4
		3.2.3 Radioactive Waste Management System..3-6
		3.2.3.1 Liquid Radioactive Waste-Management System.........................3-6
		3.2.3.2 Gaseous Radioactive Waste-Management System 3-6
		3.2.3.3 Solid Radioactive Waste-Management System 3-7
		3.2.4 Nonradioactive Waste Systems..3-7
	3.3	Power Transmission System 3-7
	3.4	References...3-8
4.0	Environmental Impacts of Construction..4-1
	4.1	Land-Use Impacts...4-1
		4.1.1 The Site and Vicinity 4-2
		4.1.2 Transmission Line Rights-of-Way...4-3
	4.2	Meteorological and Air-Quality Impacts...4-3
	4.3	Water-Related Impacts 4-4
	4.4	Ecology...4-5
		4.4.1 Terrestrial Impacts 4-6
		4.4.2 Summary of Terrestrial Impacts 4-13
		4.4.3 Aquatic Ecosystem Impacts 4-14
	4.5	Socioeconomic Impacts...4-16
		4.5.1 Physical Impacts 4-16
		4.5.2 Demography 4-16
		4.5.3 Economic Impacts to the Community 4-17
		4.5.4 Infrastructure and Community Service Impacts 4-18
		4.5.5 Summary of Socioeconomic Impacts 4-20
	4.6	Historic and Cultural Resources..4-20

4.7	Environmental Justice Impacts	4-22
4.8	Nonradiological Health Impacts	4-23
	4.8.1 Public and Occupational Health	4-23
	4.8.2 Impacts of Transporting Construction Materials and Construction Personnel to the VEGP Site	4-24
	4.8.3 Summary of Nonradiological Health Impacts	4-26
4.9	Radiological Health Impacts	4-26
	4.9.1 Direct Radiation Exposures	4-26
	4.9.2 Radiation Exposures from Gaseous Effluents	4-27
	4.9.3 Radiation Exposures from Liquid Effluents	4-28
	4.9.4 Total Dose to Site-Preparation Workers	4-29
	4.9.5 Summary of Radiological Health Impacts	4-30
4.10	Measures and Controls to Limit Adverse Impacts During Site-Preparation Activities and Construction	4-30
4.11	Site Redress Plan	4-31
4.12	Summary of Construction Impacts	4-33
4.13	References	4-33
5.0	**Environmental Impacts of Operation**	**5-1**
5.1	Land-Use Impacts	5-1
5.2	Meteorological and Air-Quality Impacts	5-2
5.3	Water-Related Impacts	5-2
5.4	Terrestrial and Aquatic Ecosystems	5-3
	5.4.1 Terrestrial Impacts	5-3
	5.4.2 Aquatic Impacts	5-8
5.5	Socioeconomic Impacts	5-10
5.6	Historic and Cultural Resource Impacts	5-10
5.7	Environmental Justice	5-11
5.8	Nonradiological Health Impacts	5-11
5.9	Radiological Impacts of Normal Operations	5-12
	5.9.1 Exposure Pathways	5-12
	5.9.2 Radiation Doses to Members of the Public	5-13
	5.9.3 Impacts to Members of the Public	5-14
	5.9.4 Occupational Doses to Workers	5-15

		5.9.5 Impacts to Biota Other than Members of the Public	5-15
		5.9.6 Radiological Monitoring	5-15
	5.10	Environmental Impacts of Postulated Accidents	5-16
		5.10.1 Design Basis Accidents	5-16
		5.10.2 Severe Accidents	5-18
		5.10.3 Severe Accident Mitigation Alternatives	5-19
		5.10.4 Summary of Postulated Accident Impacts	5-20
	5.11	Measures and Controls to Limit Adverse Impacts During Operation	5-20
	5.12	Summary of Operation Impacts	5-21
	5.13	References	5-21
6.0	Fuel Cycle, Transportation, and Decommissioning		6-1
	6.1	Fuel Cycle Impacts and Solid Waste Management	6-1
	6.2	Transportation Impacts	6-4
	6.3	Decommissioning Impacts	6-5
	6.4	References	6-5
7.0	Cumulative Impacts		7-1
	7.1	Land Use	7-1
	7.2	Air Quality	7-1
	7.3	Water Use and Quality	7-4
	7.4	Terrestrial Ecosystem	7-5
	7.5	Aquatic Ecosystem	7-6
	7.6	Socioeconomics, Historic and Cultural Resources, Environmental Justice	7-7
	7.7	Nonradiological Health	7-9
	7.8	Radiological Impacts of Normal Operation	7-9
	7.9	Severe Accidents	7-10
	7.10	Fuel Cycle, Transportation, and Decommissioning	7-10
	7.11	NRC Staff Conclusions and Recommendations	7-11
	7.12	References	7-11
8.0	Need for Power		8-1
	8.1	References	8-2
9.0	Environmental Impacts of Alternatives		9-1
	9.1	No-Action Alternative	9-1
	9.2	Energy Alternatives	9-2

  9.3 System Design Alternatives 9-3
  9.4 Region of Interest and Alternative Site Selection Process 9-4
  9.5 Evaluation of Alternative Sites 9-4
  9.6 References 9-4
10.0 Comparison of the Impacts of the Proposed Action and the Alternative Sites 10-1
  10.1 References 10-2
11.0 Conclusions and Recommendations 11-1
  11.1 Impacts of the Proposed Action 11-1
  11.2 Unavoidable Adverse Environmental Impacts 11-2
  11.3 Alternatives to the Proposed Action 11-2
  11.4 Relationship Between Short-Term Uses and Long-Term Productivity of the Human Environment 11-3
  11.5 Irreversible and Irretrievable Commitments of Resources 11-4
  11.6 Benefit-Cost Balance 11-4
  11.7 Staff Conclusions and Recommendations 11-6
  11.8 References 11-6
Appendix A – Contributors to the Supplemental Environmental Impact Statement A-1
Appendix B – Organizations Contacted B-1
Appendix C – Chronology of U.S. Nuclear Regulatory Commission Staff Environmental Review Correspondence Related to the Southern Nuclear Operating Company, Inc., Application for a Combined License for Units 3 and 4 at the Vogtle Electric Generating Plant C-1
Appendix D – Scoping Comments and Responses D-1
Appendix E – Draft Supplemental Environmental Impact Statement Comments and Responses E-1
Appendix F – Key Consultation Correspondence F-1
Appendix G – Supporting Documentation for Radiological Dose Assessment G-1
Appendix H – Authorizations and Certifications H-1
Appendix I – Vogtle Electric Generating Plant Site Characteristics, AP1000 Design Parameters and Site Interface Values I-1
Appendix J – Statements Made in the Environmental Report Considered in the U.S. Nuclear Regulatory Commission Staff's Environmental Review J-1

Figures

3-1 Proposed VEGP Site Footprint with Proposed Units 3 and 4 3-2
3-2 Areas that will be Disturbed by Construction and Preconstruction Activities for VEGP Units 3 and 4 3-3
3-3 Revised Intake Structure and Surrounding Wetlands 3-5
5-1 Extent of the 2.8°C (5.0°F)-Above-Ambient Isotherm Created by the Proposed VEGP Units 3 and 4 Discharge Pipe in the Combined Effluent Analysis 5-4

Tables

2-1 2009 Average Annual Unemployment Rates 2-8
5-1 DBA Doses for an AP1000 Reactor at the VEGP Site (Southern 2009a) 5-18
7-1 Comparison of Annual CO_2 Emission Rates 7-3

Executive Summary

On March 31, 2008, the U.S. Nuclear Regulatory Commission (NRC) received an application from Southern Nuclear Operating Company, Inc. (Southern), on behalf of itself and four co-applicants (i.e., Georgia Power Company, Oglethorpe Power Corporation, Municipal Electric Authority of Georgia, and the City of Dalton, Georgia), for combined licenses (COLs) for two new nuclear units (Units 3 and 4) to be located adjacent to the existing Vogtle Electric Generating Plant (VEGP) Units 1 and 2. The VEGP site is located in Burke County, Georgia, approximately 42 km (26 mi) southeast of Augusta, Georgia.

In Early Site Permit (ESP)-004 issued on August 26, 2009, NRC approved the VEGP site as suitable for the construction and operation of Units 3 and 4. As requested in the ESP application, the VEGP ESP also included a Limited Work Authorization (LWA) that authorized certain limited construction activities at the site in accordance with Title 10 of the Code of Federal Regulations (CFR) Subparts 50.10 and 52.24(c). As permitted by NRC regulations, the COL application references the VEGP ESP.

The proposed design specified in the COL application for the two new units is the Westinghouse AP1000 pressurized reactor. An amendment to the certified AP1000 design currently is being reviewed by NRC in a separate design certification process.

On October 2, 2009, Southern supplemented its COL application to include a request for a second LWA. The second LWA, in accordance with 10 CFR 50.10 (d), would authorize installation of reinforcing steel, sumps, drain lines, and other embedded items along with placement of concrete for the nuclear island foundation base slab.

During April, May, and June 2010, Southern submitted requests for amendments to the ESP relating to the activities authorized by the ESP LWA. In response to these applications, the NRC issued three amendments to the ESP in May, June, and July 2010, respectively. These amendments authorized Southern to use Category-1 and Category-2 backfill materials from additional onsite borrow areas and to change the classification of engineered backfill over the slopes of the excavations for Units 3 and 4. The NRC staff prepared an Environmental Assessment (EA) and Finding of No Significant Impact (FONSI) for each license amendment request.

Section 102 of the National Environmental Policy Act of 1969 (NEPA) (42 USC 4321) directs that an environmental impact statement (EIS) be prepared for major Federal actions with the potential to significantly affect the quality of the human environment. NRC has implemented Section 102 of NEPA in 10 CFR Part 51. Further, in 10 CFR 51.20, NRC has determined that the issuance of a COL under 10 CFR Part 52 is an action that requires an EIS.

The purpose of Southern's requested action is to obtain from the NRC a license to construct and operate two new nuclear power units on the VEGP site as well as an LWA to allow early commencement of certain limited construction activities. A license from the NRC to construct and operate nuclear power plants is necessary but not sufficient for construction and operation of the power plant. Southern must obtain and maintain permits from other Federal, State, and local agencies and permitting authorities. Therefore, the purpose of the NRC environmental review of the Southern application is to determine if a nuclear power plant of the proposed design can be constructed and operated at the VEGP site without unacceptable adverse impacts on the human environment.

The Southern COL application incorporates information from both the ESP Site Safety Analysis Report and Southern's environmental report (ER). Subpart A of 10 CFR Part 52 contains NRC regulations related to ESPs. An ESP is an NRC approval of a site as suitable for construction and operation of one or more new nuclear units. The NRC's detailed review of the environmental impacts of constructing and operating new units at the VEGP ESP Site is documented in NUREG-1872, *Final Environmental Impact Statement for an Early Site Permit (ESP) at the Vogtle Electric Generating Plant Site*, which was published in August 2008. For a COL application that references an ESP, the NRC staff, pursuant to 10 CFR Part 51.75(c), prepares a supplement to the ESP environmental impact statement (SEIS) in accordance with 10 CFR 51.92(e).

NRC regulations related to the environmental review of COL applications are in 10 CFR Part 51 and 10 CFR 52, Subpart C. Pursuant to NRC regulations in 10 CFR 51.50(c)(1), a COL applicant referencing an ESP need not submit information or analyses regarding environmental issues that were resolved in the ESP EIS, except to the extent the COL applicant has identified new and significant information regarding such issues. In addition, pursuant to 10 CFR 52.39, matters resolved in the ESP proceedings are considered to be resolved in any subsequent proceedings, absent identification of new and significant information.

Upon acceptance of Southern's COL application, the NRC began the environmental review process by publishing in the *Federal Register* on June 11, 2008, an Acceptance for Docketing, which announced its intent to perform a detailed technical review and conduct a hearing in accordance with Subpart L, "Informal Hearing Procedures for NRC Adjudications," of 10 CFR Part 2 (73 FR 33118). Subsequent to the site visits in August 2008 and September 2009 and in accordance with the provisions of NEPA and 10 CFR Part 51, the staff identified and evaluated the potential environmental impacts of constructing and operating two new units at the VEGP site.

The draft SEIS was published in September 2010. A 75-day comment period commenced on September 3, 2010, when the U.S. Environmental Protection Agency Notice of Filing appeared in the *Federal Register* (75 FR 54146) to allow members of the public to comment on the results of the NRC staff's review. A public meeting was held in Waynesboro, Georgia, on October 7, 2010. During this public meeting, the staff described the results of the NRC environmental review, provided members of the public with information to assist them in formulating comments on the SEIS, and accepted comments. When the comment period ended on November 24, 2010, the staff considered and addressed all comments received. All comments received on the draft SEIS are included in Appendix E.

Included in this SEIS are (1) the results of the NRC staff's analyses, which consider and weigh the environmental effects of the proposed action (i.e., issuance of the COLs and LWA) and of constructing and operating two additional nuclear units at the VEGP site; (2) mitigation measures for reducing or avoiding adverse effects; (3) the environmental impacts of alternatives to the proposed action; and (4) the staff's recommendation regarding the proposed action. To guide its assessment of environmental impacts of a proposed action or alternative actions, the NRC has established a standard of significance for impacts based on guidance developed by the Council on Environmental Quality (40 CFR 1508.27). The three significance levels established by the NRC – SMALL, MODERATE, or LARGE – are defined as follows:

> SMALL – Environmental effects are not detectable are so minor that they will neither destabilize nor noticeably alter any important attribute of the resource.
>
> MODERATE – Environmental effects are sufficient to alter noticeably, but not to destabilize, important attributes of the resource.
>
> LARGE – Environmental effects are clearly noticeable and are sufficient to destabilize important attributes of the resource.

Mitigation measures were considered for each environmental issue and are discussed in the appropriate sections of the SEIS. In preparing this SEIS, the staff reviewed Southern's COL application, including the ER and responses to staff requests for additional information; reviewed the ESP EIS and the ESP license amendment EAs; reviewed Southern's process for identifying new and significant information; consulted with Federal, State, Tribal, and local agencies; reviewed other relevant literature and documents; and followed the guidance set forth in NRC NUREG-1555, *Standard Review Plans for Environmental Reviews for Nuclear Power Plants (ESRP)*.

The NRC staff's recommendation to the Commission related to the environmental aspects of the proposed action is that the COLs and LWA be issued as proposed. This recommendation is based on (1) the COL application, including the ER and responses to staff requests for additional information submitted by Southern; (2) the staff's review conducted for the ESP application and documented in the ESP EIS; (3) the staff's review conducted for the ESP license amendments as documented in the staff's Environmental Assessments; (4) consultation with Federal, State, Tribal, and local agencies; (5) the staff's own independent review of potential new and significant information available since preparation and publication of the ESP EIS; and (6) the assessments summarized in this SEIS, including the potential mitigation measures identified and consideration of public comments received on the draft SEIS.

Abbreviations/Acronyms

ac	acre(s)
ACHP	Advisory Council on Historic Preservation
ADAMS	Agencywide Document Access and Management System
APP	Avian Protection Program
ARRA	America Recovery and Reinvestment Act
AQCR	Air Quality Control Region
BA	biological assessment
BTU	British Thermal Unit
°C	degree Celsius
CAA	Clean Air Act
CCAA	Candidate Conservation Agreement with Assurances
CDC	U.S. Center for Disease Control and Prevention
CFR	Code of Federal Regulations
CO_2	carbon dioxide
COL	combined license
CWA	Clean Water Act
CWS	cooling water system
dBA	decibel(s)
DBA	design basis accident
DCD	Design Control Document
DPS	distinct population segments
DSM	demand-side management
EA	Environmental Assessment
EAB	exclusion area boundary
EIS	environmental impact statement
EPA	U.S. Environmental Protection Agency
EPP	Environmental Protection Plan
EPRI	Electric Power Research Institute
ER	Environmental Report
ESP	early site permit
ESRP	Environmental Standard Review Plan

°F	degree Fahrenheit
FONSI	Finding of No Significant Impact
FR	Federal Register
ft	foot/feet
FWS	U.S. Fish and Wildlife Service
g	gram(s)
GCRP	U.S. Global Change Research Program
GDHR	Georgia Department of Human Resources
GDNR	Georgia Department of Natural Resources
GEIS	Generic Environmental Impact Statement
GHPD	Georgia Historic Preservation Division
gpm	gallons per minute
GPC	Georgia Power Company
GPSC	Georgia Public Service Commission
GRR	General Re-Evaluation Report
ha	hectare(s)
HLW	high-level waste
hr	hour
in.	inch(es)
IRP	integrated resource plan
km	kilometer(s)
kV	kilovolt
kWh	kilowatt hour
L	liter(s)
L/s	liter(s) per second
LAR	License Amendment Request
LLW	low-level waste
LOS	level of service
LPZ	low-population zone
LWA	Limited Work Authorization
m	meter(s)
m^3	cubic meters
MOU	Memorandum of Understanding
mi	mile(s)
mSv	millisievert

mrem	millirem(s)
MTU	metric tons uranium
MW	megawatt(s)
MWd	megawatts per day
MW(e)	megawatts electric
MWh	megawatt hour(s)
MW(t)	megawatts thermal
NAAQS	National Ambient Air Quality Standards
NEPA	National Environmental Policy Act of 1969
NHPA	National Historic Preservation Act of 1966
NIOSH	National Institute for Occupational Safety and Health
NMFS	National Marine Fisheries Service
NPDES	National Pollutant Discharge Elimination System
NRHP	National Register of Historic Places
NRC	U.S. Nuclear Regulatory Commission
ppm	parts per million
PRA	probabilistic risk assessment
RAI	Request(s) for Additional Information
RDC	Representative Delineated Corridor
REMP	Radiological Environmental Monitoring Program
ROI	region of interest
ROW	right(s)-of-way
SAMA	severe accident mitigation alternatives
SAMDA	severe accident mitigation design alternatives
SCDNR	South Carolina Department of Natural Resources
sec	second/seconds
SER	Safety Evaluation Report
SERPPAS	Southeast Regional Partnership for Planning and Sustainability
SHPO	State Historic Preservation Office/Officer
SEIS	supplemental environmental impact statement
SMEs	subject matter experts
Southern	Southern Nuclear Operating Company, Inc.
Sv	sievert
SWS	service water system
TEDE	total effective dose equivalent
TLD	thermoluminescent dosimeter

USACE	U.S. Army Corps of Engineers
USC	United States Code
VEGP	Vogtle Electric Generating Plant
Westinghouse	Westinghouse Electric Company, LLC
wt	weight
χ/Q	dispersion values
yd	yard
yd^3	cubic yards
yr	year(s)

1.0 Introduction

On March 31, 2008, Southern Nuclear Operating Company, Inc. (Southern), acting on behalf of itself and several co-applicants (Georgia Power Company, Oglethorpe Power Corporation, Municipal Electric Authority of Georgia, and the City of Dalton, Georgia), submitted to the U.S. Nuclear Regulatory Commission (NRC) an application for combined licenses (COLs) for the construction and operation of two new nuclear units at the Vogtle Electric Generating Plant (VEGP) site. The VEGP site and existing facilities are owned and operated by Georgia Power Company, Oglethorpe Power Corporation, Municipal Electric Authority of Georgia, and the City of Dalton, Georgia. Southern is the licensee and operator of the existing VEGP Units 1 and 2, and has been authorized by the VEGP co-owners to apply for COLs to construct and operate two additional units (Units 3 and 4) at the VEGP site.

1.1 Background

On August 26, 2009, the NRC approved issuance to Southern and the same four co-applicants of an early site permit (ESP) and a Limited Work Authorization (LWA) for two additional nuclear units at the VEGP site (NRC 2009). This approval was supported by information contained in NUREG-1872, *Final Environmental Impact Statement for an Early Site Permit at the Vogtle Electric Generating Plant Site* (ESP EIS) (NRC 2008a) and errata. The ESP resolved many safety and environmental issues and allowed Southern to "bank" the VEGP ESP site for up to 20 years. The LWA authorized Southern to conduct certain limited construction activities at the site in accordance with Title 10 of the Code of Federal Regulations (CFR) Subparts 50.10 and 52.24(c).

As permitted by NRC regulations, the COL application references the VEGP ESP. Southern also submitted a request for a second LWA as part of its COL application. The second LWA, in accordance with 10 CFR 50.10(d), would allow for installation of reinforcing steel, sumps, drain lines, and other embedded items along with placement of concrete for the nuclear island foundation base slab that are not included in the existing LWA (Southern 2010a).
The proposed design specified in the COL application for the two new units is the Westinghouse AP1000 pressurized reactor. An amendment to the certified AP1000 design is currently being reviewed by NRC in a separate design certification process. The draft SEIS indicated that the COL application references the AP1000 plant design that has been certified by NRC (Title 10 of the Code of Federal Regulations [CFR] Part 52, Appendix D) (Westinghouse 2005), as modified by the amendment to that design that Westinghouse Electric Company, LLC (Westinghouse), the AP1000 vendor, has submitted to the NRC. The NRC staff is reviewing the design revision separately from this COL review. At the time the draft SEIS was published, Revision 17 of the *AP1000 Design Control Document* (Westinghouse 2008) was the Revision being considered in the design certification review, and the environmental review in the draft SEIS accordingly

Introduction

accounted for the environmental impacts anticipated from use of the design in that Revision. Since publication of the draft SEIS, Westinghouse has updated its design certification application with Revision 18 of the AP1000 DCD, and the VEGP COL application has been updated to reference that Revision. The NRC staff has determined that none of the changes involved in the latest Revision has the potential to affect the environmental review documented in the SEIS. For that reason, references to Revision 17 in this SEIS have been left unchanged. If a subsequent Revision to the AP1000 DCD is submitted and referenced in the COL application, the staff will determine whether the change in Revision has the potential to affect the environmental review. Depending on the environmental significance of any such design change, the staff will supplement the SEIS as appropriate.

During April, May, and June, 2010, Southern submitted requests for three ESP license amendments associated with the previously-authorized LWA construction activities. These amendment requests sought authorization to use Category-1 and Category-2 backfill materials from additional onsite sources, including three new borrow areas, as well as to change the classification of engineered backfill over the side slopes of the excavations for Units 3 and 4 (Southern 2010b, c, d, e). The NRC prepared an Environmental Assessment (EA) and Finding of No Significant Impact (FONSI) for each license amendment request (NRC 2010a, b, c). These ESP license amendments were issued in May 2010 (NRC 2010d), June 2010 (NRC 2010e), and July 2010 (NRC 2010f), respectively.

1.1.1 COL Application and Review

To construct and operate a nuclear power plant, an ESP holder must either obtain a Construction Permit and an Operating License or obtain a COL. Either approach constitutes a separate major federal action and would require that an environmental impact statement (EIS) be issued in accordance with 10 CFR Part 51. Under 10 CFR Part 52, which contains NRC's reactor licensing regulations, and in accordance with the applicable provisions of 10 CFR Part 51, which are the NRC regulations implementing the National Environmental Policy Act of 1969 (NEPA), the NRC is required to prepare a supplemental environmental impact statement (SEIS) as part of its review of a COL application referencing an ESP. As required by 10 CFR 51.26, NRC published in the *Federal Register* a Notice of Intent (74 FR 49407) to prepare and publish a draft SEIS for public comment. The SEIS for the COLs was prepared in the same manner as the final EIS for the ESP except that NRC determined that it would not conduct a formal scoping process in accordance with 10 CFR 51.26(d). A separate Safety Evaluation Report (SER) also will be prepared in accordance with 10 CFR Part 52.

If a COL application references an ESP, the NRC staff, pursuant to 10 CFR Part 51.75(c), is required to prepare a supplement to the ESP EIS (NRC 2008a). Therefore, the staff can "tier off" the ESP EIS at the COL stage and disclose the NRC conclusion for matters resolved in the ESP review. Such matters will not be subject to litigation at the combined license stage unless new and significant information is identified. Because the VEGP COL application references the

Introduction

VEGP ESP, the NRC staff relied on the analysis in the ESP EIS as the basis in preparing the SEIS. NRC's regulatory standards for a review of a COL application are listed in 10 CFR 52.81. Detailed procedures for conducting the environmental portion of the review are found in guidance set forth in NUREG-1555, *Environmental Standard Review Plan: Standard Review Plans for Environmental Review for Nuclear Power Plants* (NRC 2000) and recent updates.

According to 10 CFR 52.80(b), an application for a COL must contain an environmental report (ER), which provides the applicant's input to the NRC's EIS. NRC regulations related to the contents of the ER are found in 10 CFR Part 51.

In accordance with 10 CFR 51.45 and 10 CFR 51.50(c)(1), Southern submitted an ER as part of its COL application (Southern 2009). In accordance with 10 CFR 51.49, Southern also submitted an ER in support of its additional LWA request (Southern 2010f). The ER submitted with the COL application is not required to contain information or analysis that was previously submitted in the ER for the ESP application or address issues that were resolved in the ESP EIS and associated proceedings.

The SEIS, together with the ESP EIS (NRC 2008a), the ESP hearing proceedings, and the ESP license amendment EAs, provides the staff's evaluation of the environmental effects of constructing and operating two AP1000 reactors at the VEGP site. In addition to considering the environmental effects of the proposed action, the SEIS addresses new and significant information with respect to alternatives to the proposed action and the benefits of the proposed action (e.g., the need for power). Southern's COL application references an ESP; therefore, in accordance with 10 CFR 52.83, issues resolved as part of the ESP proceeding remain resolved except under conditions set forth in 10 CFR 52.39(a)(2). In addition, measures and controls previously identified to limit adverse impacts are evaluated along with any new or significant information that would have the potential to affect the findings or conclusions reached in the ESP EIS.

Upon acceptance of Southern's COL application, the NRC began the environmental review process by publishing an Acceptance for Docketing in the *Federal Register* on June 11, 2008 (73 FR 33118). The Acceptance for Docketing announced NRC's intent to perform a detailed technical review and conduct a hearing in accordance with Subpart L, "Informal Hearing Procedures for NRC Adjudications," of 10 CFR Part 2.

To guide its assessment of environmental impacts of a proposed action or alternative actions, the NRC has established a standard of significance for impacts based on guidance developed by the Council on Environmental Quality (40 CFR 1508.27). The three significance levels established by the NRC – SMALL, MODERATE, or LARGE – are defined as follows:

> SMALL – Environmental effects are not detectable or are so minor that they will neither destabilize nor noticeably alter any important attribute of the resource.

Introduction

> MODERATE – Environmental effects are sufficient to alter noticeably, but not to destabilize, important attributes of the resource.
>
> LARGE – Environmental effects are clearly noticeable and are sufficient to destabilize important attributes of the resource.

This SEIS presents the staff's analysis, which considers and weighs the environmental impacts of the proposed action at the VEGP site, including the environmental impacts associated with construction and operation of Units 3 and 4 at the site, the environmental impacts of alternatives to granting the COLs, and the mitigation measures available for reducing or avoiding adverse environmental effects. The SEIS also provides the NRC staff's recommendation to the Commission regarding the issuance of the COLs and LWA for the VEGP site.

The draft SEIS was published in September 2010. A 75-day comment period commenced on September 3, 2010, when the U.S. Environmental Protection Agency Notice of Filing appeared in the *Federal Register* (75 FR 54146) to allow members of the public to comment on the results of the NRC staff's review. A public meeting was held in Waynesboro, Georgia, on October 7, 2010. During this public meeting, the staff described the results of the NRC environmental review, provided members of the public with information to assist them in formulating comments on the SEIS, and accepted comments. When the comment period ended on November 24, 2010, the staff considered and addressed all comments received. All comments received on the draft SEIS are included in Appendix E.

1.1.2 Concurrent Reviews

In a review separate from the environmental review process, the NRC analyzes the safety and security aspects of construction and operation of the proposed new reactors at the site, including the applicant's emergency planning information. These analyses will be documented in an SER. The SER will present the conclusions reached by the NRC regarding whether there is reasonable assurance that two Westinghouse AP1000 light-water reactors can be constructed and operated at the VEGP site without undue risk to the health and safety of the public and whether issuance of the license will be inimical to the common defense and security.

In addition, the AP1000 reactor design referenced in the application is a standard design that is undergoing a design certification amendment review pursuant to 10 CFR Part 52, Subpart B. This review will be the subject of a later rulemaking by the NRC.

1.2 The Proposed Federal Action

The proposed Federal action is issuance of COLs, under the provisions of 10 CFR Part 52, for two AP1000 reactors at the VEGP site and an LWA for requested construction activities. The ESP EIS (NRC 2008a) disclosed the staff's analysis of the environmental impacts that could result from the construction and operation of these two new units. This SEIS for the COL application evaluates whether any new and potentially significant information has been identified that would alter the staff's conclusions regarding issues resolved in the ESP proceeding.

In the context of a COL application that references an ESP, the term "new" in the phrase "new and significant information" is defined as any information that was both (1) not considered in preparing the ESP ER or EIS (as may be evidenced by references in these documents, applicant responses to NRC Requests for Additional Information [RAIs], comment letters, etc.) and (2) not generally known or publicly available during the preparation of the ESP EIS (such as information in reports, studies, and treatises).

For new information to be "significant," it must be material to the issue being considered; that is, it must have the potential to affect the finding or conclusions of the NRC staff's evaluation of the issue. The applicant for a COL need only provide information in the application about a previously resolved environmental issue if it is both new and significant (72 FR 49352).

In this SEIS, the staff evaluates the impacts of construction and operation of two AP1000 units, with a total combined thermal power rating of 6800 megawatts thermal (MW(t)). The proposed units would use a closed-cycle cooling system and would require a single natural draft cooling tower for each unit.

1.3 The Purpose and Need for the Proposed Action

The purpose and need for the issuance of the COLs is to provide for additional base-load electrical generating capacity in the region of interest as defined in Section 9.4.1 of the ESP EIS (NRC 2008a). Southern indicated that the proposed action also will allow it to be responsive to the Georgia legislature, which urged Georgia utilities to study the feasibility of building new nuclear power plants (Senate Resolution 865). The purpose and need for the issuance of the LWA is "... to support the project schedule by assuring that [*the proposed LWA activities*] occur independent of the COL issuance schedule and contribute to maintaining a margin in the construction schedule that ensures the operation need dates will be met" (Southern 2010e).

The ultimate decision about whether or not to build a facility and the schedule for any construction are not within the purview of NRC and would be determined by the license holder if the authorization is granted. A license from NRC to construct and operate a nuclear power plant is necessary, but not sufficient for construction and operation of the power plant. Certain long lead-time activities, such as ordering and procuring certain components and materials necessary to construct the plant, may begin before the COL is granted. Southern must obtain and maintain permits or authorizations from other Federal, State, and local agencies and permitting authorities before undertaking certain activities.

1.4 Alternatives to the Proposed Action

Section 102(2)(C)(iii) of NEPA states that an EIS is to include a detailed statement on alternatives to the proposed action. This SEIS addresses the following categories of alternatives: (1) the no-action alternative, (2) energy source alternatives, and (3) system design alternatives. In accordance with 10 CFR 51.92(e)(3), the SEIS does not contain a separate discussion of alternative sites. The NRC's detailed evaluation of alternative sites is documented in Chapters 9 and 10 of the ESP EIS (NRC 2008a).

1.5 Compliance and Consultations

Prior to construction and operation of the new unit, Southern is required to hold certain Federal, State, and local environmental permits, as well as meet applicable statutory and regulatory requirements. Southern provided a list of environmental approvals and consultations associated with the VEGP proposed Units 3 and 4 (Southern 2010e). Potential authorizations and consultations relevant to the proposed COL are included in Appendix H.

Before it could obtain a COL from NRC, it was necessary for Southern to obtain a Clean Water Act Section 401 Certification. This certification, which was issued by the Georgia Department of Natural Resources (GDNR) on June 1, 2010, ensures that the project does not conflict with water quality management programs in Georgia. Southern provided a copy of the 401 Certification to NRC as a comment to the draft SEIS (Southern 2010g).

The NRC staff has contacted the appropriate Federal, State, Tribal, and local agencies to identify any compliance, permit, or significant environmental issues of concern to the reviewing agencies that relate to the construction and operation of the proposed Units 3 and 4. A list of organizations contacted is included in Appendix B.

1.6 New and Significant Information Review

As set forth in 10 CFR 51.92, an SEIS for a COL referencing an ESP shall contain an analysis of those issues related to the impacts of construction and operation that were resolved in the ESP proceeding for which new and significant information has been identified. Information is considered new if it was (1) not considered in preparing the ESP ER or ESP EIS (NRC 2008a) (as may be evidenced by references in these documents, applicant responses to NRC RAIs, comment letters, etc.) and (2) not generally known or publicly available during the preparation of the ESP EIS (such as information in studies and reports). For information to be significant, it must be material to the issue being considered; that is, it must have the potential to affect the finding or conclusions of the NRC staff's evaluation of the issue (72 FR 49352). If there is no new and significant information for matters resolved at the ESP stage, the staff may tier off of

Introduction

the ESP EIS at the COL stage and disclose the NRC conclusions for matters considered during the ESP review.

A COL applicant should have a reasonable process to ensure it becomes aware of new and significant information that may have a bearing on the earlier NRC conclusion, and should document the results of this process in an auditable form. The NRC staff will verify that the applicant's process for identifying new and significant information is effective (72 FR 49352).

1.6.1 Applicant's Process

Southern developed a process to identify new and significant information relevant to the issues and conclusions presented in the ESP EIS. This process is detailed in *Guidance for New and Significant Information* (Southern 2007) and is summarized in the COL ER (Southern 2009). The process was designed to satisfy the requirements of 10 CFR 51.50(c) and to "... provide a methodical, comprehensive review of the conclusions presented in the ESP EIS and the supporting information for those conclusions to identify any new and significant information that has the potential to change the NRC's conclusions presented in the ESP EIS" (Southern 2009). For purposes of its review, Southern adopted definitions of "new" and "significant" previously published by the NRC (72 FR 49352).

Southern's process for identifying new and significant information began with the designation of subject matter experts (SMEs) with extensive knowledge about plant systems, site environs, station environmental issues, and the regulatory issues relevant to the plant and site. The SMEs performed a line-by-line review of the ESP EIS to identify "key inputs." This review focused on the portions of the EIS where conclusions were directly supported, especially Chapters 4, 5, 6, and 7. The review also considered key assumptions that were included in Appendix J of the ESP EIS, key site characteristics, Westinghouse design parameters and site interface values that were found in Appendix I of the ESP EIS, and dose calculation assumptions provided in Appendix G of the ESP EIS.

The SMEs reviewed the key inputs to determine if any new information exists that could affect the NRC staff's findings or conclusions. This determination typically was based, as appropriate, on current construction plans and designs, site documentation, environmental monitoring and sampling programs, interviews with Federal, State, or local officials, contact with Federal, State, or local agencies, and when necessary, the SMEs' local knowledge. The SMEs conducted a review of other information sources including interviews with industry peers, academia, and Federal, State, and local resource agencies, a review of the AP1000 Design Control Document, Westinghouse Technical Reports for the AP1000, environmental monitoring reports from existing programs, and applicable scientific literature, to determine if additional information relevant to the COL application was available that was not captured in the direct review of the ESP EIS.

Introduction

The SMEs then reviewed all information that had been identified as new to determine if it might be significant. When possible, this determination was based on comparison with regulatory limits, guidelines provided in NRC review guidance such as NUREG-1555 (NRC 2000), or other applicable criteria. When such a comparison was not possible, the SMEs used their best professional judgment to determine if new information was considered significant. The results of this review, including the bases for the conclusion on new information and the rationale for determination of significance, were summarized in documents that were audited by the NRC staff during the site audit that was conducted in late September 2009.

1.6.2 Staff Evaluation

The NRC staff's evaluation of Southern's new and significant information methodology began with the review of Southern's process as described in Rev. 0 of the VEGP Units 3 and 4 COL Application (Southern 2008). In August 2008, the staff performed an assessment of Southern's process for identifying new and significant information in three specific areas: (1) aquatic ecology, (2) terrestrial ecology, and (3) hydrology. The assessment was performed at the VEGP site near Waynesboro, Georgia, and included review of documents, staff discussions with Southern, site tours, and discussions with representatives from other State and Federal agencies including the GDNR, the U.S. Army Corps of Engineers, the U.S. Environmental Protection Agency, and the U.S. Fish and Wildlife Service. The staff raised several questions about certain aspects of the methodology that Southern needed to address. The results of that assessment were documented in a trip report (NRC 2008b).

During June 2009, the staff was provided access to the information developed during Southern's implementation of its new and significant information methodology. This access was available through a reading room set up by Southern in Richland, Washington.

After the ESP was authorized in August 2009, the NRC staff performed a new and significant information audit at the VEGP site near Waynesboro, Georgia, during the period from September 28 through October 1, 2009. The focus of the staff's audit was to determine if Southern's new and significant information methodology was robust and comprehensive and had the ability to capture any new information developed since completion of the ESP EIS and authorization of the ESP, and if Southern adhered to its process set forth in the new and significant information methodology. To make these determinations, the staff examined Southern's process in detail for all the resource areas discussed in the ESP EIS, assessed the results of Southern's review for new and significant information, and participated in several site tours including potential transmission line rights-of-way, the location of the new intake structure on the Savannah River, and the locations of cultural and historic resources on the VEGP site. In addition, the appropriate Federal, State, and local agencies and officials were contacted to verify the presence or absence of new and potentially significant information. A summary of the site audit is provided in the site audit trip report (NRC 2010g). Following the audit, the staff conducted an independent assessment of other sources of new and significant information.

Introduction

During March 2010, Southern provided new information about potential new onsite borrow areas (Southern 2010h). Because these borrow sources had not been evaluated in the ESP EIS, the NRC staff performed a second site audit during the period May 3–5, 2010, to evaluate the potential environmental impacts of developing these new borrow areas. The results of the second site audit are provided in a site audit trip report (NRC 2010h).

1.6.3 Conclusion

Based on the staff's independent review of Southern's new and significant information process, the staff determined that the process was adequate to identify new and potentially significant information concerning environmental issues addressed in the ESP EIS (NRC 2008a).

1.7 Report Contents

The subsequent chapters of this SEIS are organized as follows. Chapter 2 describes the proposed site and discusses the environment that would be affected by the addition of the new units. Chapter 3 describes the power plant characteristics to be used as the basis for evaluating the environmental impacts. Chapters 4 and 5 examine the environmental impacts of construction (Chapter 4) and operation (Chapter 5) of the proposed Units 3 and 4. Chapter 6 analyzes the environmental impacts of the uranium fuel cycle, transportation of radioactive materials, and decommissioning, while Chapter 7 discusses the cumulative impacts of the proposed action as defined in 10 CFR Part 51.75(c). Chapter 8 addresses the need for power. Chapter 9 discusses alternatives to the proposed action, and Chapter 10 summarizes the conclusions regarding the impacts of the proposed action and alternatives, while Chapter 11 summarizes the findings of the preceding chapters and presents the staff's recommendation with respect to issuance of the COLs and LWA.

1.8 References

10 CFR Part 50. Code of Federal Regulations, Title 10, *Energy*, Part 50, "Domestic Licensing of Production and Utilization Facilities."

10 CFR Part 51. Code of Federal Regulations, Title 10, *Energy*, Part 51, "Environmental Protection Regulations for Domestic Licensing and Related Regulatory Functions."

10 CFR Part 52. Code of Federal Regulations, Title 10, *Energy*, Part 52, "Licenses, Certifications, and Approvals for Nuclear Power Plants."

40 CFR Part 1508. Code of Federal Regulations, Title 40, *Protection of Environment*, Part 1508, "Terminology and Index."

Introduction

72 FR 49352. August 28, 2007. "Licenses, Certifications, and Approvals for Nuclear Power Plants." *Federal Register*, U.S. Nuclear Regulatory Commission.

73 FR 33118. June 11, 2008. "Southern Nuclear Operating Company; Acceptance for Docketing of an Application for Combined License for Vogtle Electric Generating Plant Units 3 and 4." *Federal Register*, U.S. Nuclear Regulatory Commission.

74 FR 49407. September 28, 2009. "Southern Nuclear Operating Company Vogtle Electric Generating Plant, Units 3 and 4 Combined License Application; Notice of Intent to Prepare a Supplemental Environmental Impact Statement." *Federal Register*, U.S. Nuclear Regulatory Commission.

75 FR 54145. September 3, 2010. "Environmental Impact Statements; Notice of Availability;, EIS No. 20100351, Draft EIS, NRC, GA, Vogtle Electric Generating Plant Units 3 and 4, Construction and Operation, Application for Combined Licenses (COLs). NUREG 1947." Environmental Protection Agency

Clean Water Act. 33 USC 1251, et seq. (Also referred to as the Federal Water Pollution Control Act [FWPCA].)

National Environmental Policy Act of 1969 (NEPA), as amended. 42 USC 4321, et seq.

Sen. Res. 865. 2006. "A Resolution: Urging Electric Utilities to Consider Building New Nuclear Power Plants in Georgia; Urging the Public Service Commission to Encourage Such Consideration; and for Other Purposes." Georgia State Senate, Georgia General Assembly.

Southern Nuclear Operating Company, Inc. (Southern). 2007. *Guidance for New and Significant Information.* ND-ARL-012, Version 1.0, October 8, 2007, Southern Company, Birmingham, Alabama. Accession No. ML081570520.

Southern Nuclear Operating Company, Inc. (Southern). 2008. *Vogtle Electric Generating Plant, Units 3 and 4, COL Application, Part 3 Environmental Report.* Revision 0, March 28, 2008, Southern Company, Birmingham, Alabama. Accession No. ML081050181.

Southern Nuclear Operating Company, Inc. (Southern). 2009. *Vogtle Electric Generating Plant, Units 3 and 4, COL Application, Part 3 Environmental Report.* Revision 1, August 23, 2009, Southern Company, Birmingham, Alabama. Accession No. ML092740400.

Southern Nuclear Operating Company, Inc. (Southern). 2010a. Southern Nuclear Operating Company, *Vogtle Electric Generating Plant, Units 3 and 4, COL Application, Part 6 Limited Work Authorization.* Revision 2, August 6, 2010, Southern Company, Birmingham, Alabama. Accession No. ML102220380.

Southern Nuclear Operating Company, Inc. (Southern). 2010b. Letter from B.L. Ivey to NRC, "Southern Nuclear Operating Company, Early Site Permit Site Safety Analysis Report Change Request, Vogtle Electric Generating Plant Units 3 and 4, Use of Category 1 and 2 Backfill Material for Additional Onsite Areas, on an Exigent Basis for Units 3 and 4." Letter ND-10-0795. April 20, 2010, Southern Company, Birmingham, Alabama. Accession No. ML101120089.

Southern Nuclear Operating Company, Inc. (Southern). 2010c. Letter from M. Smith, Southern, to NRC, "Southern Nuclear Operating Company, Vogtle Electric Generating Plant Units 3 and 4, Early Site Permit Site Safety Analysis Report Amendment Request, Revised Site Safety Analysis Report Markup for Onsite Sources of Backfill." Letter ND-10-0960. May 13, 2010, Southern Company, Birmingham, Alabama. Accession No. ML101340649.

Southern Nuclear Operating Company, Inc. (Southern). 2010d. Letter from C.R. Pierce, Southern, to NRC, "Southern Nuclear Operating Company, Vogtle Electric Generating Plant Units 3 and 4, Early Site Permit Site Safety Analysis Report Amendment Request, Revised Site Safety Analysis Report Markup for Onsite Sources of Backfill, Part 2." Letter ND-10-1005. May 24, 2010, Southern Company, Birmingham, Alabama. Accession No. ML101470212.

Southern Nuclear Operating Company, Inc. (Southern). 2010e. Letter from C.R. Pierce, Southern, to NRC, "Southern Nuclear Operating Company, Vogtle Electric Generating Plant Units 3 and 4, Site Safety Analysis Report License Amendment Request, Revise Backfill Geometry." Letter ND-10-0964. May 24, 2010, Southern Company, Birmingham, Alabama. Accession No. ML101470213.

Southern Nuclear Operating Company, Inc. (Southern). 2010f. Letter from C.R. Pierce, Southern, to NRC, "Southern Nuclear Operating Company, Vogtle Electric Generating Plant Units 3 and 4, Combined License Application, Environmental Report to Support Revision 1 to Part 6, LWA Request." Letter ND-10-0227. February 5, 2010, Southern Company, Birmingham, Alabama. Accession No. ML100470600.

Southern Nuclear Operating Company, Inc. (Southern). 2010g. Letter from C.R. Pierce, Southern, to NRC, "Southern Nuclear Operating Company, Vogtle Electric Generating Plant Units 3 and 4, Combined License Application, Comments on Draft Supplemental Environmental Impact Statement." November 23,1010. Southern Company, Birmingham, Alabama. Accession No. ML103300035.

Southern Nuclear Operating Company, Inc. (Southern). 2010h. Letter from M. Smith, Southern, to NRC, "Southern Nuclear Operating Company, Vogtle Electric Generating Plant Units 3 and 4, Combined License Application, *Supporting Information for Environmental Report Review.*" March 12, 2010, Southern Company, Birmingham, Alabama. Accession No. ML100750038.

Introduction

U.S. Nuclear Regulatory Commission (NRC). 2000. *Standard Review Plans for Environmental Reviews for Nuclear Power Plants.* NUREG-1555, Vol. 1, Washington, D.C. Includes 2007 revisions.

U.S. Nuclear Regulatory Commission (NRC). 2008a. *Final Environmental Impact Statement for an Early Site Permit (ESP) at the Vogtle Electric Generating Plant Site.* NUREG-1872, Vols. 1, 2, and Errata, Washington, D.C. Accession Nos. ML082240145; ML082240165, ML082260203; ML082550040.

U.S. Nuclear Regulatory Commission (NRC). 2008b. *Summary of Environmental Site Audit Related to the Review of the Combined License Application for Vogtle Electric Generating Plant, Units 3 and 4.* Washington, D.C. Accession No. ML082620184.

U.S. Nuclear Regulatory Commission (NRC). 2009. *Vogtle Electric Generating Plant Early Site Permit No. ESP-004.* Washington, D.C. Accession No. ML092290157.

U.S. Nuclear Regulatory Commission (NRC). 2010a. *Vogtle Electric Generating Plant ESP Site Early Site Permit and Limited Work Authorization Environmental Assessment and Finding of No Significant Impact.* Docket No. 52-011. Accession No. ML101380114.

U.S. Nuclear Regulatory Commission (NRC). 2010b. *Vogtle Electric Generating Plant ESP Site Early Site Permit and Limited Work Authorization Environmental Assessment and Finding of No Significant Impact.* Docket No. 52-011. Accession No. ML101670592.

U.S. Nuclear Regulatory Commission (NRC). 2010c. *Vogtle Electric Generating Plant ESP Site Early Site Permit and Limited Work Authorization Environmental Assessment and Finding of no Significant Impact.* Docket No. 52-011. Accession No. ML101660076.

U.S. Nuclear Regulatory Commission (NRC). 2010d. Letter from C. Patel, Senior Project Manager, NRC, to J.A. Miller, Executive Vice President, SNC, "Subject: Vogtle Electric Generating Plant ESP Site - Issuance of Exigent Amendment Regarding Request for Changes to the Site Safety Analysis Report." May 21, 2010. Accession No. ML101400509.

U.S. Nuclear Regulatory Commission (NRC). 2010e. Letter from T. Spicher, Project Manager, NRC, to J.A. Miller, Executive Vice President, SNC, "Subject: Vogtle Electric Generating Plant ESP Site - Issuance of Amendment Regarding Request for Changes to the Site Safety Analysis Report Regarding Onsite Sources of Backfill." June 25, 2010. Accession No. ML101760370

U.S. Nuclear Regulatory Commission (NRC). 2010f. Letter from T. Spicher, Project Manager, NRC, to J.A. Miller, Executive Vice President, SNC, Subject: Vogtle Electric Generating Plant ESP Site, Subject: Issuance of Amendment RE: Request for changes to the classification of backfill over the side slopes of Units 3 and 4 excavations." July 9, 2010. Accession No. ML101870522.

U.S. Nuclear Regulatory Commission (NRC). 2010g. *Summary of Environmental New and Significant Site Audit Related to the Review of the Combined License Application for Vogtle Electric Generating Plant Site.* Package Accession No. ML093631157.

U.S. Nuclear Regulatory Commission (NRC). 2010h. Memorandum Regarding the Site Audit Summary Concerning Environmental Impacts Associated with Acquisition of Additional Backfill Material for the Vogtle Electric Generating Plant Site Combined License Application Review. Washington, D.C. Package Accession No. ML101550095.

Westinghouse Electric Company, LLC (Westinghouse). 2005. *AP1000 Design Control Document.* AP1000 Document. APP-GW-GL-700, Revision 15, Westinghouse Electric Company, Pittsburgh, Pennsylvania. Package Accession No. ML053480403.

Westinghouse Electric Company, LLC (Westinghouse). 2008. *AP1000 Design Control Document.* AP1000 Document. APP-GW-GL-700, Revision 17, Westinghouse Electric Company, Pittsburgh, Pennsylvania. Package Accession No. ML083230168.

2.0 Affected Environment

U.S. Nuclear Regulatory Commission (NRC) staff provided a description of the affected environment in the vicinity of the Vogtle Electric Generating Plant (VEGP) early site permit (ESP) site in Chapter 2 of the ESP environmental impact statement (EIS) (NRC 2008). The applicant, Southern Nuclear Operating Company, Inc. (Southern), evaluated potential new and significant information that could affect the description of the affected environment. The NRC staff reviewed Southern's process for identifying new and significant information, but also conducted its own independent review to verify whether new and significant information has been identified. The results of those reviews are presented in this chapter. The site location is described in Section 2.1, and the land, meteorology and air quality, geology, radiological environment, water, ecology, socioeconomics, historic and cultural resources, and environmental justice aspects (or conditions) of the site are presented in Sections 2.2 through 2.10, respectively. Section 2.11 examines related Federal projects, and references cited are listed in Section 2.12.

2.1 Site Location

The staff described the location of the VEGP ESP site in Sections 2.1 and 2.2 of the ESP EIS (NRC 2008). This description included the location of the proposed Units 3 and 4 on the VEGP site in relation to the regions within 10 km (6 mi) and 80 km (50 mi) of the site. The VEGP site comprises 1282.5 ha (3169 ac) in an unincorporated area of Burke County, Georgia. The site is approximately 24 km (15 mi) east-northeast of Waynesboro, the county seat of Burke County, and 42 km (26 mi) southeast of Augusta, Georgia.

In the environmental report (ER) included in its combined license (COL) application (Southern 2009a), Southern provided no new and significant information related to site location, and the NRC staff found no new and significant information during its review of Southern's process for identifying new and significant information and the staff's visit to the VEGP site.

2.2 Land

The staff described land-related issues for the ESP site in Section 2.2 of the ESP EIS (NRC 2008). This discussion included a description of the VEGP site, the vicinity and region surrounding the site, and the existing electric power transmission system supporting the site.

In its COL ER (Southern 2009a), Southern provided no new and significant information related to land-related issues, and the NRC staff found no new and significant information during its review of Southern's process for identifying new and significant information and the staff's audit visit to the VEGP site.

Affected Environment

2.3 Meteorology and Air Quality

The meteorology and air quality of the VEGP ESP site were described by NRC in Section 2.3 of the ESP EIS (NRC 2008) and by Southern in Section 2.7 of the ESP ER (Southern 2008a). These descriptions included a summary of the climatology and air quality for the region. They also included discussions of the onsite meteorological monitoring program and associated measurements that were the bases for other assessments described in the ESP EIS. For example, estimates of site-specific atmospheric relative concentration were used to assess dose from routine and accidental radiological releases in Sections 5.9 and 5.10, respectively, of the ESP EIS (NRC 2008).

In its COL ER (Southern 2009a), Southern provided no new and significant information related to meteorology and air quality. However, during the NRC staff's independent review, new information related to changes to the National Ambient Air Quality Standard (NAAQS) for ozone was identified. The staff determined that this new information warranted further review.

The VEGP site is centrally located within the Augusta (Georgia) – Aiken (South Carolina) Interstate Air Quality Control Region (AQCR) (Title 40 Code of Federal Regulations [CFR] Part 81.114). All of the counties in this AQCR currently are designated as in attainment or unclassified for all criteria pollutants for which NAAQS have been established (40 CFR 81.311). On March 12, 2008, the U.S. Environmental Protection Agency (EPA) promulgated a revision to the NAAQS for ozone. The final rule (73 FR 16436) is designed to further protect public health by reducing the standard from 0.084 parts per million (ppm) to 0.075 ppm. Section 107(d)(1) of the Clean Air Act requires each state to submit, within 1 year of the revised standard, its recommended designation (i.e., attainment, non-attainment, or unclassified) for each county. On March 12, 2009, the Georgia Department of Natural Resources (GDNR) issued a letter to the EPA providing its recommended designations; under those recommendations Burke County remains unclassified/attainment with respect to the new ozone standard (GDNR 2009a). EPA will make its final determination no later than March 2011.

2.4 Geology

The staff described the geology of the VEGP ESP site in Section 2.4 of the ESP EIS (NRC 2008). The discussion included general descriptions of the regional geology, the topography of the site area, and the regional mineral resources. Detailed descriptions of the geologic, seismic, and geotechnical engineering properties of the site, including the results of field and laboratory investigations, were provided in the ESP Site Safety Analysis Report (Southern 2008b) and the ESP Safety Evaluation Report (NRC 2009).

Affected Environment

In its COL ER (Southern 2009a), Southern provided no new and significant information related to the environmental aspects of geology, and the NRC staff found no new and significant information during its review of Southern's process for identifying new and significant information and during the audit at the VEGP site.

2.5 Radiological Environment

Detailed descriptions of the radiological environment of the VEGP ESP site were provided by NRC in Section 2.5 of the ESP EIS (NRC 2008) and by Southern in Section 6.2 of the ESP ER (Southern 2008a). These discussions included summaries of historical data from radiological environmental monitoring program annual reports for the existing VEGP Units 1 and 2. Each year, Southern issues a report entitled *Annual Radioactive Effluent Release Report for the Vogtle Power Station*, which documents gaseous and liquid releases and resulting doses from VEGP.

In its COL ER (Southern 2009a), Southern provided no new and significant information related to radiological environment, and the NRC staff found no new and significant information during its review of Southern's process for identifying new and significant information, during the audit at the VEGP site, and during its review of recent data on releases and estimated occupational and population doses regarding the radiological environment since issuance of the VEGP ESP (Southern 2006, 2007, 2008c, 2009b).

2.6 Water

The staff described the hydrology of the VEGP ESP site in Section 2.6 of the ESP EIS (NRC 2008). These discussions included the regional and site surface water features, the regional and site hydrogeology and groundwater features, consumptive and non-consumptive surface-water and groundwater use in the area affected by the site, surface-water and groundwater quality in the area affected by the site, and existing and possible future hydrological, thermal, and chemical monitoring at the site.

In its COL ER (Southern 2009a), Southern provided no new and significant information related to hydrology, and the NRC staff found no new and significant information during its review of Southern's process for identifying new and significant information and during the audit at the VEGP site.

2.7 Ecology

The staff presented detailed descriptions of the terrestrial and aquatic ecology in the vicinity of the VEGP site in Section 2.7 of the ESP EIS (NRC 2008). The following sections update these descriptions, where appropriate, with information developed since the ESP EIS was prepared,

Affected Environment

including information from the COL ER (Southern 2009a), supplemental information provided by Southern, and reviews of current information available from Federal and State agencies.

2.7.1 Terrestrial Ecology

The staff presented a detailed description of the terrestrial resources in the vicinity of the VEGP ESP site in Section 2.7.1 of the ESP EIS (NRC 2008). This discussion included wildlife habitats, wildlife usage, and terrestrial monitoring in the vicinity of the VEGP site and the proposed transmission line rights-of-way (ROW). The evaluation also included a discussion of the important species as specified by NUREG-1555, *Environmental Standard Review Plan: Standard Review Plans for Environmental Reviews for Nuclear Power Plants* (NRC 2000), including Federally and State-listed threatened and endangered species.

In its COL ER (Southern 2009a), Southern provided no new and significant information related to terrestrial resources. The NRC staff performed site audits in September 2009 and May 2010, and contacted the GDNR, the South Carolina Department of Natural Resources (SCDNR), and the U.S. Fish and Wildlife Service (FWS) to determine if new information was available, and received responses from each of these agencies (GDNR 2009b; SCDNR 2009; FWS 2010a, b).

On October 20, 2010, FWS provided the NRC staff with an update of Federally listed threatened or endangered species that can be expected to occur in the project area (FWS 2010b). FWS identified four Federally listed terrestrial plant and animal species that may occur on or in the vicinity of the VEGP site and/or in the vicinity of the Representative Delineated Corridor (RDC) (FWS 2010b). The updated list includes the red-cockaded woodpecker (*Picoides borealis*), the wood stork (*Mycteria americana*), the Canby's dropwort (*Oxypolis canbyi*), and the eastern indigo snake (*Drymarchon couperi*). In addition to the Federally listed species, FWS provided information on the bald eagle (*Haliaeetus leucocephalus*) and the gopher tortoise (*Gopherus polyphemus*) in the response letter. Impacts to the red-cockaded woodpecker, wood stork, and Canby's dropwort are discussed in the ESP EIS. FWS indicated that there are eagle nests in Jefferson and McDuffie Counties, including one nest in the Representative Delineated Corridor (RDC). The location of the eagle nest in the RDC also was discussed in the ESP EIS.

FWS indicated that the four Federally listed terrestrial plant and animal species may occur on or in the vicinity of the VEGP site as well as within the vicinity of the RDC (FWS 2010b).

The RDC is a transmission line route of sufficient width to contain the eventual ROW for the proposed new 500-kV transmission line. It is described in Sections 2.7.1 and 4.1.2 of the ESP EIS (NRC 2008) and in the "*Corridor Study: Thomson – Vogtle 500-kV Transmission Project*" (GPC 2007), and it was the focus for the staff's analysis of potential impacts from the proposed transmission line. Southern and GPC have not determined the final route for the transmission line, but as explained in the ESP EIS, the transmission line ROW would be routed northwest from the VEGP site, passing west of Fort Gordon, a U.S. Army facility west of Augusta, Georgia,

Affected Environment

and then north to the Thomson substation. It is anticipated that the transmission line would cross primarily Burke, Jefferson, McDuffie, and Warren Counties in Georgia, and would be 46 m (150 ft) wide and 97 km (60 mi) long.

Based on the October 20, 2010 FWS letter, the four Federally listed species that can be expected to occur within the project area are the red-cockaded woodpecker (*Picoides borealis*), the wood stork (*Mycteria americana*), the Canby's dropwort (*Oxypolis canbyi*), and the eastern indigo snake (*Drymarchon couperi*). In addition to the Federally listed species, FWS provided information on the bald eagle (*Haliaeetus leucocephalus*) and the gopher tortoise (*Gopherus polyphemus*) in the response letter. Impacts to the red-cockaded woodpecker, wood stork, and Canby's dropwort are discussed in the ESP EIS. FWS indicated that there are eagle nests in Jefferson and McDuffie Counties, including one nest in the RDC. The location of the eagle nest in the RDC also was discussed in the ESP EIS.

The information discussed in this section focuses on species not previously considered in the ESP EIS. This includes the eastern indigo snake and gopher tortoise, both identified by FWS in their recent letter as species that can be expected, to occur in the project area. FWS noted that the gopher tortoise is not a Federally listed species in Georgia; however, its status is under review by FWS (FWS 2010b). Sandhills habitat that could support the gopher tortoise and the eastern indigo snake is present in the project area.

The eastern indigo snake and gopher tortoise were not included in the analysis undertaken for the ESP EIS. The eastern indigo snake was not included because it was not in previous FWS lists of species within the project area. Likewise, the gopher tortoise was not included in previous GDNR species occurrence lists for the project area. Therefore, these species are discussed below. The Federally threatened eastern indigo snake also is discussed in the Biological Assessment included in Appendix F.

FWS indicated that the gopher tortoise, a Georgia state threatened species, can be expected to occur in the project area (FWS 2010b), and currently is under review by the FWS to be listed as Federally threatened (FWS 2010b). There are no known populations of the gopher tortoise on the VEGP site or within the RDC (GDNR 2009b; FWS 2010b). The gopher tortoise is a characteristic species of the longleaf pine and wiregrass community, which includes sandhills, dry flatwoods, and turkey oak scrub. Historically, this community was represented by an open-canopied forest that allows abundant sunlight penetration and conditions favorable for a rich growth of herbaceous vegetation. Sandy soil, sunlight availability, and abundant herbaceous vegetation are key habitat requirements for the gopher tortoise. The gopher tortoise digs burrows that provide winter hibernacula, retreats from the summer heat, and shelter from fire for the tortoise and also for hundreds of invertebrate and vertebrate animal species. The gopher tortoise has been termed a "keystone species" of the longleaf pine community, meaning its existence is critical to the existence of many other species (GDNR 2009c).

Affected Environment

Southern submitted a draft Candidate Conservation Agreement with Assurances (CCAA) for the gopher tortoise at the VEGP Site. This CCAA is currently under review by FWS (SERPPAS 2010). The draft CCAA does not include the offsite portions of the proposed transmission line.

The eastern indigo snake, identified by FWS as a species that can be expected to occur in the project area, but for which there are no documented occurrences in the area is Federally listed as threatened (FWS 2010b). It occurred historically throughout Florida and in the coastal plains of Georgia, Alabama, and Mississippi (43 FR 4026; FWS 2006). Most, if not all, of the remaining viable populations of the eastern indigo snake occur in Georgia and Florida. There are no historic or recent records for the upper Coastal Plain or Fall Line sandhill region of Georgia, including Burke, McDuffie, Jefferson, and Warren Counties (FWS 2006; Diemer and Speake 1983; Stevenson 2006).

The eastern indigo snake occupies a broad range of habitats, including pine flatwoods, scrubby flatwoods, high pine, dry prairie, edges of freshwater marshes, agricultural fields, and human-altered habitats (FWS 1982). In the northern parts of its range, including southeastern Georgia, eastern indigo snakes are tied to the use of gopher tortoise burrows and longleaf pine habitat (FWS 2006). The gopher tortoise burrows are used by eastern indigo snakes to protect against cold in the winter and heat in the summer, and also for foraging, nesting, mating, and shelter prior to shedding (FWS 2006). Habitat use often varies seasonally between upland and wetland areas in Georgia (FWS 2006). Movement between habitat types may relate to the needs for thermal refugia, differences in habitat use by the juveniles and adults, or seasonal differences in availability of food resources. For these reasons, the eastern indigo snake is particularly vulnerable to habitat fragmentation (FWS 2006).

During the COL application review, Southern did identify new information with respect to the proposed new borrow areas, as described in its March 12, 2010, submittal (Southern 2010a). Southern also provided information in its subsequent submittals on May 10, May 13, and May 24, 2010, in support of requested ESP license amendments to obtain backfill material from onsite borrow areas not previously identified in the ESP (Southern 2010b, c, d). The information supplied by FWS and Southern resulted in a change in the terrestrial baseline conditions considered in the ESP EIS. The eastern indigo snake, gopher tortoise, sandhills milkvetch (*Astragalus michauxii*), and the southeastern pocket gopher (*Geomys pinetis*) all are known to occur in sandhills habitat. This habitat type is present both in the RDC and onsite.

In the ESP EIS, which was completed in the summer of 2008, the NRC staff noted that, while mounds indicative of the State-threatened southeastern pocket gopher had been identified just north of the VEGP site boundary and that similar habitat occurred nearby on the VEGP site, the footprint of construction disturbance for the ESP EIS was not expected to encompass such habitat. The EIS also indicated that, while the State-threatened sandhills milkvetch, an herbaceous legume, was known to occur within 16 km (10 mi) of the VEGP site, it had not been

Affected Environment

identified as occurring within 3.2 km (2 mi) of the VEGP site. The sandhills milkvetch has since been observed on the northern section of the VEGP site (NRC 2010a). As discussed in the staff's June 2010 Environmental Assessment (EA) (NRC 2010b) prepared in connection with Southern's license amendment request (LAR) to use three additional onsite backfill borrow areas (Southern 2010d), both species were found in a proposed new borrow area west-northwest of the power-block area in the spring of 2010 during the environmental review of the LAR. Additional details concerning the distribution and habitat preferences of the southeastern pocket gopher and the sandhills milkvetch are found in the LAR EA that was issued in June 2010 (NRC 2010b). The staff incorporated that information by reference in this SEIS.

2.7.2 Aquatic Ecology

The staff presented detailed descriptions of the aquatic ecology in the vicinity of the VEGP site in Section 2.7.2 of the ESP EIS (NRC 2008). These included descriptions of onsite ponds and streams and the Savannah River in the vicinity of the VEGP site. They also included descriptions of important species as specified by NUREG-1555 (NRC 2000), including Federally and State-listed threatened and endangered species.

In its COL ER (Southern 2009a), Southern provided no new and significant information related to aquatic ecology. On October 6, 2010, the National Marine Fisheries Service (NMFS) published in the *Federal Register* (75 FR 61904) a proposed rule for listing the Carolina and South Atlantic distinct population segments of the Atlantic sturgeon (*Acipenser oxyrinchus oxyrinchus*) as endangered under the Endangered Species Act. The staff described the life history of the Atlantic sturgeon in the ESP proceedings; however, in light of the proposed listing, the staff considered the available literature and compiled additional information in a conference consultation letter to NMFS (Appendix F).

Otherwise, the NRC staff found no new and significant information during its review of Southern's process for identifying new and significant information, the audit at the VEGP site, and contacts with representatives of FWS, NMFS, GDNR, and SCDNR (see Appendix F for the letters regarding consultation).

2.8 Socioeconomics

The staff provided a detailed description of socioeconomics in the VEGP ESP region in Section 2.8 of the ESP EIS (NRC 2008). The discussion included the socioeconomic resources that could potentially be affected by the construction and operation of the proposed Units 3 and 4 at the VEGP site. This discussion is organized into two major subsections that provide details on demographics and community characteristics. New information that has become available since issuance of the VEGP ESP is described in the following sections.

Affected Environment

2.8.1 Demographics

The staff provided a detailed discussion of the community characteristics of the VEGP ESP site in Section 2.8.1 of the ESP EIS (NRC 2008). The discussion included the resident population, transient population, and migrant populations.

In its COL ER (Southern 2009a), Southern provided no new and significant information related to demographics, and the NRC staff found no new and significant information during its review of Southern's process for identifying new and significant information, the audit at the VEGP site, and contacts with county officials.

2.8.2 Community Characteristics

The staff provided a detailed discussion of the community characteristics of the VEGP ESP site in Section 2.8.2 of the ESP EIS (NRC 2008). The discussion included the economy, taxes, transportation, aesthetics, recreation, housing, public services, and education in Burke, Richmond and Columbia Counties, which are the counties most affected by activities at the VEGP site.

In its COL ER (Southern 2009a), Southern provided no new and significant information related to community characteristics. However, the NRC staff's independent review identified changes in the community characteristics of the VEGP region that warranted further investigation. In the ESP EIS, the 2005 unemployment rate for Burke County was 7.7 percent; for Columbia County, 4.4 percent; and for Richmond County, 7.1 percent. The State of Georgia's unemployment rate was 5.2 percent. The 2009 average annual unemployment rates for Burke, Richmond, and Columbia Counties and statewide in Georgia are provided in Table 2-1. The unemployment rates of all three counties and statewide in Georgia have increased, with Burke County's unemployment rate the highest at 11.5 percent. Unemployment rates are discussed further in Section 4.5.

Table 2-1. 2009 Average Annual Unemployment Rates

	Labor Force	Employment	Unemployment Number	Unemployment Rate
Burke County	9942	8802	1140	11.5
Columbia County	60,003	55,937	4066	6.8
Richmond County	90,520	82,553	8967	9.8
Georgia	4,769,000	4,312,000	457,000	9.6

Source: USBLS 2010

Affected Environment

2.9 Historic and Cultural Resources

The staff provided a detailed discussion of the historic and cultural resources of the VEGP ESP site in Section 2.9 of the ESP EIS (NRC 2008). The discussion included the cultural background of the area and sites eligible for listing under the National Historic Preservation Act of 1966 (NHPA) (NRC 2008, Table 2-24).

In its COL ER (Southern 2009a), Southern provided no new and significant information related to historic and cultural resources. The NRC staff performed a site audit in September 2009, and contacted the Georgia State Historic Preservation Office (SHPO) during December 2009 to determine if new information was available. The new information identified during the COL application review effort was the existence of a historic cemetery located on the VEGP site outside the proposed construction footprint and the proposed new borrow areas (Southern 2010a, d). A letter report dated May 14, 2007, documents an archaeological survey that was conducted to record the boundaries and features of the cemetery (New South Associates 2007). All of the proposed additional borrow areas whose use was authorized by the ESP amendments issued in May and June 2010 are within the VEGP site boundary and are within the area of potential effect for the cultural resource analysis included in the ESP EIS (NRC 2008, 2010b, c).

In accordance with Title 36 of the Code of Federal Regulations (CFR) Subpart 800.8c, the NRC staff is using the process implemented in the National Environmental Policy Act of 1969 (NEPA) to comply with the obligations defined under Section 106 of the NHPA. The area of potential effect used by the staff for this COL review is the same as that used for the ESP review (NRC 2008).

During December 2009, NRC initiated contact with the Georgia SHPO and the Advisory Council on Historic Preservation (ACHP), and sent 25 letters to Tribes (see Appendix C for a complete listing) to begin consultations on the proposed COL action. NRC requested the participation of the SHPO, the ACHP, and the Tribes in identifying new and significant information concerning historic properties that may be impacted by this COL action.

2.10 Environmental Justice

The staff provided a discussion of environmental justice issues in the vicinity of the VEGP ESP site in Section 2.10 of the ESP EIS (NRC 2008). The discussion included analysis on the location of minority and low-income individuals, scoping and outreach performed, health preconditions and special circumstances, and migrant populations.

In its COL ER (Southern 2009a), Southern provided no new and significant information related to environmental justice, and the NRC staff found no new and significant information during its review of Southern's process for identifying new and significant information, or during the audit at the VEGP site.

Affected Environment

2.11 Related Federal Projects and Consultations

The staff discussed related Federal projects and consultations in Section 2.11 of the ESP EIS (NRC 2008). The staff reviewed the possibility that activities of other Federal agencies might impact the issuance of a COL for proposed Units 3 and 4. Any such activities could result in cumulative environmental impacts or the possible need for another Federal agency to become a cooperating or coordinating agency for preparation of this supplemental EIS (SEIS) (10 CFR 51.10(b)(2)).

In its COL ER (Southern 2009a), Southern provided no new and significant information regarding related Federal projects and consultations, and the staff found no new and significant information during its review of Southern's process for identifying new and significant information, the audits at the VEGP site, and contacts with the FWS, NMFS, ACHP, U.S. Army Corps of Engineers, and various Tribal representatives.

The NRC is required under Section 102(2)(C) of NEPA to consult with and obtain the comments of any other Federal agency that has jurisdiction by law or special expertise with respect to any environmental impact involved in the subject matter of the SEIS. During the course of preparing the SEIS, NRC consulted with the FWS, NMFS, and the ACHP. Contact correspondence is included in Appendix F.

2.12 References

10 CFR Part 51. Code of Federal Regulations. Title 10, *Energy*, Part 51, "Environmental Protection Regulations for Domestic Licensing and Related Regulatory Functions."

36 CFR Part 800. Code of Federal Regulations. Title 36, *Parks, Forests, and Public Property*, Part 800, "Protection of Historic Properties."

40 CFR Part 81. Code of Federal Regulations. Title 40, *Protection of Environment*, Part 81, "Designation of Areas for Air Quality Planning Purposes."

43 FR 4026. January 31, 1978. "Endangered and Threatened Wildlife and Plants. Listing of the Eastern Indigo Snake as a Threatened Species." *Federal Register*, U.S. Fish and Wildlife Service.

73 FR 16436. March 27, 2008. "National Ambient Air Quality Standards for Ozone." *Federal Register*, Environmental Protection Agency.

75 FR 61904. October 6, 2010. "Endangered and Threatened Wildlife and Plants; Proposed Listing Determinations for Three Distinct Population Segments of Atlantic Sturgeon in the North East Region". *Federal Register*, U.S. Department of Commerce.

Affected Environment

Clean Air Act. 42 USC 7401, et seq.

Diemer J.E. and D.W. Speake. 1983. "The Distribution of the Eastern Indigo Snake, *Drymarchon corais couperi*, in Georgia." *Journal of Herpetology* 17(3): 256-264.

Endangered Species Act of 1973. 16 USC 1531 et seq.

Georgia Department of Natural Resources (GDNR). 2009a. Letter from C.A. Couch, Georgia Department of Natural Resource, to A.S. Meiburg, U.S Environmental Protection Agency, Region 4, Atlanta, Georgia. "Recommended Designations of Ozone Non-Attainment Areas in Georgia." March 12, 2009. Accession No. ML100601088.

Georgia Department of Natural Resources (GDNR). 2009b. E-mail dated December 21, 2009 and letter dated December 17, 2009 from Katrina Morris, GDNR, to Mallecia Sutton, NRC, regarding Natural Heritage Database occurrences within the VEGP Boundary and the Transmission Line Macrocorridor. Accession No. ML100490042.

Georgia Department of Natural Resources (GDNR). 2009c. *Gopher Tortoise* (Gopherus polyphemus). Accessed December 13, 2010 at http://www.georgiawildlife.com/node/1379.

Georgia Power Company (GPC). 2007. *Corridor Study: Thomson – Vogtle 500-kV Transmission Project.* Accession No. ML071710085.

National Environmental Policy Act of 1969 (NEPA), as amended. 42 USC 4321, et seq.

National Historic Preservation Act of 1966 (NHPA). 16 USC 470, et seq.

New South Associates. 2007. *A Determination of Boundaries and Surface Features for Historic Cemetery on Plant Vogtle in Burke County, Georgia.* May 14, 2007. Not for public disclosure per Section 304 of the National Historic Preservation Act.

South Carolina Department of Natural Resources (SCDNR). 2009. E-mail from Julie Holling, SCDNR, to Mallecia Sutton, NRC, "Threatened and Endangered Species in the Vicinity of Vogtle Electric Generating Plant." December 15, 2009. Accession No. ML093491132.

Southeast Regional Partnership for Planning and Sustainability (SERPPAS). 2010. *Candidate Conservation Agreement for the Gopher Tortoise First Annual Report: October 1, 2008 – September 30, 2009.* Accessed November 15, 2010 at www.serppas.org/.../GTCCA%20First%20Annual%20Report%202008-2009.pdf.

Southern Nuclear Operating Company, Inc. (Southern). 2006. *Annual Radioactive Effluent Release Report for January 1, 2005 to December 31, 2005. Vogtle Electric Generating Plant –*

Affected Environment

Units 1 and 2, NRC Docket Nos. 50-424 and 50-425, Facility Operating License Nos. NPF-68 and NPF-81. Southern Company, Birmingham, Alabama. Accession No. ML061240254.

Southern Nuclear Operating Company, Inc. (Southern). 2007. *Annual Radioactive Effluent Release Report for January 1, 2006 to December 31, 2006. Vogtle Electric Generating Plant – Units 1 and 2*, NRC Docket Nos. 50-424 and 50-425, Facility Operating License Nos. NPF-68 and NPF-81. Southern Company, Birmingham, Alabama. Accession No. ML071220467.

Southern Nuclear Operating Company, Inc. (Southern). 2008a. *Vogtle Early Site Permit Application: Part 3. Environmental Report.* Revision 4, Southern Company, Birmingham, Alabama. Accession No. ML081020073.

Southern Nuclear Operating Company, Inc. (Southern). 2008b. "Section 2.5 Geology, Seismology, and Geotechnical Engineering, Subsections 2.5.1.1 and 2.5.1.2." In *Site Safety Analysis Report. Rev. 5.* Southern Company, Birmingham, Alabama. Accession No. ML091540908.

Southern Nuclear Operating Company, Inc. (Southern). 2008c. *Annual Radioactive Effluent Release Report for January 1, 2007 to December 31, 2007. Vogtle Electric Generating Plant – Units 1 and 2*, NRC Docket Nos. 50-424 and 50-425, Facility Operating License Nos. NPF-68 and NPF-81. Southern Company, Birmingham, Alabama. Accession No. ML081290295.

Southern Nuclear Operating Company, Inc. (Southern). 2009a. *Vogtle Electric Generating Plant, Units 3 and 4, COL Application, Part 3 Environmental Report.* Revision 1, September 23, 2009, Southern Company, Birmingham, Alabama. Accession No. ML092740400.

Southern Nuclear Operating Company, Inc. (Southern). 2009b. *Annual Radioactive Effluent Release Report for January 1, 2008 to December 31, 2008. Vogtle Electric Generating Plant – Units 1 and 2*, NRC Docket Nos. 50-424 and 50-425, Facility Operating License Nos. NPF-68 and NPF-81. Southern Company, Birmingham, Alabama. Accession No. ML091260689.

Southern Nuclear Operating Company, Inc. (Southern). 2010a. Letter from M. Smith, Southern, to NRC, "Southern Nuclear Operating Company, Vogtle Electric Generating Plant Units 3 and 4, Combined License Application, Supporting Information for Environmental Report Review." Letter ND-10-0526. March 12, 2010. Southern Company, Birmingham, Alabama. Accession No. ML100750038.

Southern Nuclear Operating Company, Inc. (Southern). 2010b. Letter from B.L. Ivey, Southern, to NRC, "Southern Nuclear Operating Company, Vogtle Electric Generating Plant Units 3 and 4, Combined License Application, Post New and Significant Audit Supporting Information." Letter ND-10-0923. May 10, 2010. Southern Company, Birmingham, Alabama. Accession Nos. ML101320256 and ML101310333.

Affected Environment

Southern Nuclear Operating Company, Inc. (Southern). 2010c. Letter from M. Smith, Southern, to NRC, "Southern Nuclear Operating Company, Vogtle Electric Generating Plant Units 3 and 4, Early Site Permit Site Safety Analysis Report Amendment Request, Revised Site Safety Analysis Report Markup for Onsite Sources of Backfill." Letter ND-10-0960. May 13, 2010, Southern Company, Birmingham, Alabama. Accession No. ML101340649.

Southern Nuclear Operating Company, Inc. (Southern). 2010d. Letter from C.R. Pierce, Southern, to NRC, "Southern Nuclear Operating Company, Vogtle Electric Generating Plant Units 3 and 4, Early Site Permit Site Safety Analysis Report Amendment Request, Revised Site Safety Analysis Report Markup for Onsite Sources of Backfill, Part 2." Letter ND-10-1005 dated May 24, 2010. Southern Company, Birmingham, Alabama. Accession No. ML101470212.

Stevenson, D.J. 2006. *Distribution and Status of the Eastern Indigo Snake (Drymarchon couperi) in Georgia: 2006*. Unpublished report to the Georgia Department of Natural Resources Nongame and Endangered Wildlife Program, Forsyth, Georgia.

U.S. Bureau of Labor Statistics (USBLS). 2010. *Local Area Unemployment Statistics*. Accessed at http://www.bls.gov/lau/ on May 12, 2010. Accession No. ML 102020520.

U.S. Fish and Wildlife Service (FWS). 1982. *Eastern Indigo Snake Recovery Plan*. Atlanta, Georgia.

U.S. Fish and Wildlife Service (FWS). 2006. *Eastern Indigo Snake (Drymarchon couperi): 5-Year Summary and Evaluation*. Southeast Region, Mississippi Ecological Services Field Office, Jackson, Mississippi.

U.S. Fish and Wildlife Service (FWS). 2010a. Letter from U.S. Fish and Wildlife Service, Athens, Georgia, to U.S. Nuclear Regulatory Commission, Washington, D.C. "Re: USFWS Log# 2009-1387." February 12, 2010. Accession No. ML100500426.

U.S. Fish and Wildlife Service (FWS). 2010b. Letter from U.S. Fish and Wildlife Service, Athens, Georgia, to U.S. Nuclear Regulatory Commission, Washington, D.C. "Re: USFWS Log# 2010-1254." October 20, 2010. Accession No. ML103010076.

U.S. Nuclear Regulatory Commission (NRC). 2000. *Environmental Standard Review Plan: Standard Review Plans for Environmental Reviews for Nuclear Power Plants*. NUREG-1555, Washington, D.C.

U.S. Nuclear Regulatory Commission (NRC). 2008. *Final Environmental Impact Statement for an Early Site Permit (ESP) at the Vogtle Electric Generating Plant Site*. NUREG-1872, Vols. 1, 2, and Errata, Washington, D.C. Accession Nos. ML082240145; ML082240165, ML082260203; ML082550040.

Affected Environment

U.S. Nuclear Regulatory Commission (NRC). 2009. "Chapter 2.0, Section 2.5 Geology, Seismology, and Geotechnical Engineering." In *Vogtle ESP Final Safety Evaluation Report*. Washington, D.C. Accession No. ML090130160.

U.S. Nuclear Regulatory Commission (NRC). 2010a. GDNR Conference Call Summaries, May 26-June 3, 2010. Washington, D.C. Accession No. ML101570079.

U.S. Nuclear Regulatory Commission (NRC). 2010b. *Vogtle Electric Generating Plant ESP Site Early Site Permit and Limited Work Authorization Environmental Assessment and Finding of No Significant Impact*. Docket No. 52-011, Washington, D.C. Accession No. ML101670592.

U.S. Nuclear Regulatory Commission (NRC). 2010c. *Vogtle Electric Generating Plant ESP Site Early Site Permit and Limited Work Authorization Environmental Assessment and Finding of no Significant Impact*. Docket No. 52-011. Accession No. ML101380114.

3.0 Site Layout and Plant Description

The U.S. Nuclear Regulatory Commission (NRC) staff provided a description of the proposed Units 3 and 4 at Vogtle Electric Generating Plant (VEGP) in Chapter 3 of the early site permit (ESP) environmental impact statement (EIS) (NRC 2008). This chapter of the combined license (COL) supplemental EIS (SEIS) provides new information relative to the key site and facility characteristics that the NRC staff used to assess the environmental impacts of the proposed action. The site layout and existing facilities are discussed in Section 3.1. The plant design and power transmission system are discussed in Sections 3.2 and 3.3, respectively. References cited in this chapter are listed in Section 3.4.

3.1 External Appearance and Plant Layout

A detailed description of the external appearance and plant layout for VEGP Units 3 and 4 and associated structures and facilities was provided in Section 3.1 of the ESP EIS (NRC 2008). The description also includes a summary of the existing VEGP Units 1 and 2 and their associated facilities and a discussion of Plant Wilson, a six-unit, oil-fueled combustion turbine facility located on the VEGP site. The ESP EIS states that the VEGP site is located on the Savannah River and that the proposed Units 3 and 4 would be located in a previously disturbed area adjacent to the existing Units 1 and 2. Figure 3-1 shows the proposed VEGP site footprint with the proposed two new units and associated facilities. Figure 3-2 shows the areas on the site that will be disturbed by construction and preconstruction activities.

3.2 Plant Description

Section 3.2 of the ESP EIS (NRC 2008) described VEGP, including information about the Westinghouse AP1000 plant design that has been certified by NRC (Title 10 of the Code of Federal Regulations [CFR] Part 52, Appendix D) (Westinghouse 2005) and that has been selected by Southern Nuclear Operating Company, Inc. (Southern), as the reactor design for the proposed Units 3 and 4. Westinghouse Electric Company, LLC (Westinghouse), the AP1000 vendor, submitted Revision 17 of the *AP1000 Design Control Document* to the NRC for review (Westinghouse 2008), and the NRC staff is reviewing the design revision separately from this proposed action.

Section 3.2 of the ESP EIS also discussed the proposed cooling system and power output for proposed Units 3 and 4. The proposed cooling system would consist of one concrete natural-draft hyperbolic cooling tower for each unit, and each unit would operate at an estimated net electrical power output of approximately 1117 MW(e) (NRC 2008).

Site Layout and Plant Description

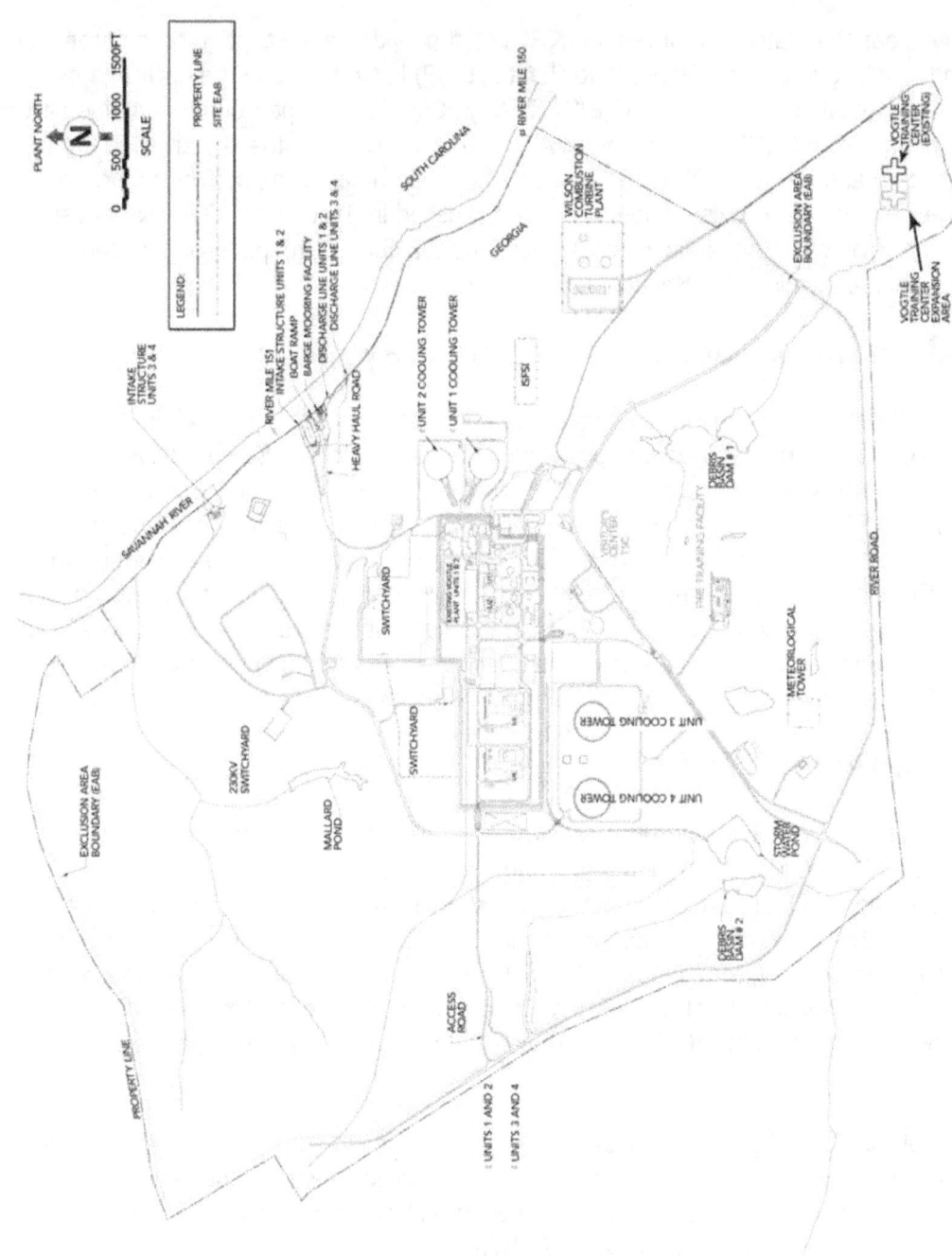

Figure 3-1. Proposed VEGP Site Footprint with Proposed Units 3 and 4

Site Layout and Plant Description

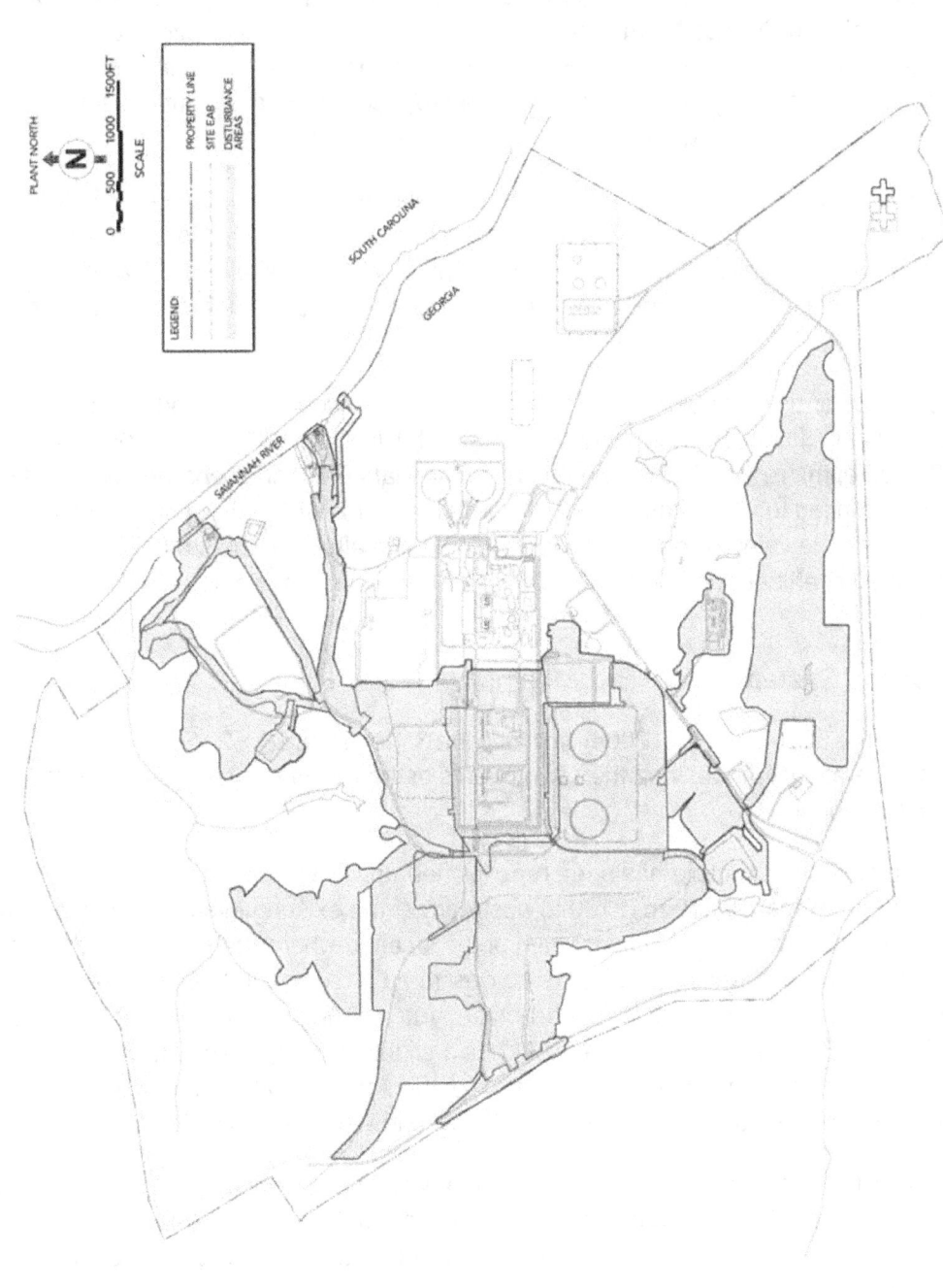

Figure 3-2. Areas that will be Disturbed by Construction and Preconstruction Activities for VEGP Units 3 and 4

Site Layout and Plant Description

3.2.1 Plant Water Use

Sections 3.2.1 and 3.2.2 of the ESP EIS (NRC 2008) and Section 3.3 of the ESP ER (Southern 2008) described plant water use for the proposed Units 3 and 4. These sections described the surface-water and groundwater withdrawals required for operation of the facility, the consumptive and nonconsumptive water uses of the proposed units, the plant effluent streams, and the plant water-treatment systems.

Southern provided no new and significant information related to plant water use in the COL ER, and the staff found no new and significant information during its review of Southern's process for identifying new and significant information and during the VEGP site audit. However, the NRC staff's review did identify the following information that warranted further staff analysis in this SEIS.

Estimated plant water use for operation of Units 3 and 4 is provided in Appendix I. The normal and maximum plant effluent discharges to the Savannah River are 631 L/s (10,008 gpm) and 2000 L/s (31,695 gpm), respectively. The impact of the plant effluent discharge described in the ESP EIS corresponded to a maximum discharge rate of 1941 L/s (30,761 gpm), which is 3 percent less than the value given above. Accordingly, the effect on the staff's ESP EIS conclusion of a plant effluent discharge of 2000 L/s (31,695 gpm) is evaluated in Section 5.3 of this document.

3.2.2 Cooling System

Section 3.2.2 of the ESP EIS (NRC 2008) and Section 3.4 of the ESP ER (Southern 2008) described the operational modes and the components of the cooling water system for the proposed Units 3 and 4.

The cooling water intake structure has been repositioned upstream approximately 46 m (150 ft), which places it approximately 650 m (2130 ft) upstream of the existing intakes for Units 1 and 2, and approximately 427 m (1400 ft) downstream of the location where the stream from Mallard Pond enters the Savannah River. Southern also described a change in the dimensions of the intake structure (Southern 2010), lowering the intake structure floor from elevation 38.1 m to 32.0 m (125 ft to 105 ft). In addition, there would be a slight bend (approximately 30 degrees) roughly halfway down the canal to orient the mouth of the intake canal perpendicular to the river. Figure 3-3 illustrates the revised intake structure and wetlands in its vicinity.

Southern determined the information in the preceding paragraph to be new but not significant information, and provided no other new information related to the cooling system in the COL ER. During its review of Southern's process for identifying new and significant information and during the audit at the VEGP site, the staff found no additional new information that warranted further analysis.

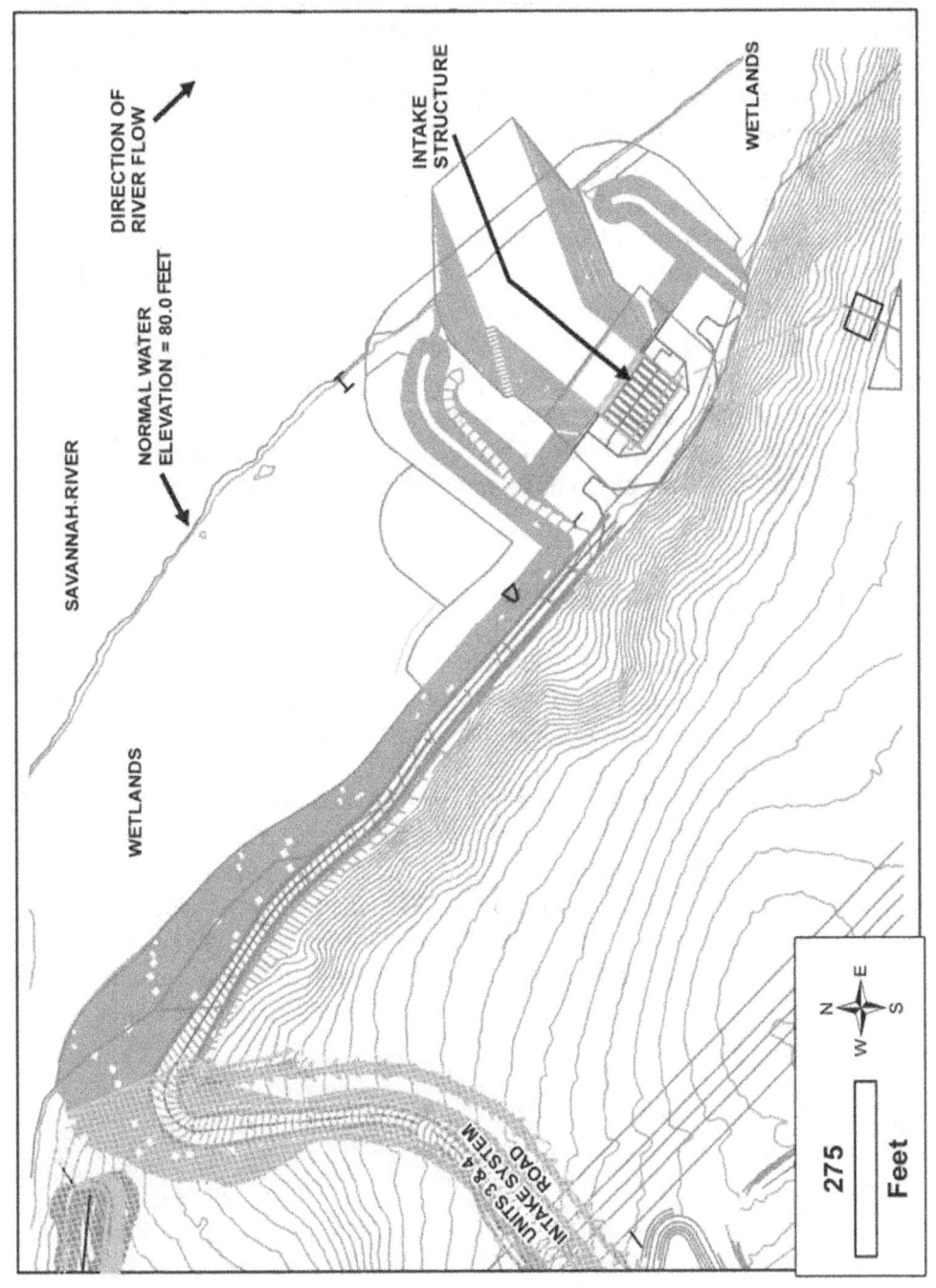

Figure 3-3. Revised Intake Structure and Surrounding Wetlands

Site Layout and Plant Description

3.2.3 Radioactive Waste Management System

Section 3.2.3 of the ESP EIS (NRC 2008) and Section 3.5 of the ESP ER (Southern 2008) provided summary descriptions of the liquid, gaseous, and solid radioactive waste-management systems for the AP1000 reactor, based on Revision 15 of the *AP1000 Design Control Document* (Westinghouse 2005). The summaries of the radioactive waste-management system presented in the ESP EIS are augmented below where additional descriptive information was provided by Southern in its COL application (Southern 2009). A more detailed description of these systems can be found in Chapter 11 of Revision 17 of the *AP1000 Design Control Document* (Westinghouse 2008). The description of the radioactive waste-management system provided in the COL ER is based on information from Revision 17 of the *AP1000 Design Control Document* (Westinghouse 2008). None of the changes in the description of the radioactive waste-management system from Revision 15 to Revision 17 of the *AP1000 Design Control Document* is considered to be significant for the purposes of the environmental review. In particular, the radioactive effluent release source terms are identical for Revision 15 and Revision 17 of the *AP1000 Design Control Document*. Therefore, there is no change in the design characteristic that is most relevant to dose and other environmental impacts associated with radioactive waste.

3.2.3.1 Liquid Radioactive Waste-Management System

The liquid radioactive waste-management system functions to control, collect, process, handle, store, and dispose of liquids containing radioactive material. Section 3.2.3.1 of the ESP EIS (NRC 2008) described the liquid radioactive waste-management system.

The liquid radioactive effluent source term for the proposed Units 3 and 4, taken from Revision 15 of the *AP1000 Design Control Document* (Westinghouse 2005), was presented in Appendix G, Table G-1 of the ESP EIS (NRC 2008). The liquid radioactive effluent source term presented in Revision 17 of the *AP1000 Design Control Document* (Westinghouse 2008) is unchanged from Revision 15 of the *AP1000 Design Control Document* (Westinghouse 2005). Dose calculation results presented in Section 5.9 of the ESP EIS (NRC 2008) remain valid and show that all the dose projected to the maximally exposed individual is within the design objectives identified in 10 CFR Part 50, Appendix I.

3.2.3.2 Gaseous Radioactive Waste-Management System

The gaseous radioactive waste-management system functions to collect, process, and discharge radioactive or hydrogen-bearing gaseous wastes. Section 3.2.3.2 of the ESP EIS (NRC 2008) described the gaseous radioactive waste-management system.

Site Layout and Plant Description

The gaseous radioactive effluent release source term for proposed Units 3 and 4, taken from Revision 15 of the *AP1000 Design Control Document* (Westinghouse 2005), was presented in Appendix G, Table G-4 of the ESP EIS (NRC 2008). The gaseous radioactive effluent source term presented in Revision 17 of the *AP1000 Design Control Document* (Westinghouse 2008) is unchanged from Revision 15 of the *AP1000 Design Control Document* (Westinghouse 2005). The results of calculations presented in Section 5.9 of the ESP EIS (NRC 2008) remain valid and show that all the projected dose to the maximally exposed individual is within the design objectives identified in 10 CFR Part 50, Appendix I.

3.2.3.3   Solid Radioactive Waste-Management System

The solid radioactive waste-management system functions to treat, store, package, and dispose of dry or wet solids. Section 3.2.3.3 of the ESP EIS (NRC 2008) described the solid radioactive waste-management system. Southern provided no new and significant information related to radioactive waste systems in the COL ER (Southern 2009), and the staff found no new and significant information during its review of Southern's process for identifying new and significant information and during the audit at the VEGP site. However, Section 6.1 of this SEIS describes the NRC staff's assessment of the potential environmental impacts that might occur if permanent disposal facilities for low-level solid radioactive waste remain unavailable to VEGP and Southern's contingency plans for interim management of such waste need to be implemented.

3.2.4   Nonradioactive Waste Systems

Section 3.2.4 of the ESP EIS (NRC 2008) and Section 3.6 of the ESP ER (Southern 2008) described the nonradioactive waste systems for the VEGP site. Southern provided no new and significant information related to nonradioactive waste systems in the COL ER (Southern 2009), and the staff found no new and significant information during its review of Southern's process for identifying new and significant information and during the audit at the VEGP site.

3.3 Power Transmission System

Section 3.3 of the ESP EIS (NRC 2008) described Southern's proposed system for transmitting the power produced by the proposed Units 3 and 4 to the regional distribution grid.

As described in Section 3.3 of the ESP EIS, Southern determined that one additional 500-kV transmission line in a new transmission line right-of-way would be required. The new transmission line would connect the substation for the proposed Units 3 and 4 to the Thomson substation located west of Augusta, Georgia. The precise route of the new transmission line right-of-way has yet to be determined, but it would be within a previously defined Representative Delineated Corridor, as summarized in Section 2.7.1 of this document.

Site Layout and Plant Description

Southern provided no new and significant information regarding the route of the new transmission line right-of-way in its COL ER (Southern 2009), and the staff found no additional new and significant information during its review of Southern's process for identifying new and significant information and during the audit at the VEGP site.

3.4 References

10 CFR Part 50. Code of Federal Regulations, Title 10, *Energy*, Part 50, "Domestic Licensing of Production and Utilization Facilities."

10 CFR Part 52. Code of Federal Regulations, Title 10, *Energy*, Part 52, "Licenses, Certifications, and Approvals for Nuclear Power Plants."

Southern Nuclear Operating Company, Inc. (Southern). 2008. *Vogtle Early Site Permit Application: Part 3. Environmental Report.* Revision 5, Southern Company, Birmingham, Alabama. Package Accession No. ML091550858.

Southern Nuclear Operating Company, Inc. (Southern). 2009. *Vogtle Electric Generating Plant, Units 3 and 4, COL Application: Part 3 Environmental Report.* Revision 1, September 23, 2009. Southern Company, Birmingham, Alabama. Accession No. ML092740400.

Southern Nuclear Operating Company, Inc. (Southern). 2010. Letter from C.R. Pierce, Southern, to NRC, "Southern Nuclear Operating Company Vogtle Electric Generating Plant Units 3 and 4 Combined License Application, Response to Request for Additional Information Letter on Environmental Issues." ND-10-0023 dated January 8, 2010. Southern Company, Birmingham, Alabama. Accession No. ML100120479.

U.S. Nuclear Regulatory Commission (NRC). 2008. *Final Environmental Impact Statement for an Early Site Permit (ESP) at the Vogtle Electric Generating Plant Site.* NUREG-1872, Vols. 1, 2, and Errata, Washington, D.C. Accession Nos. ML082240145; ML082240165, ML082260203; ML082550040.

Westinghouse Electric Company, LLC (Westinghouse). 2005. *AP1000 Design Control Document.* AP1000 Document. APP-GW-GL-700, Revision 15, Westinghouse Electric Company, Pittsburgh, Pennsylvania. Package Accession No. ML053480403.

Westinghouse Electric Company, LLC (Westinghouse). 2008. *AP1000 Design Control Document.* AP1000 Document. APP-GW-GL-700, Revision 17, Westinghouse Electric Company, Pittsburgh, Pennsylvania. Package Accession No. ML083230168.

4.0 Environmental Impacts of Construction

In Chapter 4 of the early site permit (ESP) environmental impact statement (EIS) (NRC 2008a), the U.S. Nuclear Regulatory Commission (NRC) staff provided an analysis of the environmental impacts of constructing the proposed Units 3 and 4 at the Vogtle Electric Generating Plant (VEGP) site. The applicant, Southern Nuclear Operating Company, Inc. (Southern), in its environmental report (ER) evaluated new and potentially significant information related to the impacts of construction in as part of its combined license (COL) application (Southern 2009a). The NRC staff reviewed Southern's process for identifying new and significant information, but also conducted its own independent review to verify whether new and significant information had been identified. The results of that review are presented in this chapter. Sections 4.1 through 4.9 discuss the potential new and significant information regarding the impacts on land use; meteorology and air quality; water use and quality; terrestrial and aquatic ecosystems; socioeconomics; historic and cultural resources; environmental justice; nonradiological health effects; and radiological health effects. Section 4.10 describes the applicable measures and controls that would limit the adverse impacts of construction of the proposed Units 3 and 4. An overview of the site redress plan that is applicable to both the Limited Work Authorization (LWA) issued concurrently with the ESP and the second LWA requested by Southern as part of its COL application is provided in Section 4.11. A summary of the construction-related impacts is presented in Section 4.12. References cited in this chapter are listed in Section 4.13. Cumulative impacts of construction and other past, present, and future actions are discussed in Chapter 7. The technical analyses provided in this chapter support the results, conclusions, and recommendations presented in Chapter 11.

Because the VEGP COL application references an approved ESP, the significance levels of the potential adverse impacts for the various areas evaluated will remain the same as documented in the ESP EIS (NRC 2008a) unless new and significant information has been identified that would change the original significance level. The definition of new and significant information is documented in a 2007 *Federal Register* notice (72 FR 49352) and is described in Chapter 1 of this supplemental EIS (SEIS).

4.1 Land-Use Impacts

This section provides information on land-use impacts associated with construction of proposed Units 3 and 4 at the VEGP site. Topics discussed are land-use impacts at the VEGP site and in the vicinity of the site (Section 4.1.1) and land-use impacts in transmission line rights-of-way (ROW) and offsite areas (Section 4.1.2).

Environmental Impacts of Construction

4.1.1 The Site and Vicinity

The NRC staff's assessment of the land-use impacts related to construction of the proposed Units 3 and 4 was provided in Section 4.1.1 of the ESP EIS (NRC 2008a). The assessment addressed the land area that would be impacted by various construction activities. Based on the staff's analysis in the ESP proceeding, the staff concluded that the land-use impacts of construction would be SMALL.

In the ER included in its COL application, Southern indicated that there is no new and significant information regarding construction-related impacts on land use (Southern 2009a, 2010a). During its review of the COL application, the NRC staff performed an independent review of potential new and significant information related to land use by reviewing Southern's ER, auditing Southern's process for identifying new and significant information, examining other information available at the site audit, and considering applicable regulations and reference documents. This review identified the following new information that warranted further review:

- The VEGP site land area impacted on a long-term basis would increase from the 131 ha (324 ac) stated in the ESP EIS to approximately 153 ha (379 ac) (Southern 2009b). The revised area includes land for the fire training facility and the simulator building.

- The VEGP site land area impacted on a short-term basis would increase by approximately 108 ha (267 ac) to a total of 200 ha (494 ac). The additional land area consists of three onsite locations that would be used as a source of Category 1 and Category 2 backfill. The staff analyzed the environmental impacts associated with this additional land in an Environmental Assessment (EA) and finding of no significant impact (NRC 2010a).

- The entire VEGP site has been designated an Energy Production District in the Burke County Comprehensive Plan (MACTEC 2007).

The NRC staff determined that the new information does not have the potential to change the staff's impact characterization in the ESP EIS. The reasons for this determination are (1) the additional affected acreage is on the VEGP site and (2) the entire VEGP site is designated an Energy Production District in the Burke County Comprehensive Plan (MACTEC 2007). Based on this review, the staff determined that the conclusion presented in Section 4.1.1 of the ESP EIS remains valid.

Southern indicated in a new and significant information evaluation (Southern 2010b) that it may subsequently seek to obtain engineering grade backfill materials from an existing, permitted, offsite borrow source. Southern stated that it has not made a final decision on whether to submit an ESP license amendment request (LAR) to the NRC to use this borrow source, and will not make the decision until it determines whether the already-approved onsite sources will

Environmental Impacts of Construction

be sufficient for its construction needs. The staff recognizes that the use (or possible expansion) of an offsite borrow source could have adverse impacts to land-use; however, because the extent to which such an offsite source would be disturbed or expanded, if it is even needed at all, is not presently known, and the potential significance of those land use impacts cannot be evaluated until a LAR (to use offsite borrow sources) is submitted. If Southern submits a LAR to use an offsite source, the staff would conduct an environmental review as part of its determination on that LAR. .

4.1.2 Transmission Line Rights-of-Way

The NRC staff's assessment of the land-use impacts related to the construction of the planned new transmission lines and ROW to serve proposed Units 3 and 4 was provided in Section 4.1.2 of the ESP EIS (NRC 2008a). Based on the staff's analysis, impacts to land use were considered to be MODERATE.

In its COL ER (Southern 2009a), Southern indicated that there is no new and significant information regarding construction-related impacts on the transmission line ROW. During its review of the COL application, the NRC staff independently verified that no new and significant information was available related to construction impacts on the transmission line ROW by reviewing Southern's ER, auditing Southern's process for identifying new and significant information, examining other information available at the site audit, and considering applicable regulations and reference documents. Based on this review, the staff determined that the conclusion presented in Section 4.1.2 of the ESP EIS that the impacts would be MODERATE remains bounding and valid.

4.2 Meteorological and Air-Quality Impacts

The NRC staff's assessment of meteorological and air quality construction-related impacts, including dust generation during ground clearing and emissions from construction equipment and workers' vehicles, was provided in Section 4.2 of the ESP EIS (NRC 2008a). Based on the staff's analysis, construction-related impacts to meteorology and air quality were considered to be SMALL.

In its COL ER (Southern 2009a), Southern indicated that there is no new and significant information regarding construction-related impacts on meteorology and air quality. During its review of the COL application, the NRC staff performed an independent review of potential new and significant information related to meteorology and air quality by reviewing Southern's ER, auditing Southern's process for identifying new and significant information, examining other information available at the site audit, and considering applicable regulations and reference

Environmental Impacts of Construction

documents. The review identified new information related to potential changes in construction traffic as well as changes to the National Ambient Air Quality Standard (NAAQS) for ozone that warranted further review.

During the September 2009 site audit, Southern indicated that a traffic study had been completed in July 2009 (Neel-Schaffer 2009). The traffic study uses different workforce and shift assumptions than were used in the ESP EIS (NRC 2008a); however, the staff determined that these assumptions are reasonable and the results remain consistent with the ESP EIS. In addition to the vehicle traffic analyzed in the traffic study, Southern has indicated the potential need for additional truck deliveries if more backfill material is needed than could be obtained onsite (Southern 2010b). Southern stated that traffic impacts would be minimized by using different routes near the site for inbound and outbound trucks. Although the potential truck traffic would result in more air emissions, these emissions would be temporary and would be completed before peak construction begins (Southern 2010b).). Therefore, the staff, after analyzing the new information identified in Southern's traffic study, finds that the air quality conclusions reached in the ESP EIS remain unaffected because the changes have a marginal effect on the staff's previous conclusions.

As discussed in Section 2.3, the U.S. Environmental Protection Agency (EPA) promulgated a revision to the NAAQS for ozone on March 12, 2008. The final rule (73 FR 16436) reduced the ozone standard from 0.084 ppm to 0.075 ppm. Section 107(d)(1) of the Clean Air Act requires each state to submit, within 1 year of the revised standard, its recommended designation (i.e., attainment, non-attainment, or unclassified) for each county. On March 12, 2009, the Georgia Department of Natural Resources (GDNR) issued a letter to the EPA providing its recommended designations. Under those recommendations, Burke County remains unclassified/attainment with respect to the new ozone standard (GDNR 2009a). EPA will make its final determination on attainment status no later than March 2011. Based upon on the staff's review of new and significant information and the fact that GDNR has determined that Burke County will remain designated as an attainment area with respect to the NAAQS standard, the NRC staff determined that the new information was not significant the conclusions presented in the ESP EIS remain valid.

4.3 Water-Related Impacts

The NRC staff's assessment of the water-related impacts associated with construction of the proposed Units 3 and 4 at the VEGP site were provided in Section 4.3 of the ESP EIS (NRC 2008a). Based on the staff's analysis, construction-related impacts of hydrological alterations and on water use and water quality were considered to be SMALL.

In its COL ER (Southern 2009a) and RAI responses (Southern 2010c), Southern provided new information on the proposed intake structure design, as described in Section 3.2.2. Changes to

Environmental Impacts of Construction

the design (Southern 2010c) do not substantially modify the width of the intake canal or the length of the canal extending beyond the existing river bank. The impacts of hydrological alterations resulting from construction activities would thus remain localized and temporary as concluded in the ESP EIS (NRC 2008a).In support of its recent requests to amend the ESP site safety analysis report, Southern provided new information regarding additional onsite borrow areas from which it sought to obtain backfill material, including three new borrow areas in previously undeveloped portions of the VEGP site (Southern 2010d, e). The NRC staff, as part of its review of hydrological alterations associated with the three new borrow areas, relied on the environmental assessments supporting the amendments to the ESP (NRC 2010a, b). Southern stated in its May 24, 2010, submittal that these borrow areas are included in their National Pollutant Discharge Elimination System (NPDES) permit for construction stormwater. Southern also indicated that all excavations would be redressed according to the site-specific Erosion Sedimentation and Control Plan of the NPDES permit. Additionally, Southern stated that,the excavations would neither intersect the water table nor require dewatering. Just as important, the NPDES permit along with the recently acquired Clean Water Act Section 401 certification, and U.S. Army Corps of Engineers (USACE)an individual Department of the Army Clean Water Act Section 404/Rivers and Harbors Act Section 10 permit, will ensure impacts from the additional excavations are minimized. Based on the above, the NRC staff determined that the conclusions reached in the ESP EIS with respect to surface water and groundwater remains valid for excavations from the new borrow areas (NRC 2010a, b).

During its review of the COL application, the NRC staff performed an independent review of potential new and significant information regarding water-related impacts of construction by reviewing Southern's ER, auditing Southern's process for identifying new and significant information, examining other information available at the site audit (including permits for groundwater withdrawal and dewatering of the surficial aquifer during construction) and provided by Southern subsequent to the site audit, reviewing information submitted as part of Southern's ESP license amendment requests, and considering applicable regulations and reference documents. Beyond the information identified by Southern and discussed above, the staff's review identified no additional information requiring further staff consideration. Based on this review, the staff determined that the conclusions presented in the ESP EIS, that impacts would be SMALL, remain valid.

4.4 Ecology

This section provides information on terrestrial and aquatic resource impacts associated with construction of the proposed Units 3 and 4 at the VEGP site. Topics discussed are terrestrial and aquatic resource impacts at the VEGP site and in the vicinity of the site (Sections 4.4.1 and 4.4.2).

Environmental Impacts of Construction

4.4.1 Terrestrial Impacts

The NRC staff's assessment of the potential construction impacts to terrestrial resources, including impacts to Federal and State-listed threatened and endangered species, was provided in Sections 4.4.1 and 4.4.3 of the ESP EIS (NRC 2008a). Terrestrial-resource-related impacts of construction, including impacts on Federal and State-listed species that are discussed in the ESP EIS include wildlife habitat removal during ground clearing, direct and indirect impacts to wetlands during construction, wildlife displacement and mortality related to construction activities and increased traffic, avian collisions with tall structures during construction, and noise from construction activities. Based on the staff's analysis, construction-related impacts to terrestrial resources were considered to be SMALL in the vicinity of the VEGP site. The construction-related impacts on terrestrial resources in the vicinity of the new transmission line were considered to be SMALL to MODERATE because of the uncertainty regarding the actual transmission line route, as well as the uncertainty regarding the distribution of State-protected species along and within the ROW.

In its COL ER (Southern 2009a), Southern indicated that there is no new and significant information regarding construction-related impacts on terrestrial resources. During its review of the COL application, the NRC staff performed an independent review of potential new and significant information related to terrestrial resources by reviewing Southern's ER, reviewing information submitted as part of the ESP LAR activities to obtain backfill from additional onsite borrow areas, auditing Southern's process for identifying new and significant information, examining other information available at the site audit, considering applicable regulations and reference documents, and contacting the South Carolina Department of Natural Resources (SCDNR), U.S. Fish and Wildlife Service (FWS), and GDNR (NRC 2010c, d, e; SCDNR 2009; GDNR 2009b, c; FWS 2010a, b). This review identified new information related to construction-related impacts to wildlife habitat, wetlands, and Federal and State-listed species that warranted additional staff analysis.

Information relating to additional proposed onsite borrow areas was submitted by Southern on March 12, 2010, as part of the new and significant evaluation for the COL (Southern 2010a). Southern also submitted information pertaining to these borrow areas in subsequent submittals supporting its LAR to obtain backfill material from areas not previously identified in the ESP (Southern 2010e, f).

The borrow areas requested under Amendment 1 were located in onsite areas whose disturbance had already been evaluated in the ESP EIS, thus staff's Amendment 1 EA concluded that terrestrial resource impacts associated with these locations would be consistent with the impacts previously evaluated in the ESP EIS and found not to be significant (NRC 2010b). With respect to the borrow locations requested under Amendment 2, which were not previously evaluated in the ESP EIS, the NRC staff described and evaluated the associated

Environmental Impacts of Construction

potential impacts on terrestrial resources within these areas in the Amendment 2 EA issued in June 2010 (NRC 2010a). Accordingly, as described further below, the staff incorporates the description and analysis in the Amendment 2 EA by reference in this SEIS.

As discussed in the ESP EIS (NRC 2008a), approximately 225 ha (556 ac) would be disturbed during construction of proposed Units 3 and 4, including 131 ha (324 ac) that would be permanently disturbed and an additional 94 ha (232 ac) that could be temporarily disturbed. Southern updated the estimated acreage needed for construction of proposed Units 3 and 4 and currently estimates that approximately 353 ha (873 ac) would be disturbed by construction of the proposed Units 3 and 4, including approximately 153 ha (379 ac) that could be permanently disturbed for facilities and onsite infrastructure; 92 ha (227 ac) that would be temporarily disturbed for parking, laydown areas, and spoils piles; and 108 ha (267 ac) that have been cleared and excavated for backfill material (Southern 2009a, b; 2010a, c, d).

The additional 22 ha (55 ac) impacted for permanent facilities would result in a change in habitat types impacted for some facilities. An additional 1.2 ha (3.0 ac) of planted pines, previously disturbed areas, and open fields would be cleared during construction of permanent facilities. An estimated 21 ha (52 ac) of hardwood habitat would be lost to permanent structures and facilities, representing an increase from the 2 ha (5 ac) that was estimated in the ESP EIS. This additional acreage is a fragmented mosaic of hardwood remnants interspersed among planted pine and previously disturbed areas. The updated onsite hardwood disturbance estimates are still a small fraction (less than 0.1 percent) of the total acres of hardwood habitat available (31,669 ha [78,253 ac]) within 16 km (10 mi) of the site (USGS 2001).

Hardwood habitats have much greater plant species and structural diversity than upland fields, planted pine forests, and previously disturbed areas, and are thus assumed to be much more important as wildlife habitat. However, as noted above, the updated onsite hardwood habitat lost to permanent structures and facilities represents a small percentage of the total available hardwood habitat available onsite and in the vicinity of the VEGP site. In addition, as discussed in the Amendment 2 EA issued in June 2010 (NRC 2010a), approximately 108 ha (267 ac) in three locations composed of planted longleaf (*Pinus palustris*), loblolly (*P. taeda*), and slash pines (*P. elliottii*) will be cleared to obtain backfill material. The areas would be stabilized with permanent vegetation when land-disturbing activities have been completed. Southern has committed to replanting all the areas in longleaf pine, if possible. Two sandhills species, the southeastern pocket gopher (*Geomys pinetis*) and the sandhills milkvetch (*Astragalus michauxii*), both of which are listed as State-threatened by GDNR, were found in one of the proposed borrow areas. The NRC staff discussed the loss of sandhills habitat with GDNR. GDNR indicated that there is a general concern for the loss of sandhills habitat. However, sandhills habitat quality in the areas being affected by obtaining the additional backfill material authorized by the ESP amendments is considered to be marginal compared to the quality of sandhills habitat located on the northern section of the VEGP site, which would not be disturbed

Environmental Impacts of Construction

(NRC 2010e, GDNR 2009d). Southern has voluntarily collaborated with GDNR and the Georgia Plant Conservation Alliance to mitigate impacts to the southeastern pocket gopher and the sandhills milkvetch. In the Amendment 1 and Amendment 2 EAs, issued in May and June 2010, the staff also described, among other matters, the applicable stormwater permitting provisions and the best management practices Southern intends to follow for erosion and sediment control (NRC 2010a, b). In the Amendment 2, EA the staff also evaluated the impacts to habitat from relocation of the State-threatened species associated with obtaining the additional backfill material and determined that there would not be any destabilizing effect on terrestrial resources. With respect to the EAs for ESP Amendments 1, 2, and 3, the staff determined that approval of the ESP amendments would have no significant impact (NRC 2010a, b, f).

On September 30, 2010, Southern received an individual Department of the Army Clean Water Act Section 404/Rivers and Harbors Act Section 10 permit for the VEGP site (USACE 2010). This permit authorizes impacts to 3.75 ha (9.23 ac) of jurisdictional wetland area, which represents approximately 5 percent of the 69 ha (170 ac) of wetlands that occur on the VEGP site. As discussed in the ESP EIS, Southern originally estimated that approximately 8.5 ha (21.0 ac) of wetlands would be directly affected by Units 3 and 4 construction activities (NRC 2008a). The updated wetlands information reflects a decrease in the amount of wetland habitat that would be impacted during construction. On March 3, 2011, Southern provided an update to the NRC regarding a change in its September 30, 2010 Department of the Army permit. This amendment gave Southern permission to acquire additional wetland credits from other approved banks. The compensatory mitigation will consist of the purchase of 45.53 and 24.87 wetland mitigation credits from Phinizy Swamp Mitigation Bank and Brushy Creek Mitigation Bank, respectively; both are approved U.S. Army Corps of Engineers (USACE) mitigation banks (Southern 2011). No new information was identified regarding potential impacts to wetlands within the new transmission line right-of-way. To satisfy the remainder of the wetland mitigation requirements, Southern will purchase the wetland mitigation credits at the Margin Bay Mitigation Bank or the Wilhelmina Morgan Mitigation Bank; both of these banks are also USACE approved mitigation banks in the secondary service area. Accordingly, the staff's conclusion in the ESP EIS with respect to impacts to wetlands remains bounding.

During its review, the NRC staff also identified new information related to onsite and offsite infrastructure alterations in connection with how the large reactor components and other materials would be delivered to the site.

Southern submitted a letter to the NRC in February 2010 stating that large components and other construction materials would be transported to the VEGP site via rail, using the Norfolk-Southern rail line from Savannah, Georgia, to Waynesboro, Georgia, where the line connects with the spur to VEGP (Southern 2010g). The letter states that there would be no substantive changes made to either the Norfolk-Southern rail line or to the private spur line to VEGP to support the shipment of an estimated 70 components and pieces of heavy equipment that will require special cars or size considerations. Some routine track maintenance, (e.g., replacement

Environmental Impacts of Construction

of cross ties and/or ballast) may be necessary, but no land disturbing activities or modifications of bridges, overpasses, or other structures would be needed. Southern stated that modifications would be needed for the onsite rail yard and rail spur to support storage and unloading of equipment and materials delivered by rail. The rail yard is located in an area previously disturbed by construction of VEGP Units 1 and 2 and is within the current disturbance footprint.

Based on the information in Southern's February 2010 letter (Southern 2010g) and in the information received in Southern's RAI response (Southern 2010c), which indicates that no significant land-disturbing activities will be needed to support rail transport and delivery of large components to the site, the staff does not expect either the transportation of reactor components to the site or modifications to the onsite rail yard and spur to adversely impact terrestrial resources, including threatened and endangered species.

The combined loss of sandhills habitat, hardwood forest and bottomland wetlands, planted pine habitat, and open field habitat during the construction of Units 3 and 4 and the clearing of the new borrow areas for backfill material would reduce available habitat for wildlife, including two State-threatened species, the southeastern pocket gopher and sandhills milkvetch. However, Georgia is currently working to restore sandhills habitat across the state, which includes planting longleaf pine. Southern has committed to replant the disturbed onsite borrow areas in longleaf pine, if possible (Southern 2010h). In addition, the areas that have been disturbed are of marginal quality compared to the remaining higher quality habitat available onsite. Planted pine, open field, and bottomland hardwood wetland habitats are available in other locations onsite and in the region. Furthermore, as explained in the Amendment 2 EA, the potential losses to the southeastern pocket gopher and sandhills milkvetch are isolated and will not jeopardize the stability or viability of any of the remaining populations in Georgia. These populations occur in different locations throughout the state and each population is not dependent on the success of others. Therefore, and for the reasons discussed above and in more detail in the Amendment 2 EA (NRC 2010a), construction activities associated with the proposed action are not expected to destabilize terrestrial resources, including the State-threatened southeastern pocket gopher and sandhills milkvetch.

As part of the NRC's responsibilities under Section 7 of the Endangered Species Act, the NRC staff prepared a (BA) documenting potential impacts on the Federally listed threatened or endangered species as a result of the limited site preparation activities at the VEGP site (including construction of the onsite portion of the new 500-kV transmission line). The BA was submitted to FWS on January 25, 2008 (NRC 2008b), and FWS concurred with the findings on September 19, 2008 (FWS 2008).

In a letter dated January 7, 2010, NRC requested that the FWS Field Office in Brunswick, Georgia, provide information regarding Federally listed species and critical habitat that may

Environmental Impacts of Construction

have changed since the 2008 consultation (NRC 2010c). On February 12, 2010, FWS provided a response letter indicating listed species under FWS purview had been adequately addressed for limited site-preparation activities on the VEGP site (FWS 2010a). On October 20, 2010, FWS provided an updated list of Federally listed threatened or endangered species that can be expected to occur in the project area (FWS 2010b). FWS identified four Federally listed terrestrial plant and animal species that may occur on or in the vicinity of the VEGP site as well as within the vicinity of the Representative Delineated Corridor (RDC) (FWS 2010b). These four species are the red-cockaded woodpecker (*Picoides borealis*), wood stork (*Mycteria americana*), Canby's dropwort (*Oxypolis canbyi*), and eastern indigo snake (*Drymarchon couperi*).

In addition to the Federally listed species, FWS provided information on the bald eagle (*Haliaeetus leucocephalus*) and the gopher tortoise (*Gopherus polyphemus*) in the response letter. FWS indicated there are eagle nests in Jefferson and McDuffie Counties, including one nest in the RDC (FWS 2010b). The location of the eagle nest in the RDC was discussed in the ESP EIS. Further, the impacts to the red-cockaded woodpecker, wood stork, and Canby's dropwort were discussed in the ESP EIS.

The eastern indigo snake and gopher tortoise were not included in the analysis in the ESP EIS. The eastern indigo snake was not included because it was not previously listed in FWS species lists for the counties within the project area (Burke, Jefferson, McDuffie or Warren Counties) (NRC 2008a). Likewise, GDNR indicated there have been no known occurrences of the gopher tortoise in the project area (GDNR 2009b, c).

The information discussed in this section focuses on species not previously considered in the ESP. This includes the eastern indigo snake and gopher tortoise, both identified by FWS in its recent letter as species that can be expected, to occur in the project area. FWS noted that the gopher tortoise is not a Federally listed species in Georgia; however, it is under review by FWS (FWS 2010b). Sandhills habitat that could support the gopher tortoise and the eastern indigo snake is present in the project area (GDNR 2009b). Therefore, these species are discussed below.

NRC submitted a biological assessment (BA) to FWS on February 24, 2011 to document potential impacts on Federally listed threatened or endangered terrestrial species resulting from operation of Units 3 and 4 and ancillary facilities, as well as construction and operation of the proposed transmission line ROW. This BA is included in Appendix F. A BA documenting potential impacts on the Federally listed threatened or endangered species as a result of the site preparation and preliminary construction of the nonsafety-related structures, systems, or components on the VEGP site was submitted to FWS on January 25, 2008 (NRC 2008b), and FWS concurred with the findings on September 19, 2008 (FWS 2008) Appendix F.

Environmental Impacts of Construction

The eastern indigo snake was Federally listed as threatened by FWS in 1978 (43 FR 4026). In its October 20, 2010, letter to NRC, FWS noted that there are no documented occurrences of the eastern indigo snake on the VEGP site or in the RDC ; however, FWS recommends that any pedestrian surveys of sandhill habitats, especially those with gopher tortoise burrows, should include cursory surveys to determine the presence of the eastern indigo snake (FWS 2010b). The eastern indigo snake is not documented in Burke County or any of the counties crossed by the proposed transmission line ROW. Suitable habitat may occur in the RDC, and gopher tortoise burrows are in the vicinity. However, because the project area is outside the historic and current range of the eastern indigo snake and because no further impacts to sandhills habitat are projected to occur on the VEGP site, the staff determined that it is unlikely that either building activities at the VEGP site or the construction of the proposed transmission line will adversely affect this species.

There are no known Federally threatened or endangered terrestrial species on the VEGP site and/or in the RDC, with the exception of the American alligator (*Alligator mississippiensis*). As explained in the ESP EIS and Amendments 1 and 2 of the EA, while an alligator has previously been observed in Mallard Pond on the VEGP site (See Figures 3-1 and 3-2), alligators appear to be relatively common in the Savannah River near and on the VEGP site, and construction impacts on alligators would be negligible because any displacement would be temporary and ample habitat exists in the region. Furthermore, there are no adequate nesting and foraging locations for the Federally endangered red-cockaded woodpecker in the additional onsite areas that have been and would be disturbed. Details on the 21 ha (52 ac) currently enrolled in the Red-Cockaded Woodpecker Safe Harbor Agreement acreage that would be impacted are discussed in the EA for ESP Amendment 2 (NRC 2010a); Southern intends to retain the this area under the agreement and to replant it in longleaf pine, if possible, once the areas have been stabilized and closed out.

As noted above, the October 20, 2010, FWS letter included information on the gopher tortoise and the bald eagle. The gopher tortoise is a Georgia state-threatened species and is currently under review by the FWS to be listed as a Federally threatened species (FWS 2010b). There are no known populations of the gopher tortoise on the VEGP site as well as within the RDC (GDNR 2009c; FWS 2010b). Southern submitted a draft Candidate Conservation Agreement with Assurances (CCAA) for the gopher tortoise at the VEGP site. This CCAA is currently under review by FWS (SERPPAS 2010). In light of the CCAA and because no further impacts to sandhills habitat are projected to occur on the VEGP site, the staff considers it unlikely that the gopher tortoise will be affected onsite. The draft CCAA does not include the offsite portions of the proposed transmission line. In the October 20, 2010, letter to NRC, FWS recommended that tortoise surveys be included in surveys that are conducted where sandhills habitat exists. FWS stated that there are several areas within the RDC that have sandhills habitat that may contain gopher tortoises (FWS 2010b). Georgia Power Company (GPC) would site the transmission line ROW in accordance with Georgia Code Title 22, Section 22-3-161

Environmental Impacts of Construction

(Ga. Code Ann. 2004). GPC's procedures for implementing this code include consultation with GDNR as well as an evaluation of impacts to special habitats (including wetlands) and threatened and endangered species. Impacts to State-protected species are likely to be minimal provided that adequate surveys are conducted prior to commencement of transmission line construction and that consultation with GDNR is initiated, as needed. However, without proper surveys, consultation, and appropriate mitigation, the impact could be greater than negligible in the RDC, which is consistent with the staff's analysis in the ESP EIS.

The bald eagle, a state-threatened species, was Federally delisted under the Endangered Species Act in August 2007. There are bald eagle nests in Jefferson and McDuffie counties in Georgia, and one known location of an active nest within the RDC (FWS 2010b). Potential impacts to the bald eagle were discussed in the ESP EIS. For example, as noted in the ESP EIS, GPC would ensure that the new transmission line ROW would not come within 180 m (600 ft) of this known bald eagle nesting site (GPC 2007).

NRC received comments on the COL draft SEIS from the U.S. Department of Interior expressing concern about avian collisions with tall structures and transmission lines and what mitigative measures GPC will use to minimize impacts (see Appendix F). The ESP EIS included an analysis of construction-related avian collisions with structures, including transmission lines in Section 4.4.1.2. However, additional information on the mitigation measures to minimize impacts to avian species is provided below.

GPC has developed an Avian Protection Program that includes guidelines for siting new transmission lines. When siting new transmission lines, substations, or other GPC facilities, available information on migratory and resident bird populations will be taken into account to ensure that the lines or facilities will have as little adverse impact as practicable on these bird species (GPC 2006).

The Avian Protection Plan states that, in areas where agencies are concerned about the safety of protected birds, consideration of appropriate siting and placement will reduce the likelihood of collisions. When possible, areas with known bird concentrations will be avoided, and vegetation or topographic characteristics that would naturally lead to shielding the birds from collision would be used. If this practice is not possible, installing visibility devices also may reduce the risk of collision. Examples of these devices are marker balls or other line-visibility devices placed in varying configurations, depending on the line or location. The effectiveness of these devices has been validated by Federal and State agencies in conjunction with Edison Electric Institute (GPC 2006).

When designing power transmission lines in high bird-use areas or on Federal land, GPC construction standards for transmission, distribution, and substation equipment and facilities will reflect the most appropriate and practicable "raptor-safe" specifications for new construction consistent with available information. The objective is to provide a spacing of 1.5 m (60 in.)

Environmental Impacts of Construction

between energized conductors and grounded hardware, or to insulate energized hardware if such spacing is not possible. The design standards are consistent with raptor-safe specifications recommended by Federal wildlife agencies (GPC 2006).

No critical habitat for threatened or endangered species is present on the VEGP site as well as within the RDC. Other than the consideration of the indigo snake discussed above, the new information did not reveal impacts that may affect Federally listed species or critical habitat in a manner not previously considered in the ESP EIS. There are no anticipated adverse impacts to Federally listed species as a result of construction on the VEGP site, including within the RDC.

4.4.2 Summary of Terrestrial Impacts

In summary, the staff has reviewed the COL application and subsequent submittals, has performed an independent review of potential new and significant information related to terrestrial resources, has reviewed information submitted in conjunction with the ESP license amendments, has audited Southern's process for identifying new and significant information, has examined information provided at the site audits, has considered applicable regulations and reference documents, and has contacted the GDNR, SCDNR, and FWS.

Southern is required to comply with conditions of the NPDES construction storm water general permit issued by GDNR's Environmental Protection Division, and Southern has committed to using best management practices to minimize impacts from erosion. Southern has voluntarily mitigated impacts to the southeastern pocket gopher and the sandhills milkvetch, both of which are State-threatened species. Southern also has committed to replant longleaf pine in areas that would be disturbed, if possible (Southern 2010f). Longleaf pine is a fundamental component of sandhills habitat and a species ideally suited to the soil type and regional topography.

Based on the total acres of habitat that would be disturbed for the proposed project and Southern's efforts to mitigate impacts to State-threatened species in connection with the use of onsite borrow areas, the NRC staff concludes that site preparation and construction activities related to building VEGP Units 3 and 4 could have a MODERATE impact on local terrestrial resources through the loss of habitat and the displacement of localized populations of two State-threatened species, the southeastern pocket gopher and the sandhills milkvetch, but would not have a destabilizing effect either on wildlife habitats or on the populations of these two State-listed species in Georgia.

The staff also reviewed the information provided above regarding construction-related impacts on terrestrial resources in the vicinity of the new transmission line ROW. This review included consideration of the new information on eastern indigo snake and the gopher tortoise. Although sandhills habitat that could support these species is present, neither species is known to occur in the RDC. Because of the uncertainty regarding the actual transmission line route, as well as

Environmental Impacts of Construction

the uncertainty regarding the distribution of wetlands and State-protected species along and within the ROW, the staff's conclusion that these impacts would be SMALL to MODERATE remains bounding and valid.

Southern indicated in a new and significant information evaluation (Southern 2010b) that it may subsequently seek to obtain engineering grade backfill materials from an existing, permitted, offsite borrow source. Southern stated that it has not made a final decision on whether to submit an ESP LAR to the NRC to use this borrow source, and will not make that determination until it determines whether the already-approved onsite sources will be sufficient for its construction needs. The staff recognizes that the use (or possible expansion) of an offsite borrow source could have adverse impacts to terrestrial resources; however, because the extent to which such an offsite source would be disturbed or expanded, if it is even needed at all, is not presently known, the potential significance of those ecological impacts cannot be evaluated until a LAR for use of offsite borrow sources is submitted. If Southern submits a LAR to use an offsite source, the staff would conduct an environmental review as part of its determination on that LAR.

4.4.3 Aquatic Ecosystem Impacts

The NRC staff's assessment of the aquatic ecology related impacts, including the impacts to aquatic biota in onsite ponds and streams from soil-disturbing activities, to aquatic biota in the Savannah River from construction of the cooling water intake structure, the barge structure, and the discharge structure, and from construction of the proposed transmission line, was provided in Section 4.4.2 of the ESP EIS (NRC 2008a). The impacts to important species, including Federally and State-listed threatened and endangered species, were discussed in Sections 4.4.2 and 4.4.3.2 of the ESP EIS. Based on the staff's analysis in the ESP EIS, construction-related impacts to the aquatic biota in the onsite water bodies and the Savannah River were considered to be SMALL.

In its COL ER (Southern 2009a), Southern indicated that there is no new and significant information regarding construction related impacts on aquatic ecology. The NRC staff independently reviewed Southern's ER, audited Southern's process for identifying new and significant information, examined other information available at the site audit, and discussed potential construction impacts with resource agencies (i.e., FWS, SCDNR, and GDNR; see Appendix F). Southern subsequently provided new information on three additional onsite borrow areas from which it sought to obtain backfill material via license amendment (Southern 2010d, e, f). Based on the information provided by Southern and the NRC analysis in the ESP EIS, the staff concluded in the LAR EAs for Amendments 1 and 2 (NRC 2010a, b) that site preparation and construction activities at the additional onsite borrow locations are similar to those that have been previously analyzed and documented in the ESP EIS, and that the aquatic resource impacts of activities which would be conducted at the borrow areas are consistent with

Environmental Impacts of Construction

the impacts previously examined and found not to be significant. Accordingly, the staff incorporates by reference its analysis in the LAR EAs (NRC 2010a, b).

As part of the NRC's responsibilities under Section 7 of the Endangered Species Act, the staff prepared a BA in connection with the Vogtle ESP review, documenting potential impacts on the shortnose sturgeon (*Acipenser brevirostrum*) as a result of preconstruction activities including constructing the intake and discharge systems and modifying the barge slip for the proposed Units 3 and 4. That BA, which was submitted to the National Marine Fisheries Service on January 25, 2008 (NRC 2008b), concluded that the proposed action is not likely to adversely affect the shortnose sturgeon. The NMFS concurred with that determination (NMFS 2008). In a letter dated September 3, 2010, NRC confirmed with NMFS that the ESP stage consultation encompassed the proposed actions included in the COL application (NRC 2010g).

On October 6, 2010, NMFS published in the *Federal Register* (75 FR 61904) a proposed rule for listing the Carolina and South Atlantic distinct population segments of the Atlantic sturgeon (*Acipenser oxyrinchus oxyrinchus*) as endangered under the Endangered Species Act. In the ESP proceeding, the NRC staff determined that impacts to the Atlantic sturgeon would be SMALL. The staff has determined that the project has not been modified in a way that would cause an effect to the Atlantic sturgeon not previously considered in the ESP proceeding. Nevertheless, because of the listing proposal, the staff compiled information in a conference consultation letter to NMFS on March 2, 2011. A copy of this letter is provided in Appendix F. None of the information compiled by the staff for the Atlantic sturgeon resulted in a change to the conclusions in Chapter 4 of the ESP EIS because none of the contemplated shoreline construction activities will prevent the Atlantic sturgeon from migrating past the site.

On September 30, 2010, Southern received an individual Department of the Army Clean Water Act Section 404/Rivers and Harbors Act Section 10 permit for the VEGP site (USACE 2010). This permit authorizes impacts to 224 m (734 ft) of stream (the Georgia side of the Savannah River), which is equivalent to 0.57 ha (1.42 ac) of open water. In addition, it authorizes impacts to 0.03 ha (0.07 ac) of ephemeral stream in the southeast corner of the site near the debris basins discussed in Section 4.4.1.1 and 4.4.2.2 of the ESP EIS. Compensatory mitigation will consist of the purchase of 2224 stream mitigation credits from the Bath Branch Mitigation Bank, an approved USACE mitigation bank that services the project area (USACE 2010). Southern also received a Section 401 Water Quality Certification from the GDNR dated June 1, 2010 (USACE 2010). Although the amount of affected shoreline described in the Department of the Army permit represents an increase over the 155 m (510 ft) of shoreline disturbance cited in the ESP EIS, it remains a small fraction of the shoreline that bounds the VEGP site and a small fraction of the shoreline habitat on this stretch of the Savannah River. Accordingly, the staff determined that this change does not alter its impact conclusion in the ESP EIS.

Environmental Impacts of Construction

The staff has not identified any additional new information that warranted further analysis in the SEIS. Based on this review, the staff determined that the conclusions presented in the ESP EIS remain valid.

4.5 Socioeconomic Impacts

This section evaluates the social and economic impacts to the surrounding region as a result of constructing the proposed Units 3 and 4 at the VEGP site. Topics discussed are the socioeconomic impacts at the VEGP site and in the 80-km (50-mi) region of the site with an emphasis on Burke, Columbia, and Richmond Counties (Section 4.5).

4.5.1 Physical Impacts

The NRC staff's assessment of the physical impacts, including noise, odor, vehicle exhaust emissions, aesthetics, and dust, were provided in Section 4.5.1 of the ESP EIS (NRC 2008a). Based on the staff's analysis and Southern's representation that it would undertake mitigation measures, construction-related physical impacts on workers and the local public, buildings, roads, and aesthetics were considered to be SMALL, with the exception of a MODERATE impact on aesthetics as a result on construction of new transmission lines.

In its COL ER (Southern 2009a), Southern indicated that there is no new and significant information regarding construction-related physical impacts on workers and the local public, buildings, roads, and aesthetics. During its review of the COL application, the NRC staff independently verified that there is no new and significant information related to physical impacts by reviewing Southern's ER, auditing Southern's process for identifying new and significant information, examining other information available at the site audit, and considering applicable regulations and reference documents.

Based on this review, the staff determined that the conclusions presented in the ESP EIS that impacts would be SMALL, with the exception of MODERATE aesthetic impacts related to transmission lines, remain bounding and valid.

4.5.2 Demography

The NRC staff's assessment of the demographic impacts was provided in Section 4.5.2 of the ESP EIS (NRC 2008a). Based on the staff's analysis described in the ESP EIS, the regional impacts from the in-migration of workers as a result of construction activities were projected to be SMALL in most of the region, but MODERATE in Burke County. Based on information from Southern, the ESP EIS estimated that approximately 2500 construction workers would be expected to in-migrate into the region.

Environmental Impacts of Construction

In its COL ER (Southern 2009a), Southern indicated that there is no new and significant information regarding construction-related demographic impacts on the 80-km (50-mi) region. During its review of the COL application, the NRC staff performed an independent review of potential new and significant information related to demographic impacts by reviewing Southern's ER, auditing Southern's process for identifying new and significant information, examining other information available at the site audit, and considering applicable regulations and reference documents. This review identified new information related to the need for additional onsite and offsite backfill material that warranted evaluation. As explained in the Amendment 1 and 2 EAs (NRC 2010a, b), backfill activities would occur concurrently with other site preparation activities and would not require additional workers beyond the workforce evaluated in the ESP EIS. The staff analyzed the environmental impacts associated with onsite backfill activities in two EAs, both of which resulted in findings of no significant impact (NRC 2010a, b). Accordingly, the staff incorporates by reference its analysis in the LAR EAs. The staff has not identified any additional new information that warranted further analysis in the SEIS.

Based on this review, the staff determined that the conclusions presented in the ESP EIS remain bounding and valid.

4.5.3 Economic Impacts to the Community

The staff's assessment of the economic and tax-related impacts was provided in Section 4.5.3 of the ESP EIS (NRC 2008a). Based on the staff's analysis described in the ESP EIS, construction impacts to the regional economy were considered to be SMALL, with the exception of a possible MODERATE and beneficial impact in Burke County.

In its COL ER (Southern 2009a), Southern indicated that there is no new and significant information regarding construction-related economic impacts to the community. During its review of the COL application, the staff performed an independent review of potential new and significant information related to economic impacts by reviewing Southern's ER, auditing Southern's process for identifying new and significant information, examining other information available at the site audit, considering applicable regulations and reference documents, and discussions with Burke County officials. This review identified new information related to the local unemployment rate that warranted additional evaluation.

As shown in Table 2-1, unemployment rates for Burke, Richmond, and Columbia Counties and statewide in Georgia have risen recently. This development is consistent with the current economic slowdown throughout the United States and is not unique to the VEGP region. In the short term, higher unemployment could lead to an increased demand for social services, a decrease in income tax to the state, and to an extent, a decrease in sales tax to the counties. However, construction of the proposed Units 3 and 4 could alleviate these impacts by providing

Environmental Impacts of Construction

jobs to unemployed individuals either directly at the site or through multiplier-induced, indirect jobs described in Section 4.5.3 of the ESP EIS (NRC 2008a). Construction of the proposed Units 3 and 4 would also provide additional tax revenue for Burke County that would provide funding for any additional social services needed due to the higher unemployment. In the long term, by the time construction peaks, unemployment will likely have had time to adjust and adverse impacts from decreased tax revenue or increased social service demands will have subsided. Based on this review, the NRC staff determined that the conclusions presented in the ESP EIS remain valid.

4.5.4 Infrastructure and Community Service Impacts

The NRC staff's assessment of the infrastructure and community-service impacts was provided in Section 4.5.4 of the ESP EIS (NRC 2008a). Based on the staff's ESP analysis, the infrastructure and community-service impacts from the relocation of workers as a result of construction activities were projected to be SMALL in most of the region with two exceptions. The staff found in the EIS that there remains a possibility of a MODERATE impact on transportation during peak construction if mitigation strategies are not implemented and a MODERATE impact on housing and public services if the less-populated counties see a larger than expected number of in-migrating construction workers.

During the September 2009 site audit, Southern indicated that a traffic study had been completed in July 2009 (Neel-Schaffer 2009). The traffic study uses different workforce and shift assumptions than were used in the ESP EIS (NRC 2008a); however, the staff determined that these assumptions are reasonable and the results remain consistent with the ESP EIS. The traffic study is based on assumptions that 25 percent of workers will carpool during both the day shift, which will consist of 75 percent of the construction workforce, and the nightshift, which will consist of the remaining 25 percent of the workforce. The traffic study does not account for outage workers or truck deliveries. The two scenarios used in the traffic study are the construction ramp-up in January 2011 and the peak construction stage in March 2013. Approximately 1200 construction workers are expected to be present in January 2011. Assuming 75 percent of the workers are on the day shift, 25 percent on the night shift, and that 25 percent of workers would carpool, approximately 675 vehicles will be on the day shift and 225 on the night shift. Approximately 4300 construction workers are expected to be present in March 2013 with approximately 2419 vehicles on the day shift and 806 vehicles on the night shift. In the January 2011 projections, most intersections near VEGP would range from a level of service (LOS) of A to an LOS of C. However, the eastbound and westbound sections of the intersection of River Road and Hancock Road would have LOS D and LOS F ratings, respectively. LOS A is the best rating, corresponding to no wait times at an intersection, and LOS F is the worst rating, corresponding to long wait times at an intersection. According to the new traffic study, intersection ratings during the peak construction period occurring in 2013

Environmental Impacts of Construction

would include as many as five LOS F ratings, with considerable wait times at several intersections.

Recommendations from the traffic study for the 2011 scenario were minor improvements such as restriping affected lanes. The traffic study's recommendations to Southern for the 2013 scenario were more extensive, proposing several additional turn lanes, as well as rerouting existing plant traffic and the realignment of Ebenezer Church Road with the entrance to the VEGP gate. Staggering construction shifts also would alleviate traffic congestion on heavily impacted intersections.

In addition to the vehicle traffic analyzed in the traffic study, Southern has indicated the potential need for additional truck deliveries if additional backfill material is needed that would be obtained offsite. In its analysis of the impact of the truck deliveries, Southern assumed all deliveries would be made during the 10-hour day shift coinciding with Units 1 and 2 operations shift change, but not during the Units 3 and 4 construction shift change. Southern also assumed deliveries would consist of approximately 250 trucks a day. Each truck is the equivalent of 3.5 vehicles on the road by Georgia Department of Transportation definitions. The additional 250 truck deliveries are equivalent to 875 vehicles a day (which equals 87.5 vehicles one way per hour during the 10 hour shift). The additional 87.5 vehicles one way an hour are within the design capacity limits for the roads near the VEGP site even during the current shift changes for the existing Units 1 and 2. Design capacity limits on Georgia roads are 1700 (2-lane roadway) and 2000 (4-lane roadway) vehicles each way per hour. Georgia capacity limits were used for analysis on South Carolina roads too. Impacts would be minimized by using different routes near the site for inbound (SR 23) and outbound (SR 56) trucks. Deliveries are expected to last 7 months and would be completed before the peak of construction begins (Southern 2010b).

Although the July 2009 traffic study uses different (more conservative) assumptions than the ESP EIS (NRC 2008a), the impacts and recommendations are similar. In the ESP EIS (NRC 2008a), the NRC staff concluded that impacts to transportation would be SMALL to MODERATE for local highways and River Road in the vicinity of VEGP. The 2009 traffic study commissioned by Southern and the potential additional backfill truck deliveries further support the MODERATE impact on River Road and other nearby intersections. The traffic study and potential additional backfill truck deliveries confirm that traffic impacts will noticeably alter the local roads during shift changes, but the recommendations also demonstrate that, by implementing mitigating measures, the impacts could be managed. Therefore, the NRC staff determined that the MODERATE conclusion presented in the ESP EIS with respect to transportation impacts remains valid.

In regard to other infrastructure and community-service impacts, there is no new and significant information regarding construction-related impacts in the region within an 80-km (50-mi) radius

Environmental Impacts of Construction

of the VEGP site. During its review of the COL application, the NRC staff independently verified that no new and significant information was available related to infrastructure and community-service impacts by reviewing Southern's ER, auditing Southern's process for identifying new and significant information, examining other information available at the site audit, and considering applicable regulations, reference documents, and discussions with county officials. Based on this review, the staff determined that the conclusions presented in the ESP EIS remain bounding and valid.

4.5.5 Summary of Socioeconomic Impacts

As described in the ESP EIS (NRC 2008a), adverse socioeconomic impacts resulting from construction of proposed Units 3 and 4 range from SMALL to MODERATE, and beneficial impacts range from SMALL to MODERATE. For the reasons described above, these conclusions remain valid.

4.6 Historic and Cultural Resources

The NRC staff's assessment of the construction-related impacts to historic and cultural resources, including sites that are listed or eligible for listing under the National Historic Preservation Act of 1966 (NHPA), was provided in Section 4.6 of the ESP EIS (NRC 2008a). Based on the staff's analysis, construction-related impacts to historic and cultural resources were considered to be MODERATE.

In its COL ER (Southern 2009a), Southern indicated that there is no new and significant information regarding construction-related impacts to historic and cultural resources. During its review of the COL application, the NRC staff performed an independent review of potential new and significant information related to historic and cultural resources by reviewing Southern's ER, auditing Southern's process for identifying new and significant information, examining other information available at the site audit, considering applicable regulations and reference documents, and contact with the Georgia State Historic Preservation Officer (SHPO) Advisory Council, and Tribes (see Appendix C for complete listing).

This review identified new information related to the presence of a historic cemetery on the VEGP site (New South Associates 2007) and mitigation for impacts to a site eligible to the National Register of Historic Places (NRHP), which warranted further staff consideration. Southern has installed a fence around the cemetery, determined that the planned construction actions will not impact the site, and has consulted with the SHPO regarding protection and mitigation of the site. Archaeological site 9BK416 is a large multicomponent prehistoric site and is described the ESP EIS (NRC 2008a). Archaeological site 9BK416 is eligible for listing in the National Register of Historic Places.

Environmental Impacts of Construction

Southern signed a Memorandum of Understanding (MOU) with the Georgia SHPO for "... the preservation of the remaining balance of site 9BK416 from physical disturbance and performance of additional archaeological surveys as directed by the SHPO" (GHPD 2010). In the MOU, Southern states, "The proposed project will disturb approximately 2.5 acres of the estimated 29 total acres of site 9BK416. The disturbance constitutes approximately 8.5 percent of the total estimated site and results from the installation of the river water intake piping, an electrical duct bank and associated ROW clearings. Based on consultation and supporting field surveys, the SHPO determined the proposed project will impact site 9BK416, but will not adversely impact the site." The new information provides further indication that Southern will protect historic and cultural resources on the VEGP site, or mitigate impacts in coordination with the SHPO. As a result of these protective measures proposed by Southern and consultation with the SHPO, the staff concludes that the identification of the historic cemetery and the signed MOU does not change its conclusion that the construction activities will alter but not destabilize the cultural resources in the vicinity of the VEGP site.

The staff's review also identified new information related to Southern's use of backfill from three additional onsite borrow areas as authorized by amendment of the ESP (Southern 2010e). All of the new borrow areas are within the VEGP site and also are within the area of potential effect for the cultural resource analysis included in the ESP EIS (NRC 2008a). The known cultural resources located within the additional borrow areas were recommended as not eligible for inclusion in the NRHP. The Georgia SHPO concurred with this finding by letter (GDNR 2007). In June 2010, NRC consulted with the SHPO regarding the use of the onsite borrow areas and the SHPO "... agreed with NRC that the backfill operations will have no effect to properties listed on or eligible for listing on the National Register of Historic Places..." (GDNR 2010). The staff, incorporates in completing its analysis for the SEIS, relied on the results of its Environmental Assessments completed for a few amendments related to the ESP by reference in this SEIS (NRC 2010a). As a result of the cultural resources analysis, field investigations, procedures Southern has in place for unanticipated cultural resources discoveries, and the consultation with the SHPO, the NRC staff concludes that the use of the additional onsite backfill areas (Southern 2010e) will not change its conclusions in the ESP EIS. Further, the staff found that while the construction activities will likely alter cultural resources in the vicinity of the VEGP site, the resource will not not destabilized.

Southern indicated in a new and significant information evaluation (Southern 2010b) that it may subsequently seek to obtain engineering grade backfill materials from an existing, permitted, offsite borrow source. Southern stated that it has not made a final decision on whether to submit an ESP LAR to the NRC to use this borrow source, and will not make that determination until it determines whether the already-approved onsite sources will be sufficient for its construction needs. The staff recognizes that the use (or possible expansion) of an offsite borrow source could have adverse impacts to cultural and historic resources; however, because the extent to which such an offsite source would be disturbed or expanded, if it is even needed

Environmental Impacts of Construction

at all, is not presently known, the potential significance of those historic and cultural resource impacts cannot be evaluated until an LAR is submitted. If Southern submits an LAR to use an offsite source, the staff would conduct an environmental review as part of its determination on that LAR.

Based on this review, the NRC staff determined that the conclusions presented in the ESP EIS remain valid.

4.7 Environmental Justice Impacts

The NRC staff's assessment of environmental justice impacts, including environmental pathways, socioeconomic impacts, and subsistence and special conditions, was provided in Section 4.7.1 of the ESP EIS (NRC 2008a). Based on the staff's analysis, construction impacts to environmental justice were considered to be SMALL.

In its COL ER (Southern 2009a), Southern indicated that there is no new and significant information regarding construction-related impacts on environmental justice. During its review of the COL application, the NRC staff performed an independent review of potential new and significant information related to environmental justice by reviewing Southern's ER, auditing Southern's process for identifying new and significant information, examining other information available at the site audit, and considering applicable regulations and reference documents. This review identified new information related to the impacts on traffic that warranted evaluation. As described in Section 4.5.4, Southern has completed a new traffic study and has indicated the potential for additional truck deliveries for offsite backfill. In regards to the new study, the assumptions are different, but the conclusions are similar and still lead to a MODERATE impact on roads near the VEGP site and a SMALL impact elsewhere. As stated in the traffic study, Southern plans to mitigate potentially adverse impacts via roadway and traffic control improvements. With respect to the potential need for offsite backfill, the hypothetical truck delivery routes identified by Southern would likely run through a small number of additional minority or low-income communities north of the VEGP site in South Carolina. However, the delivery routes would not be concentrated in minority or low-income communities nor are there likely to be noticeable adverse impacts (such as from traffic or air emissions) to these communities because the additional vehicles related to deliveries would remain within the design capacity of the roads. Therefore the staff determined that the SMALL conclusion presented in the ESP EIS with respect to environmental justice impacts remains valid.

Based on this review, the NRC staff determined that the conclusions presented in the ESP EIS remain bounding and valid.

4.8 Nonradiological Health Impacts

The NRC staff provided a description of the nonradiological health impacts for construction of the proposed Units 3 and 4 in Section 4.8 of the ESP EIS (NRC 2008a). Physical impacts of construction on public and occupational health, including dust, vehicle emissions, noise, and transportation of materials and personnel, were summarized. Public and occupational health is discussed in Section 4.8.1, while the impacts of transporting construction materials and construction personnel to the VEGP site are discussed in Section 4.8.2.

4.8.1 Public and Occupational Health

The NRC staff's assessment of the public and occupational health-related impacts, including air quality, site-preparation and construction worker health, and noise impacts, were provided in Section 4.8.1 of the ESP EIS (NRC 2008a). Based on the staff's analysis, construction-related impacts to public and occupational health were considered to be SMALL.

In its COL ER (Southern 2009a), Southern indicated that there is no new and significant information regarding construction-related impacts on public and occupational health. During its initial review of the COL application, the NRC staff independently verified that there was no new and significant information related to public and occupational health by reviewing Southern's ER, auditing Southern's process for identifying new and significant information, examining other information available at the site audit, and considering applicable regulations and reference documents. Subsequently, Southern also provided new information on three additional onsite borrow areas from which it sought to obtain backfill material via license amendment (Southern 2010e). Based on the information provided by Southern and the NRC analysis in the ESP EIS, the staff concluded in its EA for Amendment 2 (NRC 2010a) that site preparation and construction activities at the additional onsite borrow locations are similar to those that have been previously analyzed and documented in the ESP EIS, and that the nonradiological health impacts on workers and the public from activities conducted at the borrow areas are consistent with the impacts previously examined and found not to be significant. Accordingly, the staff incorporates by reference its analysis in the LAR EAs (NRC 2010a, b). The staff has not identified any additional new information that warranted further analysis in the SEIS.

Based on this review, the staff determined that the conclusions presented in the ESP EIS remain bounding and valid.

Environmental Impacts of Construction

4.8.2 Impacts of Transporting Construction Materials and Construction Personnel to the VEGP Site

The NRC staff's assessment of the nonradiological impacts associated with transporting construction materials and personnel to and from the VEGP site was presented in Section 4.8.2 of the ESP EIS (NRC 2008a). These impacts include the damage, injuries, and fatalities associated with vehicular accidents. Based on the staff's analysis, the transportation-related impacts on human health were considered to be SMALL.

In its COL ER (Southern 2009a), Southern provided no new or significant information related to transportation accidents. During its initial review of the COL application, the NRC staff independently verified that there was no new and significant information related to transportation of construction materials and personnel through its evaluation of Southern's process for identifying new and significant information, additional information provided by Southern at the site audit, and the staff's independent review of available information. However, subsequent to the site audit, Southern determined that it would need to obtain backfill material from onsite borrow areas other than those previously specified in the ESP site safety analysis report. Accordingly, Southern submitted LARs to obtain approval for the use of backfill from additional onsite borrow areas. The NRC staff evaluated the nonradiological impacts associated with truck transport of backfill material from these additional locations (NRC 2010a) and determined that the additional truck shipments would not significantly increase the nonradiological impacts presented in the ESP EIS (NRC 2008a). Accordingly, the staff incorporates by reference its analysis in the Amendment 2 EA (NRC 2010a).

Additionally, Southern indicated in a new and significant information evaluation (Southern 2010b) that it may subsequently seek to obtain engineering grade backfill materials from an offsite borrow source. Although Southern has not made a final decision on whether to submit an ESP LAR to do so, and thus a final plan is not before the NRC, the NRC staff conducted an evaluation of the nonradiological impacts of transporting backfill material from offsite borrow areas to the VEGP site, to assess whether such a development could potentially affect the staff's conclusions in the ESP EIS regarding nonradiological impacts associated with building Units 3 and 4.

The nonradiological impacts of transporting backfill material from offsite borrow areas to the VEGP site were calculated using the same general approach and data that were used in the ESP EIS and in the Amendment 2 EA (NRC 2010a). To calculate nonradiological impacts, shipping distances are multiplied by unit rates (i.e., accidents, injuries, and fatalities per unit distance). The bases and assumptions for these calculations are listed below:

Environmental Impacts of Construction

- The NRC staff assumed that a total of 611,644 m^3 (800,000 yd^3) of backfill would be transported by truck from a nearby borrow source to the power-block area of the Units 3 and 4 site (Southern 2010b).

- Southern assumed that shipment capacities for backfill material are approximately 15 m^3 (20 yd^3) per truck load (Southern 2010a).

- The NRC staff assumed that the average one-way shipping distance for backfill material to be about 96.6 km (60 mi) based on information provided by the Southern (Southern 2010b). This distance was doubled to account for the empty return trip.

- Accident, injury, and fatality rates for transporting building materials were taken from Table 4 in ANL/ESD/TM-150, *State-level Accident Rates for Surface Freight Transportation: A Reexamination* (Saricks and Tompkins 1999). Rates for the State of Georgia were used for backfill material shipments. The data provided in Saricks and Tompkins (1999) are representative of heavy-truck accident rates.

- The DOT Federal Motor Carrier Safety Administration evaluated the data underlying the Saricks and Tompkins (1999) rates, which were taken from the Motor Carrier Management Information System, and determined that the rates were under-reported. Therefore, the accident, injury, and fatality rates from Saricks and Tompkins (1999) were adjusted using factors derived from data provided by the University of Michigan Transportation Research Institute (UMTRI 2003). The University of Michigan Transportation Research Institute data indicate that accident rates for the period from 1994 to 1996, which are the same data used by Saricks and Tompkins (1999), were under-reported by about 39 percent. Injury and fatality rates were under-reported by 16 percent and 36 percent, respectively. As a result, the accident, injury, and fatality rates were increased by factors of 1.64, 1.20, and 1.57, respectively, to account for the apparent under-reporting. These adjustments were applied to the construction materials, which are transported by heavy truck shipments similar to those evaluated by Saricks and Tompkins (1999), but not to commuter traffic accidents.

The estimated nonradiological impacts of transporting backfill materials to the power-block area of the VEGP site from an offsite source are approximately 8.5 accidents, 4.1 injuries, and 0.2 fatalities. The estimated total annual nonradiological fatalities related to transporting backfill material represents about a 2.4 percent increase above the average 9.8 traffic fatalities per year that occurred in Burke County, Georgia, from 2004 to 2008 (DOT 2010). Even when considered in combination with the minor increase in traffic fatality risk analyzed in the ESP EIS, this increase remains small relative to the current traffic fatality risks in the area surrounding the proposed VEGP site.

Environmental Impacts of Construction

Based on this review and on information analyzed in the Amendment 2 EA for additional onsite borrow areas (NRC 2010a), the NRC staff determined that the conclusions related to the nonradiological impacts of transporting construction materials and personnel to and from the proposed Units 3 and 4 presented in the ESP EIS remain valid.

4.8.3 Summary of Nonradiological Health Impacts

The NRC staff concluded in the ESP EIS that nonradiological health impacts to construction and operational workers at the VEGP site and to the local population from fugitive dust, occupational injuries, noise, and transport of materials and personnel would be SMALL. During its review of the COL application, the NRC staff independently examined information related to public and occupational health by reviewing Southern's ER, auditing Southern's process for identifying new and significant information, examining other information available at the site audit, considering the information provided in conjunction with the Amendment 2 LAR (Southern 2010e) and information regarding the potential LAR for use of offsite backfill, and considering applicable regulations and reference documents.

Based on this review and information in the EA (NRC 2010a), the staff determined that the conclusions presented in the ESP EIS remain bounding and valid.

4.9 Radiological Health Impacts

The NRC staff provided a description of the radiological health impacts for construction of the proposed Units 3 and 4 at the VEGP site in Section 4.9 of the ESP EIS (NRC 2008a). The sources of radiation exposure for construction workers included exposures from direct radiation, gaseous radioactive effluents, and liquid radioactive waste discharges from routine operations at the existing VEGP Units 1 and 2 during construction of proposed Units 3 and 4. For the purposes of this discussion, construction and site-preparation workers were assumed to be members of the public; therefore, the dose estimates were compared to the dose limits for the public, pursuant to Title 10 of the Code of Federal Regulations (CFR) Part 20, Subpart D. Southern noted that all major construction activities are expected to occur outside the protected area boundary for the existing VEGP Units 1 and 2 but inside the restricted area boundary (Southern 2008a). The impact of direct radiation exposure is discussed in Section 4.9.1, gaseous effluents in Section 4.9.2, and liquid effluents in Section 4.9.3, while total dose to site preparation workers is discussed in Section 4.9.4.

4.9.1 Direct Radiation Exposures

The NRC staff's assessment of direct radiation exposures was provided in Section 4.9.1 of the ESP EIS (NRC 2008a). Based on the staff's analysis, construction-related impacts resulting from direct radiation exposure were considered to be SMALL.

Environmental Impacts of Construction

In its COL ER (Southern 2009a), Southern indicated that there is no new and significant information regarding construction-related impacts resulting from direct radiation exposure. During its initial review of the COL application, the NRC staff independently verified that there was no new and significant information related to direct radiation exposure by reviewing Southern's ER, auditing Southern's process for identifying new and significant information, examining other information available at the site audit, and considering applicable regulations, reference documents, and recent data on direct radiation sources that have become available since issuance of the VEGP ESP (Southern 2006, 2007, 2008b, 2009c). Southern subsequently provided new information on three additional borrow areas from which it sought to obtain backfill material via license amendment (Southern 2010e). Based on the information provided by Southern and the NRC analysis in the ESP EIS, the staff concluded in its EA (NRC 2010a) that site preparation and construction activities at the additional onsite borrow locations are similar to those that have been previously analyzed and documented in the ESP EIS, and that the radiological health impacts of direct radiation exposure of workers conducting activities at the borrow areas are consistent with the impacts previously examined and found not to be significant. Accordingly, the staff incorporates by reference its analysis in the Amendment 2 EA (NRC 2010a). As discussed in Section 2.5 of this SEIS, the data and analysis showed that direct radiation exposure rates remained within trends indentified in the ESP EIS. Also, in the COL ER (Southern 2009), Southern indicated that a new low-level waste (LLW) storage area had been developed northwest of the existing Unit 2 cooling tower to accommodate wastes from the existing units as well as Units 3 and 4. Because of the distance between the LLW storage area and the proposed construction area, Southern determined and the staff agrees that the LLW storage area would provide a negligible contribution to direct radiation dose to construction workers.

In addition, at certain times during construction, Southern would receive, possess, and use specific radioactive byproduct, source, and special nuclear material in support of construction and preparations for operation. These sources of low-level radiation are required to be controlled by the applicant's radiation protection program and have very specific uses under controlled conditions. The dose to construction workers from these sources of byproduct, source, and special nuclear material is expected to result in a negligible contribution to this estimate of construction worker doses in the ESP EIS.

The staff has not identified any additional new information that warranted further analysis in the SEIS. Based 2 EA (NRC 2010a), the staff determined that the conclusions presented in the ESP EIS remain valid.

4.9.2 Radiation Exposures from Gaseous Effluents

The NRC staff's assessment of radiation exposures resulting from gaseous effluents was provided in Section 4.9.2 of the ESP EIS (NRC 2008a). Based on the staff's analysis,

Environmental Impacts of Construction

construction-related impacts resulting from radiation exposure to gaseous effluents were considered to be SMALL.

In its COL ER (Southern 2009a), Southern indicated that there is no new and significant information regarding construction-related impacts resulting from radiation exposure to gaseous effluents. During its initial review of the COL application, the NRC staff independently verified that there was no new and significant information related to gaseous effluents by reviewing Southern's ER, auditing Southern's process for identifying new and significant information, examining other information available at the site audit, and considering applicable regulations, reference documents, and recent data on gaseous effluents that have become available since issuance of the VEGP ESP (Southern 2006, 2007, 2008b, 2009c). Southern subsequently provided new information on three additional borrow areas from which it sought to obtain backfill material via license amendment (Southern 2010e). Based on the information provided by Southern and the NRC analysis in the ESP EIS, the staff concluded in its EA (NRC 2010a) that site preparation and construction activities at the additional onsite borrow locations are similar to those that have been previously analyzed and documented in the ESP EIS, and that the radiological health impacts of exposure of workers to gaseous effluents while conducting activities at the borrow areas are consistent with the impacts previously examined and found not to be significant. Accordingly, the staff incorporates by reference its analysis in the Amendment 2 EA (NRC 2010a). The staff has not identified any additional new information that warranted further analysis in the SEIS. As discussed in Section 2.5 of this SEIS, the data and analysis showed that radiation exposure rates resulting from gaseous effluents remained within trends identified in the ESP EIS.

Based on this review and information in the EA (NRC 2010a), the staff determined that the conclusions presented in the ESP EIS remain valid.

4.9.3 Radiation Exposures from Liquid Effluents

The NRC staff's assessment of radiation exposures resulting from liquid effluents was provided in Section 4.9.3 of the ESP EIS (NRC 2008a). Based on the staff's analysis, construction-related impacts resulting from radiation exposure to liquid effluents were considered to be SMALL.

In its COL ER (Southern 2009a), Southern indicated that there is no new and significant information regarding construction-related impacts resulting from radiation exposure to liquid effluents. During its initial review of the COL application, the NRC staff independently verified that there was no new and significant information related to liquid effluents by reviewing Southern's ER, auditing Southern's process for identifying new and significant information, examining other information available at the site audit, and considering applicable regulations, reference documents, and recent data on liquid effluents that have become available since

Environmental Impacts of Construction

issuance of the VEGP ESP (Southern 2006, 2007, 2008b, 2009c). Southern subsequently provided new information on three additional borrow areas from which it sought to obtain backfill material via license amendment (Southern 2010e). Based on the information provided by Southern and the NRC analysis in the ESP EIS, the staff concluded in its EA (NRC 2010a) that site preparation and construction activities at the additional onsite borrow locations are similar to those that have been previously analyzed and documented in the ESP EIS, and that the radiological health impacts of exposure of workers to liquid effluents while conducting activities at the borrow areas are consistent with the impacts previously examined and found not to be significant. Accordingly, the staff incorporates by reference its analysis in the Amendment 2 EA (NRC 2010a). The staff has not identified any additional new information that warranted further analysis in the SEIS. As discussed in Section 2.5 of this SEIS, the data and analysis showed that radiation exposure rates resulting from liquid effluents remained within trends identified in the ESP EIS.

Based on this review, the staff determined that the conclusions presented in the ESP EIS remain valid.

4.9.4 Total Dose to Site-Preparation Workers

The NRC staff's assessment of total dose to site-preparation workers was provided in Section 4.9.3 of the ESP EIS (NRC 2008a). Here, the term site preparation workers refers to workers performing either preconstruction or construction activities. Based on the staff's analysis, construction-related impacts resulting from total dose to site-preparation workers were considered to be SMALL.

In its COL ER (Southern 2009a), Southern indicated that there is no new and significant information regarding construction-related impacts resulting from total dose to site-preparation workers. During its initial review of the COL application, the NRC staff independently verified that there was no new and significant information related to total dose to site-preparation workers by reviewing Southern's ER, auditing Southern's process for identifying new and significant information, examining other information available at the site audit, and considering applicable regulations, reference documents, and recent data on direct radiation sources and radiological effluents that have become available since issuance of the ESP (Southern 2006, 2007, 2008b, 2009c). Southern subsequently provided new information on three additional borrow areas from which it sought to obtain backfill material via license amendment (Southern 2010e). Based on the information provided by Southern and the NRC analysis in the ESP EIS, the staff concluded in its EA (NRC 2010a) that site preparation and construction activities at the additional onsite borrow locations are similar to those that have been previously analyzed and documented in the ESP EIS, and that the radiological health impacts of exposures of workers while conducting activities at the borrow areas are consistent with the impacts previously examined and found not to be significant. Accordingly, the staff incorporates by reference its

Environmental Impacts of Construction

analysis in the Amendment 2 EA (NRC 2010a). The staff has not identified any additional new information that warranted further analysis in the SEIS. As discussed in Section 2.5 of this SEIS, the data and analysis showed that total dose to site preparation workers remained within trends identified in the ESP EIS.

Based on this review and information in the EA (NRC 2010a), the staff determined that total dose to site-preparation workers at the VEGP site remained within the limits specified in Federal environmental radiation standards – 10 CFR Part 20; 10 CFR Part 50, Appendix I; and 40 CFR Part 190 – and that the conclusions presented in the ESP EIS remain valid.

4.9.5 Summary of Radiological Health Impacts

The NRC staff concluded in the ESP EIS that radiological health impacts to construction workers at the VEGP site would be SMALL. During its review of the COL application, the staff independently examined information related to radiological exposure by reviewing Southern's ER, auditing Southern's process for identifying new and significant information, examining other information available at the site audit, considering information Southern submitted in conjunction with the Amendment 2 LAR for additional onsite borrow sources, and considering applicable regulations, reference documents, and recent data on direct radiation sources and radiological effluents that have become available since issuance of the VEGP ESP (Southern 2006, 2007, 2008b, 2009c).

Based on this review and information in the Amendment 2 EA (NRC 2010a), the staff determined that total dose to construction workers at the VEGP site remained within the limits specified in Federal environmental radiation standards – 10 CFR Part 20; 10 CFR Part 50, Appendix I; and 40 CFR Part 190 – and that the conclusions presented in the ESP EIS remain valid.

4.10 Measures and Controls to Limit Adverse Impacts During Site-Preparation Activities and Construction

The staff's assessment of the measures and controls to limit adverse impacts during site-preparation and construction were addressed in Section 4.10 of the ESP EIS (NRC 2008a). Part 10 of Southern's COL application includes a draft Environmental Protection Plan (EPP) for the site, which identifies proposed conditions, monitoring, reporting, and record keeping for environmental data during construction. The draft EPP provided with the COL application is substantively similar to the EPP attached as Appendix G to ESP-004 (NRC 2009).

In its COL ER (Southern 2009a), Southern indicated that there is no new and significant information regarding measures and controls to limit adverse impacts, but that it remains committed to the mitigation measures described in Section 4.10 of the ESP EIS. During its

Environmental Impacts of Construction

review of the COL application, the NRC staff identified an MOU between the Georgia SHPO and Southern (GHPD 2010) that related to measures and controls to limit adverse impacts to cultural resources. Additionally, the staff identified new information (Southern 2010e) indicating that prior to developing the additional onsite backfill borrow sources associated with its second ESP LAR, Southern implemented rare plant and animal relocation programs in an attempt to minimize impacts (NRC 2010a). The NRC staff discussed these measures in the EA for Amendment 2 and incorporates that discussion by reference in this SEIS. With respect to the COL review, the NRC staff performed an independent analysis by reviewing Southern's ER, auditing Southern's process for identifying new and significant information, examining other information available at the site audit, information submitted in conjunction with the ESP LARs, and considering applicable regulations and reference documents.

With respect to historic and cultural resources, the MOU between the SHPO and Southern is for the preservation of the remaining balance of site 9BK416 from physical disturbance and performance of additional archaeological surveys as directed by the Georgia Historic Preservation Division (GHPD). The proposed project would disturb approximately 1 ha (2.5 ac) of the estimated 11.7 ha (29 ac) of site 9BK416. The SHPO determined that based on consultation and supporting field surveys the proposed project would impact site 9BK416, but not adversely impact the site (GHPD 2010). As described in Section 4.6, the staff considered these measures and controls in reaching its impact conclusion.

As noted above, regarding rare species, Southern implemented voluntary programs to relocate the southeastern pocket gopher and the sandhills milkvetch prior to development of a new borrow source in an area with populations of both of these species. These efforts have resulted in the relocation of both southeastern pocket gophers and sandhills milkvetch plants to an area on the northern part of the VEGP site. The relocation programs were developed in consultation with GDNR.

Based on this review, with the addition of the MOU and the species relocation programs, the staff determined that the measures and controls identified to limit adverse impacts during site preparation activities and construction presented in the ESP EIS remain valid, and also that Southern's proposed EPP is appropriate. If the COLs are issued, the staff would include the EPP as part of the licenses.

4.11 Site Redress Plan

Southern submitted a revised site redress plan as part of its ESP application (Southern 2008c), and the NRC staff described and evaluated that plan in Section 4.11 of the ESP EIS (NRC 2008a). The purpose of the site redress plan was to ensure that the VEGP site would be returned to an environmentally stable and aesthetically acceptable condition if the proposed Units 3 and 4 were not fully developed to generate electricity. The site redress plan is

Environmental Impacts of Construction

applicable specifically to those actions allowed under the LWA that was issued concurrently with the ESP in August 2009 (NRC 2009).

In its COL ER (Southern 2009a), Southern indicated that there is no new and significant information regarding the current site redress plan. In October 2009, Southern submitted an application for a second LWA that, if approved by the NRC, would allow for additional construction-related activities to be conducted prior to issuance of the COLs for Units 3 and 4. The second LWA, in accordance with 10 CFR 50.10(d), would authorize installation of reinforcing steel, sumps, drain lines, and other embedded items along with placement of concrete for the nuclear island foundation base slab. The second LWA application indicates that the existing site redress plan would be applicable to the additional LWA activities. During its review of the COL application, the NRC staff independently verified that no new and significant information was available related to the site redress plan by reviewing Southern's ER, auditing Southern's process for identifying new and significant information, examining other information available at the site audit, and considering applicable regulations and reference documents. In its ER submitted in support of the second LWA request, Southern explained why, in each resource area evaluated in Chapter 4 of the ESP EIS, the requested LWA activities would involve no additional impacts beyond those presented in the ESP EIS (Southern 2010h). The staff reviewed and independently assessed Southern's evaluation of the LWA impacts.

In the ESP EIS, the staff examined the construction activities requested in Southern's ESP LWA application and determined that the environmental impacts of those activities would be a small proportion of the impacts of the combined construction and site preparation activities. The staff determined that the LWA impacts would be bounded by the analysis of those overall impacts, and would be SMALL. As Southern's ER in support of its second LWA explains, that is also true of the subset of construction activities requested in the second LWA, in that they represent a small proportion of the planned construction and preconstruction activities and would occur entirely within the footprint of the nuclear island. Accordingly, the ESP conclusion regarding the impacts of the ESP LWA reinforces a conclusion that construction impacts specifically attributable to the October 2009 LWA request would likewise be SMALL.

Based on this review, the staff verified that the site redress plan discussed in the ESP EIS would adequately redress the impacts of the activities requested under the second LWA in the event construction is terminated by Southern or its successor, the COL application is withdrawn by Southern or denied by the NRC, or the second LWA is revoked by the NRC. As a result, the staff's conclusion in accordance with 10 CFR 50.10(c) that the LWA activities requested in the October 2009 submittal would not result in any significant adverse environmental impacts that could not be redressed is bounding and valid.

Environmental Impacts of Construction

4.12 Summary of Construction Impacts

Impact level characterizations identified by the NRC staff during the evaluation of the ESP application were documented in Table 4-7 of the ESP EIS (NRC 2008a). In addition to impact characterizations, environmental impacts categories were listed in Table 4-7 along with the specific measures and controls Southern proposed to implement in connection with those impact categories. For the reasons stated in this chapter, the NRC staff's review of information available during the site audit and from other information sources did not identify any information that would change the impact characterization for any of the categories in Table 4-7 of the ESP EIS (NRC 2008a), with the exception of the impact level for onsite terrestrial resources, which changed from SMALL to MODERATE for reasons described in Section 4.4.1 of this SEIS. The staff determined that the activities associated with the second LWA are a small subset of the overall construction activities that would occur entirely within the footprint of the nuclear island. Therefore, impacts from the activities requested in the second LWA would be SMALL for all resource areas.

4.13 References

10 CFR Part 20. Code of Federal Regulations. Title 10, *Energy*, Part 20, "Standards for Protection Against Radiation."

10 CFR Part 50. Code of Federal Regulations, Title 10, *Energy*, Part 50, "Domestic Licensing of Production and Utilization Facilities."

40 CFR Part 190. Code of Federal Regulations, Title 40, *Protection of Environment*, Part 190, "Environmental Radiation Protection Standards for Nuclear Power Operation."

50 CFR Part 402. Code of Federal Regulations, Title 50, *Wildlife and Fisheries*, Part 402, "Interagency Cooperation—Endangered Species Act of 1973, as Amended."

43 FR 4026. January 31, 1978. "Endangered and Threatened Wildlife and Plants. Listing of the Eastern Indigo Snake as a Threatened Species." *Federal Register*, U.S. Fish and Wildlife Service.

72 FR 49352. August 28, 2007. "Licenses, Certifications, and Approvals for Nuclear Power Plants; Final Rule." *Federal Register*, U.S. Nuclear Regulatory Commission.

73 FR 16436. March 27, 2008. "National Ambient Air Quality Standards for Ozone." *Federal Register*, Environmental Protection Agency.

Environmental Impacts of Construction

75 FR 61904. October 6, 2010. "Endangered and Threatened Wildlife and Plants; Proposed Listings for Two Distinct Population Segments of Atlantic Sturgeon (*Acipenser oxyrinchus oxyrinchus*) in the Southeast." *Federal Register*. U.S. Department of Commerce.

Bald and Golden Eagle Protection Act. 1992. 16 USC 668 et seq.

Clean Air Act. 42 USC 7401, et seq.

Clean Water Act (CWA). 33 USC 1251, et seq. (also called the Federal Water Pollution Control Act [FWPCA]).

Endangered Species Act of 1973. 16 USC 1531, et seq.

Ga. Code Ann. 22-3-161. 2004. "Selection of route for electric transmission line; settlement negotiations with property owners." *Georgia Code Annotated*.

Georgia Historic Preservation Division (GHPD). 2010. Letter from D. Crass, GHPD, to T.C. Moorer, Southern, "Memorandum of Understanding - Archaeological Site 9BK416 Vogtle Electric Generating Plant Expansion Burke County, Georgia, HP-060428-001." January 20, 2010. Accession No. ML100500302.

Georgia Department of Natural Resources (GDNR). 2007. Letter from K. Anderson-Cordova, GDNR, to M.D. Notich, NRC, "Early Site Permit – Draft EIS, Vogtle Electric Generating Plant Expansion, Burke County, Georgia, HP-060428-001." December 27, 2007. Accession No. ML080070095.

Georgia Department of Natural Resources (GDNR). 2009a. Letter from C.A. Couch, Georgia Department of Natural Resources, to A.S. Meiburg, U.S Environmental Protection Agency, Region 4, Atlanta, Georgia. "Recommended Designations of Ozone Non-Attainment Areas in Georgia." March 12, 2009. Accession No. ML100601088.

Georgia Department of Natural Resources (GDNR). 2009b. E-mail from Matt Elliott, GDNR, to Mallecia Sutton, NRC. December 16, 2009. Accession No. ML093500211.

Georgia Department of Natural Resources (GDNR). 2009c. E-mail dated December 21, 2009 and letter dated December 17, 2009 from Katrina Morris, GDNR to Mallecia Sutton, NRC, regarding Natural Heritage Database occurrences within the VEGP Boundary and the Transmission Line Macrocorridor.
Accession No. ML100490042.

Georgia Department of Natural Resources (GDNR). 2009d. *Sandhills Inventory*. Accession No. ML101690125.

Environmental Impacts of Construction

Georgia Department of Natural Resources (GDNR). 2010. E-mail from Elizabeth Shirk (GDNR, Historic Preservation Division) to Mallecia Sutton, NRC. June 17, 2010. Accession No. ML101940268.

Georgia Power Company (GPC). 2006. *Avian Protection Program for Georgia Power Company, Rev 1.* March 14, 2006. Accession No. ML063000228.

Georgia Power Company (GPC). 2007. *Corridor Study - Thomson Vogtle 500-kV Transmission Project.* Atlanta, Georgia. Accession No. ML070460368.

MACTEC. 2007. *Burke County Comprehensive Plan.* Accessed February 4, 2010 at http://www.burkechamber.org/burke-county/department-profile.php?Department_ID=6. Accession No. ML100600797.

National Historic Preservation Act of 1966 (NHPA). 16 USC 470, et seq.

Neel-Schaffer. 2009. *Traffic Study, Plant Vogtle Construction of Two New Units.* July 2009. Accession No. ML101940362.

New South Associates. 2007. *A Determination of Boundaries and Surface Features for Historic Cemetery on Plant Vogtle in Burke County, Georgia. May 14, 2007.* Not for public disclosure per Section 304 of the National Historic Preservation Act and Supplemental Staff Guidance to NUREG-1555, "Environmental Standard Review Plan," for Cultural and Historic Reviews (Accession No. ML10055073013).

Rivers and Harbors Appropriations Act of 1889. 33 USC 403, Section 10.

Saricks C.L. and M.M. Tompkins. 1999. *State-Level Accident Rates for Surface Freight Transportation: A Reexamination.* ANL/ESD/TM-150. Argonne National Laboratory, Argonne, Illinois.

South Carolina Department of Natural Resources (SCDNR). 2009. E-mail from Julie Holling, SCDNR, Data Manager, to Mallecia Sutton, NRC, December 15, 2009. Columbia, South Carolina. Accession No. ML093491132.

Southeast Regional Partnership for Planning and Sustainability (SERPPAS). 2010. *Candidate Conservation Agreement for the Gopher Tortoise First Annual Report: October 1, 2008 – September 30, 2009.* February 2010. Accessed at www.serppas.org/.../GTCCA%20First%20Annual%20Report%202008-2009.pdf on November 15, 2010.

Environmental Impacts of Construction

Southern Nuclear Operating Company, Inc. (Southern). 2006. *Annual Radioactive Effluent Release Report for January 1, 2005 to December 31, 2005. Vogtle Electric Generating Plant – Units 1 and 2*, NRC Docket Nos. 50-424 and 50-425, Facility Operating License Nos. NPF-68 and NPF-81. Southern Company, Birmingham, Alabama. Accession No. ML061240254.

Southern Nuclear Operating Company, Inc. (Southern). 2007. *Annual Radioactive Effluent Release Report for January 1, 2006 to December 31, 2006. Vogtle Electric Generating Plant – Units 1 and 2*, NRC Docket Nos. 50-424 and 50-425, Facility Operating License Nos. NPF-68 and NPF-81. Southern Company, Birmingham, Alabama. Package Accession No. ML071220467.

Southern Nuclear Operating Company, Inc. (Southern). 2008a. *Vogtle Early Site Permit Application: Part 3. Environmental Report.* Revision 4, Southern Company, Birmingham, Alabama. Package Accession No. ML081020073.

Southern Nuclear Operating Company, Inc. (Southern). 2008b. *Annual Radioactive Effluent Release Report for January 1, 2007 to December 31, 2007. Vogtle Electric Generating Plant – Units 1 and 2*, NRC Docket Nos. 50-424 and 50-425, Facility Operating License Nos. NPF-68 and NPF-81. Southern Company, Birmingham, Alabama. Accession No. ML081290295.

Southern Nuclear Operating Company, Inc. (Southern). 2008c. *Vogtle Early Site Permit Application: Part 4. Site Redress Plan.* Revision 4, Southern Company, Birmingham, Alabama. Accession No. ML081020073.

Southern Nuclear Operating Company (Southern). 2009a. *Vogtle Electric Generating Plant, Units 3 and 4, COL Application, Part 3 Environmental Report.* Revision 1, September 23, 2009. Southern Company, Birmingham, Alabama. Accession No. ML092740400.

Southern Nuclear Operating Company (Southern). 2009b. *Vogtle Units 3 and 4 Revised Disturbance Table.* Accession No. ML101960587.

Southern Nuclear Operating Company, Inc. (Southern). 2009c. *Annual Radioactive Effluent Release Report for January 1, 2008 to December 31, 2008. Vogtle Electric Generating Plant – Units 1 and 2*, NRC Docket Nos. 50-424 and 50-425, Facility Operating License Nos. NPF-68 and NPF-81. Southern Company, Birmingham, Alabama. Accession No. ML091260689.

Southern Nuclear Operating Company, Inc. (Southern). 2010a. *Supporting Information for Environmental Report Review.* March 12, 2010. Southern Company, Birmingham, Alabama. Accession No. ML100750038.

Southern Nuclear Operating Company, Inc. (Southern). 2010b. Letter from C.R. Pierce, Southern, to NRC, "Southern Nuclear Operating Company, Vogtle Electric Generating Plant

Environmental Impacts of Construction

Units 3 and 4, Combined License Application, New and Significant Information Evaluation For the Transportation of Backfill From an Offsite Source." Letter ND-10-1389. July 16, 2010. Southern Company, Birmingham, Alabama. Accession No. ML102010031.

Southern Nuclear Operating Company, Inc. (Southern). 2010c. Letter from C.R. Pierce, Southern, to NRC, "Southern Nuclear Operating Company, Vogtle Electric Generating Plant Units 3 and 4 Combined License Application, Response to Request for Additional Information Letter on Environmental Issues." Letter ND-10-0023. January 8, 2010. Southern Company, Birmingham, Alabama. Accession No. ML100120479.

Southern Nuclear Operating Company, Inc. (Southern). 2010d. Letter from M. Smith, Southern, to NRC, "Southern Nuclear Operating Company, Vogtle Electric Generating Plant Units 3 and 4, Early Site Permit Site Safety Analysis Report Amendment Request, Revised Site Safety Analysis Report Markup for Onsite Sources of Backfill." Letter ND-10-0960. May 13, 2010, Southern Company, Birmingham, Alabama. Accession No. ML101340649.

Southern Nuclear Operating Company, Inc. (Southern). 2010e. Letter from C.R. Pierce, Southern, to NRC, "Southern Nuclear Operating Company, Vogtle Electric Generating Plant Units 3 and 4, Early Site Permit Site Safety Analysis Report Amendment Request, Revised Site Safety Analysis Report Markup for Onsite Sources of Backfill, Part 2." Letter ND-10-1005. May 24, 2010. Southern Company, Birmingham, Alabama. Accession No. ML101470212.

Southern Nuclear Operating Company, Inc. (Southern). 2010f. Letter from B.L. Ivey, Southern, to NRC, "Southern Nuclear Operating Company, Vogtle Electric Generating Plant Units 3 and 4 Combined License Application, Post New and Significant Audit Supporting Information." Letter ND-10-0923. May 10, 2010. Southern Company, Birmingham, Alabama. Accession Nos. ML101320256 and ML101310333.

Southern Nuclear Operating Company, Inc. (Southern). 2010g. Letter from Charles R. Pierce, Southern Nuclear Operating Company, Inc., to NRC, "Southern Nuclear Operating Company, Vogtle Electric Generating Plant Units 3 and 4 Combined License Applications, Large Component Transportation Method Decision." February 19, 2010. Southern Company, Birmingham, Alabama. Accession No. ML100550033.

Southern Nuclear Operating Company, Inc. (Southern). 2010h. Letter from C.R. Pierce, Southern, to NRC, "Southern Nuclear Operating Company, Vogtle Electric Generating Plant Units 3 and 4 Combined License Application, Environmental Report to Support Revision 1 to Part 6, LWA Request." Letter ND-10-227. February 5, 2010. Southern Company, Birmingham, Alabama. Accession No. ML100470600.

U.S. Army Corps of Engineers (USACE). 2010. Letter from Carol Bernstein (USACE Chief, Coastal Branch) to Thomas Moorer (Southern), "Subject: Signed Department of the Army Permit for the expansion of the existing Vogtle Electric Generating Plant." SAS-2007-01837.

Environmental Impacts of Construction

U.S. Department of Commerce, National Oceanic and Atmospheric Administration, National Marine Fisheries Service (NMFS). 2008. Letter from Roy E. Crabtree, Ph.D., Regional Administrator to William Burton, NRC, "A Biological Assessment for the Shortnose Sturgeon for the Vogtle Electric Generating Plant Early Site Permit Application." August 11, 2008. Accession No. ML082480450.

U.S. Department of Transportation (DOT). 2010. *Traffic Safety Facts, Burke County, Georgia, 2004-2008*. Accessed June 1, 2010 at http://www-nrd.nhtsa.dot.gov/departments/nrd-30/ncsa/STSI/13_GA/2008/Counties/Georgia_Burke%20County_2008.HTM.

U.S. Fish and Wildlife Service (FWS). 2008. Letter from U.S. Fish and Wildlife Service, Athens, Georgia to U.S. Nuclear Regulatory Commission, Washington, DC. Re: USFWS Log# 08-FA-0473. September 19, 2008. Accession No. ML082760694.

U.S. Fish and Wildlife Service (FWS). 2010a. Letter from U.S. Fish and Wildlife Service, Athens, Georgia, to U.S. Nuclear Regulatory Commission, Washington, D.C. Re: USFWS Log# 2009-1387. February 12, 2010. Accession No. ML100500426.

U.S. Fish and Wildlife Service (FWS). 2010b. Letter from U.S. Fish and Wildlife Service, Athens, Georgia, to U.S. Nuclear Regulatory Commission, Washington, D.C. Re: USFWS Log# 2010-1254. October 20, 2010. Accession No. ML103010076.

U.S. Geological Survey (USGS). 2001. *National Land Cover Database Zone 55 Land Cover Layer*. Accessed on February 11, 2010 at http://www.mrlc.gov/nlcd.php. Accession No. ML100601081.

U.S. Nuclear Regulatory Commission (NRC). 2008a. *Final Environmental Impact Statement for an Early Site Permit (ESP) at the Vogtle Electric Generating Plant Site.* NUREG-1872, Vols. 1, 2, and Errata, Washington, D.C. Accession Nos. ML082240145; ML082240165, ML082260203; ML082550040.

U.S. Nuclear Regulatory Commission (NRC). 2008b. Letter from NRC to FWS dated January 25, 2008. Subject: Biological Assessment for Threatened and Endangered Species and Designated Critical Habitat for the Vogtle Electric Generating Plant Early Site Permit Application. Accession No. ML080100512.

U.S. Nuclear Regulatory Commission (NRC). 2008c. Letter from W. Burton, NRC, to D. Bernhart, NMFS, "Subject: Biological Assessment for the Shortnose Sturgeon for the Vogtle Electric Generating Plant Early Site Permit Application." January 25, 2008. Accession No. ML080070538 and ML080100588.

U.S. Nuclear Regulatory Commission (NRC). 2009. *Southern Nuclear Operating Company, Vogtle Electric Generating Plant ESP Site, Docket No. 52-011, Early Site Permit and Limited Work Authorization. Early Site Permit No. ESP-004.* Washington. D.C. Accession No. ML092290157.

U.S. Nuclear Regulatory Commission (NRC). 2010a. Vogtle Electric Generating Plant ESP Site Early Site Permit and Limited Work Authorization EA and Finding of No Significant Impact. Docket No. 52-011, Washington, D.C. Accession No. ML101670592.

U.S. Nuclear Regulatory Commission (NRC). 2010b. Vogtle Electric Generating Plant ESP Site Early Site Permit and Limited Work Authorization EA and Finding of No Significant Impact. Docket No. 52-011, Washington, D.C. Accession No. ML101380114.

U.S. Nuclear Regulatory Commission (NRC). 2010c. Letter from NRC to FWS dated January 7, 2010. Washington, D.C. Accession No. ML092600684.

U.S. Nuclear Regulatory Commission (NRC). 2010d. Site Audit Summary. Washington, D.C. Accession No. ML101550095.

U.S. Nuclear Regulatory Commission (NRC). 2010e. GDNR Conference Call Summaries, May 26-June 3, 2010. Washington, D.C. Accession No. ML101670079.

U.S. Nuclear Regulatory Commission (NRC). 2010f. *Vogtle Electric Generating Plant ESP Site Early Site Permit and Limited Work Authorization Environmental Assessment and Finding of No Significant Impact.* Docket No. 52-011. Accession No. ML101660076.

U.S. Nuclear Regulatory Commission. 2010g. Letter from G. Hatchett, NRC, to D. Bernhart, NMFS, "Notification of the Issuance and Request for Comments on the Draft Supplemental Environmental Impact Statement for Vogtle Electric Generating Plant, Units 3 and 4 Combined License Application." September 3, 2010. Accession No. ML102320162.

University of Michigan Transportation Research Institute (UMTRI). 2003. *Evaluation of the Motor Carrier Management Information System Crash File, Phase One.* UMTRI, Ann Arbor, Michigan.

Southern Nuclear Operating Company, Inc. (Southern). 2010h. Letter from C.R. Pierce, Southern, to NRC, "Southern Nuclear Operating Company, Vogtle Electric Generating Plant Units 3 and 4 Combined License Application, Notification of Approved Change to Environmental Permit SAS-2007-01837." Letter ND-11-0424. March 3, 2011. Southern Company, Birmingham, Alabama. Accession No. ML110660152.

5.0 Environmental Impacts of Operation

In Chapter 5 of the early site permit (ESP) environmental impact statement (EIS) (NRC 2008a), the U.S. Nuclear Regulatory Commission (NRC) staff provided a description of the environmental impacts of operating the proposed Units 3 and 4 at the Vogtle Electric Generating Plant (VEGP) site. The applicant, Southern Nuclear Operating Company, Inc. (Southern), evaluated the potential new and significant information that could affect impacts of operation. The NRC staff reviewed Southern's process for identifying new and significant information, but also conducted its own independent review to verify whether new and significant information had been identified. The results of that review are presented in the following sections. Sections 5.1 through 5.10 discuss the potential operational impacts on land use, meteorology and air quality; water use and quality; terrestrial and aquatic ecosystems; socioeconomics; historic and cultural resources; environmental justice; nonradiological health effects; radiological health effects; and postulated accidents. Applicable measures and controls that would limit the adverse impacts during the 40-year operating period for the proposed Units 3 and 4 are described in Section 5.11. A summary of the operational impact is presented in Section 5.12. The references cited in this chapter are listed in Section 5.13.

5.1 Land-Use Impacts

The NRC staff's assessment of the land-use impacts related to the operation of proposed Units 3 and 4 and the planned new transmission line right-of-way was provided in Section 5.1 of the ESP EIS (NRC 2008a). Based on the staff's analysis, impacts to land use were considered to be SMALL.

In the environmental report (ER) included in its combined license (COL) application (Southern 2009), Southern indicated that there is no new and significant information regarding impacts of the operation of Units 3 and 4 and the planned new transmission line right-of-way (ROW) on land use. During its review of the COL application, the NRC staff independently verified that no new and significant information was available related to the land-use impacts of operating Units 3 and 4 and the planned new transmission line ROW by reviewing Southern's ER, auditing Southern's process for identifying new and significant information, examining other information available at the site audit, and considering applicable regulations and reference documents.

Based on this review, the staff determined that the conclusions presented in Section 5.1 of the ESP EIS remain bounding and valid.

Environmental Impacts of Operation

5.2 Meteorological and Air-Quality Impacts

The NRC staff's assessment of meteorology and air-quality impacts, including impacts from the cooling tower plumes and emissions from the operation of auxiliary generators and boilers, was provided in Section 5.2 of the ESP EIS (NRC 2008a). Based on the staff's analysis, operation-related impacts to meteorology and air quality were considered to be SMALL.

In its COL ER (Southern 2009), Southern indicated that there is no new and significant information regarding construction-related impacts on meteorology and air quality. During its review of the COL application, the NRC staff performed an independent review of potential new and significant information related to meteorology and air quality by reviewing Southern's ER, auditing Southern's process for identifying new and significant information, examining other information available at the site audit, and considering applicable regulations and reference documents. During this review, the staff identified new information related to changes to the National Ambient Air Quality Standard (NAAQS) for ozone that warranted further review.

As discussed in Chapter 2.3, the U.S. Environmental Protection Agency (EPA) promulgated a revision to the NAAQS for ozone on March 12, 2008. The final rule (73 FR 16436) reduced the ozone standard from 0.084 parts per million (ppm) to 0.075 ppm. Section 107(d)(1) of the Clean Air Act (CAA) requires each state to submit, within 1 year of the revised standard, its recommended designation (i.e., attainment, non-attainment, or unclassified) for each county. On March 12, 2009, the Georgia Department of Natural Resources (GDNR) issued a letter to EPA providing GDNR's recommended designations; Burke County remains unclassified/ attainment for the new ozone standard (GDNR 2009). EPA will make its final determination regarding attainment status no later than March 2011.

Based on this review and the fact that Burke County has been proposed to remain in attainment, the NRC staff determined that the conclusions presented in the ESP EIS (NRC 2008a) remain bounding and valid.

5.3 Water-Related Impacts

The NRC staff's assessment of the water-related impacts associated with operation of the proposed Units 3 and 4 was provided in Section 5.3 of the ESP EIS (NRC 2008a). Based on the staff's analysis, operations-related impacts of hydrological alterations on water use and water quality were considered to be SMALL.

During its review of the COL application, the NRC staff performed an independent review of potential new and significant information regarding water-related impacts of operation by reviewing Southern's ER, auditing Southern's process for identifying new and significant information, examining other information available at the site audit (including permits for

Environmental Impacts of Operation

groundwater withdrawal and dewatering of the surficial aquifer during construction), and considering applicable regulations and reference documents.

In its COL ER (Southern 2009) and request for additional information (RAI) responses (Southern 2010), Southern provided new information on the proposed intake structure design, as described in Section 3.2.2. These design changes would have no impact on water use and water quality during operation and therefore do not change the assessment of operations-related impacts described in the ESP EIS.

As described in Section 3.2.1, during its review, the staff identified information on the total effluent discharge to the Savannah River that warranted further staff analysis in the SEIS. The discharge estimate is 2000 L/s (31,695 gpm) (Southern 2010), which is 3 percent more than the value of 1941 L/s (30,761 gpm) used in the ESP EIS to evaluate water-quality impacts of operations. The NRC staff performed an independent assessment of the thermal effluent plume's extent using a total discharge of 2000 L/s (31,695 gpm) and assuming the same conservative conditions described in ESP EIS Section 5.3.3. The extent of the thermal plume was estimated as the 2.8°C (5.0°F)-above-ambient isotherm using CORMIX Version 6.0 (Doneker and Jirka 2007). The 3-percent increase in discharge resulted in an increase in the estimated thermal plume extent from 29.6 m (97 ft) to 33.6 m (110 ft) in length and from 4.6 m (15 ft) to 5.2 m (17 ft) in width. The extent of the 2.8°C (5.0°F)-above-ambient isotherm is shown in Figure 5-1. Because the estimated extent of the thermal plume remains small in relation to the width of the river, the 3 percent increase in the discharge does not result in a change to the staff's impact conclusion in the ESP EIS.

Based on this review, the staff determined that the conclusion presented in the ESP EIS, that impacts would be SMALL, remains valid.

5.4 Terrestrial and Aquatic Ecosystems

5.4.1 Terrestrial Impacts

The NRC staff's assessments of the potential operational impacts to terrestrial resources, including impacts to Federally and State-listed threatened and endangered species, were provided in Sections 5.4.1 and 5.4.3 of the ESP EIS (NRC 2008a). Terrestrial-resource-related impacts of operations that are discussed in the ESP EIS include impacts on vegetation related to cooling tower drift, icing, fogging, or increased humidity; bird collisions with cooling towers and transmission lines; cooling tower noise; shoreline habitat; transmission line ROW management; electromagnetic fields; transmission line ROW maintenance on floodplains and wetlands; and Federal and State-listed species. Based on the staff's analysis, operations-related impacts to terrestrial resources were considered to be SMALL.

Environmental Impacts of Operation

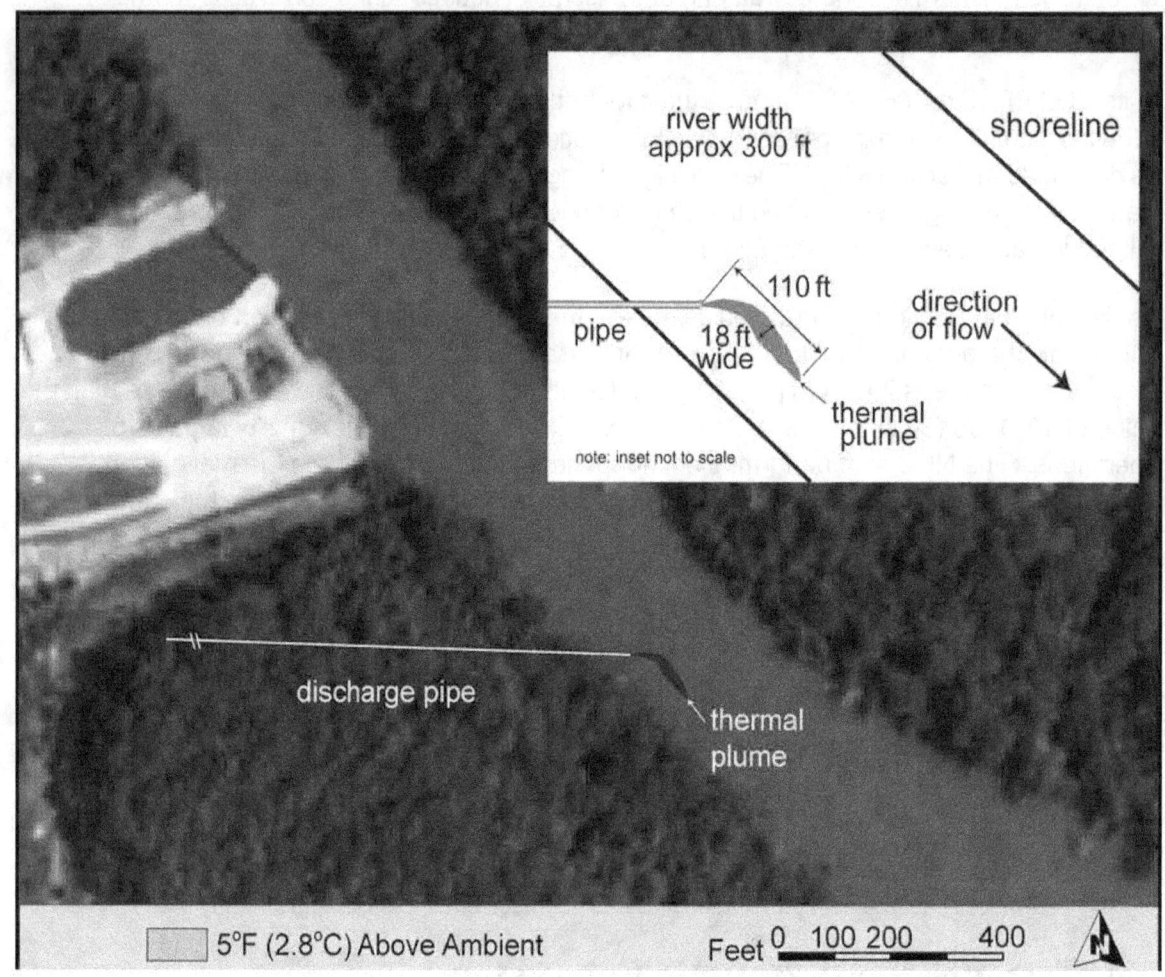

Figure 5-1. Extent of the 2.8°C (5.0°F)-Above-Ambient Isotherm Created by the Proposed VEGP Units 3 and 4 Discharge Pipe in the Combined Effluent Analysis

In its COL ER (Southern 2009), Southern indicated that there is no new and significant information regarding operations-related impacts on terrestrial resources. During its review of the COL application, the NRC staff independently verified that there is no new and significant information related to terrestrial ecology by reviewing Southern's ER, auditing Southern's process for identifying new and significant information, examining other information available at the site audit, considering applicable regulations and reference documents, and contacting representatives of the South Carolina Department of Natural Resources (SCDNR), U.S. Fish and Wildlife Service (FWS), and GDNR (see Appendix F).

Environmental Impacts of Operation

In a letter dated January 7, 2010, NRC requested that the FWS Field Office in Brunswick, Georgia, provide information regarding Federally listed species and critical habitat that may have changed since the 2008 consultation (NRC 2010a). On October 20, 2010, FWS provided an updated list of Federally listed threatened or endangered species that can be expected to occur in the project area (FWS 2010). FWS identified four Federally listed terrestrial plant and animal species that may occur on or in the vicinity of the VEGP site as well as within in the vicinity of the Representative Delineated Corridor (RDC) (FWS 2010). These four species are the red-cockaded woodpecker (*Picoides borealis*), the wood stork (*Mycteria americana*), the Canby's dropwort (*Oxypolis canbyi*), and the eastern indigo snake (*Drymarchon couperi*). Impacts to the red-cockaded woodpecker, wood stork, and Canby's dropwort are discussed in the ESP EIS.

In addition to the Federally listed species, FWS provided information on the bald eagle (*Haliaeetus leucocephalus*) and the gopher tortoise (*Gopherus polyphemus*) in the response letter FWS indicated there are eagle nests in Jefferson and McDuffie Counties, including one nest in the RDC (FWS 2010). The location of the eagle nest in the RDC also was discussed in the ESP EIS.

The eastern indigo snake and gopher tortoise were not included in the analysis in the ESP EIS. The eastern indigo snake was not included because it was not previously listed in FWS species lists for the counties within the project area (Burke, Jefferson, McDuffie, or Warren Counties) (NRC 2008a). Likewise, GDNR indicated there are no known occurrences of the gopher tortoise in the project area (GDNR 2009).

The information discussed in this section focuses on species not previously considered in the ESP EIS. This includes the eastern indigo snake and gopher tortoise, both identified by FWS in its October 20, 2010 letter as species that can be expected to, but not known to, occur in the project area. FWS noted that the gopher tortoise is not a Federally listed species in Georgia; however, FWS is reviewing its status (FWS 2010). Sandhills habitat that could support the gopher tortoise and the eastern indigo snake is present in the project area (GDNR 2009). Therefore, these species are discussed below.

NRC submitted a biological assessment (BA) to FWS on February 24, 2011 to document potential impacts on Federally listed threatened or endangered terrestrial species resulting from operation of Units 3 and 4 and ancillary facilities, as well as construction and operation of the proposed transmission line ROW. This BA is included in Appendix F. A BA documenting potential impacts on the Federally listed threatened or endangered species as a result of the site preparation and preliminary construction of the nonsafety-related structures, systems, or components on the VEGP site was submitted to FWS on January 25, 2008 (NRC 2008b), and FWS concurred with the findings on September 19, 2008 (FWS 2008).

Environmental Impacts of Operation

The eastern indigo snake was Federally listed as threatened by FWS in 1978 (FWS 1978). The eastern indigo snake is not documented in Burke County or any of the counties crossed by the proposed transmission line ROW. Suitable habitat may occur in the RDC, and gopher tortoise burrows, which are used by the eastern indigo snake, are in the vicinity. However, the project area is outside the historic and current range of the eastern indigo snake. Therefore, the NRC staff determined it is unlikely that operation of the proposed Units 3 and 4 and operation of the proposed transmission line will affect this species.

As noted above, the October 20, 2010, FWS letter included information on the gopher tortoise and the bald eagle (FWS 2010). The gopher tortoise is a Georgia state threatened species, and currently is under review by the FWS to be listed as Federally threatened (FWS 2010). There are no known populations of the gopher tortoise on the VEGP site or within the RDC (GDNR 2009; FWS 2010). Southern submitted a draft Candidate Conservation Agreement with Assurances (CCAA) for the gopher tortoise at the VEGP Site. This CCAA currently is under review by FWS (SERPPAS 2010). In light of the CCAA, and because no further impacts to sandhills habitat are projected to occur on the VEGP site, the staff considers it unlikely that the gopher tortoise will be affected onsite. The draft CCAA does not include the offsite portions of the proposed transmission line. In its October 20, 2010, letter to NRC, FWS recommends that tortoise surveys be included in surveys that are conducted where sandhills habitat exists. FWS also states that there are several areas within the RDC that have sandhills habitat that may contain gopher tortoises (FWS 2010). The impact on the gopher tortoise in the ROW due to ROW maintenance activities is not known because of the uncertainty of the final routing of the transmission line. However, there are no known tortoise locations within the RDC, and Georgia Power Company (GPC) has established maintenance practices and procedures to protect sensitive areas and species along existing transmission line ROWs. Therefore, the staff has determined the impacts to the gopher tortoise would likely be minimal.

The bald eagle, a State-threatened species, was Federally delisted under the Endangered Species Act in August 2007. There are bald eagle nests in Jefferson and McDuffie Counties in Georgia, and one known location of an active nest in McDuffie County in the vicinity of the proposed new transmission line ROW (FWS 2010). Potential impacts to the bald eagle were discussed in the ESP EIS. For example, as noted in the ESP EIS, the proposed 180-m (600-ft) buffer around the known bald eagle nest site would minimize any potential impacts from transmission line maintenance.

NRC received comments on the COL draft SEIS from the U.S. Department of Interior expressing concern about avian collisions with tall structures and transmission lines and what mitigative measures GPC will use to minimize impacts (see Appendix F). In Sections 5.4.1.2 and 5.4.1.6 of the ESP EIS (NRC 2008a), NRC included an analysis of operation-related avian collisions with structures, including cooling towers and transmission lines. However, additional

Environmental Impacts of Operation

information on the mitigation measures to minimize impacts to avian species during operation is provided below.

As discussed in the ESP EIS, the natural draft cooling towers associated with the proposed Units 3 and 4 would be 180-m (600-ft) high (NRC 2008a). The VEGP site is located adjacent to the Savannah River, and although migratory birds pass through the vicinity of the site, it is not located on a major American flyway. No formal bird collision surveys have been conducted at the VEGP site; however, the Environmental Protection Plan (EPP) for VEGP Units 1 and 2 stipulates that any excessive bird-impact events be reported to NRC within 24 hours (Southern 1989). No excessive bird-impact events have been reported onsite. The conclusion presented in the *Generic Environmental Impact Statement for License Renewal of Nuclear Plants* (GEIS) (NRC 1996) for nuclear power plant license renewals is that bird collisions with natural draft cooling towers are of small significance at all operating nuclear plants, including those with multiple cooling towers. Consequently, the incremental number of bird collisions, if any, associated with the operation of the two new natural draft cooling towers for the proposed Units 3 and 4 at the VEGP site, would be minimal.

Avian mortalities resulting from collisions with conductors, guy wires, and overhead ground (static) wires have not been specifically documented on GPC system components, but are known to occur on other utilities and communication systems. GPC has installed spiral vibration dampers to increase visibility on some of the transmission lines, especially along the coastal areas where the wood stork is known to nest and forage (GPC 2006). As noted above, of the EPP for the existing Units 1 and 2 stipulates that any excessive bird-impact events be reported to NRC within 24 hours (Southern 1989). Transmission line and ROW maintenance personnel have not reported bird deaths attributed to collisions or contact with Unit 1 and 2 transmission lines (Southern 2008).

The Electric Power Research Institute (EPRI) (1993) notes that factors appearing to influence the rate of avian impacts with structures are diverse and related to bird behavior, weather, and the attributes of the structure. Structure height, location, configuration, and lighting all appear to play a role in avian mortality. Weather, such as low cloud ceilings, advancing weather fronts, and fog also contribute to this phenomenon. Larger birds, such as waterfowl, are more prone to collisions with transmission lines, especially transmission lines that cross wetland areas used by large concentrations of birds (EPRI 1993).

EPRI (1993) documents electrocution of large birds, particularly eagles, as a source of mortality that could be significant to listed species. Electrocutions do not normally occur on lines where voltages are greater than 69 kV because the distance between lines is too great to be spanned by birds (EPRI 1993). The voltage of the proposed new transmission line is greater than 69 kV; therefore, bald eagles and other large bird populations should not be noticeably affected by

Environmental Impacts of Operation

transmission-line electrocutions. GPC has implemented an Avian Protection Program to monitor and address the impacts of transmission lines on birds.

The addition of the proposed transmission line likely would present new opportunities for bird collisions. However, the additional number of bird collisions, if any, would not be expected to cause a measurable reduction in local bird populations. Any impact events would be coordinated with GPC's Environmental Field Services and, if necessary, coordination also would involve FWS (GPC 2006). Consequently, the incremental number of bird collisions posed by the operation of the new transmission line for the proposed Units 3 and 4 at the VEGP site is anticipated to be negligible.

Based on the review of the new information presented above regarding operation-related impacts on terrestrial resources, the staff determined that the conclusion presented in the ESP EIS, that operational impacts would be SMALL, remains bounding and valid.

5.4.2 Aquatic Impacts

The NRC staff's assessments of aquatic-ecology-related impacts were provided in Section 5.4.2 and 5.4.3 of the ESP EIS (NRC 2008a). The staff assessed impacts to onsite streams and ponds and to the Savannah River from operation of the cooling-water system, including impacts from entrainment and impingement resulting from the operation of the intake system; impacts from operation of the discharge including thermal, chemical, and physical impacts; and impacts from transmission-line maintenance. Impacts to important species, including Federally and State-listed threatened and endangered species, also are discussed in Sections 5.4.2 and 5.4.3 of the ESP EIS (NRC 2008a). Based on the staff's analysis, operations-related impacts to the aquatic resources were considered to be SMALL.

In its COL ER (Southern 2009), Southern indicated that there is no new and significant information regarding operations-related impacts on aquatic biota. However, Southern indicated that there would be a 3 percent increase in the discharge flow. As explained in Section 5.3 of this SEIS, using the same conservative assumptions employed in the ESP EIS analysis, this change would result in only a small increase in the size of the thermal plume as defined by the 2.8°C (5.0°F)-above-ambient isotherm – from 29.6 m (97 ft) to 33.6 m (110 ft) in length and from 4.6 m (15 ft) to 5.2 m (17 ft) in width. The NRC staff reviewed this information and determined that consistent with the reasoning identified by the ESP EIS analysis, the thermal plume would remain small compared to the width of the Savannah River at that location, and it still would not impede fish passage up and down the river. Accordingly, this minor change would not affect the conclusion in the ESP EIS related to the impacts to aquatic biota from thermal discharges resulting from operation of two additional units. In addition to independently reviewing the ER, the NRC staff audited Southern's process for identifying new and significant information, examined other information available at the site audit, and discussed potential operational

Environmental Impacts of Operation

impacts with resource agencies (i.e., FWS, the National Marine Fisheries Service [NMFS], SCDNR, and GDNR; see Appendix F for the consultation letters).

During the site audit, Southern informed the NRC staff that the design and location of the cooling water intake structure for proposed Units 3 and 4 had changed. As a result, the staff requested further information on the design and location to determine whether any of these changes might affect the entrainment and/or impingement of aquatic organisms. In response to requests for additional information from the NRC staff, Southern (2010) indicated the intake structure would be located 46 m (150 ft) upstream of its previously designated location. The staff determined that this new location would not alter the basis for the staff's analysis and conclusion in the ESP EIS because the orientation of the mouth of the intake canal in relation to the river (perpendicular) has not changed, and because the new location of the intake canal is in habitat similar to that in the previous location (on a straight portion of the river and in the same floodplain). In addition, Southern described the changes to the intake design (Southern 2010) and indicated that no changes had been made to the water withdrawal rates, through-screen velocities, traveling screen mesh size, or to the hydraulic zone of influence, which are the main factors that would impact the entrainment or impingement rate of aquatic biota during operation of the cooling water intake structure. As a result, the staff determined there was no change to the impact on aquatic biota from entrainment or impingement as discussed in the ESP EIS.

As part of the NRC's responsibilities under Section 7 of the Endangered Species Act, the staff prepared a BA in connection with the Vogtle ESP review, documenting potential impacts on the shortnose sturgeon (*Acipenser brevirostrum*) as a result of construction of two new units at the VEGP site. That BA was submitted to NMFS (NRC 2008c). In its response (NMFS 2008), NFMS stated its conclusion that the proposed action, including the risk of sturgeon impingement with the intake structure and the potential effect from thermal discharge and chemical effluents, is not likely to adversely affect the shortnose sturgeon. The staff has determined that the project has not been modified in a way that was not previously considered in the ESP EIS or that would cause an effect to the shortnose sturgeon. In a letter dated September 3, 2010, the NRC confirmed with NMFS that the ESP stage consultation encompassed the proposed actions included in the COL application (NRC 2010b).

On October 6, 2010, NMFS published in the *Federal Register* (75 FR 61904) a proposed rule for listing the Carolina and South Atlantic distinct population segments of the Atlantic sturgeon (*Acipenser oxyrinchus oxyrinchus*) as endangered under the Endangered Species Act. In the ESP proceeding, the staff determined that impacts to the Atlantic sturgeon would be SMALL. The NRC staff has determined that the project has not been modified in a way that would cause an effect to the Atlantic sturgeon that was not previously considered in the ESP proceeding. Nevertheless, because of the listing proposal, the staff compiled information regarding the Atlantic sturgeon distribution and life history in a conference consultation letter to NMFS on March 2, 2011. A copy of this letter is provided in Appendix F. None of the information

Environmental Impacts of Operation

examined by the staff resulted in a change to the conclusions in Chapter 5 of the ESP EIS because it remained fully consistent with the staff's assessment that the species' demersal eggs and migratory behavior of larval sturgeon, as well as the design features of the intake structure and the anticipated extent of the thermal plume, would all minimize the potential impacts of plant operation to the Atlantic sturgeon.

Based on this review, the staff determined that the conclusions presented in the ESP EIS and the hearing proceedings remain valid.

5.5 Socioeconomic Impacts

The NRC staff's assessments of the socioeconomic-related impacts, including physical impacts, demographic impacts, economic impacts, and infrastructure and community-service impacts, were provided in Section 5.5 of the ESP EIS (NRC 2008a). Based on the staff's analysis, operations-related impacts to socioeconomics were considered to be SMALL, with the following three exceptions: (1) a MODERATE impact associated with the aesthetics of the transmission lines, (2) a MODERATE beneficial impact on the economy of Burke County, and (3) a LARGE beneficial property tax impact in Burke County.

In its COL ER (Southern 2009), Southern indicated that there is no new and significant information regarding operations-related impacts on socioeconomics. During its review of the COL application, the NRC staff independently verified that there is no new and significant information related to socioeconomics by reviewing Southern's ER, auditing Southern's process for identifying new and significant information, examining other information available at the site audit, considering applicable regulations and reference documents, and contacts with county officials.

Based on this review, the staff determined that the conclusions presented in the ESP EIS remain bounding and valid.

5.6 Historic and Cultural Resource Impacts

The NRC staff's assessment of impacts from operation of Units 3 and 4 to historic and cultural resources was provided in Section 5.6 of the ESP EIS (NRC 2008a). Based on the staff's analysis, operational impacts related to historic and cultural resources were considered to be SMALL.

In its COL ER (Southern 2009), Southern indicated that there is no new and significant information regarding operations-related impacts on historic and cultural resources. During its review of the COL application, the NRC staff independently verified that there is no new and significant information regarding operational impacts related to historic and cultural resources by

Environmental Impacts of Operation

reviewing Southern's ER, auditing Southern's process for identifying new and significant information, examining other information available at the site audit, considering applicable regulations and reference documents, and contact with the Georgia State Historic Preservation Officer (SHPO), Advisory Council on Historic Preservation, and Tribes (see Appendix C for the complete listing). The staff notes that, as described in Section 4.6, Southern has signed a memorandum of understanding with the Georgia SHPO (GHPD 2010). This action further indicates that Southern will protect historic and cultural resources on the VEGP site or mitigate impacts in consultation with the Georgia SHPO.

Based on this review, the NRC staff determined that the conclusions presented in the ESP EIS remain valid.

5.7 Environmental Justice

The NRC staff's assessment of the environmental justice-related impacts, including health and environmental impacts, socioeconomic impacts, and subsistence and special conditions, was provided in Section 5.7 of the ESP EIS (NRC 2008a). Based on the staff's analysis, operations-related environmental justice impacts were considered to be SMALL.

In its COL ER (Southern 2009), Southern indicated that there is no new and significant information regarding operations-related impacts on environmental justice. During its review of the COL application, the NRC staff independently verified that there is no new and significant information related to environmental justice by reviewing Southern's ER, auditing Southern's process for identifying new and significant information, examining other information available at the site audit, and considering applicable regulations and reference documents.

Based on this review, the staff determined that the conclusions presented in the ESP EIS remain bounding and valid.

5.8 Nonradiological Health Impacts

The NRC staff's assessment of the nonradiological health impacts for operation of the proposed Units 3 and 4 at the VEGP site was provided in Section 5.8 of the ESP EIS (NRC 2008a). Health impacts to the public from the cooling system, noise generated by operations, electromagnetic fields, other occupational health concerns, and transporting operations and outage workers were summarized. Health impacts from the same sources also were evaluated for workers at the proposed Units 3 and 4.

The NRC staff concluded in the ESP EIS that nonradiological health impacts to the public and the workers from the cooling system (e.g., exposure to thermophilic organisms), noise generated by unit operations, acute effects of electromagnetic fields at the higher power levels,

Environmental Impacts of Operation

occupational health-related impacts (e.g., falls, electric shock, etc.), and transporting operations and outage workers to/from the two additional units would be SMALL.

In the ESP EIS, the staff did not reach a conclusion on the chronic effects of electromagnetic fields. The staff found that available information was not sufficient to cause the staff to consider the potential impacts of electromagnetic fields as significant to the public.

In its COL ER (Southern 2009), Southern indicated that there is no new and significant information regarding operations-related impacts to nonradiological health. During its review of the COL application, the NRC staff independently verified that there is no new and significant information by reviewing Southern's ER, auditing Southern's process for identifying new and significant information, examining other information available at the site audit, and considering applicable regulations and reference documents, including recent data from the U.S. Centers for Disease Control and Prevention (CDC 2009), Georgia Department of Human Resources (GDHR 2009), and South Carolina Department of Health and Environmental Control (SCDHEC 2008, 2009, 2010).

Based on this review, the staff determined that the conclusions presented in the ESP EIS remain bounding and valid.

5.9 Radiological Impacts of Normal Operations

The NRC staff's assessment of the radiological health impacts resulting from normal operation of the proposed Units 3 and 4 at the VEGP site was provided in Section 5.9 of the ESP EIS (NRC 2008a). The discussion included the estimated radiation dose to a member of the public and to the biota in the vicinity of the VEGP site. Estimated doses to workers at the proposed units also were discussed.

This section considers whether new and significant information has been identified relative to the radiological health impacts during operation of the proposed Units 3 and 4. Exposure pathways are discussed in Section 5.9.1, radiological doses to members of the public are discussed in Section 5.9.2, impacts to members of the public are discussed in Section 5.9.3, occupational doses to workers are discussed in Section 5.9.4, impacts to biota other than members of the public are discussed in Section 5.9.5, and radiological monitoring is discussed in Section 5.9.6.

5.9.1 Exposure Pathways

The staff provided a summary of exposure pathways considered in its assessment of radiological impacts of normal operations in Section 5.9.1 of the ESP EIS (NRC 2008a).

Environmental Impacts of Operation

In its COL ER (Southern 2009), Southern indicated that there is no new and significant information regarding the exposure pathways considered in the analyses. During its review of the COL application, the NRC staff independently verified that there is no new and significant information related to exposure pathways by reviewing Southern's ER, auditing Southern's process for identifying new and significant information, examining other information available at the site audit, considering applicable regulations and reference documents, and reviewing the most recent offsite dose calculation manual for the existing Units 1 and 2. Although the new dairy being developed near Girard, Georgia, (approximately 9.6 km [6 mi] south of the VEGP site) is not considered in the analysis because it is greater than 8 km (5 mi) from the existing and proposed units, milk from the dairy will be monitored by Southern for radionuclides. Monitoring of milk from local dairies is carried out as part of the radiological monitoring program for the existing Units 1 and 2. Southern staff indicated during the site audit, and the NRC staff verified, that no previous samples had indicated the presence of radionuclides. The new dairy in Girard, Georgia, will become the nearest dairy being monitored.

Based on this review, the staff determined that the exposure pathways considered in the ESP EIS remain bounding and valid.

5.9.2 Radiation Doses to Members of the Public

The NRC staff's assessment of radiation doses to members of the public was provided in Section 5.9.2 of the ESP EIS (NRC 2008a).

In its COL ER (Southern 2009), Southern indicated that there is no new and significant information regarding the radiation doses to members of the public. During its review of the COL application, the NRC staff independently verified that there is no new and significant information related to radiation doses to members of the public by reviewing Southern's ER, auditing Southern's process for identifying new and significant information, examining other information available at the site audit, considering applicable regulations and reference documents, and reviewing the most recent offsite dose calculation manual for the existing Units 1 and 2.

For the ESP EIS (NRC 2008a), radiological impacts were determined using data from Revision 15 of the Westinghouse AP1000 reactor design (Westinghouse 2005) with expected direct radiation and liquid and gaseous radiological effluent rates. The Southern ESP application referenced Revision 15 of the AP1000 standard reactor design, and Revision 15 is certified by rule in Title 10 of the Code of Federal Regulations (CFR) Part 52, Appendix D. Prior to publication of the ESP EIS, Westinghouse submitted Revision 16 (Westinghouse 2007) to the AP1000 reactor design to the NRC for review. The staff noted this submission in the ESP EIS, but did not update the analyses with respect to radiological impacts because the staff review of Revision 16 was not complete. Subsequently, Westinghouse submitted Revision 17

Environmental Impacts of Operation

(Westinghouse 2008) to the AP1000 reactor design. Although Revision 17 remains under a separate design certification review pursuant to 10 CFR Part 52, the NRC staff has considered the impact of this latest revision in its evaluation of potential impacts for normal operations in this SEIS. For normal operations, the staff has not found any changes in estimated direct radiation, gaseous radiological effluent releases, or liquid radiological effluent releases based on data in Revisions 15, 16, and 17.

In its COL ER (Southern 2009), Southern indicated that a new low-level waste (LLW) storage area had been developed northwest of the existing Unit 2 cooling tower to accommodate wastes from the existing units as well as Units 3 and 4. Because of the distance between the LLW storage area and the proposed construction area, Southern determined and the staff agrees that the LLW storage area would provide negligible contribution to direct radiation dose to construction workers. Likewise, because of distances, occupancy factors, and the lack of effluents from the facility, doses to members of the public, operations personnel, and other biota would also be negligible.

Based on this review, the staff determined that the radiation doses to members of the public described in the ESP EIS remain valid.

5.9.3   Impacts to Members of the Public

The NRC staff's assessment of the estimated impacts to members of the public was provided in Section 5.9.3 of the ESP EIS (NRC 2008a), including to a maximally exposed individual near the VEGP site and a population dose (collective dose to the population within 80 km [50 mi]) in the vicinity of the VEGP site. Based on the NRC staff's analysis, operation-related health impacts to individual members of the public and the population resulting from radiation exposure were considered to be SMALL.

In its COL ER (Southern 2009), Southern indicated that there is no new and significant information regarding the impacts to members of the public. During its review of the COL application, the staff independently verified that there is no new and significant information related to impacts to members of the public by reviewing Southern's ER, auditing Southern's process for identifying new and significant information, examining other information available at the site audit, considering applicable regulations and reference documents, and reviewing the most recent offsite dose calculation manual for the existing Units 1 and 2.

Based on this review, the staff determined that the radiation doses to members of the public described in the ESP EIS remain valid.

Environmental Impacts of Operation

5.9.4 Occupational Doses to Workers

The staff's assessment of the estimated impacts to occupational workers was provided in Section 5.9.4 of the ESP EIS (NRC 2008a). Based on the staff's analysis, operation-related health impacts to occupational workers resulting from radiation exposure were considered to be SMALL.

In its COL ER, Southern indicated that there is no new and significant information regarding the impacts to occupational workers. During its review of the COL application, the NRC staff independently verified that there is no new and significant information related to impacts to members of the public by reviewing Southern's ER, auditing Southern's process for identifying new and significant information, examining other information available at the site audit, and considering applicable regulations and reference documents.

Based on this review, the staff determined that the radiation doses to occupational workers described in the ESP EIS remain valid.

5.9.5 Impacts to Biota Other than Members of the Public

The NRC staff's assessment of the estimated impacts to biota other than members of the public was provided in Section 5.9.5 of the ESP EIS (NRC 2008a). Based on the staff's analysis, operation-related health impacts to biota from radiation exposure were considered to be SMALL.

In its COL ER (Southern 2009), Southern indicated that there is no new and significant information regarding the impacts to biota. During its review of the COL application, the NRC staff independently verified that there is no new and significant information related to impacts to biota by reviewing Southern's ER, auditing Southern's process for identifying new and significant information, examining other information available at the site audit, considering applicable regulations and reference documents, and reviewing the most recent offsite dose calculation manual for the existing Units 1 and 2.

Based on this review, the staff determined that the radiation doses to biota other than members of the public described in the ESP EIS remain valid.

5.9.6 Radiological Monitoring

In Section 5.9.6 of the ESP EIS (NRC 2008a), the NRC staff provided a summary of radiological monitoring performed at and near the VEGP site.

In its COL ER (Southern 2009), Southern indicated that there is no new and significant information regarding radiological monitoring. During its review of the COL application, the NRC staff independently verified that there is no new and significant information related to radiological monitoring by reviewing Southern's ER, auditing Southern's process for identifying new and significant information, examining other information available at the site audit, considering applicable regulations and reference documents, and reviewing the most recent offsite dose calculation manual for the existing Units 1 and 2.

Based on this review, the staff determined that the radiological monitoring described in the ESP EIS remains valid.

5.10 Environmental Impacts of Postulated Accidents

The NRC staff's assessment of the environmental impacts of postulated design basis accidents and severe accidents for AP1000 reactors at the VEGP ESP site was provided in Section 5.10 of the ESP EIS (NRC 2008a). Based on the staff's analysis, the environmental impacts of design-basis and severe accidents were considered to be SMALL.

The Southern ESP application referenced Revision 15 of the *AP1000 Design Control Document* for the AP1000 standard reactor design (Westinghouse 2005). Revision 15 is certified by rule in 10 CFR Part 52, Appendix D. Prior to publication of the ESP EIS, Westinghouse submitted Revision 16 to the *AP1000 Design Control Document* (Westinghouse 2007) for NRC staff review. The staff noted this submission in the ESP EIS, but did not update the accident analyses because the staff review of Revision 16 was not complete. Subsequently, Westinghouse submitted Revision 17 of the *AP1000 Design Control Document* (Westinghouse 2008). Consequently, Southern updated its review of potential impacts for postulated accidents based on Revision 17 of the *AP1000 Design Control Document*, which is under separate review by the NRC staff pursuant to 10 CFR Part 52.

The term "accident," as used in this section, refers to any off-normal event not addressed in Section 5.9 that results in release of radioactive materials into the environment. The focus of this review is on events that could lead to releases substantially in excess of permissible limits for normal operations. Normal release limits are specified in 10 CFR Part 20, Appendix B, Table 2.

5.10.1 Design Basis Accidents

The NRC staff's review of Design Basis Accidents (DBAs) was provided in Section 5.10.1 of the ESP EIS (NRC 2008a). The review of environmental impacts of postulated accidents in the ESP EIS assumed the location of two new nuclear units at the VEGP ESP site. The calculation approach used by Southern for its COL application is consistent with the approach described in the ESP EIS (NRC 2008a) and is summarized below.

Environmental Impacts of Operation

Southern evaluated the potential consequences of postulated accidents to demonstrate that an AP1000 reactor could be constructed and operated at the VEGP site without undue risk to the health and safety of the public (Southern 2008). These evaluations used a set of DBAs that are representative for the AP1000 reactor design and site-specific meteorological data. The set of accidents covers events that range from a relatively high probability of occurrence with relatively low consequences to a relatively low probability with high consequences.

The DBA analyses in the ESP EIS (NRC 2008a) assumed that the postulated releases would occur from the location on an imaginary border of an area surrounding all release points for the two proposed units that would result in the greatest doses at the exclusion area and low population zone boundaries. The units proposed in the COL application are situated entirely within the area assumed in the ESP application, so the previous exclusion area boundary and low-population zone distances remain valid for the COL application. The staff evaluated potential consequences of DBAs following procedures outlined in regulatory guides and standard review plans. Potential consequences of accidental releases depend on characteristics of the specific radionuclides released, radionuclide release rates, and meteorological conditions. Methods for evaluating potential accidents are based on guidance in Regulatory Guide 1.183 (NRC 2000).

Based on the ESP review and having found no new and significant information applicable to this analysis, the NRC staff concludes that the atmospheric dispersion factors (χ/Qs) for the VEGP site are still applicable for evaluating potential environmental consequences of postulated DBAs for Revision 17 of the *AP1000 Design Control Document* (Westinghouse 2008) at the VEGP site.

Table 5-1 lists the set of DBAs considered and presents estimates of the environmental consequences of each accident in terms of total effective dose equivalent (TEDE), which is the sum of the committed effective dose equivalent from inhalation and the effective dose equivalent from external exposure. The DBAs listed in the table are the same as those being considered in the design certification and those that were considered in the ESP review. The NRC staff independently reviewed the calculation of the consequences of the DBAs in Revision 17 of the *AP1000 Design Control Document* and found the calculations to be correct. There are no environmental criteria related to the potential consequences of DBAs. Consequently, the review criteria used in the staff's safety review of DBA doses are included in Table 5-1 to illustrate the magnitude of the calculated environmental consequences (TEDE doses). In all cases, the calculated TEDE values are considerably smaller than the TEDE doses used as safety review criteria. Further, in no case is the consequence estimate significantly different than the corresponding estimate presented in the ESP EIS (NRC 2008a). Therefore, the staff determined that the conclusion in the ESP EIS that the environmental consequences of DBAs for an AP1000 reactor at the VEGP site are SMALL remains valid.

Environmental Impacts of Operation

Table 5-1. DBA Doses for an AP1000 Reactor at the VEGP Site (Southern 2009a)

Accident	Standard Review Plan Section[b]	TEDE in rem[a]		
		Exclusion Area Boundary	Low-Population Zone	Safety Review Criterion
Main steam line break	15.0.3			
Pre-incident iodine spike		0.07	0.03	25[c]
Equilibrium iodine activity		0.08	0.08	2.5[d]
Loss-of-coolant accident	15.0.3	3.6	1.5	25[c]
Steam generator tube rupture	15.0.3			
Pre-incident iodine spike		0.16	0.04	25[c]
Equilibrium iodine activity		0.08	0.02	2.5[d]
Locked rotor	15.0.3			
No feedwater		0.06	0.01	2.5[d]
Feedwater available		0.04	0.02	2.5[d]
Failure of small lines carrying primary coolant outside containment	15.0.3	0.15	0.03	2.5[d]
Rod ejection accident	15.0.3	0.27	0.17	6.3[d]
Fuel handling	15.0.3	0.38	0.07	6.3[d]

(a) To convert rem to Sv, divide rem by 100.
(b) NUREG-0800 (NRC 2007).
(c) 10 CFR 52.79(a)(2) and 10 CFR 100.21.
(d) Standard Review Plan criterion.

5.10.2 Severe Accidents

The staff's analysis of the potential consequences of severe accidents was provided in Section 5.10.2 of the ESP EIS (NRC 2008a). The staff concluded that the probability-weighted consequences of the severe accidents for an AP1000 reactor at the VEGP ESP site were SMALL and that the issue was resolved.

Southern conducted a search for new information related to severe accidents and states that there have been no significant changes in either the reactor-specific or site-specific information used in the severe accident consequence assessment (Southern 2009). The NRC staff has reviewed the process that Southern used to search for new information and has conducted its own search. The staff concurs that there is no new and significant information related to the site-specific input to the severe accident consequence assessment in Section 5.10.2 of the ESP EIS.

The NRC staff evaluated the significance of the new information related to the AP1000 design. Westinghouse reviewed the AP1000 Probabilistic Risk Analysis (PRA) for Revision 15 of the *AP1000 Design Control Document* (Westinghouse 2005) and concluded that the PRA remained

Environmental Impacts of Operation

valid for a proposed Revision 16 of the *AP1000 Design Control Document* (Westinghouse 2007); the PRA is unchanged for Revision 17 (Westinghouse 2008). The NRC staff also evaluated the current PRA using DC/COL-ISG-3, *Probabilistic Risk Assessment Information to Support Design Certification and Combined License Applications*, (NRC 2008c), and concluded that the PRA submitted with Revision 15 is a conservative and acceptable basis for evaluating severe accident consequences for the current revision.

Because the NRC staff is not aware of any new and significant site-specific or reactor-specific information, the NRC staff determined that its conclusion set forth in Section 5.10.2 of the ESP EIS that the probability-weighted consequences of severe accidents at the VEGP site would be SMALL remains valid.

5.10.3 Severe Accident Mitigation Alternatives

The NRC staff provided a review of Severe Accident Mitigation Alternatives (SAMAs) for Revision 15 of the AP1000 reactor design at the VEGP site in Section 5.10.3 of the ESP EIS (NRC 2008a). The staff found that the VEGP site characteristics are within the site characteristics considered in the severe accident design mitigation alternatives (SAMDA) review conducted for certification of the AP1000 design (10 CFR 52, Appendix D). Consequently, further SAMDA review was precluded by rule. The other attributes of the SAMA review, namely procedures and training, were also addressed in the ESP EIS.

In its COL ER, Southern states that there is no new and significant information related to postulated accidents (Southern 2009). However, the NRC staff notes that the ER did contain an update of information on DBAs associated with the proposed revision to the AP1000 design. In the previous section of this SEIS, the staff reviewed the information used in the severe accident consequence assessment included in the staff's ESP EIS and determined that the revised reactor design did not change any of the input to the severe accident consequence assessment.

Westinghouse reviewed the AP1000 PRA for Revision 15 and concluded that the PRA remains valid for a proposed revision of the design control document (Westinghouse 2007); the PRA is unchanged for Revision 17. Furthermore, the NRC staff evaluated the current PRA using DC/COL-ISG-3 (NRC 2008c) and concluded that the PRA submitted with Revision 15 is a conservative and acceptable basis for evaluating strategies for mitigating severe accidents. Therefore, the NRC staff considers the PRA for Revision 15 of the design control document to be an adequate basis for a SAMDA analysis for an application referencing Revision 17. Consequently, the NRC staff incorporates, by reference, the environmental assessment accompanying the design certification rulemaking for 10 CFR Part 52, Appendix D (NRC 2005).

Environmental Impacts of Operation

Because there is no new and significant information related to either the site-specific data used in the ESP EIS to conclude that the characteristics of the VEGP site are bounded by those considered in the generic SAMDA review or to the AP1000 PRA, the NRC staff reaffirms and adopts the ESP EIS conclusions that there are no cost-effective SAMDAs for an AP1000 at the VEGP site.

Other attributes of the SAMA review, namely procedures and training, have been addressed by Southern's statement that "…appropriate administrative controls on plant operations would be incorporated into the plants' management systems as part of its baseline…." (Southern 2008). Further, the staff notes that, pursuant to regulatory requirements, procedures and training, programs are being developed. The staff has a reasonable expectation that risk mitigation measures will be considered when procedures would be in place and training would be completed prior to loading fuel. Therefore, the NRC staff concludes that SAMAs were appropriately considered in the ESP EIS.

5.10.4 Summary of Postulated Accident Impacts

In the ESP EIS (NRC 2008a), the staff evaluated the environmental impacts from DBAs and severe accidents for an AP1000 at the VEGP site and considered SAMAs. Based on the information provided by Southern and NRC's own independent review, the staff concluded that the potential environmental impacts (risks) from postulated accidents from the operation of the proposed AP1000 reactors would be SMALL and that additional mitigation is not warranted. Staff from Southern and NRC have considered new information, including changes to the certified AP1000 reactor design, and determined that there is no new and significant information. Therefore, the staff concludes that ESP EIS conclusions related to DBAs, severe accidents, and SAMAs remain valid.

5.11 Measures and Controls to Limit Adverse Impacts During Operation

The staff's assessment of measures and controls to limit adverse impacts during operation are provided in Section 5.11 of the ESP EIS (NRC 2008a).

In its COL ER (Southern 2009), Southern indicated that there is no new and significant information regarding measures and controls to limit adverse impacts during construction, but did indicate that it remains committed to the mitigation measures included in Section 5.11 of the ESP EIS. During its independent review of the COL application, the NRC staff evaluated new and significant information related to the measures and controls by reviewing Southern's ER, auditing Southern's process for identifying new and significant information, examining other information available at the site audit, and considering applicable regulations and reference documents. As discussed in Section 5.6, a memorandum of understanding (GHPD 2010) has

Environmental Impacts of Operation

been signed between Southern and the Georgia SHPO concerning protection of archaeological site 9BK416. The staff determined that this agreement constitutes a new measure and control.

Additionally, Part 10 of the COL application includes a draft EPP for the site, which identifies proposed conditions, monitoring, reporting, and record keeping for environmental data during operations.

Based on its review, the staff determined that, with the addition of the Memorandum of Understanding that was identified, the measures and controls to limit adverse impacts during operation as presented in the ESP EIS remain valid, and also that Southern's proposed EPP is appropriate. If the COL is issued, the staff will include the EPP as part of the license.

5.12 Summary of Operation Impacts

Impact level categories identified during the evaluation of the ESP application are documented in Table 5-19 of the ESP EIS (NRC 2008a). These levels are designated as SMALL, MODERATE, or LARGE as a measure of their expected adverse impacts. The NRC staff's review of information available during both site audits and from other information sources did not identify any information that would change the designation for any of the categories in Table 5-19 of the ESP EIS.

5.13 References

10 CFR Part 20. Code of Federal Regulations, Title 10, *Energy*, Part 20, "Standards for Protection Against Radiation."

10 CFR Part 52. Code of Federal Regulations, Title 10, *Energy*, Part 52, "Licenses, Certifications, and Approvals for Nuclear Power Plants."

10 CFR Part 100. Code of Federal Regulations, Title 10, *Energy*, Part 100 "Reactor Site Criteria."

73 FR 16436. March 27, 2008. "National Ambient Air Quality Standards for Ozone." *Federal Register*, Environmental Protection Agency.

75 FR 61904. October 6, 2010. "Endangered and Threatened Wildlife and Plants; Proposed Listings for Two Distinct Population Segments of Atlantic Sturgeon (*Acipenser oxyrinchus oxyrinchus*) in the Southeast." *Federal Register*. U.S. Department of Commerce.

Bald and Golden Eagle Protection Act. 1991. 16 USC 668 et seq.

Centers for Disease Control and Prevention (CDC). 2009. "Summary of Notifiable Diseases — United States, 2007." *Morbidity and Mortality Weekly Report* 56(53). July 9, 2009. Atlanta, Georgia. Accession No. ML100470938.

Clean Air Act (CAA). 42 USC 7401, et seq.

Doneker, R.L. and G.H. Jirka. 2007. *Cormix User Manual, A Hydrodynamic Mixing Zone Model and Decision Support System for Pollutant Discharges into Surface Waters.* EPA#823/K-07-001. Accessed at http://www.cormix.info. (No-cost site, but registration required.)

Electric Power Research Institute (EPRI). 1993. *Proceedings: Avian Interactions with Utility Structure, International Workshop, September 13-16, 1992, Miami, Florida.* EPRI TR-103268, Palo Alto, California.

Endangered Species Act of 1973. 16 USC 1531 et seq.

Georgia Department of Natural Resources (GDNR). 2009. E-mail dated December 21, 2009 and letter dated December 17, 2009 from Katrina Morris, GDNR, to Mallecia Sutton, NRC, regarding Natural Heritage Database occurrences within the VEGP Boundary and the Transmission Line Macrocorridor. Accession No. ML100490042.

Georgia Historic Preservation Division (GHPD). 2010. Letter from D. Crass, GHPD, to T.C. Morrer, Southern, "Memorandum of Understanding, Archaeological Site 9BK416 Vogtle Electric Generating Plant Expansion, Burke County, Georgia," Southern Company, Atlanta, Georgia. Accession No. ML100500302.

Georgia Power Company (GPC). 2006. *Avian Protection Program for Georgia Power Company, Rev 1.* March 14, 2006. Accession No. ML063000228.

South Carolina Department of Health and Environmental Control (SCDHEC). 2008. *2006 South Carolina Annual Report on Reportable Conditions.* Columbia, South Carolina. Accession No. ML100481115.

South Carolina Department of Health and Environmental Control (SCDHEC). 2009. *Epi Notes Disease Prevention and Epidemiology Newsletter* XXVI(2), Columbia, South Carolina. Accession No. ML100470958.

South Carolina Department of Health and Environmental Control (SCDHEC). 2010. *Epi Notes Disease Prevention and Epidemiology Newsletter* XXVII(1):Winter 2010, Columbia, South Carolina. Accession No. ML100470979.

Southeast Regional Partnership for Planning and Sustainability (SERPPAS). 2010. *Candidate Conservation Agreement for the Gopher Tortoise First Annual Report: October 1, 2008 – September 30, 2009.* Accessed at www.serppas.org/.../GTCCA%20First%20Annual%20Report%202008-2009.pdf on November 15, 2010.

Southern Nuclear Operating Company, Inc. (Southern). 1989. *Environmental Protection Plan, Appendix B to the Facility Operating License No. NPF-68 and Facility Operating License No. NPF-81, Vogtle Electric Generating Plant, Units 1 and 2.* Southern Nuclear, Docket Nos. 50-424 and 50-425. March 31, 1989. Southern Company, Birmingham, Alabama. Accession No. ML012350369.

Southern Nuclear Operating Company, Inc. (Southern). 2008. *Southern Nuclear Operating Company, Vogtle Early Site Permit Application*, Revision 4, Southern Company, Birmingham, Alabama. Package Accession No. ML081020073.

Southern Nuclear Operating Company, Inc. (Southern). 2009. *Vogtle Electric Generating Plant, Units 3 and 4, COL Application: Part 3. Environmental Report.* Revision 1, August 23, 2009, Southern Company, Birmingham, Alabama. Accession No. ML092740400.

Southern Nuclear Operating Company, Inc. (Southern). 2010. Letter from C.R. Pierce, Southern, to NRC, "Southern Nuclear Operating Company Vogtle Electric Generating Plant Units 3 and 4 Combined License Application, Response to Request for Additional Information Letter on Environmental Issues," January 8, 2010. Southern Company, Birmingham, Alabama. Accession No. ML100120479.

U.S. Department of Commerce, National Oceanic and Atmospheric Administration, National Marine Fisheries Service (NMFS). 2008. Letter from Roy E. Crabtree, Ph.D., Regional Administrator to William Burton, NRC, dated August 11, 2008, "A Biological Assessment for the Shortnose Sturgeon for the Vogtle Electric Generating Plant Early Site Permit Application." Accession No. ML082480450.

U.S. Fish and Wildlife Service (FWS). 1978. "Endangered and Threatened Wildlife and Plants. Listing of the Eastern Indigo Snake as a Threatened Species." *Federal Register* 43:4026.

U.S. Fish and Wildlife Service (FWS). 2008. Letter from U.S. Fish and Wildlife Service, Athens, Georgia to U.S. Nuclear Regulatory Commission, Washington, DC. Re: USFWS Log# 08-FA-0473. September 19, 2008. Accession No. ML082760694.

U.S. Fish and Wildlife Service (FWS). 2010. Letter from S. Tucker, FWS, Athens, Georgia, to G.P. Hatchett, NRC, Washington, D.C., "Re: USFWS Log# 2010-1254." October 20, 2010. Accession No. ML103010076.

Environmental Impacts of Operation

U.S. Nuclear Regulatory Commission (NRC). 1996. *Generic Environmental Impact Statement for License Renewal of Nuclear Plants.* NUREG-1437, Volumes 1 and 2, U.S. Nuclear Regulatory Commission, Washington, D.C.

U.S. Nuclear Regulatory Commission (NRC). 2000. *Alternative Radiological Source Terms for Evaluating Design Basis Accidents at Nuclear Power Plants.* Regulatory Guide 1.183, Washington, D.C.

U.S. Nuclear Regulatory Commission (NRC). 2005. *Environmental Assessment Relating to the Certification of the AP1000 Standard Plant Design. Docket No. 52-006.* Washington, D.C. Accession No. ML053250292.

U.S. Nuclear Regulatory Commission (NRC). 2007. *Standard Review Plan for the Review of Safety Analysis Reports for Nuclear Power Plants.* NUREG-0800, Washington, D.C.

U.S. Nuclear Regulatory Commission (NRC). 2008a. *Final Environmental Impact Statement for an Early Site Permit (ESP) at the Vogtle Electric Generating Plant Site.* NUREG-1872, Vols. 1, 2, and Errata, Washington, D.C. Accession Nos. ML082240145; ML082240165, ML082260203; ML082550040.

U.S. Nuclear Regulatory Commission (NRC). 2008b. Letter from NRC to FWS dated January 25, 2008. Subject: Biological Assessment for Threatened and Endangered Species and Designated Critical Habitat for the Vogtle Electric Generating Plant Early Site Permit Application. Accession No. ML080100512.

U.S. Nuclear Regulatory Commission (NRC). 2008c. Letter from W. Burton, NRC, to D. Bernhart, NMFS, "Subject: Biological Assessment for the Shortnose Sturgeon for the Vogtle Electric Generating Plant Early Site Permit Application." dated January 25, 2008. Accession Nos. ML080070538, ML080100588.

U.S. Nuclear Regulatory Commission (NRC). 2008c. *Interim Staff Guidance, Probabilistic Risk Assessment Information to Support Design Certification and Combined License Applications.* DC/COL-ISG-3, Washington, D.C. Accession No. ML081430675.

U.S. Nuclear Regulatory Commission (NRC). 2010a. Letter from G.P. Hatchett, NRC, to S. Tucker, FWS, "Request for a List of Protected Species Within the Area Under Evaluation for the Vogtle Electric Generating Plant, Units 3 and 4 Combined License Applications," January 7, 2010. Washington, D.C. Accession No. ML092600684.

Environmental Impacts of Operation

U.S. Nuclear Regulatory Commission (NRC). 2010b. Letter from G.P. Hatchett, NRC, to D. Bernhart, NMFS, "Notification of the Issuance and Request for Comments on the Draft Supplemental Environmental impact Statement for Vogtle Electric Generating Plant, Units 3 and 4 Combined License Application." September 3, 2010. Accession No. ML102320162.

Westinghouse Electric Company, LLC (Westinghouse). 2005. *AP1000 Design Control Document.* AP1000 Document. APP-GW-GL-700, Revision 15, Westinghouse Electric Company, Pittsburgh, Pennsylvania. Package Accession No. ML053480403.

Westinghouse Electric Company, LLC (Westinghouse). 2007. *AP1000 Design Control Document.* AP1000 Document. APP-GW-GL-700, Revision 16, Westinghouse Electric Company, Pittsburgh, Pennsylvania. Package Accession No. ML071580939.

Westinghouse Electric Company, LLC (Westinghouse). 2008. *AP1000 Design Control Document.* AP1000 Document. APP-GW-GL-700, Revision 17, Westinghouse Electric Company, Pittsburgh, Pennsylvania. Package Accession No. ML083230168.

6.0 Fuel Cycle, Transportation, and Decommissioning

In Chapter 6 of the Vogtle Electric Generating Plant (VEGP) early site permit (ESP) environmental impact statement (EIS) (NRC 2008), the U.S. Nuclear Regulatory Commission (NRC) staff provided a description of the environmental impacts from (1) the uranium fuel cycle and solid waste management, (2) the transportation of radioactive material, and (3) the decommissioning of two new nuclear units at the VEGP site. Fuel cycle impacts and solid waste management are discussed in Section 6.1. Transportation impacts are discussed in Section 6.2. Decommissioning impacts are discussed in Section 6.3. The list of references cited is in Section 6.4.

6.1 Fuel Cycle Impacts and Solid Waste Management

The NRC staff's assessment of fuel cycle and solid waste-management-related impacts was provided in Section 6.1 of the ESP EIS (NRC 2008). Based on the staff's analysis, environmental impacts were considered to be SMALL.

Southern Nuclear Operating Company, Inc. (Southern) stated in the environmental report (ER) included in its combined license (COL) application that there is no new and significant information regarding fuel cycle and solid-waste management-related environmental impacts (Southern 2009a). During its review of the COL application, the staff independently verified that there is no new and significant information related to fuel cycle and solid-waste management by reviewing Southern's ER, auditing Southern's process for identifying new and significant information, examining other information available at the site audit, and considering applicable regulations and reference documents, including Southern's response to the staff's request for additional information regarding the proposed solid-waste-management system (Southern 2009b). However, because of additional information submitted by Southern regarding its low-level waste (LLW) disposal options and associated contingency plans, the staff assessed the significance of this information for its analysis in the ESP EIS of the environmental impacts of the uranium fuel cycle regarding LLW management.

The quantities of buried radioactive waste material (i.e., LLW, high-level waste [HLW], and transuranic waste) are specified in Table S–3 (Title 10 of the Code of Federal Regulations (CFR) Subpart 51.51(b)). For LLW disposal at land burial facilities, the Commission notes in Table S–3 that there would be no significant radioactive releases to the environment.

Southern indicated in its response to the staff's request for additional information (ND-09-1540) that the Barnwell LLW disposal facility in Barnwell, South Carolina, no longer accepts Class-B and Class-C wastes from sources in states outside of the Atlantic Compact (Southern 2009b). By the time Units 3 and 4 begin operations, Southern stated that it expects to enter into an

Fuel Cycle, Transportation, and Decommissioning

agreement with an NRC-licensed facility that would accept LLW from VEGP. If that expectation is not met, Southern indicated it could implement measures to limit the generation of Class-B and Class-C wastes, extending the capacity of the onsite Auxiliary Building to store such wastes. Southern noted that it also could construct additional storage facilities onsite and has indicated that such facilities would be designed and operated to meet the guidance standards in Appendix 11.4-A of NUREG-0800, *Standard Review Plan for the Review of Safety Analysis Reports for Nuclear Power* (NRC 1987). Finally, Southern indicated that it could enter into an agreement with a third-party contractor to process, store, own, and ultimately dispose of LLW from VEGP. Because Southern indicates that it would choose one or a combination of these options, the staff considered the environmental impacts of each of these three options.

Table S–3 addresses the environmental impacts expected if Southern enters into an agreement with an NRC-licensed facility for disposal of LLW, and Table S–4 addresses the environmental impacts from transportation of LLW as discussed in the ESP EIS (NRC 2008). The use of third-party contractors was not explicitly addressed in Tables S–3 and S–4; however, such third-party contractors are currently licensed by the NRC and are required to comply with 10 CFR Part 20 dose limits. The impacts from onsite storage or use of a third-party contractor are therefore expected to be similar, and the additional environmental impacts are not significant compared to the impacts described in Tables S–3 and S–4.

The measures to reduce the generation of Class-B and Class-C wastes described by Southern, such as mixing spent resins to limit radioactivity concentrations, could increase the volume of LLW, but would not increase the total curies of radioactive material in the waste. The volume of waste would still be bounded by or very similar to the estimates shown in Table S–3, and the environmental impacts would not be significantly different.

When applicable criteria are met, the NRC's regulations (10 CFR 50.59) allow licensees operating nuclear power plants to construct and operate additional onsite LLW storage facilities without seeking approval from the NRC. Licensees are required to evaluate the safety and environmental impacts before constructing the facility and make those evaluations available to NRC inspectors. A number of nuclear power plant licensees have constructed and operate such facilities in the United States, including Southern, which currently maintains an onsite LLW storage area for VEGP Units 1 and 2. These facilities have available storage capacity for 6 to 8 years of accumulated waste and adequate room for expansion (Southern 2008). Typically, these facilities are constructed near the power block inside the security fence on land that has already been disturbed during initial plant construction. Therefore, the impacts on environmental resources (e.g., land use and aquatic and terrestrial biota) of such additional storage would be very small. All of the NRC (10 CFR Part 20) and U.S. Environmental Protection Agency (EPA) (40 CFR Part 190) dose limitations would apply both for public and occupational radiation exposure and the radiation doses continue to be below 0.25 mSv/yr (25 mrem/yr), which is the dose limit stated in 40 CFR Part 190. The NRC staff concludes that

Fuel Cycle, Transportation, and Decommissioning

doses to members of the public within the NRC and EPA regulations are a small impact. Therefore, the staff concludes the environmental impacts from any additional or expanded LLW storage facilities that Southern might construct and operate would be SMALL.

In addition, NUREG-1437, *Generic Environmental Impact Statement for License Renewal of Nuclear Plants, Main Report, Final Report*, assessed the impacts of LLW storage onsite at currently operating nuclear power plants and concluded that the radiation doses to offsite individuals from interim LLW storage are insignificant (NRC 1996). The types and amounts of LLW generated by the proposed Units 3 and 4 would be very similar to those generated by currently operating nuclear power plants, and the construction and operation of these interim LLW storage facilities would be very similar to the construction and operation of the currently operating facilities.

The Commission notes that HLW and transuranic waste are to be buried at a repository, such as the proposed geologic HLW repository at Yucca Mountain, Nevada, and that no release to the environment is expected to be associated with such disposal because it has been assumed that all of the gaseous and volatile radionuclides contained in the spent fuel are released to the atmosphere before the disposal of the waste. In NUREG-0116, *Environmental Survey of the Reprocessing and Waste Management Portions of the LWR Fuel Cycle* (NRC 1976), which provides background and context for the Table S–3 values established by the Commission, the staff indicates that HLW and transuranic waste will be buried and will not be released to the environment.

As part of the Table S–3 rulemaking, the staff evaluated, along with more conservative assumptions, this zero-release assumption associated with waste burial in a repository, and the NRC reached an overall generic determination that fuel cycle impacts would not be significant. In 1983, the U.S. Supreme Court affirmed the NRC's position that the zero-release assumption was reasonable in the context of the Table S–3 rulemaking to address generically the impacts of the uranium fuel cycle in individual reactor licensing proceedings (Baltimore Gas and Electric Company vs. Natural Resources Defense Council, Inc. 1983).

Furthermore, in the Commission's Waste Confidence Decision, 10 CFR 51.23(a), the Commission has made the generic determination that "… if necessary, spent fuel generated in any reactor can be stored safely and without significant environmental impacts for at least 60 years beyond the licensed life for operation (which may include the term of a revised or renewed license) of that reactor in a combination of storage in its spent fuel storage basin and at either onsite or offsite independent spent fuel storage installations. Further, the Commission believes there is reasonable assurance that sufficient mined geologic repository capacity will be available to dispose of the commercial high-level radioactive waste and spent fuel generated in any reactor when necessary." In addition, 10 CFR 51.23(b) applies the generic determination in section 51.23(a) to provide that "… no discussion of any environmental impact of spent fuel

Fuel Cycle, Transportation, and Decommissioning

storage in reactor facility storage pools or independent spent fuel storage installations (ISFSI) for the period following the term of the....reactor combined license or amendment....is required in any....environmental impact statement....prepared in connection with the ....issuance or amendment of a combined license for a nuclear power reactor under parts 52 or 54 of this chapter."

In the context of operating license renewal, Sections 6.2 and 6.4 of NUREG-1437, *Generic Environmental Impact Statement for License Renewal of Nuclear Plants, Main Report, Final Report* (NRC 1996), provide additional descriptions of the generation, storage, and ultimate disposal of LLW, mixed waste, and spent fuel from power reactors, concluding that environmental impacts from these activities are SMALL. For the reasons stated above, the NRC staff concludes that the environmental impacts of radioactive waste storage and disposal associated with Units 3 and 4 would be minor, and that the conclusions presented in the ESP EIS remain valid.

6.2 Transportation Impacts

The staff's assessment of the impacts to public health from transporting unirradiated fuel, spent fuel, and radioactive waste to and from the VEGP site was provided in Section 6.2 of the ESP EIS (NRC 2008). The staff concluded in the ESP EIS that the radiological and nonradiological impacts on human health would be SMALL.

Southern indicated in its COL ER (Southern 2009a) that there is no new and significant information regarding transportation-related impacts. During its review of the COL application, the staff independently verified that there is no new and significant information regarding transportation-related impacts. This was performed by reviewing Southern's ER and supporting documentation, auditing Southern's process for identifying new and significant information, examining other information available at the site audit, and considering applicable regulations and updates to reference documents cited in this SEIS.

The NRC staff notes that, on March 3, 2010, the U.S. Department of Energy submitted a motion to the Atomic Safety and Licensing Board to withdraw with prejudice its application for a permanent geologic repository at Yucca Mountain, Nevada (DOE 2010). The motion was subsequently denied by the Atomic Safety and Licensing Board (NRC 2010). Regardless of the final outcome of this proceeding, the staff concludes that transportation impacts are roughly proportional to the distance from the reactor site to the repository site, in this case Georgia to Nevada. The distance from the VEGP site to any new planned repository in the contiguous United States would be no more than double the distance from the VEGP site to Yucca Mountain. Doubling the environmental impact estimates from the transportation of spent reactor fuel, as presented in the ESP EIS (NRC 2008), would provide a reasonable bounding estimate

Fuel Cycle, Transportation, and Decommissioning

of the impacts for NEPA purposes. The staff concludes that the environmental impacts of these doubled estimates would still be SMALL.

Based on this review, the staff determined that the conclusions presented in the ESP EIS regarding transportation-related impacts remain valid.

6.3 Decommissioning Impacts

The NRC staff's assessment of the decommissioning-related impacts was provided in Section 6.3 of the ESP EIS. Based on the staff's analysis, these environmental impacts were considered to be SMALL.

Southern indicated in its COL ER (Southern 2009a) that there is no new and significant information regarding decommissioning-related impacts. During its review of the COL application, the staff independently verified that there is no new and significant information related to decommissioning by reviewing Southern's ER, auditing Southern's process for identifying new and significant information, examining other information available at the site audit, and considering applicable regulations and reference documents.

Based on this review, the staff determined that the conclusions presented in the ESP EIS remain bounding and valid.

6.4 References

10 CFR Part 20. Code of Federal Regulations, Title 10, *Energy*, Part 20, "Standards for Protection against Radiation."

10 CFR Part 50. Code of Federal Regulations, Title 10, *Energy*, Part 50, "Domestic Licensing of Production and Utilization Facilities."

10 CFR Part 51. Code of Federal Regulations, Title 10, *Energy*, Part 51, "Environmental Protection Regulations for Domestic Licensing and Related Regulatory Functions."

40 CFR Part 190. Code of Federal Regulations, Title 40, *Protection of Environment*, Part 190, "Environmental Radiation Protection Standards for Nuclear Power Operations."

75 FR 81032. December 23, 2010. "Consideration of Environmental Impacts of Temporary Storage of Spent Fuel After Cessation of Reactor Operation." *Federal Register*. U.S. Nuclear Regulatory Commission.

Baltimore Gas & Electric Company vs. Natural Resources Defense Council, Inc. June 6, 1983. *United States Reports*, Vol. 460, p. 1034.

Fuel Cycle, Transportation, and Decommissioning

Southern Nuclear Operating Company (Southern). 2008. *10 CFR 50.59 Screening/Evaluation.* Activity/Document No. RER A071430001, Version No. 1.0, Birmingham, Alabama. Accession No. ML102100523.

Southern Nuclear Operating Company (Southern). 2009a. *Vogtle Electric Generating Plant, Units 3 and 4, COL Application, Part 3 Environmental Report.* Revision 1, September 23, 2009, Birmingham, Alabama. Accession No. ML092740400.

Southern Nuclear Operating Company (Southern). 2009b. *Response to NRC RAI Letter No. 039 on the VEGP Units 3 & 4 COL Application Involving the Solid Waste Management System*, ND-09-1540. September 23, 2009, Birmingham, Alabama. Accession No. ML092680023.

U.S. Department of Energy (DOE). 2010. *In the Matter of U.S. Department of Energy (High-Level Waste Repository), Docket No. 63-001. U.S. Department of Energy's Motion to Withdraw.* March 3, 2010, Washington, D.C. Accession No. ML100621397.

U.S. Nuclear Regulatory Commission (NRC). 1976. *Environmental Survey of the Reprocessing and Waste Management Portions of the LWR Fuel Cycle.* NUREG-0116 (Supplement 1 to WASH-1248), Washington, D.C.

U.S. Nuclear Regulatory Commission (NRC). 1987. *Standard Review Plan for the Review of Safety Analysis Reports for Nuclear Power Plants.* NUREG-0800, Washington, D.C.

U.S. Nuclear Regulatory Commission (NRC). 1996. *Generic Environmental Impact Statement for License Renewal of Nuclear Plants, Main Report, Final Report.* NUREG-1437 Vol. 1, Washington, D.C.

U.S. Nuclear Regulatory Commission (NRC). 2008. *Final Environmental Impact Statement for an Early Site Permit (ESP) at the Vogtle Electric Generating Plant Site.* NUREG-1872, Vols. 1, 2, and Errata, Washington, D.C. Accession Nos. ML082240145; ML082240165, ML082260203; ML082550040.

U.S. Nuclear Regulatory Commission (NRC). 2010. Atomic Safety and Licensing Board, in the Matter of U.S. Department of Energy (High Level Waste Repository), Memorandum and Order (Granting intervention to Petitioners and Denying Withdrawal Motion), June 29, 2010, Washington D.C. Accession No. ML101800299.

7.0 Cumulative Impacts

In Chapter 7 of the Vogtle Electric Generating Plant (VEGP) early site permit (ESP) environmental impact statement (EIS) (NRC 2008), the U.S. Nuclear Regulatory Commission (NRC) staff provided a description of the potential cumulative impacts that could result from construction and operation of the proposed Units 3 and 4. The discussions in the ESP EIS included past, present, and reasonably foreseeable actions, and the geographical area over which the past, present, and reasonably foreseeable actions could contribute to cumulative impacts. This chapter of the supplemental EIS (SEIS) provides new information relative to cumulative impacts. Land use, air quality, water use and quality, terrestrial and aquatic ecosystems, socioeconomics and historic and cultural resources, nonradiological health, radiological impacts, severe accidents, fuel cycle, transportation, and decommissioning are discussed in Sections 7.1 through 7.10 of this chapter. The staff's conclusions are summarized in Section 7.11, and references are listed in Section 7.12.

7.1 Land Use

The NRC staff's assessment of the cumulative land-use impacts related to the construction and operation of the proposed Units 3 and 4 was provided in Section 7.1 of the ESP EIS (NRC 2008). Based on its analysis in the ESP EIS, the staff determined that cumulative land-use impacts would be SMALL.

In the environmental report (ER) included in its combined license (COL) application (Southern 2009), Southern Nuclear Operating Company, Inc. (Southern) indicated that there is no new and significant information regarding cumulative impacts related to the construction and operation of the proposed Units 3 and 4. During its review of the COL application, the NRC staff independently verified that there is no new and significant information related to the cumulative land-use impacts of constructing and operating Units 3 and 4 by reviewing Southern's ER, information submitted in support of ESP license amendment requests, auditing Southern's process for identifying new and significant information, examining other information available at the site audit, and considering applicable regulations and reference documents. Based on this review, the staff determined that the conclusion presented in Section 7.1 of the ESP EIS remains valid.

7.2 Air Quality

The NRC staff's assessment of cumulative air-quality impacts from criteria air pollutants was provided in Section 7.2 of the ESP EIS (NRC 2008). Permitted air-emission sources in the vicinity of the VEGP site include the Allen B. Wilson Combustion Turbine Plant (Plant Wilson) located on the VEGP site and the U.S. Department of Energy's Savannah River Site in South

Cumulative Impacts

Carolina. In addition, a mixed-oxide nuclear fuel facility has been proposed for development on the Savannah River Site. Based on the staff's analysis, cumulative impacts to air quality were considered to be SMALL.

In its COL ER (Southern 2009), Southern indicated that there is no new and significant information regarding cumulative impacts on air quality. During its review of Southern's COL application, the NRC staff performed an independent review of potential new and significant information related to meteorology and air quality by reviewing Southern's ER, auditing Southern's process for identifying new and significant information, examining other information available at the site audit, and considering applicable regulations and reference documents. This review identified new information related to potential changes in construction traffic as well as changes to the National Ambient Air Quality Standard (NAAQS) for ozone that warranted further staff analysis.

As discussed in Section 2.3, the U.S. Environmental Protection Agency (EPA) promulgated a revision to the NAAQS for ozone on March 12, 2008. The final rule (73 FR 16436) reduced the ozone standard from 0.084 ppm to 0.075 ppm. Section 107(d)(1) of the Clean Air Act (CAA) requires each state to submit, within one year of the revised standard, its recommended designation (i.e., attainment, non-attainment, or unclassified) for each county. On March 12, 2009, the Georgia Department of Natural Resources (GDNR) issued a letter to the EPA providing its recommended designations; Burke County remains unclassified/attainment for the new ozone standard (GDNR 2009). EPA will make its final determination on attainment status no later than March 2011. Based on this review and the fact that GDNR has determined that Burke County will remain in attainment, the NRC staff determined that the conclusions presented in the ESP EIS remain bounding and valid.

In Section 4.2, it was noted that Southern has indicated the potential need for additional truck deliveries if more backfill material is needed than could be obtained onsite; this would result in additional truck traffic to and from the site (Southern 2010a). Traffic impacts would be minimized by using different routes for inbound and outbound trucks. Although the potential truck traffic would result in more air emissions, these emissions would be temporary and would be completed before the peak of construction begins (Southern 2010a). The staff therefore expects the air quality conclusions presented in the ESP EIS related to construction traffic would remain valid.

In November 2009, the Commission issued Commission Order CLI-09-21 (NRC 2009), which provided guidance to the NRC staff to "... include consideration of carbon dioxide and other greenhouse gas emissions in its environmental reviews for major licensing actions under the National Environmental Policy Act." Although the staff considered greenhouse gas emissions in the ESP EIS and the issue therefore is not new, the staff has nevertheless re-examined its previous analysis to show conformance with the Commission's instructions in CLI-09-21.

Cumulative Impacts

While there are some carbon dioxide (CO_2) emissions associated with the construction and operation of a nuclear power plant, the life-cycle contributions are dominated by emissions associated with the uranium fuel cycle. These emissions primarily result from the operation of fossil-fueled power plants that provide the electricity needed to manufacture the fuel. Published estimates of life-cycle CO_2 emission rates from operating nuclear power plants worldwide average around 0.066 metric tons[a] (0.073 short tons) of CO_2 for each megawatt hour (MWh) generated, with a large fraction of these emissions associated with the fuel cycle (Sovacool 2008). For comparison, a coal-fired power plant emits about 1.02 metric tons (1.12 short tons[b]) of CO_2 for each MWh generated (EPA 2009a).

For consistency with Table S–3 of 10 CFR 51.51, the NRC staff has estimated the fuel cycle CO_2 emissions as 0.05 metric tons (0.055 short tons) of CO_2 per MWh generated. For a 1000 MW nuclear power reactor, the resulting annual CO_2 emission rate is approximately 447,000 metric tons (492,733 short tons). For context, Table 7-1 compares this value to other CO_2 emission estimates, including other sources of base-load power generation.

Table 7-1. Comparison of Annual CO_2 Emission Rates

Source	Metric Tons per Year	Short Tons per Year
Global Emissions	28,000,000,000[a]	30,865,000,000
United States	6,000,000,000[a]	6,614,000,000
1000 MW Coal-Fired Power Plant	8,939,000[b]	9,854,000
1000 MW Natural-Gas-Fired Power Plant	4,511,000[b]	4,973,000
1000 MW Nuclear Power Plant[c]	447,000	492,733
Average U.S. Passenger Vehicle	5[d]	5.5

(a) EPA 2009b
(b) EPA 2009a
(c) Including emissions from fuel cycle processes and operations; 90 percent capacity factor.
(d) EPA 2009c

As discussed in the state-of-the-science report issued by the U.S. Global Change Research Program (GCRP), it is the "... production and use of energy that is the primary cause of global warming, and in turn, climate change will eventually affect our production and use of energy. The vast majority of U.S. greenhouse gas emissions, about 87 percent, come from energy production and use...." Approximately one-third of the greenhouse gas emissions result from generating electricity and heat (GCRP 2009).

(a) The published emission estimates are reported in terms of grams (g) of CO_2 per kilowatt hour (kWh). The metric tons and short-ton (U.S.) values shown in this section are conversions from the published values.
(b) The published emission estimates are reported in terms of metric tons. The short-ton (U.S.) values shown in this section are conversions from the published values.

Cumulative Impacts

For the following reasons, it is difficult to evaluate cumulative impacts of a single or combination of greenhouse gas sources.

- The impact is global rather than local or regional.
- The impact is not particularly sensitive to location of the release point.
- The magnitude of individual greenhouse gas sources related to human activity, no matter how large compared to other sources, are small when compared to the total mass of greenhouse gases in the atmosphere.
- The total number and variety of greenhouse gas sources are extremely large, and they are located everywhere.

These points are illustrated by the magnitude and comparison of annual CO_2 emission rates listed in Table 7-1.

Evaluation of cumulative impacts of greenhouse gas emissions requires the use of a global climate model. The GCRP report (GCRP 2009) provides a synthesis of the results of numerous climate modeling studies. The NRC staff concludes that the cumulative impacts of greenhouse emissions around the world as presented in the GCRP report are the appropriate basis for its evaluation of cumulative impacts. Based on the impacts set forth in the GCRP report, the staff concludes that the cumulative impacts of greenhouse gas emissions are significant at the global level. The staff further concludes that the cumulative impact level would be significant, either with or without the greenhouse gas emissions of the proposed project.

Consequently, the NRC staff has determined that the proper approach to addressing the cumulative impacts of greenhouse gases, including CO_2, is to recognize that they are important contributors to climate change and that the carbon footprint is a relevant factor in evaluating energy alternatives. Among the viable energy generation sources for base-load power listed in Table 7-1, the CO_2 emissions from nuclear power plants (including the associated fuel cycle processes and operations) are considerably less than emissions from natural-gas-fired and coal-fired power plants, and the staff considers these emissions and their impacts to be SMALL both in isolation and cumulatively when compared to these other viable sources of base-load energy. Accordingly, the staff determined that the conclusions presented in Section 7.2 of the ESP EIS remain valid.

7.3 Water Use and Quality

The NRC staff's assessment of the water-related cumulative impacts of the proposed Units 3 and 4, the existing Units 1 and 2, the U.S. Department of Energy's Savannah River Site directly across the Savannah River from the VEGP site, and other water users in the region was provided in Section 7.3 of the ESP EIS (NRC 2008). The staff considered saltwater intrusion in

Cumulative Impacts

the State of Georgia, tritium that has been found in the unconfined aquifer, and contamination in the environment surrounding the Savannah River Site. Based on the staff's analysis, cumulative impacts to water use and water quality were considered to be SMALL.

In its COL ER (Southern 2009), Southern indicated that there is no new and significant information regarding cumulative impacts on water use and water quality. During its review of the COL application, the staff independently verified that there is no new and significant information related to water use and water quality by reviewing Southern's ER, information submitted in support of ESP license amendment requests, auditing Southern's process for identifying new and significant information, examining other information available at the site audit, and considering applicable regulations and reference documents.

Based on this review, the NRC staff determined that the conclusions presented in the ESP EIS remain valid.

7.4 Terrestrial Ecosystem

The NRC staff's cumulative impact assessment of the terrestrial resources in the vicinity of the VEGP site and the proposed transmission line right-of-way was provided in Section 7.4 of the ESP EIS (NRC 2008). Based on the staff's analysis, cumulative impacts to terrestrial resources were considered to be SMALL.

In its COL ER (Southern 2009), Southern indicated that there is no new and significant information regarding cumulative impacts on terrestrial resources. During its review of the COL application, the staff independently verified that there is no new and significant information related to the cumulative impact assessment of terrestrial resources by reviewing Southern's ER, reviewing information submitted as part of the license amendment request (LAR) activities to obtain backfill from additional onsite borrow areas, auditing Southern's process for identifying new and significant information, examining other information available at the site audit, considering applicable regulations and reference documents, and contacting the U.S. Fish and Wildlife Service (FWS), the Georgia Department of Natural Resources (GDNR), and the South Carolina Department of Natural Resources (SCDNR).

The land that would be disturbed for permanent structures and land that has been cleared for additional backfill material is composed of hardwood forest and bottomland wetlands, planted pine, sandhills, and open field habitats. The sandhills habitat that has been disturbed is of marginal quality compared to the remaining higher quality sandhills habitat available onsite. Planted pine, open field, and bottomland hardwood wetland habitats are available in other locations onsite and in the region. Furthermore, as explained in the Environmental Assessment for ESP Amendment 2 (NRC 2010), the potential losses to the southeastern pocket gopher (*Geomys pinetis*) and sandhills milkvetch (*Astragalus michauxii*) are isolated and will not

Cumulative Impacts

jeopardize the stability or viability of any of the remaining populations in Georgia. These populations occur in different locations throughout the state and each population is not dependent on the success of others. Staff did not identify new and significant information concerning any activities or projects in the geographic region of interest that would result in an adverse cumulative effect on terrestrial resources, including wildlife habitats and the State-threatened southeastern pocket gopher and sandhills milkvetch. Based on this review, the NRC staff determined that, while the localized impact has increased, the conclusions presented in the ESP EIS, that cumulative impacts to terrestrial resources would be SMALL, remain valid.

7.5 Aquatic Ecosystem

The staff's assessment of the cumulative impacts to aquatic resources in the Savannah River from upstream of the VEGP site to the mouth of the river was provided in Section 7.5 of the ESP EIS (NRC 2008). Based on the staff's analysis, cumulative impacts to aquatic resources were considered to be SMALL.

One of the sources of cumulative impact discussed in the ESP EIS and subsequent hearing proceedings was the potential for impacts from dredging the Federal navigation channel to facilitate shipment of large components to the site. In February 2010, Southern submitted a letter to NRC stating that large components and other construction materials would be transported to the VEGP site via rail using the Norfolk-Southern rail line from Savannah, Georgia, to Waynesboro, Georgia, where the line connects with the spur to VEGP (Southern 2010b). The letter also states that Southern will not construct a barge slip or seek maintenance dredging of the Savannah River navigation channel. Thus, in the absence of these shoreline construction or dredging activities, the cumulative impacts to aquatic resources would not include any impacts from these sources and thus would be bounded by the potential impacts described in Section 7.5 of the ESP EIS.

In its COL ER (Southern 2009), Southern indicated that there is no new and significant information regarding cumulative impacts on aquatic ecology. During the review of the COL application, the staff identified new, warranted further staff review information related to cumulative impacts.

On November 15, 2010, the U.S. Army Corps of Engineers (USACE) published a draft General Re-Evaluation Report (GRR) (USACE 2010a) and a Tier II EIS (USACE 2010b) related to determining the feasibility of improvements to the Federal navigation project at Savannah Harbor. The GRR and EIS assess mitigation plans for alternative channel depths from -42 to -48 feet mean lower low water. The Savannah Harbor expansion project has the potential to result in the loss of several hundred acres of habitat for fish, including essential fish habitat for shortnose sturgeon and striped bass, and habitat for other fish species in the Savannah River estuary. Many mitigation measures are being considered in connection with this project, including building a fish-way around the New Savannah Bluff Lock and Dam at Augusta, Georgia, which would open up an additional 32 km (20 mi) of habitat upstream of the dam (USACE 2010a). As explained in the ESP EIS, construction of the proposed units at the VEGP site would temporarily affect less than 0.6 ha (1.5 ac) of sturgeon migratory habitat (NRC 2008). Water withdrawal rates during operation would be less than 1 percent of Savannah River flow during average flow conditions, and the small zone of influence would have a negligible impact on pelagic spawning (NRC 2008). Furthermore, the proposed activities associated with the VEGP expansion would not impede the mitigation measures being considered for the Savannah River expansion project. Accordingly, construction and operation of the proposed VEGP units would not have an adverse cumulative impact on important fish species when considered together with the potential Savannah Harbor expansion project.

No other cumulative impacts were identified by the staff following review of Southern's ER, information submitted in support of ESP license amendment requests, auditing Southern's process for identifying new and significant information, examining other information available at the site audit, considering applicable reference documents, and contacts with the FWS, National Marine Fisheries Service, GDNR, USACE, and SCDNR.

Based on this review, the staff determined that the conclusions presented in the ESP EIS remain valid.

7.6 Socioeconomics, Historic and Cultural Resources, Environmental Justice

The NRC staff's assessment of the cumulative socioeconomic impacts related to the construction and operation of the proposed Units 3 and 4 was provided in Section 7.6 of the ESP EIS (NRC 2008). Based on the staff's analysis, impacts to socioeconomics were considered to be SMALL, with the exception for a possible MODERATE impact on roads, housing, and public services in Burke County during construction and a LARGE beneficial impact from property taxes collected in Burke County during operations. Based on the staff's analyses, cumulative impacts to historic and cultural resources were considered to be MODERATE, and Environmental Justice Impacts were considered to be SMALL.

In its COL ER (Southern 2009), Southern indicated that there is no new and significant information regarding cumulative impacts related to the construction and operation of the proposed Units 3 and 4. During its review of the COL application, the staff reviewed Southern's ER, audited Southern's process for identifying new and significant information, examined other information available at the site audit, and considered applicable regulations, reference documents, and discussions with state and county officials, Georgia State Historic Preservation Division, Advisory Council on Historic Preservation, and potentially interested Tribes (see

Cumulative Impacts

Appendix C for complete listing). This independent review identified new information in the areas of historic and cultural resources and socioeconomics that warranted further staff review.

As described in Section 4.6 of this SEIS, the staff identified a historic cemetery located on the VEGP site outside the proposed construction footprint. Southern has installed a fence around the cemetery, determined that the planned construction actions would not impact the site, and has consulted with the State Historic Preservation Office (SHPO) regarding protection and mitigation of the site. As a result of these protective measures proposed by Southern and consultation with the SHPO, the staff concludes that the identification of the historic cemetery does not change its conclusion regarding the cumulative impacts to historic and cultural resources in the vicinity of the VEGP site. The staff evaluated new proposed onsite borrow areas as a result of the LAR (Southern 2010b). The impacts to historic and cultural resources associated with the new proposed onsite borrow areas are previously described in Section 4.6. There are no NRHP eligible properties located in the vicinity of the proposed onsite borrow areas. As a result of the cultural resources analysis, field investigations, procedures Southern has in place for unanticipated cultural resources discoveries, and the consultation with the SHPO, the staff concludes that the proposed new onsite borrow areas do not change its conclusions regarding cumulative impacts to historic and cultural resources in the vicinity of the VEGP site.

This independent review also identified new information related to funding provided by the American Recovery and Reinvestment Act (ARRA), which warranted further staff consideration. A significant amount of the ARRA funding that could have potential socioeconomic impacts on Columbia and Richmond Counties in Georgia has been allocated to the nearby Savannah River Site. The ARRA funding has saved and created thousands of jobs at the Savannah River Site, which is near the VEGP site (DOE 2009). However, ARRA is not a renewable source of funding, and ARRA-related employment will diminish before construction of the proposed Units 3 and 4 peaks; therefore, the staff does not expect any increase in cumulative impacts. The NRC staff's independent review found no new and significant information regarding environmental justice.

Section 4.5 of this SEIS described the possibility of Southern needing additional backfill material delivered by truck from an offsite source, thus adding additional vehicles to the roadways (Southern 2010a). Traffic impacts would be minimized by using different routes for inbound and outbound trucks. As discussed in Section 4.5, although the truck deliveries would increase the amount of traffic on the roadways, the increases would remain within the design capacities of the roads, and the increased traffic would be temporary and completed before the peak of construction begins (Southern 2010a). Based on this review, the staff determined that the conclusions presented in the ESP EIS remain valid.

7.7 Nonradiological Health

The NRC staff's assessment of cumulative nonradiological, health-related impacts was provided in Section 7.7 of the ESP EIS (NRC 2008). Based on the staff's analysis, cumulative impacts to nonradiological health were considered to be SMALL.

In its COL ER (Southern 2009), Southern indicated that there is no new and significant information regarding cumulative impacts on nonradiological health. During its review of the COL application, the staff independently verified that there is no new and significant information related to nonradiological health by reviewing Southern's ER, auditing Southern's process for identifying new and significant information, examining other information available at the site audit, and considering applicable regulations and reference documents. However, subsequent to the site audit, Southern determined that it would need to obtain backfill material from onsite borrow areas other than those previously specified in the ESP site safety analysis report. Accordingly, Southern submitted license amendment requests to obtain approval of the use of backfill from additional onsite and offsite borrow areas. The NRC staff evaluated the nonradiological impacts associated with truck transport of backfill material from these additional locations (NRC 2010) and determined that the additional truck shipments would not significantly increase the nonradiological impacts presented in the ESP EIS (NRC 2008). Furthermore, in Section 4.8.2 of this SEIS, the staff examined the potential increase in traffic fatality risk in the event Southern were to need to obtain additional backfill material from an offsite source. As explained in Section 4.8.2, even when considered in combination with the minor increase in traffic fatality risk analyzed in the ESP EIS, this increase remains small relative to the current traffic fatality risks in the area surrounding the proposed VEGP site.

Based on this review, the staff determined that the conclusions presented in the ESP EIS remain valid.

7.8 Radiological Impacts of Normal Operation

The NRC staff's assessment of cumulative radiological, health-related impacts was provided in Section 7.8 of the ESP EIS (NRC 2008). Based on the staff's analysis, cumulative impacts to radiological health were considered to be SMALL.

In its COL ER (Southern 2009), Southern indicated that there is no new and significant information regarding cumulative impacts on radiological health. During its review of the COL application, the staff independently verified that there is no new and significant information related to radiological health by reviewing Southern's ER, information submitted in support of ESP license amendment requests, auditing Southern's process for identifying new and significant information, examining other information available at the site audit, and considering applicable regulations and reference documents.

Cumulative Impacts

In Section 6.1 of this SEIS, the staff analyzed the potential environmental impacts of additional onsite or offsite storage of low-level radioactive waste, if it becomes necessary for Southern to implement one or more of the contingency options it has described. For the reasons described in those sections, implementation of those contingencies would not result in doses in excess of the applicable 10 CFR Part 20 limits, and thus any cumulative impacts would be SMALL.

Based on this review, the staff determined that the conclusions presented in the ESP EIS remain bounding and valid.

7.9 Severe Accidents

The NRC staff's assessment of cumulative, severe-accident-related impacts was provided in Section 7.9 of the ESP EIS (NRC 2008). Based on the staff's analysis, cumulative impacts of severe accidents were considered to be SMALL.

In its COL ER (Southern 2009), Southern indicated that there is no new and significant information regarding cumulative impacts related to severe accidents. During its review of the COL application, the NRC staff independently verified that there is no new and significant information related to radiological health by reviewing Southern's ER, auditing Southern's process for identifying new and significant information, examining other information available at the site audit, and considering applicable regulations and reference documents.

Based on this review, the staff determined that the conclusions presented in the ESP EIS remain bounding and valid.

7.10 Fuel Cycle, Transportation, and Decommissioning

The NRC staff's assessment of impacts related to the fuel cycle, transportation, and decommissioning was provided in Section 7.10 of the ESP EIS (NRC 2008). Based on the staff's analysis, cumulative impacts related to the fuel cycle, transportation, and decommissioning were considered to be SMALL.

In its COL ER (Southern 2009), Southern indicated that there is no new and significant information regarding cumulative impacts related to the fuel cycle, transportation, and decommissioning. During its review of the COL application, the staff independently verified that there is no new and significant information related to the fuel cycle, transportation, and decommissioning by reviewing Southern's ER, auditing Southern's process for identifying new and significant information, examining other information available at the site audit, and considering applicable regulations and reference documents.

Based on this review, the NRC determined that the conclusions presented in the ESP EIS remain bounding and valid.

7.11 NRC Staff Conclusions and Recommendations

The NRC staff considered the potential impacts resulting from constructing and operating the proposed Units 3 and 4 together with the past, present, and reasonably foreseeable future actions in the VEGP site area. The staff summarized its conclusions in Section 7.11 of the ESP EIS and found that all potential cumulative impacts resulting from construction and operation generally would be SMALL, and additional mitigation was not warranted. The staff's review of Southern's process for identifying new and significant information results from the VEGP site audit, and contacts with various Federal, State, and Tribal agencies identified no information that would change these cumulative impact designations.

7.12 References

10 CFR Part 20. Code of Federal Regulations, Title 10, *Energy*, Part 20, "Standards for the Protection Against Radiation."

10 CFR Part 51. Code of Federal Regulations, Title 10, *Energy*, Part 51, "Environmental Protection Regulations for Domestic Licensing and Related Regulatory Functions."

73 FR 16436. March 27, 2008. "National Ambient Air Quality Standards for Ozone; Final Standard." *Federal Register*, U.S. Environmental Protection Agency.

American Recovery and Reinvestment Act of 2009. Public Law 111-5.

Clean Air Act. 42 USC 7401, et seq,

Georgia Department of Natural Resources (GDNR). 2009. Letter from C.A. Couch, Georgia Department of Natural Resources, to A.S. Meiburg, U.S Environmental Protection Agency, Region 4, Atlanta, Georgia. "Recommended Designations of Ozone Non-Attainment Areas in Georgia." March 12, 2009. Accession No. ML100601088.

Southern Nuclear Operating Company, Inc. (Southern). 2009. *Vogtle Electric Generating Plant, Units 3 and 4, COL Application, Part 3 Environmental Report*. Revision 1, September 23, 2009. Southern Company, Birmingham, Alabama. Accession No. ML092740400.

Cumulative Impacts

Southern Nuclear Operating Company, Inc. (Southern). 2010a. Letter from C.R. Pierce, Southern, to NRC, "Southern Nuclear Operating Company, Vogtle Electric Generating Plant Units 3 and 4, Combined License Application, New and Significant Information Evaluation For the Transportation of Backfill From an Offsite Source." Letter ND-10-1389. July 16, 2010. Southern Company, Birmingham, Alabama. Accession No. ML102010031.

Southern Nuclear Operating Company, Inc. (Southern). 2010b. Letter from Mr. Charles R. Pierce, Southern to NRC, "Southern Nuclear Operating Company, Vogtle Electric Generating Plant Units 3 and 4, Combined License Application, Large Component Transportation Method Decision." Letter ND-10-0261. February 19, 2010. Southern Company, Birmingham, Alabama. Accession No. ML100550033.

Sovacool, B.K. 2008. "Valuing the Greenhouse Gas Emissions from Nuclear Power: A Critical Survey." *Energy Policy* 36(2008):2940–2953.

U.S. Army Corps of Engineers. 2010a. *Draft General Re-Evaluation Report for Savannah Harbor Expansion Project. Chatham County, Georgia and Jasper County, South Carolina.* Dated 15 November 2010. Accessed January 6, 2010 at http://www.sas.usace.army.mil/shexpan/SHEPreport.html.

U.S. Army Corps of Engineers. 2010b. *Draft Tier II Environmental Impact Statement for the Savannah Harbor Expansion Project. Chatham County, Georgia and Jasper County, South Carolina.* Accessed January 6, 2010 at http://www.sas.usace.army.mil/shexpan/SHEPTierII.html.

U.S. Department of Energy (DOE). 2009. "Recovery and Reinvestment." Accessed January 7, 2010, at http://www.energy.gov/recovery/. Accession No. ML100601103.

U.S. Environmental Protection Agency (EPA). 2009a. *Air Emissions*. Accessed January 12, 2010, at http://www.epa.gov/RDEE/energy-and-you/affect/air-emissions.html. Accession No. ML100600773.

U.S. Environmental Protection Agency (EPA). 2009b. *Global Greenhouse Gas Data*. Accessed January 22, 2010, at http://www.epa.gov/climatechange/emissions/globalghg.html. Accession No. ML100221499.

U.S. Environmental Protection Agency (EPA). 2009c. Emission Facts: Greenhouse Gas Emissions for a Typical Passenger Vehicle. Accessed January 22, 2010 at http://www.epa.gov/OMS/climate/42f05004.htm. Accession No. ML100600790.

Cumulative Impacts

U.S. Global Change Research Program (GCRP). 2009. *Global Climate Change Impacts in the United States*. T.R. Karl, J.M. Melillo, and T.C. Peterson, eds, Cambridge University Press, New York.

U.S. Nuclear Regulatory Commission (NRC). 2008. *Final Environmental Impact Statement for an Early Site Permit (ESP) at the Vogtle Electric Generating Plant Site.* NUREG-1872, Vols. 1, 2, and Errata, Washington, D.C. Accession Nos. ML082240145; ML082240165, ML082260203; ML082550040.

U.S. Nuclear Regulatory Commission (NRC). 2009. Memorandum and Order, Duke Energy Carolinas, LLC, Combined License Application for William States Lee Nuclear Station, Units 1 and 2, and Tennessee Valley Authority, Bellefonte Nuclear Power Plan, Units 3 and 4, CLI-09-21, November 3, 2009. Washington, D.C. Accession No. ML093070689.

U.S. Nuclear Regulatory Commission (NRC). 2010. *Vogtle Electric Generating Plant ESP Site Early Site Permit and Limited Work Authorization Environmental Assessment and Finding of No Significant Impact.* Docket No. 52-011. Accession No. ML101670592.

8.0 Need for Power

A discussion of the need for power from proposed Units 3 and 4 was provided in Chapter 8 of the Vogtle Electric Generating Plant (VEGP) early site permit (ESP) environmental impact statement (EIS) (NRC 2008). This section describes the need for power assessment for the proposed units. The discussion in the ESP EIS is organized into four major subsections that provide details on the power system, power demand, power supply, and the assessment of need for power.

Southern Nuclear Operating Company, Inc. (Southern) indicated in its combined licenses (COL) environmental report (ER) that there is no new and significant information regarding need for power (Southern 2009). During its review of the COL application, the U.S. Nuclear Regulatory Commission (NRC) staff performed an independent review of potential new and significant information related to need for power that included reviewing Southern's ER, auditing Southern's process for identifying new and significant information, examining other information available at the site audit, and considering applicable regulations and reference documents including the Georgia Power Company (GPC) Integrated Resource Plan (IRP) (GPC 2010) which was approved by the Georgia Public Service Commission (GPSC) on July 13, 2010 (GPSC 2010a).

A certification for construction of the proposed Units 3 and 4 was approved by GPSC in March 2009 (GPSC 2009) and was amended in June 2010 (GPSC 2010b) with additional information concerning the need for power and other issues after the original certification was remanded back to the GPSC by the Fulton County Superior Court. In its June 2010 decision, GPSC specifically found that:

- There will be a need for new base-load generation in Georgia during the 2016 to 2017 timeframe.
- Demand side management programs do not eliminate the need for new base-load generation.

A certification is issued if GPSC finds there is a need for new capacity and the resource being used is economical and reliable. That GPSC has found that a need for power exists and decided to issue the Certification further supports the conclusions presented in the ESP EIS (NRC 2008) that a need for power in the region of interest exists. Based on this review, the staff determined that the conclusions regarding need for power presented in the ESP EIS remain valid.

Need For Power

8.1 References

Georgia Power Company (GPC). 2010. *Integrated Resource Plan.* Georgia Public Service Commission. Docket No. 31081. Accessed March 10, 2010 at ftp://www.psc.state.ga.us/Dockets/31081/125981.zip. Accession No. ML100760714.

Georgia Public Service Commission (GPSC). 2009. *Docket No. 27800. Georgia Power's Application for the Certification of Units 3 and 4 at Plant Vogtle and Updated Integrated Resource Plan. Amended Certification Order.* March 17, 2009. Accession No. ML100600818.

Georgia Public Service Commission (GPSC). 2010a. *Docket 31081. Georgia Power Company's Application for Approval of its 2010 Integrated Resource Plan, Docket 31082. Georgia Power Company's application for the Certification of Demand-side Management Programs for its 2010 Integrated Resource Plan.* July 13, 2010, Atlanta, Georgia. Accession No. ML101960367

Georgia Public Service Commission (GPSC). 2010b. *Docket No. 27800. Georgia Power's Application for the Certification of Units 3 and 4 at Plant Vogtle and Updated Integrated Resource Plan. Order on Remand.* June 17, 2010, Atlanta, Georgia. Accession No. ML101960603

Southern Nuclear Operating Company (Southern). 2009. *Vogtle Electric Generating Plant, Units 3 and 4, COL Application, Part 3 Environmental Report,* Revision 1, September 23, 2009. Southern Company, Birmingham, Alabama. Accession No. ML092740400.

U.S. Nuclear Regulatory Commission (NRC). 2008. *Final Environmental Impact Statement for an Early Site Permit (ESP) at the Vogtle Electric Generating Plant Site.* NUREG-1872, Vols. 1, 2, and Errata, Washington, D.C. Accession Nos. ML082240145; ML082240165, ML082260203; ML082550040.

9.0 Environmental Impacts of Alternatives

The environmental impacts of alternatives to the proposed action were evaluated in Chapter 9 of the early site permit (ESP) environmental impact statement (EIS) (NRC 2008). This chapter discusses new and significant information, where applicable, concerning alternatives to the proposed action. Topics discussed are the no-action alternative (Section 9.1), energy alternatives (Section 9.2), system design alternatives (Section 9.3), Southern's region of interest (ROI) and site selection process (Section 9.4), and evaluation of alternative sites (Section 9.5).

9.1 No-Action Alternative

For purposes of a combined license (COL) application, the no-action alternative refers to a scenario in which the U.S. Nuclear Regulatory Commission (NRC) would deny Southern Nuclear Operating Company's (Southern's) application for COLs and a second limited work authorization (LWA). Upon such a denial, the construction and operation of new nuclear generating units at the Vogtle Electric Generating Plant (VEGP) ESP site in accordance with Title 10 of the Code of Federal Regulations (CFR) Part 52, including performance of the LWA construction activities requested pursuant to 10 CFR 50.10(d), would not occur. There would be no environmental impacts at the VEGP site associated with not issuing the COLs, except the impacts associated with (1) any activities not within the definition of construction at 10 CFR 51.4, (2) activities authorized by the LWA included in the ESP (NRC 2009) issued to Southern and conducted prior to the time the COLs are denied, and/or (3) activities performed under the second LWA that Southern requested in conjunction with its COL application (if the second LWA were granted by the NRC prior to denial of the COLs) and conducted prior to the time the COL requests are denied. At the same time, the benefits associated with the proposed action would not occur. If the Commission approved the COLs but denied the requested LWA, the construction activities associated with the LWA would still occur, but at a somewhat later time. In that scenario, the benefits of the LWA – for example, potentially earlier completion of construction and, accordingly, earlier commencement of power production – would not be realized.

If the COL requests (including the second LWA request) are denied, the power will still be needed as discussed in Chapter 8 of the ESP EIS (NRC 2008). As described in Section 9.2 of the ESP EIS, Southern would have a variety of options for meeting power needs including constructing a new nuclear power plant at another site, constructing a coal-fired or natural-gas-fired plant at the VEGP site or at another site, and pursuing one or more of the other energy alternatives discussed in Section 9.2. There would be environmental impacts associated with each of these options that would occur at the site of implementation.

Environmental Impacts of Alternatives

9.2 Energy Alternatives

In Section 9.2 of the ESP EIS, the NRC staff evaluated alternative energy sources (NRC 2008). Based on its analysis in the ESP EIS, the staff concluded in Section 9.2.5 of the ESP EIS that from an environmental perspective, none of the viable energy alternatives would be clearly preferable to construction of a new base-load nuclear power generation plant. The basis for this conclusion is summarized in Table 9-4 in the ESP EIS (NRC 2008).

During its review of Southern's COL application, the NRC staff performed an independent review of potential new and significant information related to energy alternatives by reviewing Southern's environmental report and supporting information, responses to requests for additional information (Southern 2010), auditing Southern's process for identifying new and significant information, examining other information available at the site audit, and considering applicable regulations and reference documents. This review identified the following new information that warranted further review:

- Georgia Power Company (GPC) expects to achieve approximately 900 MW(e) of demand reduction by 2013 through the implementation of existing and new demand-side management (DSM) programs. This load reduction represents more than 5 percent of GPC's current load (GPC 2010). The 900 MW(e) is already accounted for (partly as a load reduction and partly as a capacity resource) in GPC's Integrated Resource Plan (IRP) and is therefore not available to offset the need for two new nuclear generating units that would generate base-load power.

- Southern has no plans to reactivate any retired power plants in its ROI.

The staff determined that the new DSM information does not have the potential to change the staff's conclusion in Section 9.3.5 of the ESP EIS. The reasons for this determination are (1) the additional 900 MW(e) attributable to DSM programs is accounted for in GPC's IRP (GPC 2010) and is, therefore, not available to offset the need for two new nuclear generating units that would generate base-load power and (2) none of Southern's retired power plants would be available to offset the need for the new nuclear units.

On June 3, 2010, EPA issued a rule tailoring the applicability criteria that determine which stationary sources and modification to existing projects become subject to permitting requirements for GHG emissions under the Prevention of Significant Deterioration (PSD) and Title V programs of the Clean Air Act (75 FR 31514). According to the Tailoring Rule, GHGs are a regulated new source review (NSR) pollutant under the PSD major source permitting program if the source (1) is otherwise subject to PSD (for another regulated NSR pollutant) and (2) has a GHG potential to emit equal to or greater than 75,000 tons per year of CO_2 equivalent ("carbon

Environmental Impacts of Alternatives

dioxide equivalent" adjusting for different global warming potentials for different GHGs). Such sources would be subject to best available control technology (BACT). The use of BACT has the potential to reduce the amount of GHGs emitted from stationary source facilities. The implementation of this rule could reduce the amount of GHGs from the values indicated in Table 7-1 for coal and natural gas, as well as from other alternative energy sources that would otherwise have appreciable uncontrolled GHG emissions. The GHG emissions from the production of electricity from a nuclear power source are primarily from the fuel cycle and could be reduced further if the electricity from a fossil fuel source powering the fuel cycle was subject to BACT controls. The emission of GHGs from the production of electrical energy from a nuclear power source is orders of magnitude less than those of the reasonable alternative energy sources. Accordingly, the comparative relationship between the energy sources listed in Table 7-1 would not change meaningfully, even if the reduction of GHG emissions from the nuclear fuel cycle are ignored, because GHG emissions from the other energy source alternatives would not be sufficiently reduced to make them environmentally preferable to the proposed project.

In addition, as discussed in Chapters 4, 5, and 7 of this COL supplemental environmental impact statement (SEIS), the staff did not identify any information that would change any of the entries in the nuclear column of Table 9-4 in the ESP EIS (NRC 2008). As discussed in Section 4.4.1 of this SEIS, although the staff's conclusion with respect to magnitude of the onsite terrestrial impacts increased, the staff determined that the overall conclusion for Ecology in Table 9-4 of the ESP EIS would still be the range of SMALL to MODERATE; thus, the overall comparison of impacts with other energy alternatives would not change. Accordingly, the staff affirms its conclusion in Section 9.2.5 of the ESP EIS (NRC 2008) that, from an environmental perspective, none of the viable energy alternatives would be clearly preferable to construction of a new base-load nuclear power generation plant at the VEGP ESP site.

9.3 System Design Alternatives

The information and associated impacts for this section are provided and resolved in Section 9.3 of the ESP EIS (NRC 2008). Once-through cooling and dry or hybrid wet/dry cooling towers were evaluated by the staff as alternatives to the proposed wet cooling tower design. The NRC staff concluded that none of the alternatives would be preferable to the proposed wet cooling towers for proposed Units 3 and 4. For the reasons discussed in earlier chapters of this SEIS, the new information available since completion of the ESP EIS does not significantly affect the impact on the environment of the proposed cooling towers as analyzed in the ESP EIS, and the staff concludes that those impacts remain SMALL. Accordingly, the staff concludes that the wet cooling tower design remains preferable to the alternatives considered in the ESP EIS.

Environmental Impacts of Alternatives

9.4 Region of Interest and Alternative Site Selection Process

The staff's review of Southern's ROI and site selection process was provided in Section 9.4 of the ESP EIS (NRC 2008). No additional discussion of this topic is required in a supplement to an ESP EIS that is prepared for a COL application (10 CFR 51.92(e)(3)).

9.5 Evaluation of Alternative Sites

The staff's evaluation of alternative sites was provided in Section 9.5 of the ESP EIS (NRC 2008). That review determined that none of the alternative sites would be environmentally preferable or obviously superior to the proposed VEGP site. No additional discussion of this topic is required in a supplement to an ESP EIS that is prepared for a COL application (10 CFR 51.92(e)(3)).

9.6 References

10 CFR Part 50. Code of Federal Regulations. Title 10, *Energy*, Part 50, "Domestic Licensing of Production and Utilization Facilities."

10 CFR Part 51. Code of Federal Regulations, Title 10, *Energy*, Part 51, "Environmental Protection Regulations for Domestic Licensing and Related Regulatory Functions."

10 CFR Part 52. Code of Federal Regulations, Title 10, *Energy*, Part 52, "Licenses, Certifications, and Approvals for Nuclear Power Plants."

Georgia Power Company (GPC). 2010. *Integrated Resource Plan*. Georgia Public Service Commission. Docket No. 31081. Accessed March 10, 2010 at ftp://www.psc.state.ga.us/Dockets/31081/125981.zip. Accession No. ML100760714.

Southern Nuclear Operating Company, Inc. (Southern). 2010. Letter from M. Smith, Southern, to NRC, "Southern Nuclear Operating Company, Vogtle Electric Generating Plant Units 3 and 4, Combined License Application, Supporting Information for Environmental Report Review." Letter ND-10-0526. March 12, 2010, Southern Company, Birmingham, Alabama. Accession No. ML100750038.

U.S. Nuclear Regulatory Commission (NRC). 2008. *Final Environmental Impact Statement for an Early Site Permit (ESP) at the Vogtle Electric Generating Plant Site*. NUREG-1872, Vols. 1, 2, and Errata, Washington, D.C. Accession Nos. ML082240145; ML082240165, ML082260203; ML082550040.

U.S. Nuclear Regulatory Commission (NRC). 2009. *Southern Nuclear Operating Company Vogtle Electric Generating Plant ESP Site, Docket No. 52-011, Early Site Permit and Limited Work Authorization.* Early Site Permit No. ESP-004, Washington, D.C. Accession No. ML0902290157.

10.0 Comparison of the Impacts of the Proposed Action and the Alternative Sites

A comparison of the proposed action at Vogtle Electric Generating Plant (VEGP) and at three alternative sites was provided in Chapter 10 of the early site permit (ESP) environmental impact statement (EIS) (NRC 2008). The U.S. Nuclear Regulatory Commission (NRC) staff concluded that none of the alternative sites was environmentally preferable or obviously superior to the proposed VEGP ESP site. As set out at Title 10 of the Code of Federal Regulations (CFR) Part 51 (10 CFR 51.92(e)(3)), no additional discussion of alternative sites is required in a supplemental EIS (SEIS) that is prepared for a combined license (COL) application referencing an ESP.

Chapter 10 of the ESP EIS also compares the proposed action with the no-action alternative, which in this SEIS refers to a scenario in which the NRC would deny Southern Nuclear Operating Company's (Southern's) application for COLs and a second limited work authorization (LWA). As described in Section 9.1 of this SEIS, if the COLs and second LWA applications were denied, the construction and operation of new nuclear generating units at the VEGP ESP site would not occur. There would be no environmental impacts at the VEGP site associated with not issuing the COLs, except the impacts associated with (1) activities conducted by Southern that are not within the definition of construction at 10 CFR 51.4, (2) activities performed under the LWA that was granted concurrently with the ESP, and conducted prior to the time the COLs were denied, and/or (3) activities performed under the second LWA that Southern requested in conjunction with its COL application (if the second LWA were granted by NRC prior to denial of the COLs). Under the no-action alternative, the benefits associated with the proposed action would not occur. The power would still be needed as discussed in Chapter 8 of the ESP EIS (NRC 2008). Southern would have a variety of options for meeting power needs, as discussed in Section 9.2 of the ESP EIS. There would be environmental impacts associated with each of these options that would occur at the site of implementation.

Redress would be required for any actions performed pursuant to the first LWA and second LWA (if issued prior to denial of the COLs) in accordance with the Site Redress Plan in Appendix F of the ESP issued to Southern (NRC 2009). As discussed in Sections 4.11 and 10.4 of the ESP EIS (NRC 2008), the staff concluded that LWA activities would not result in any significant adverse impacts that could not be redressed. The NRC staff affirms this conclusion for activities conducted under the LWA granted with the ESP and any activities that would be conducted under the second LWA request if the request is granted prior to issuance of the COLs. There also would be impacts associated with activities performed by Southern that are not within the definition of construction at 10 CFR 51.4. Redress for these activities would be conducted according to the laws and regulations of Burke County and the State of Georgia.

Comparison of the Impacts of the Proposed Action and the Alternative Sites

10.1 References

10 CFR Part 51. Code of Federal Regulations, Title 10, *Energy*, Part 51, "Environmental Protection Regulations for Domestic Licensing and Related Regulatory Functions."

U.S. Nuclear Regulatory Commission (NRC). 2008. *Final Environmental Impact Statement for an Early Site Permit (ESP) at the Vogtle Electric Generating Plant Site.* NUREG-1872, Vols. 1, 2, and Errata, Washington, D.C. Accession Nos. ML082240145; ML082240165, ML082260203; ML082550040.

U.S. Nuclear Regulatory Commission (NRC). 2009. *Southern Nuclear Operating Company Vogtle Electric Generating Plant ESP Site, Docket No. 52-011, Early Site Permit and Limited Work Authorization.* Early Site Permit No. ESP-004, Washington, D.C. Accession No. ML092290157.

11.0 Conclusions and Recommendations

The U.S. Nuclear Regulatory Commission (NRC) staff's conclusions and recommendations for the Vogtle Electric Generating Plant (VEGP) early site permit (ESP) environmental impact statement (EIS) were provided in Chapter 11 of the ESP EIS (NRC 2008). As described in Chapter 1 of this supplemental EIS (SEIS), Southern Nuclear Operating Company, Inc. (Southern) evaluated, and the NRC staff independently reviewed, the potential new and significant information with respect to environmental impacts that could occur if combined licenses (COLs) and a second Limited Work Authorization (LWA) were issued to Southern for proposed Units 3 and 4 at the VEGP ESP site. The results of the NRC staff review are presented in Chapters 1 though 10 of this SEIS. Southern's COL application, and accompanying environmental report (ER) (Southern 2009), reference an ESP, so where appropriate, this SEIS adopts the analysis and the results of the environmental review conducted in support of the ESP application and incorporates by reference the analyses and results presented in the ESP EIS.

Mitigation measures were considered for each environmental issue and are discussed in the appropriate sections. During its environmental review, the NRC staff considered planned activities and actions that Southern indicated it and others would likely take should Southern receive two COLs and an LWA.

Impacts of the proposed action are summarized in Section 11.1. Unavoidable adverse environmental impacts, alternatives to the proposed action, the relationship between short-term uses and long-term productivity of the human environment, irreversible and irretrievable commitments of resources, benefit-cost balance, and the staff conclusions and recommendations are described in Sections 11.2 through 11.7, respectively. The references cited are listed in Section 11.8.

11.1 Impacts of the Proposed Action

A summary of the impacts associated with issuance of the ESP and the first LWA was given in Section 11.1 of the ESP EIS (NRC 2008). This information, as supplemented by this SEIS, provides the basis for an informed decision concerning the environmental impacts of issuance of COLs and a second LWA by the NRC. In the staff's review of new and significant information for the COL review, with the exception of terrestrial ecology as described in Section 4.4.1, no new and significant information was identified that would change any of the conclusions stated in the ESP EIS.

Conclusions and Recommendations

11.2 Unavoidable Adverse Environmental Impacts

The NRC staff's' assessment of unavoidable adverse environmental impacts during construction and operation of the proposed Units 3 and 4 was provided in Section 11.2 of the ESP EIS (NRC 2008). That assessment explained whether adverse impacts had been identified, listed actions anticipated to mitigate impacts, and noted which impacts would be unavoidable. In its COL ER (Southern 2009), Southern concluded that there is no new and significant information related to unavoidable adverse environmental impacts, but it did note there would be an increase in the permanently disturbed land area, from 131 ha (324 ac) to 153 ha (379 ac). The development of additional onsite borrow areas also increased the amount of additional land disturbance (Southern 2010a) from 92 ha (227 ac) to 200 ha (494 ac). These changes in land area were noted and evaluated by the NRC staff in Chapter 4 of this SEIS. Development of the new borrow areas also resulted in the loss or diminishment of populations of two State-listed species (the southeastern pocket gopher [*Geomys pinetis*] and the sandhills milkvetch [*Astragalus michauxii*]). These impacts were noted and evaluated in Section 4.4 of this SEIS.

While these land use and terrestrial resource impacts would be adverse and unavoidable, the staff's review identified actions to mitigate these impacts. These mitigating actions are consistent with those described in Section 11.2 of the ESP EIS, and include compliance with the requirements of applicable Federal, State, Tribal, and local permits, and observance of best management practices. With respect to the impacts to the State-listed species, the staff's analysis in Section 4.4.1 of this SEIS also describes Southern's efforts to relocate the onsite populations of these species and to replant the disturbed areas with longleaf pine, if possible. These developments do not alter the staff's conclusions in Section 11.2 of the ESP EIS, and in the staff's review of new and significant information, as described throughout this SEIS, no other information was identified that would change the conclusions stated in Section 11.2 of the ESP EIS regarding unavoidable adverse environmental impacts.

If the second LWA requested by Southern were granted by NRC and the COLs subsequently denied, there would be some environmental impacts at the VEGP site from the conduct of those activities. However, the staff concluded in Chapter 10 of this SEIS that any such impacts related to NRC authorized activities could be adequately redressed.

11.3 Alternatives to the Proposed Action

The proposed action for this SEIS is identified in Section 1.2. A summary of the alternatives to the proposed action at the ESP stage was presented in Section 11.3 of the ESP EIS (NRC 2008). Alternatives to the proposed action discussed in this SEIS are the no-action alternative, energy alternatives, and system design alternatives. As described in Sections 9.2 and 9.3 of this SEIS, no new and significant information was identified in the areas of energy alternatives

Conclusions and Recommendations

or system design alternatives. Therefore, the staff determined that the conclusions regarding these alternatives in the ESP EIS remain valid.

The no-action alternative is discussed in Section 9.1 of this SEIS. Under the no-action alternative, the NRC would not issue the COLs or second LWA to Southern. There would be no environmental impacts associated with not issuing the COLs, except the impacts associated with activities not within the definition of construction at 10 CFR 50.10(a) and 10 CFR 51.4 and any activities performed under an LWA prior to the time the COLs were denied. At the same time, the benefits associated with the proposed action would not occur. If the COL application is denied, the power would still be needed as discussed in Chapter 8 of this SEIS. Southern would have a variety of options for meeting power needs, including constructing a new nuclear power plant at another site, constructing a coal- or natural-gas-fired plant at the VEGP site or at another site, and pursuing one or more of the energy alternatives discussed in Sections 9.2.1 and 9.2.2 of the ESP EIS. There would be environmental impacts associated with each of these options that would occur at the site of implementation. For reasons explained in Chapter 9 of the ESP EIS, however, the options evaluated in Sections 9.2.1 and 9.2.3 were determined not to be reasonable alternatives to providing new baseload power generation capacity.

11.4 Relationship Between Short-Term Uses and Long-Term Productivity of the Human Environment

The staff's review of the relationship between local short-term uses of the environment and the long-term productivity of the environment for the ESP and first LWA application was provided in Section 11.4 of the ESP EIS (NRC 2008). As stated in the ESP EIS, the evaluation of the relationship between local short-term uses of the environment and the maintenance and enhancement of long-term productivity for the construction and operation of proposed COL units can be performed by discussing the benefits of operating the units. The principal benefit is the production of electricity. The analysis of the benefit-cost balance was presented in Section 11.6 of the ESP EIS. If new nuclear power plants are constructed on the VEGP site, power production would continue until the COLs expire or the licensee chooses to cease operation. Once the plants are shut down, they would be decommissioned according to NRC regulations. Once decommissioning is complete and the NRC license is terminated, the site would be available for other uses.

In its COL ER (Southern 2009), Southern indicated that it had identified no new and significant information relative to this topic. In the NRC staff's review of new and significant information for the COL review, no information was identified that would change the conclusions in the ESP EIS, for the proposed action identified in Section 1.2 of this SEIS, regarding short-term uses and long-term productivity.

Conclusions and Recommendations

11.5 Irreversible and Irretrievable Commitments of Resources

The NRC staff's review of the irreversible and irretrievable commitments of resources associated with the proposed action at the ESP stage was provided in Section 11.5 of the ESP EIS (NRC 2008). As stated in the ESP EIS, irretrievable commitments of resources during construction of the proposed new units generally would be similar to that of any major construction project. The staff expects that the use of construction materials in the quantities associated with those expected for proposed Units 3 and 4, while irretrievable, would be of small consequence with respect to the availability of such resources. Likewise, as stated in the ESP EIS, the main resource that would be irretrievably committed during operation of the new nuclear units would be uranium. However, the availability of uranium ore and existing stockpiles of highly enriched uranium in the United States and Russia that could be processed into fuel is sufficient, so the irreversible and irretrievable commitment would be of small consequence.

In its COL ER (Southern 2009), Southern indicated that there is no new and significant information relative to the irreversible and irretrievable commitment of resources related to its request for COLs and a second LWA. In the NRC staff's independent evaluation and review of the COL ER and Southern's process for identifying new and significant information, and supplemental information provided by Southern (Southern 2010b), no new and significant information was identified that would change the conclusions identified in the ESP EIS regarding irreversible and irretrievable commitments of resources.

11.6 Benefit-Cost Balance

A benefit-cost balance discussion is provided in Section 11.6 of the ESP EIS (NRC 2008). Southern indicated in its COL ER (Southern 2009) that there is no new and significant information regarding benefits and costs related to the proposed Units 3 and 4. During its review of the COL application, the NRC staff independently reviewed Southern's ER, audited Southern's process for identifying new and significant information, examined other information available at the site audit, and considered applicable regulations and reference documents. In doing so, the NRC staff identified new information in the areas of project benefits and ecological costs that warranted further analysis in the SEIS.

In March 2009, the Georgia Public Service Commission (GPSC) issued a certification to Southern for construction of the proposed Units 3 and 4 (GPSC 2009). This certification was amended in June 2010 (GPSC 2010) after being remanded back the GPSC by the Fulton County Superior Court. The amended certification (GPSC 2010) further substantiates the conclusions in the ESP EIS concerning the benefits of the proposed action, especially

Conclusions and Recommendations

concerning price stability and fuel diversity in Georgia. Specifically, the GPSC found in its June 2010 decision that:

- Fuel diversity is necessary to protect ratepayers from fuel cost and environmental cost risks.

- The addition of base-load nuclear generation will preserve the diversity of fuel sources necessary to assure reliable and economical supply of electric power and energy for the Georgia retail consumers of GPSC.

- The fuel cost savings likely to result from adding nuclear base-load capacity offer substantial assurance of reliable and economical supply of power and energy to GPSC's Georgia retail consumers.

As described in Section 4.4.1 of this SEIS, the development of additional onsite borrow sources that were not considered in the ESP EIS resulted in the loss or diminishment of populations of two species that are listed as State-threatened by the Georgia Department of Natural Resources. However, although the staff's conclusion with respect to the magnitude of the onsite terrestrial impacts increased, the staff determined that the overall conclusion for Ecology in Table 11-3 of the ESP EIS would continue to be the range of SMALL to MODERATE. The staff did not identify any other new information in the areas of project benefits and environmental costs that has the potential to affect its conclusions in the EIS with respect to the cost-benefit analysis.

Southern has requested a second LWA along with two COLs. The second LWA would allow Southern to perform certain construction activities before the COLs are issued. The economic and environmental costs associated with the second LWA would be a small fraction of the overall costs of construction and operating the proposed facility. The primary benefit from authorizing the LWA activities in the second LWA request in advance of issuing the COLs is that it would enable Southern to maintain the overall project schedule of construction and operation-need dates, thereby decreasing the chance for cost overruns.

Based on this review, including consideration of the benefits and costs of the construction activities requested in the second LWA, the staff determined that the assessment of costs and benefits presented in the ESP EIS remains valid. The potential societal benefits to the local economy and the electricity generated appear to be larger in comparison to the overall external socio-environmental costs, including the increase in terrestrial ecology impact. Consequently, the staff continues to conclude that the construction and operation of the proposed Units 3 and 4, with mitigation measures identified by the staff, would have accrued benefits that most likely would outweigh the economic, environmental, and social costs associated with constructing and operating two new units at the VEGP site.

Conclusions and Recommendations

11.7 Staff Conclusions and Recommendations

The NRC staff's recommendation to the Commission related to the environmental aspects of the proposed action is that the COLs and the LWA be issued. The staff's evaluation of the safety and security aspects of the proposed action will be addressed in the staff's Safety Evaluation Report. This recommendation is based on (1) Southern's COL ER (Southern 2009) and responses to staff requests for additional information; (2) the staff's review conducted for the ESP application (Southern 2008) and the assessment documented in the ESP EIS (NRC 2008); (3) consultation with Federal, State, and Tribal agencies; (4) the staff's own independent review of potential new and significant information available since preparation and publication of the ESP EIS; and (5) the assessments summarized in this SEIS, including the potential mitigation measures identified and consideration of public comments received on the draft SEIS. Finally, the staff concludes that the requested LWA construction activities defined at 10 CFR 50.10(a) and described in the site redress plan would not result in any significant adverse environmental impacts that cannot be redressed.

11.8 References

10 CFR Part 50. Code of Federal Regulations, Title 10, *Energy*, Part 50, "Domestic Licensing of Production and Utilization Facilities."

10 CFR Part 51. Code of Federal Regulations, Title 10, *Energy*, Part 51, "Environmental Protection Regulations for Domestic Licensing and Related Regulatory Functions."

Georgia Public Service Commission (GPSC). 2009. *Docket No. 27800. Georgia Power's Application for the Certification of Units 3 and 4 at Plant Vogtle and Updated Integrated Resource Plan. Amended Certification Order.* March 17, 2009. Accession No. ML100600818.

Georgia Public Service Commission (GPSC). 2010. *Docket No. 27800. Georgia Power's Application for the Certification of Units 3 and 4 at Plant Vogtle and Updated Integrated Resource Plan. Order on Remand.* June 17, 2010. Accession No. ML101960603.

Southern Nuclear Operating Company (Southern). 2008. *Vogtle Early Site Permit Application. Part 3: Environmental Report.* Revision 4. Southern Company, Birmingham, Alabama. Package Accession No. ML081020073.

Southern Nuclear Operating Company (Southern). 2009. *Vogtle Electric Generating Plant, Units 3 and 4, COL Application, Part 3 Environmental Report*, Revision 1, September 23, 2009. Southern Company, Birmingham, Alabama. Accession No. ML092740400.

Southern Nuclear Operating Company (Southern). 2010a. Letter from C.R. Pierce, Southern, to NRC, "Southern Nuclear Operating Company, Vogtle Electric Generating Plant Units 3 and 4, Early Site Permit Site Safety Analysis Report Amendment Request, Revised Site Safety Analysis Report Markup for Onsite Sources of Backfill, Part 2." Letter ND-10-1005. May 24, 2010. Southern Company, Birmingham, Alabama. Accession No. ML101470212.

Southern Nuclear Operating Company (Southern). 2010b. Letter from C.R. Pierce, Southern, to NRC, "Southern Nuclear Operating Company Vogtle Electric Generating Plant Units 3 and 4 Combined License Application, Response to Request for Additional Information Letter on Environmental Issues." Letter ND-10-0023. January 8, 2010. Southern Company, Birmingham, Alabama. Accession No. ML100120479.

U.S. Nuclear Regulatory Commission (NRC). 2008. *Final Environmental Impact Statement for an Early Site Permit (ESP) at the Vogtle Electric Generating Plant Site.* NUREG-1872, Vols. 1, 2, and Errata, Washington, D.C. Accession Nos. ML082240145; ML082240165, ML082260203; ML082550040.

Appendix A

Contributors to the Supplemental Environmental Impact Statement

Appendix A

Appendix A

Contributors to the Supplemental Environmental Impact Statement

The overall responsibility for the preparation of this supplemental environmental impact statement was assigned to the Office of New Reactors, U.S. Nuclear Regulatory Commission (NRC). The statement was prepared by members of the Office of New Reactors with assistance from other NRC organizations and the Pacific Northwest National Laboratory.

Name	Affiliation	Function or Expertise
NUCLEAR REGULATORY COMMISSION		
Mallecia Sutton	Office of New Reactors	Environmental Project Manager
Mark Notich	Office of New Reactors	Co-Environmental Project Manager
Gregory Hatchett	Office of New Reactors	Branch Chief
Steve Shaffer	Office of New Reactors	Radiological Health
Richard Emch	Office of New Reactors	Radiological Health
Jill Caverly	Office of New Reactors	Hydrology
Daniel Mussatti	Office of New Reactors	Socioeconomics, Environmental Justice, Cost-Benefit Balance, Need for Power
Nancy Kuntzleman	Office of New Reactors	Terrestrial/Aquatic Ecology
Jennifer Davis	Office of Federal State Materials and Environmental Management Programs	Cultural Resources
John Fringer	Office of New Reactors	Land Use, Transmission Lines, Alternatives, Nonradiological Health
Michelle Hart	Office of New Reactors	Design Basis and Severe Accidents
Brad Harvey	Office of New Reactors	Meteorology and Air Quality
Norma Garcia-Santos	Office of Nuclear Material Safety and Safeguards	Transportation
Lucieann Vechioli	Office of Nuclear Material Safety and Safeguards	Transportation
Stan Echols	Office of Nuclear Material Safety and Safeguards	Fuel Cycle
James Shephard	Office of Federal, State, Environmental Management	Decommissioning

Appendix A

Name	Affiliation	Function or Expertise
PACIFIC NORTHWEST NATIONAL LABORATORY[a]		
Michael Sackschewsky		Task Leader
Kimberly Leigh		Deputy Task Leader
Amanda Stegen		Terrestrial Ecology
Michael Smith		Radiological and Nonradiological Health, Decommissioning
Jeremy Rishel		Meteorology and Air Quality
Michelle Niemeyer		Socioeconomics, Environmental Justice, Benefit-Cost Balance, Need for Power
Katherine Cort		Socioeconomics, Environmental Justice, Benefit-Cost Balance, Need for Power
Lance Vail		Water Use, Hydrology, Plant System Alternatives
Philip Meyer		Water Use, Hydrology, Plant System Alternatives
Rebekah Krieg		Aquatic Ecology
Beverly Miller		Aquatic Ecology
Paul Hendrickson		Energy and Site Alternatives, Land Use
Daniel Strom		Radiological and Nonradiological Health, Decommissioning
James V. Ramsdell, Jr.		Meteorology and Air Quality, Design Basis and Severe Accidents, Nonradiological Health
Phil Daling		Transportation
Ellen Prendergast-Kennedy		Cultural Resources
Tara O'Neil		Cultural Resources
Andre Coleman		Mapping and Spatial Analysis
Cary Counts		Technical Editing
Cornelia Brim		Technical Editing
Dave Payson		Technical Editing
Tomiann Parker		References
Susan Gulley		References, Review
Michael Parker		Document Design
Elaine Schneider		Graphics
Donna Austin-Workman		Graphics
Rose Zanders		Graphics

(a) Pacific Northwest National Laboratory is operated by Battelle for the U.S. Department of Energy.

Appendix B

Organizations Contacted

Appendix B

Appendix B

Organizations Contacted

This appendix lists the Federal, State, regional, Tribal, and local organizations that were contacted during the course of the U.S. Nuclear Regulatory Commission staff's independent review of new and significant information potential environmental impacts from the construction and operation of new nuclear units at the Vogtle Electric Generating Plant in Burke County, Georgia. See Appendix B of the early site permit (ESP) environmental impact statement, dated August 2008, for a listing of organizations contacted during the ESP review (NRC 2008).

Absentee-Shawnee Tribe of Oklahoma, Shawnee, Oklahoma

Advisory Council on Historic Preservation, Washington, D.C.

Alabama-Coushatta Tribe of Texas, Livingston, Texas

Alabama-Quassarte Tribal Town, Wetumka, Oklahoma

Catawba Indian Tribe, Catawba, South Carolina

Cherokee Nation of Oklahoma, Tahlequah, Oklahoma

Chickasaw Nation of Oklahoma, Ada, Oklahoma

Coushatta Tribe of Louisiana, Elton, Louisiana

Burke County Board of Commissioners, Waynesboro, Georgia

Eastern Band of Cherokee Indians, Cherokee, North Carolina

Georgia Department of Natural Resources, Atlanta, Georgia

Georgia Department of Natural Resources, Social Circle, Georgia

Georgia Tribe of Eastern Cherokee, Clayton, Georgia

Kialegee Tribal Town, Wetumka, Oklahoma

Miccosukee Tribe of Indians of Florida, Miami, Florida

Mississippi Band of Choctaw Indians, Choctaw, Mississippi

Muscogee (Creek) Nation of Oklahoma, Okmulgee, Oklahoma

Poarch Band of Creek Indians, Atmore, Alabama

Seminole Nation of Oklahoma, Wewoka, Oklahoma

South Carolina Department of Natural Resources, Columbia, South Carolina

Seminole Tribe of Florida, Clewiston, Florida

Thlopthlocco Tribal Town, Okemah, Oklahoma

United Keetoowah Band of Cherokee Indians, Tahlequah, Oklahoma

U.S. Army Corps of Engineers, Savannah, Georgia

U.S. Fish and Wildlife Service, Brunswick, Georgia

U.S. National Marine Fisheries Service, Southeast Regional Office, St. Petersburg, Florida

B.1 Reference

U.S. Nuclear Regulatory Commission (NRC). 2008a. *Final Environmental Impact Statement for an Early Site Permit (ESP) at the Vogtle Electric Generating Plant Site.* NUREG-1872, Vols. 1, 2, and Errata, Washington, D.C. Accession Nos. ML082240145; ML082240165, ML082260203; ML082550040.

Appendix C

Chronology of U.S. Nuclear Regulatory Commission Staff Environmental Review Correspondence Related to the Southern Nuclear Operating Company, Inc., Application for Combined Licenses for Units 3 and 4 at the Vogtle Electric Generating Plant

Appendix C

Chronology of U.S. Nuclear Regulatory Commission Staff Environmental Review Correspondence Related to the Southern Nuclear Operating Company, Inc., Application for Combined Licenses for Units 3 and 4 at the Vogtle Electric Generating Plant

This appendix contains a chronological listing of correspondence between the U.S. Nuclear Regulatory Commission (NRC) and Southern Nuclear Operating Company, Inc. (Southern), and other correspondence related to the NRC staff's environmental review of Southern's combined license (COL) application for two AP1000 reactors at the Vogtle Electric Generating Plant (VEGP). Correspondence information pertinent to the early site permit (ESP) review of Units 3 and 4 can be found in Appendix C of the ESP environmental impact statement dated August 2008. All documents, with the exception of those containing proprietary or sensitive information, have been placed in the Commission's Public Document Room, at One White Flint North, 11555 Rockville Pike (first floor), Rockville, Maryland. Such documents are also available electronically from the Public Electronic Reading Room found on the Internet at the following Web address: <http://www.nrc.gov/reading-rm/adams.html>. From this site, the public can gain access to the NRC's Agencywide Documents Access and Management System (ADAMS), which provides text and image files of NRC's public documents in the publicly available records component of ADAMS. The ADAMS accession number for each document is included below:

May 5, 2008	*Federal Register* Notice of Receipt and Availability of Application for a Combined License (Accession No. ML081780052)
May 30, 2008	Letter from Mr. J.A. (Buzz) Miller, Sr. Vice President, Nuclear Development, Southern Nuclear Operating Company, regarding the Acceptance Review for the Vogtle Electric Generating Plant Units 3 and 4 Combined License Application (Accession No. ML081480138)
June 11, 2008	*Federal Register* Notice regarding Acceptance for Docketing of an Application for Combined License for Vogtle Electric Generation Plant Units 3 and 4 (Accession No. ML081770650)

Appendix C

July 2, 2008	Note to File: Public Outreach Meeting on the Vogtle Electric Generating Plant Units 3 and 4 Combined License Application (Accession No. ML081850263)
July 9, 2008	Letter from NRC to Ms. Gwen Jackson, Burke County Library, regarding Application by Southern Nuclear Operating Company for a Combined License for Units 3 and 4 at the Vogtle Electric Generating Plant (Accession No. ML081780805)
July 16, 2008	Letter from NRC to Mr. J.A. (Buzz) Miller, Southern Nuclear Operating Company, regarding the Notice of Intent to Prepare a Supplement to the Environmental Impact Statement in Relation to the Combined License Application for the Vogtle Electric Generating Plant (Accession No. ML081500677)
August 8, 2008	Letter from NRC to the U.S. Environmental Protection Agency, regarding the Final Environmental Impact Statement for an Early Site Permit at the Vogtle Electric Generating Plant Site (Accession No. ML081910396)
August 11, 2008	Note to File: Summary of Public Outreach Meeting to Discuss the Review of the Vogtle Combined License Application (Accession No. ML082190977)
October 14, 2008	Trip Report – August 11 through August 12, 2008, VEGP, Units 3 and 4, COL Site Audit (Accession No. ML082620184)
September 21, 2009	Note to File: Audit Execution Plan for New and Significant Information Audit and Plant Vogtle Combined License Supplement Environmental Impact Statement (Accession No. ML092600338)
September 23, 2009	Letter from Mr. Charles R. Pierce, Southern Nuclear Operating Company, to NRC, regarding Revision 1 to the Environmental Report for Vogtle Electric Generating Plant Units 3 and 4 Combined License Application Package (Accession No. ML092740396)
September 28, 2009	*Federal Register* Notice of Intent to Prepare a Supplemental Environmental Impact Statement (Accession No. ML092650823)
October 2, 2009	Letter from Mr. Michael K. Smith, Southern Nuclear Operating Company, to NRC, regarding Revision 1 to Part 6, Limited Work Authorization Request for the Vogtle Electric Generating Plant Units 3 and 4 Combined License Application (Accession No. ML092960549)

Appendix C

October 15, 2009	Letter from Mr. J.A. (Buzz) Miller, Vice President, Nuclear Development, Southern Nuclear Operating Company, to NRC, regarding Post New and Significant Audit Supporting Information (Accession No. ML092960312)
October 28, 2009	Letter from NRC to Ms. Gwen Jackson, Burke County Library, regarding Environmental Revision 1 by Southern Nuclear Operating Company for a Combined License for Units 3 and 4 at the Vogtle Electric Generating Plant (Accession No. ML093000052)
December 7, 2009	E-mail from NRC to Ms. Julie Holling, South Carolina Department of Natural Resources, regarding South Carolina State Threatened and Endangered Species in the Vicinity of Vogtle Electric Generating Plant (Accession No. ML093491061)
December 9, 2009	Letter from NRC to Dr. Dave Crass, Acting Division Director and Deputy SHPO, State of Georgia Historic Preservation Officer, Historic Preservation Division, Department of Natural Resources, regarding Vogtle, Units 3 and 4 COL Review, Atlanta, GA SHPO (Accession No. ML092600744)
December 10, 2009	Letter from NRC to Mr. J.A. (Buzz) Miller, Sr. Vice President, Nuclear Development, Southern Nuclear Operating Company, regarding Request for Additional Information Regarding the Environmental Review of the Combined License Application for Vogtle Electric General Plant, Units 3 and 4 (Accession No. ML093140059)
December 10, 2009	Letter from NRC to Ms. Stephanie Rolin, NAGPRA Contact, Poarch Band of Creek Indians initiating Consultation to the Tribes for Vogtle COLA (Accession No. ML092730038)
December 10, 2009	Letter from NRC to Ms. Emma Sue Holland, United Keetoowah Band of Cherokee Indians, initiating Consultation to the Tribes for Vogtle COLA (Accession No. ML092740546)
December 10, 2009	Letter from NRC to Mr. Eddie Tullis, Chairperson, Poarch Band of Creek Indians, initiating Consultation to the Tribes for Vogtle COLA (Accession No. ML092670288)
December 10, 2009	Letter from NRC to Ms. Kathy McCoy, NAGPRA Contact, Eastern Band of Cherokee Indians, initiating Consultation to the Tribes for Vogtle COLA (Accession No. ML092730317)

Appendix C

December 10, 2009   Letter from NRC to Mr. John Zachary, Attorney-at-Law, c/o Coushatta Tribe of Louisiana, initiating Consultation to the Tribes for Vogtle COLA (Accession No. ML092730292)

December 10, 2009   Letter from NRC to Ms. Evelyn Bucktrot, Town King, Kialegee Tribal Town, initiating Consultation to the Tribes for Vogtle COLA (Accession No. ML092740388)

December 10, 2009   Letter from NRC to Mr. Steven Terry, Land Resource Manager, Miccosukee Tribe of Indians of Florida, initiating Consultation to the Tribes for Vogtle COLA (Accession No. ML092740375)

December 10, 2009   Letter from NRC to Ms. Gale Thrower, NAGPRA Contact, Poarch Band of Creek Indians, initiating Consultation to the Tribes for Vogtle COLA (Accession No. ML092710241)

December 10, 2009   Letter from NRC to Mr. Louis McGertt, Town King, Thlopthlocco Tribal Town initiating, Consultation to the Tribes for Vogtle COLA (Accession No. ML092740554)

December 10, 2009   Letter from NRC to Mr. A. D. Ellis, Principal Chief, Muscogee (Creek) Nation, initiating Consultation to the Tribes for Vogtle COLA (Accession No. ML092730350)

December 10, 2009   Letter from NRC to Mr. Richard L. Allen, NAGPRA Contact, Cherokee Nation of Oklahoma, initiating Consultation to the Tribes for Vogtle COLA (Accession No. ML092730092)

December 10, 2009   Letter from NRC to Ms. Gingy Nail, NAGPRA Contact, Chickasaw Nation, initiating Consultation to the Tribes for Vogtle COLA (Accession No. ML092730177)

December 10, 2009   Letter from NRC to Mr. Bill Anoatubby, Governor, Chickasaw Nation of Oklahoma, initiating Consultation to the Tribes for Vogtle COLA (Accession No. ML092730147)

December 10, 2009   Letter from NRC to Mr. Charles Thurmond, NAGPRA Contact, Georgia Tribe of Eastern Cherokee, initiating Consultation to the Tribes for Vogtle COLA (Accession No. ML092730371)

Appendix C

December 10, 2009	Letter from NRC to Mr. Tarpie Yargee, Alabama-Quassarte Tribal Town, initiating Consultation to the Tribes for Vogtle COLA (Accession No. ML092730274)
December 10, 2009	Letter from NRC to Mr. Pare Bowlegs, Seminole Nation of Oklahoma, initiating Consultation to the Tribes for Vogtle COLA (Accession No. ML092930629)
December 10, 2009	Letter from NRC to Mr. Michell Hicks, Principal Chief, Eastern Band of Cherokee Indians, initiating Consultation to the Tribes for Vogtle COLA (Accession No. ML092940250)
December 10, 2009	Letter from NRC to Mr. Dallas Proctor, Chief, United Keetoowah Band of Cherokee Indians, initiating Consultation to the Tribes for Vogtle COLA (Accession No. ML092740393)
December 10, 2009	Letter from NRC to Ms. Karen Kaniatobe, Director of the Cultural/ Historical Preservation Department, initiating Consultation to the Tribes for Vogtle COLA (Accession No. ML092730283)
December 10, 2009	Letter from NRC to Ms. Debbie Thomas, Tribal Historic Preservation Officer, NAGPRA Coordinator, Alabama- Coushatta Tribe of Texas, initiating Consultation to the Tribes for Vogtle COLA (Accession No. ML092730252)
December 10, 2009	Letter from NRC to Mrs. Joyce Bear, NAGPRA Contact, Muscogee (Creek) Nation of Oklahoma, initiating Consultation to the Tribes for Vogtle COLA (Accession No. ML092920490)
December 10, 2009	Letter from NRC to Mr. Chadwick Smith, Principal Chief, Cherokee Nation of Oklahoma, initiating Consultation to the Tribes for Vogtle COLA (Accession No. ML092730059)
December 10, 2009	Letter from NRC to Mr. Gilbert Blue, Chairperson, Catawba Indian Tribe, initiating Consultation to the Tribes for Vogtle COLA (Accession No. ML092730321)
December 10, 2009	Letter from NRC to Mr. Willard Steele, Deputy THPO, Seminole Tribe of Florida, initiating Consultation to the Tribes for Vogtle COLA (Accession No. ML092920488)

Appendix C

December 10, 2009	Letter from NRC to Mr. Kenneth Carleton, THPO/ Tribal Archaeologist, Mississippi Band of Choctaw Indians, initiating Consultation to the Tribes for Vogtle COLA (Accession No. ML092730208)
December 15, 2009	E-mail from Ms. Julie Holling, South Carolina Department of Natural Resources, to NRC, regarding South Carolina State Threatened and Endangered Species in the Vicinity of Vogtle Electric Generating Plant (Accession No. ML093491132)
December 15, 2009	E-mail from NRC to Mr. Matt Elliot, Georgia Department of Natural Resources, regarding updated Georgia state-listed species information (Accession No. ML093491138)
December 16, 2009	E-mail from Mr. Matt Elliot, Georgia Department of Natural Resources, to NRC, regarding GDNR e-mail Vogtle COL (Accession No. ML093500211)
December 17, 2009	Letter from Ms. Katrina Morris, Georgia Department of Natural Resources, to NRC, regarding known occurrences of natural communities, plants and animals of highest priority conversation status on or near Vogtle COL, Burke County, Georgia (Accession No. ML100490042)
December 23, 2009	Letter to Mr. Don Klima, Office of Federal Agency Programs, Advisory Council on Historic Preservation, regarding Request for Information on Historic Properties within the Area Under Evaluation for the VEGP, Units 3 and 4 COL (Accession No. ML092600785)
December 23, 2009	Summary of Teleconference Held with Southern Nuclear Operating Company regarding Vogtle Electric Generating Plant Site for a COL (Accession No. ML093410022)
January 7, 2010	Letter from NRC to Ms. Sandra S. Tucker, Field Supervisor, U.S Fish and Wildlife Service Coastal Sub Office, regarding Request for List of Protected Species (Accession No. ML092600684)
January 7, 2010	Letter from NRC to Mr. Donald Rodgers, Catawba Indian Nation, regarding the U.S. Nuclear Regulatory Commission's Supplemental Environmental Impact Statement for Southern Nuclear Operating Company's Combined License Application for the Proposed Construction and Operation of Units 3 and 4 at the Vogtle Electric Generating Plant in Waynesboro, Georgia (Accession No. ML100060777)

Appendix C

January 8, 2010	Letter from Mr. Charles R. Pierce, Southern Nuclear Operating Company, to NRC, regarding the Response to Request for Additional Information Letter on Environmental Issues (Accession No. ML100120479)
January 20, 2010	Letter from Dr. David Crass, Georgia Department of Natural Resources, Historic Preservation Division, to Mr. Thomas Moorer, Southern Nuclear Operating Company, Memorandum of Understanding – Archaeological Site 9BK416 Vogtle Electric Generating Plant Expansion, Burke County, Georgia, HP-060428-001 (Accession No. ML100500302)
January 28, 2010	Letter from NRC to Mr. J.A. (Buzz) Miller, Southern Nuclear Operating Company, regarding Issuance of the Environmental Review Schedule for the Combined License Application Review for Vogtle Electric Generating Plant, Units 3 and 4 (Accession No. ML092630002)
January 29, 2010	Letter from Mr. J.A. (Buzz) Miller, Southern Nuclear Operating Company, to NRC, regarding Replacement DVD for Letter ND-09-1673 (10/15/09) (Accession No. ML100300006)
February 4, 2010	Letter from NRC to Mr. J.A. (Buzz) Miller, Southern Nuclear Operating Company, regarding Request for Additional Information Regarding the Environmental Review of the Limited Work Authorization for the Vogtle Electric Generating Plant, Units 3 and 4 (Accession No. ML100280034)
February 5, 2010	Letter from Mr. Charles R. Pierce, Southern Nuclear Operating Company, to NRC, regarding the Environmental Report to Support Revision 1 to Part 6, Limited Work Authorization Request, of the Vogtle Electric Generating Plant Units 3 and 4 Combined License Application (Accession No. ML100470600)
February 12, 2010	Letter from Ms. Sandra Tucker, U.S. Fish and Wildlife Service, to NRC, regarding USFWS Log Number 2009-1387 (Accession No. ML100500426)
February 19, 2010	Letter from Mr. Charles R. Pierce, Southern Nuclear Operating Company, to NRC, regarding the Large Component Transportation Decision (Accession No. ML100550033)

Appendix C

March 1, 2010	Note to File, Discussion with the U.S. Army Corps of Engineers, Savannah District, concerning their participation in the development of the supplemental environmental impact statement for the combined operating license for the Vogtle Electric Generating Plant, Units 3 and 4 (Accession No. ML100570038)
March 11, 2010	Letter from Mr. Brian L. (Pete) Ivey, Southern Nuclear Operating Company, to NRC, regarding Supplement to Environmental Report in Support of Revision 1 to Part 6, LWA Request (Accession No. ML100750657)
March 12, 2010	Letter from Mr. Michael K. Smith, Southern Nuclear Operating Company, to NRC, regarding Vogtle Electric Generating Plant Units 3 and 4 Combined License Application Supporting Information for Environmental Report Review (Accession No. ML100750038)
April 6, 2010	Memorandum regarding Summary of the Environmental Site Audit Related to the Review of the Combined License Application for Vogtle Electric Generating Plant Site (Package Accession No. ML093631157)
April 28, 2010	Summary of Meeting to Discuss Southern Nuclear Operating Company's Plans for Potential License Amendments Regarding Safety-Related Backfill for its Early Site Permit for the Vogtle Site (Accession No. ML101160362)
May 10, 2010	Summary of Meeting with Southern Nuclear Operating Company to Discuss Plans to Request Exemption from Requirements of 10 CFR 50, Section 10(d) for its Combined License Application for the Vogtle Electric Generating Plant Proposed Units 3 and 4 (Accession No. ML101250259)
May 10, 2010	Letter from Mr. B.L. (Pete) Ivey, Southern Nuclear Operating Company, to NRC, regarding Post New and Significant Audit Support Information (Accession No. ML101310333)
May 18, 2010	Letter from NRC to Mr. J.A. (Buzz) Miller, Southern Nuclear Operating Company, regarding Revision of the Environmental Review Schedule for the Combined License Application Review for Vogtle Electric Generating Plant, Units 3 and 4 (Accession No. ML101330353)
June 16, 2010	Memorandum Regarding the Site Audit Summary Concerning Environmental Impacts Associated with Acquisition of Additional Backfill

Appendix C

	Material for the Vogtle Electric Generating Plant Site Combined License Application Review (Package Accession No. ML101550095)
June 17, 2010	Summary of Teleconference Calls Held with the Georgia Department of Natural Resources for the Vogtle Electric Generating Plant, Units 3 and 4 Onsite Backfill Amendment (Accession No. ML101670079)
June 18, 2010	E-mail from Ms. Elizabeth Shirk, Georgia Department of Natural Resources, Historic Preservation Division, to NRC, regarding Vogtle Electric Generating Plant, Burke County, Georgia, Units 3 and 4 Supplement (Accession No. ML101940268)
July 14, 2010	Summary of the teleconference held with Southern Nuclear Operating Company regarding the Vogtle Electric Generating Plant, Units 3 and 4 Combined License Application (Accession No. ML 100620862)
July 16, 2010	Letter from Mr. Charles R. Pierce, Southern Nuclear Operating Company, to NRC, regarding New and Significant information evaluation for the transportation of backfill from an offsite source (Accession No. ML102010031)
August 3, 2010	Letter from NRC to Mr. J.A. (Buzz) Miller, Southern Nuclear Operating Company, regarding Revision of the Environmental Review Schedule for the Combined License Application Review for Vogtle Electric Generating Plant, Units 3 and 4 (Accession No. ML102100311)
August 6, 2010	Letter from Mr. B.L. Ivey, Southern Nuclear Operating Company, regarding Vogtle Electric Generating Plant, Units 3 and 4, COL Application, Part 6, Limited Work Authorization Request, Revision 2 (Accession No. ML102220380)
August 26, 2010	Letter from NRC to U.S. Environmental Protection Agency, regarding Notice of Availability of the Draft Supplemental Environmental Impact Statement for the Vogtle Electric Generating Plant, Units 3 and 4 Combined Licenses Application (Accession No. ML102070018)

Appendix C

August 26, 2010	Letter from NRC to Mr. J.A. (Buzz) Miller, Southern Nuclear Operating Company, regarding Notice of Availability of the Draft Supplemental Environmental Impact Statement for Combined Licenses for the Vogtle Electric Generating Plant, Units 3 and 4 (Accession No. ML102080062)
August 27, 2010	Letter from NRC to Ms. Gwen Jackson, Burke County Public Library, regarding Maintenance of Reference Materials at the Burke County Library for the Draft Supplemental [Environmental] Impact Statement for Combined Licenses for the Vogtle Electric Generating Plant, Units 3 and 4 (Accession No. ML102170028)
September 2, 2010	Letter from NRC to Mr. A.D. Ellis, Principal Chief of the Muscogee (Creek) Nation, regarding Section 106 Consultation and Notification of the Issuance and Request for Comments on the Draft Supplemental Environmental Impact Statement for Vogtle Electric Generating Plant, Units 3 and 4 Combined License Application (Accession No. ML102000264)
September 2, 2010	Letter from NRC to Mr. Dallas Proctor, Chief of the United Keetoowah Band of Cherokee Indians, regarding Section 106 Consultation and Notification of the Issuance and Request for Comments on the Draft Supplemental Environmental Impact Statement for Vogtle Electric Generating Plant, Units 3 and 4 (Accession No. ML102000360)
September 2, 2010	Letter from NRC to Mr. Eddie Tullis, Chairperson Poarch Band of Creek Indians, regarding Section 106 Consultation and Notification of the Issuance and Request for Comments on the Draft Supplemental Environmental Impact Statement for Vogtle Electric Generating Plant, Units 3 and 4 (Accession No. ML102000149)
September 2, 2010	Letter from NRC to Ms. Emma Sue Holland, NAGPRA Contact for the United Keetoowah Band of Cherokee Indians, regarding Section 106 Consultation and Notification of the Issuance and Request for Comments on the Draft Supplemental Environmental Impact Statement for Vogtle Electric Generating Plant, Units 3 and 4 (Accession No. ML102000191)
September 2, 2010	Letter from NRC to Ms. Kathy McCoy, NAGPRA Contact for the Eastern Band of Cherokee Indians, regarding Section 106 Consultation and Notification of the Issuance and Request for Comments on the Draft Supplemental Environmental Impact Statement for Vogtle Electric Generating Plant, Units 3 and 4 (Accession No. ML102000210)

Appendix C

September 2, 2010	Letter from NRC to Mr. John Zachary, Coushatta Tribe of Louisiana, regarding Section 106 Consultation and Notification of the Issuance and Request for Comments on the Draft Supplemental Environmental Impact Statement for Vogtle Electric Generating Plant, Units 3 and 4 (Accession No. ML102000219)
September 2, 2010	Letter from NRC to Ms. Evelyn Bucktrot, Town King of the Kialegee Tribal Town, regarding Section 106 Consultation and Notification of the Issuance and Request for Comments on the Draft Supplemental Environmental Impact Statement for Vogtle Electric Generating Plant, Units 3 and 4 (Accession No. ML102000224)
September 2, 2010	Letter from NRC to Mr. Steven Terry, Miccosukee Tribe of Indians of Florida, regarding Section 106 Consultation and Notification of the Issuance and Request for Comments on the Draft Supplemental Environmental Impact Statement for Vogtle Electric Generating Plant, Units 3 and 4 (Accession No. ML102000228)
September 2, 2010	Letter from NRC to Ms. Gale Thrower, NAGPRA Contact for the Poarch Band of Creek Indians, regarding Section 106 Consultation and Notification of the Issuance and Request for Comments on the Draft Supplemental Environmental Impact Statement for Vogtle Electric Generating Plant, Units 3 and 4 (Accession No. ML102000233)
September 2, 2010	Letter from NRC to Mr. Louis McGertt, Town King of the Thlopthlocco Tribal Town, regarding Section 106 Consultation and Notification of the Issuance and Request for Comments on the Draft Supplemental Environmental Impact Statement for Vogtle Electric Generating Plant, Units 3 and 4 (Accession No. ML102000240)
September 2, 2010	Letter from NRC to Mr. Richard L. Allen, NAGPRA Contact for the Cherokee Nation of Oklahoma, regarding Section 106 Consultation and Notification of the Issuance and Request for Comments on the Draft Supplemental Environmental Impact Statement for Vogtle Electric Generating Plant, Units 3 and 4 (Accession No. ML102000287)
September 2, 2010	Letter from NRC to Ms. Gingy (Virginia) Hail, NAGPRA Contact for the Chickasaw Nation, regarding Section 106 Consultation and Notification of the Issuance and Request for Comments on the Draft Supplemental Environmental Impact Statement for Vogtle Electric Generating Plant, Units 3 and 4 (Accession No. ML102000331)

Appendix C

September 2, 2010	Letter from NRC to Mr. Bill Anoatubby, Governor of the Chickasaw Nation of Oklahoma, regarding Section 106 Consultation and Notification of the Issuance and Request for Comments on the Draft Supplemental Environmental Impact Statement for Vogtle Electric Generating Plant, Units 3 and 4 (Accession No. ML102000335)
September 2, 2010	Letter from NRC to Mr. Charles Thurmond, NAGPRA Contact for the Georgia Tribe of Eastern Cherokee, regarding Section 106 Consultation and Notification of the Issuance and Request for Comments on the Draft Supplemental Environmental Impact Statement for Vogtle Electric Generating Plant, Units 3 and 4 (Accession No. ML102000345)
September 2, 2010	Letter from NRC to Mr. Tarpie Yargee, Alabama-Quassarte Tribal Town, regarding Section 106 Consultation and Notification of the Issuance and Request for Comments on the Draft Supplemental Environmental Impact Statement for Vogtle Electric Generating Plant, Units 3 and 4 (Accession No. ML102000349)
September 2, 2010	Letter from NRC to Mr. Pare Bowlegs, Seminole Nation of Oklahoma, regarding Section 106 Consultation and Notification of the Issuance and Request for Comments on the Draft Supplemental Environmental Impact Statement for Vogtle Electric Generating Plant, Units 3 and 4 (Accession No. ML102000355)
September 2, 2010	Letter from NRC to Mr. Michell Hicks, Principal Chief of the Eastern Band of Cherokee Indians, regarding Section 106 Consultation and Notification of the Issuance and Request for Comments on the Draft Supplemental Environmental Impact Statement for Vogtle Electric Generating Plant, Units 3 and 4 (Accession No. ML102000358)
September 2, 2010	Letter from NRC to Ms. Karen Kaniatobe, Absentee-Shawnee Tribe of Oklahoma, regarding Section 106 Consultation and Notification of the Issuance and Request for Comments on the Draft Supplemental Environmental Impact Statement for Vogtle Electric Generating Plant, Units 3 and 4 (Accession No. ML102000365)
September 2, 2010	Letter from NRC to Ms. Debbie Thomas, NAGPRA Coordinator of the Alabama-Coushatta Tribe of Texas, regarding Section 106 Consultation and Notification of the Issuance and Request for Comments on the Draft Supplemental Environmental Impact Statement for Vogtle Electric Generating Plant, Units 3 and 4 (Accession No. ML102000367)

Appendix C

September 2, 2010   Letter from NRC to Ms. Joyce A. Bear, NAGPRA Contact of the Muscogee (Creek) Nation of Oklahoma, regarding Section 106 Consultation and Notification of the Issuance and Request for Comments on the Draft Supplemental Environmental Impact Statement for Vogtle Electric Generating Plant, Units 3 and 4 (Accession No. ML102000368)

September 2, 2010   Letter from NRC to Mr. Chadwick Smith, Principal Chief of the Cherokee Nation of Oklahoma, regarding Section 106 Consultation and Notification of the Issuance and Request for Comments on the Draft Supplemental Environmental Impact Statement for Vogtle Electric Generating Plant, Units 3 and 4 (Accession No. ML102000375)

September 2, 2010   Letter from NRC to Mr. Willard Steele, Seminole Tribe of Florida, regarding Section 106 Consultation and Notification of the Issuance and Request for Comments on the Draft Supplemental Environmental Impact Statement for Vogtle Electric Generating Plant, Units 3 and 4 (Accession No. ML102000382)

September 2, 2010   Letter from NRC to Mr. Kenneth H. Carleton, Mississippi Band of Choctaw Indians, regarding Section 106 Consultation and Notification of the Issuance and Request for Comments on the Draft Supplemental Environmental Impact Statement for Vogtle Electric Generating Plant, Units 3 and 4 (Accession No. ML102000384)

September 2, 2010   Letter from NRC to Ms. Stephanie Rolin, NAGPRA Contact of the Poarch Band of Creek Indians, regarding Section 106 Consultation and Notification of the Issuance and Request for Comments on the Draft Supplemental Environmental Impact Statement for Vogtle Electric Generating Plant, Units 3 and 4 (Accession No. ML102000390)

September 2, 2010   Letter from NRC to Ms. Carol Bernstein, U.S. Army Corps of Engineers Savannah District, regarding Notification of the Issuance and Request for Comments on the Draft Supplemental Environmental Impact Statement for Vogtle Electric Generating Plant, Units 3 and 4 (Accession No. ML102320187)

September 3, 2010   Letter from NRC to Mr. David Bernhart, National Marine Fisheries Service, regarding Notification of the Issuance and Request for Comments on the Draft Supplemental Environmental Impact Statement for Vogtle Electric Generating Plant, Units 3 and 4 (Accession No. ML102320162)

Appendix C

September 3, 2010	Letter from NRC to Mr. Robert D. Perry, South Carolina Department of Natural Resources, regarding Notification of the Issuance and Request for Comments on the Draft Supplemental Environmental Impact Statement for Vogtle Electric Generating Plant, Units 3 and 4 (Accession No. ML102320174)
September 3, 2010	Letter from NRC to Ms. Sandra Tucker, U.S. Fish and Wildlife Service, regarding Notification of the Issuance and Request for Comments on the Draft Supplemental Environmental Impact Statement for Vogtle Electric Generating Plant, Units 3 and 4 (Accession No. ML102320222)
September 21, 2010	October 7, 2010 Public Meeting Notice for the Draft Supplemental Environmental Impact Statement for the Combined Licenses for Vogtle Electric Generating Plant, Units 3 and 4 (Accession No. ML102070021)
September 27, 2010	Letter from NRC to Commissioner Andrews, Burke County, regarding an Invitation to a Meeting with the U.S. Nuclear Regulatory Commission to Discuss the Draft Supplemental Environmental Impact Statement for Combined Licenses for Vogtle Electric Generating Plant Units 3 and 4 (Accession No. ML102710050)
September 27, 2010	Letter from NRC to Commissioner Crockett, Burke County, regarding an Invitation to a Meeting with the U.S. Nuclear Regulatory Commission to Discuss the Draft Supplemental Environmental Impact Statement for Combined Licenses for Vogtle Electric Generating Plant Units 3 and 4 (Accession No. ML102710053)
September 27, 2010	Letter from NRC to Commissioner Delaigle, Burke County, regarding an Invitation to a Meeting with the U.S. Nuclear Regulatory Commission to Discuss the Draft Supplemental Environmental Impact Statement for Combined Licenses for Vogtle Electric Generating Plant Units 3 and 4 (Accession No. ML102710057)
September 29, 2010	Letter from NRC to Commissioner Tinley, Burke County, regarding an Invitation to a Meeting with the U.S. Nuclear Regulatory Commission to Discuss the Draft Supplemental Environmental Impact Statement for Combined Licenses for Vogtle Electric Generating Plant Units 3 and 4 (Accession No. ML102710064)

Appendix C

September 29, 2010   Letter from NRC to Commissioner Lucious Abrams, Burke County, regarding an Invitation to a Meeting with the U.S. Nuclear Regulatory Commission to Discuss the Draft Supplemental Environmental Impact Statement for Combined Licenses for Vogtle Electric Generating Plant Units 3 and 4 (Accession No. ML102700514)

September 29, 2010   Letter from NRC to Mayor Deloach, City of Waynesboro, regarding an Invitation to a Meeting with the U.S. Nuclear Regulatory Commission to Discuss the Draft Supplemental Environmental Impact Statement for Combined Licenses for Vogtle Electric Generating Plant Units 3 and 4 (Accession No. ML102710083)

September 29, 2010   Letter from NRC to Mr. Bill Tinley, Waynesboro City Council, regarding an Invitation to a Meeting with the U.S. Nuclear Regulatory Commission to Discuss the Draft Supplemental Environmental Impact Statement for Combined Licenses for Vogtle Electric Generating Plant Units 3 and 4 (Accession No. ML102710115)

September 29, 2010   Letter from NRC to Mr. Herman Brown, Waynesboro City Council, regarding an Invitation to a Meeting with the U.S. Nuclear Regulatory Commission to Discuss the Draft Supplemental Environmental Impact Statement for Combined Licenses for Vogtle Electric Generating Plant Units 3 and 4 (Accession No. ML102710096)

September 29, 2010   Letter from NRC to Mr. James Jones, Waynesboro City Council, regarding an Invitation to a Meeting with the U.S. Nuclear Regulatory Commission to Discuss the Draft Supplemental Environmental Impact Statement for Combined Licenses for Vogtle Electric Generating Plant Units 3 and 4 (Accession No. ML102710137)

September 29, 2010   Letter from NRC to Mr. Richard Byne, Waynesboro City Council, regarding an Invitation to a Meeting with the U.S. Nuclear Regulatory Commission to Discuss the Draft Supplemental Environmental Impact Statement for Combined Licenses for Vogtle Electric Generating Plant Units 3 and 4 (Accession No. ML102710147)

September 29, 2010   Letter from NRC to Mr. Willie Williams, Waynesboro City Council, regarding an Invitation to a Meeting with the U.S. Nuclear Regulatory Commission to Discuss the Draft Supplemental Environmental Impact Statement for Combined Licenses for Vogtle Electric Generating Plant Units 3 and 4 (Accession No. ML102710182)

Appendix C

September 29, 2010   Letter from NRC to Ms. Linda Bailey, Superintendent Burke County Public Schools, regarding an Invitation to a Meeting with the U.S. Nuclear Regulatory Commission to Discuss the Draft Supplemental Environmental Impact Statement for Combined Licenses for Vogtle Electric Generating Plant Units 3 and 4 (Accession No. ML102710201)

September 29, 2010   Letter from NRC to Ms. Portia Lodge Washington, Waynesboro City Council, regarding an Invitation to a Meeting with the U.S. Nuclear Regulatory Commission to Discuss the Draft Supplemental Environmental Impact Statement for Combined Licenses for Vogtle Electric Generating Plant Units 3 and 4 (Accession No. ML102710169)

October 6, 2010   E-mail from Mr. Bryant J. Celestine, Alabama-Coushatta Tribe of Texas, to NRC, regarding the Draft SEIS (Accession No. ML102940055)

October 20, 2010   Letter from Ms. Sandra Tucker, U.S. Fish and Wildlife Service, to NRC regarding FWS Log #2010-1254 (Accession No. ML103010076)

October 28, 2010   Summary of Teleconference with the U.S. Fish and Wildlife Service to Discuss Species List for the Vogtle Combined Licenses Biological Assessment (Accession No. ML102990317)

October 29, 2010   Letter from NRC to Mr. J.A. (Buzz) Miller, Southern Nuclear Operating Company, regarding the Revised Review Schedule (Accession No. ML102310362)

November 10, 2010   Summary of the Public Meeting for the Draft Supplemental Environmental Impact Statement for the Combined Licenses for Vogtle Electric Generating Plant Units 3 and 4 (Package Accession No. ML103130518)

November 15, 2010   Letter from Mr. Heinz J. Mueller, U.S. Environmental Protection Agency, to NRC, regarding Draft Supplemental Environmental Impact Statement for Combined Licenses Vogtle, Units 3 and 4, Construction and Operation, Application, NUREG-1947, CEQ No. 20100351 (Accession No. ML103370044)

November 29, 2010   Letter from Mr. Gregory Hogue, U.S. Department of the Interior, to NRC, regarding Comments for the Draft Environmental Impact Statement (DEIS) for Vogtle Nuclear Plant Units 3 and 4, Application for Combined Licenses (COLs), NUREG-1947, Burke County, Georgia (Accession No. ML103330069)

Appendix C

November 23, 2010   Letter from Mr. C. R. Pierce, Southern Nuclear Operating Company, to NRC, regarding Comments on Draft Supplemental Environmental Impact Statement (Accession No. ML103300035)

February 24, 2011   Letter from NRC to Ms. Sandra Tucker, U.S. Fish and Wildlife Service, regarding Biological Assessment for Threatened and Endangered Species and Designated Critical Habitat (Accession No. ML103410237)

March 2, 2011   Letter from NRC to Mr. David Bernhart, National Fisheries Service, regarding Conference Consultation for the Atlantic Sturgeon (Accession No. ML110460152)

March 3, 2011   Letter from Mr. Charles R. Pierce, Southern Nuclear Operating Company, to NRC, regarding Notification of Approved Change to Environmental Permit SAS-2007-01837
(Accession No. ML110660152)

Appendix D

Scoping Comments and Responses

Appendix D

Scoping Comments and Responses

Appendix D of the Vogtle Electric Generating Plant (VEGP) early site permit (ESP) environmental impact statement (EIS) details the scoping comments received under that review process. The ESP was granted in August 2009. The combined operating license (COL) application, revision 0, was submitted in March 2008, while revision 1 of the COL application was submitted on September 23, 2009. In accordance with 10 CFR Part 51.26, the U.S. Nuclear Regulatory Commission (NRC) staff published a Notice of Intent to prepare an EIS related to the VEGP in the *Federal Register* on September 28, 2009. Furthermore, 10 CFR 51.26(d) states that scoping is not required for a supplement to an EIS prepared for a COL application that references an ESP. Therefore, no formal scoping comment period occurred. A public outreach meeting was held on July 17, 2008. A summary of that meeting can be found at the NRC Public Electronic Reading Room found on the Internet at the following Web address: http://www.nrc.gov/reading-rm/adams.html, Accession No. ML082190977.

Appendix E

Draft Supplemental Environmental Impact Statement Comments and Responses

Appendix E

Draft Supplemental Environmental Impact Statement Comments and Responses

As part of the U.S. Nuclear Regulatory Commission (NRC) review of the Southern Nuclear Operating Company, Inc. (Southern) application for combined licenses (COLs) for proposed Units 3 and 4 at the Vogtle Electric Generating Plant (VEGP) site, the NRC solicited comments from the public on a draft of this supplemental environmental impact statement (SEIS). NRC regulations related to the environmental review of COL applications are contained in Title 10 of the Code of Federal Regulations (CFR) Part 51 and 10 CFR Part 52, Subpart C. Pursuant to NRC regulations in 10 CFR 51.50(c)(1), a COL applicant referencing an early site permit (ESP) need not submit information or analyses regarding environmental issues that were resolved in the ESP final environmental impact statement (EIS), except to the extent the COL applicant has identified new and significant information regarding such issues. In addition, pursuant to 10 CFR 52.39, matters resolved in the ESP proceedings are considered to be resolved in any subsequent proceedings, absent identification of new and significant information. The NRC staff prepared this SEIS to the ESP EIS (NRC 2008a), in support of the COL application for the proposed Units 3 and 4 at the VEGP site.

The draft SEIS was published in September 2010 (NRC 2010a). A 75-day comment period began on September 3, 2010, when the U.S. Environmental Protection Agency (EPA) issued a Notice of Availability (75 FR 54190) of the draft SEIS to allow members of the public to comment on the results of the environmental review. On October 7, 2010, a public meeting was held at the Augusta Technical College in Waynesboro, Georgia. At the meeting, NRC staff described the results of the environmental review, answered questions related to the review, and provided members of the public with information to assist them in formulating their comments.

As part of the process to solicit public comments on the draft SEIS, the staff

- placed a copy of the draft SEIS at the Burke County Public Library in Waynesboro, Georgia
- made the draft SEIS available in the NRC's Public Document Room in Rockville, Maryland
- placed a copy of the draft SEIS on the NRC website at http://www.nrc.gov/reading-rm/doc-collections/nuregs/staff/sr1947/
- provided a copy of the draft SEIS to any member of the public that requested one
- sent copies of the draft SEIS to certain Federal, State, and local agencies

Appendix E

- published a notice of availability of the draft SEIS in the *Federal Register* on September 3, 2010 (75 FR 54190)
- filed the draft SEIS with the EPA
- announced and held a public meeting at the Augusta Technical College in Waynesboro, Georgia, to describe the results of the environmental review, answer any related questions, and take public comments.

Approximately 80 people attended the meeting and 22 attendees provided oral comments. A certified court reporter recorded the oral comments and prepared a written transcript of the meeting. The transcript of the public meeting is part of the public record for the proposed project and was used to establish correspondence between comments contained in this volume of the SEIS and oral comments received at the public meeting. In addition to the comments received at the public meeting, the NRC received 37 letters and e-mail messages with comments. The comment period closed on November 24, 2010; however, the NRC did, to the degree permitted by the schedule, consider comments submitted after the comment period ended.

A meeting summary is available from the Publicly Available Records component of NRC's Agency-wide Document Access and Management System (ADAMS); its accession number is ML103130579 (NRC 2010c). The transcript of the public meeting, and the letters and e-mail messages providing comments on the draft SEIS, are also available in ADAMS; accession numbers are provided in Table E-1. ADAMS is accessible at www.nrc.gov/reading-rm/adams.html, which provides access through the NRC's Public Electronic Reading Room link. Persons who do not have access to ADAMS or who encounter problems in accessing the documents located in ADAMS should contact the NRC's Public Document Room reference staff at 1-800-397-4209 or 301-415-4737, or by e-mail at pdr@nrc.gov.

E.1 Disposition of Comments

This appendix contains all of the comments extracted from the comment letters and e-mail messages provided to the review team during the comment period, as well as the comments from the transcript. Each set of comments from a given commenter was given a unique alpha identifier (commenter ID), allowing each set of comments from a commenter to be traced back to the transcript, letter, or e-mail in which the comments were submitted.

After the comment period, the staff considered and dispositioned all comments received. To identify each individual comment, the team reviewed the transcript of the public meeting and each letter and e-mail received related to the draft SEIS. Table E-1 lists commenters identified by name, affiliation (if given), comment number, and the source of the comment including its ADAMS accession number. As part of the review, the staff identified statements that they

Appendix E

believed were related to the proposed action and recorded the statements as comments. Each comment was assigned to a specific subject area, and similar comments were grouped together. Finally, responses were prepared for each comment or group of comments.

This appendix presents the comments and the staff responses to them grouped by similar issues in the following order:

- Comments Concerning Process – COL
- Comments Concerning Process – NEPA
- Comments Concerning Land Use – Transmission Lines
- Comments Concerning Meteorology and Air Quality
- Comments Concerning Hydrology – Surface Water
- Comments Concerning Hydrology – Groundwater
- Comments Concerning Ecology – Terrestrial
- Comments Concerning Ecology – Aquatic
- Comments Concerning Socioeconomics
- Comments Concerning Historic and Cultural Resources
- Comments Concerning Environmental Justice
- Comments Concerning Health – Nonradiological
- Comments Concerning Health – Radiological
- Comments Concerning Accidents – Design Basis
- Comments Concerning Accidents – Severe
- Comments Concerning the Uranium Fuel Cycle
- Comments Concerning Transportation
- Comments Concerning the Need for Power
- Comments Concerning Alternatives – Energy
- Comments Concerning Benefit-Cost Balance
- General Comments in Support of the Licensing Action
- General Comments in Support of the Licensing Process
- General Comments in Support of Nuclear Power
- General Comments in Support of the Existing Plant
- General Comments in Opposition to the Licensing Action
- General Comments in Opposition to the Licensing Process
- General Comments in Opposition to Nuclear Power
- General Comments in Opposition to the Existing Plant
- Comments Concerning Issues Outside Scope – Emergency Preparedness
- Comments Concerning Issues Outside Scope – Miscellaneous
- Comments Concerning Issues Outside Scope – NRC Oversight
- Comments Concerning Issues Outside Scope – Safety
- Comments Concerning Issues Outside Scope – Security and Terrorism

Appendix E

When the comments resulted in a change in the text of the draft SEIS, the corresponding response refers the reader to the appropriate section of the final SEIS where the change was made. Throughout the final SEIS, with the exception of this new Appendix E, revisions to the text from the draft SEIS are indicated by vertical lines (change bars) in the margin beside the text.

Some comments addressed topics and issues that are not part of the environmental review for this proposed action. These comments included questions about the NRC's safety review, general statements of support or opposition to nuclear power, observations regarding national nuclear waste management policies, comments on the NRC regulatory process in general, and comments on NRC regulations. These comments are included, but detailed responses to such comments are not provided because they addressed issues that do not directly relate to the environmental effects of this proposed action and are thus outside the scope of the National Environmental Policy Act (NEPA) review of this proposed action. If appropriate, these comments were forwarded to the cognizant organization within the NRC for consideration.

Many comments specifically addressed the scope of the environmental review, analyses, and issues contained in the draft SEIS, including comments about potential impacts, proposed mitigation, the agency review process, and the public comment period. Detailed responses to each of these comments are provided in this appendix.

Table E-1. Individuals Providing Comments During the Comment Period

Commenter	Affiliation (if stated)	Comment Source and ADAMS Accession #	Correspondence ID
Abrams, Lucious	Self	Meeting Transcript (ML103130550)	0016-3
Arnold, Judy	Self	E-mail (ML103330061)	0003
Barczak, Sara	Southern Alliance for Clean Energy	Meeting Transcript (ML103130550)	0016-5
Barczak, Sara	Southern Alliance for Clean Energy	Meeting Transcript (ML103140538)	0017-1
Baxley, Robin	Self	Meeting Transcript (ML103130550)	0016-9
Baxter, Farouk	Self	E-mail (ML103560158)	0001
Boatenreiter, Glenn	Self	E-mail (ML103330045)	0003
Booher, Sam	Self	Meeting Transcript (ML103130550)	0016-21
Byne, Dick	Waynesboro City Council	Meeting Transcript (ML103130550)	0016-2

Appendix E

Table E-1. (contd)

Commenter	Affiliation (if stated)	Comment Source and ADAMS Accession #	Correspondence ID
Carroll, Glenn	Nuclear Watch South	E-mail (ML103330030)	0002
Carroll, Glenn	Nuclear Watch South	Meeting Transcript (ML103130550)	0016-11
Carter, Pat	Self	E-mail (ML103330034)	0003
Celestine, Bryant	Alabama-Coushatta Tribe of Texas	E-mail (ML102940055)	0012
Cumbow, Kay	Self	E-mail (ML103330053)	0003
Dawson, Daneille	Self	E-mail (ML103330064)	0003
DeLoach, George	Self	Meeting Transcript (ML103130550)	0016-1
Dooley, Gerald	Self	E-mail (ML103330054)	0003
Elam, Terry	Augusta Technical College	Meeting Transcript (ML103130550)	0016-4
Elam, Terry	Augusta Technical College	Meeting Transcript (ML103140541)	0018-1
Falconer, Kimberly	Self	E-mail (ML103330038)	0003
Hatch, Sarah	Nuke Watch South	E-mail (ML103330031)	0003
Henson, Courtney	Self	Meeting Transcript (ML103130550)	0016-15
Howard, Claude	Self	Meeting Transcript (ML103130550)	0016-17
Kasenow, Lisa	Self	E-mail (ML103330043)	0003
Kushner, Adele	Action for a Clean Environment	E-mail (ML103330037)	0005
Lewis, Marvin	Self	E-mail (ML103330048)	0003
Lomas, Judith	Self	E-mail (ML103330046)	0003
Lusk, Phil	Self	E-mail (ML103330044)	0003
McConnell, Joy	Self	E-mail (ML103330063)	0003
McNulty, Joy	Self	E-mail (ML103330055)	0003
Michetti, Susan	Self	E-mail (ML103330052)	0007
Mills, Nancy	Self	E-mail (ML103330035)	0003
Mitchell, Tommy	Self	Meeting Transcript (ML103130550)	0016-8
Mueller, Heinz J.	U.S. Environmental Protection Agency	Letter (ML103370044)	0019
Ogley-Oliver, Emma	Self	Meeting Transcript (ML103130550)	0016-16

Appendix E

Table E-1. (contd)

Commenter	Affiliation (if stated)	Comment Source and ADAMS Accession #	Correspondence ID
Parr, Sue	Augusta Metro Chamber of Commerce	Meeting Transcript (ML103130550)	0016-12
Patrie, Lewis E.	Western N. C. Physicians for Social Responsibility	E-mail (ML102940057)	0013
Paul, Bobbie	Women's Action for New Directions	Meeting Transcript (ML103130550)	0016-7
Pierce, Charles	Southern Nuclear Operating Company	Letter (ML103330035)	0014
Rivard, Betsy	Self	Meeting Transcript (ML103130550)	0016-20
Roberts, Ashley	Burke Co. Chamber of Commerce	Meeting Transcript (ML103130550)	0016-13
Sardi, David	Self	Meeting Transcript (ML103130550)	0016-22
Sheppard, Deborah	Self	E-mail (ML103330033)	0003
Smith, Nathan	Self	E-mail (ML103330047)	0006
Hogue, Gregory	U.S. Department of the Interior	E-mail (ML103330069)	0010
Stephens, Annie Laura	Self	Meeting Transcript (ML103130550)	0016-10
Stone, Jesse	Self	Meeting Transcript (ML103130550)	0016-14
Taylor, F	Self	E-mail (ML103330032)	0004
Thomas, Ellen	Self	E-mail (ML103330060)	0008
Thomas, Russel	Self	E-mail (ML103330040)	0003
Trujillo, Dianne	Self	E-mail (ML103330058)	0003
Utley, Charles	Self	Meeting Transcript (ML103130550)	0016-6
Valentin, Dianne	Self	Meeting Transcript (ML103130550)	0016-19
Vejdani, Vivianne	South Carolina Department of Natural Resources	E-mail (ML102940054)	0011
Villarreal, Tasha	Self	E-mail (ML103330039)	0003
Vincent, Patricia	Self	Meeting Transcript (ML103130550)	0016-18
Zeller, Lou	Blue Ridge Environmental Defense League	E-mail (ML103330070)	0009

Appendix E

E.2 Comments and Responses

Table E-2 is an alphabetical index to the comment categories and lists the commenter names and comment identification numbers that were included in each category. The balance of this document presents the comments and responses organized by topic category. References appear in Section E.3 at the end of the appendix.

Table E-2. Comment Categories

Comment Category	Commenter (Comment ID)
Accidents-Design Basis	• Sardi, David (0016-22-8)
Accidents-Severe	• Arnold, Judy (0003-8) • Boatenreiter, Glenn (0003-8) • Carroll, Glenn (0002-8) • Carter, Pat (0003-8) • Cumbow, Kay (0003-8) • Dawson, Daneille (0003-8) • Dooley, Gerald (0003-8) • Falconer, Kimberly (0003-8) • Hatch, Sarah (0003-8) • Kasenow, Lisa (0003-8) • Kushner, Adele (0005-8) • Lewis, Marvin (0003-8) • Lomas, Judith (0003-8) • Lusk, Phil (0003-8) • McConnell, Joy (0003-8) • McNulty, Joy (0003-8) • Mills, Nancy (0003-8) • Sheppard, Deborah (0003-8) • Thomas, Russel (0003-8) • Trujillo, Dianne (0003-8) • Villarreal, Tasha (0003-8)
Alternatives-Energy	• Arnold, Judy (0003-2) (0003-6) • Barczak, Sara (0016-5-7) (0017-1-6) • Boatenreiter, Glenn (0003-2) (0003-6) • Carroll, Glenn (0002-2) (0002-6) (0016-11-4) (0016-11-5) (0016-11-6) • Carter, Pat (0003-2) (0003-6) • Cumbow, Kay (0003-2) (0003-6) • Dawson, Daneille (0003-2) (0003-6) • Dooley, Gerald (0003-2) (0003-6) • Falconer, Kimberly (0003-2) (0003-6) • Hatch, Sarah (0003-2) (0003-6) • Kasenow, Lisa (0003-2) (0003-6)

Appendix E

Table E-2. (contd)

Comment Category	Commenter (Comment ID)
	• Kushner, Adele (0005-2) (0005-6) • Lewis, Marvin (0003-2) (0003-6) • Lomas, Judith (0003-2) (0003-6) • Lusk, Phil (0003-2) (0003-6) • McConnell, Joy (0003-2) (0003-6) • McNulty, Joy (0003-2) (0003-6) • Michetti, Susan (0007-3) (0007-4) • Mills, Nancy (0003-2) (0003-6) • Patrie, Lewis E. (0013-3) • Rivard, Betsy (0016-20-12) • Sardi, David (0016-22-4) (0016-22-14) • Sheppard, Deborah (0003-2) (0003-6) • Taylor, F (0004-2) • Thomas, Ellen (0008-1) (0008-5) • Thomas, Russel (0003-2) (0003-6) • Trujillo, Dianne (0003-2) (0003-6) • Villarreal, Tasha (0003-2) (0003-6) • Vincent, Patricia (0016-18-3)
Benefit-Cost Balance	• Arnold, Judy (0003-5) • Barczak, Sara (0016-5-2) (0016-5-6) (0017-1-2) (0017-1-5) • Boatenreiter, Glenn (0003-5) • Carroll, Glenn (0002-5) (0016-11-2) • Carter, Pat (0003-5) • Cumbow, Kay (0003-5) • Dawson, Daneille (0003-5) • Dooley, Gerald (0003-5) • Falconer, Kimberly (0003-5) • Hatch, Sarah (0003-5) • Henson, Courtney (0016-15-3) (0016-15-5) • Kasenow, Lisa (0003-5) • Kushner, Adele (0005-5) • Lewis, Marvin (0003-5) • Lomas, Judith (0003-5) • Lusk, Phil (0003-5) • McConnell, Joy (0003-5) • McNulty, Joy (0003-5) • Michetti, Susan (0007-2) (0007-5) (0007-10) • Mills, Nancy (0003-5) • Ogley-Oliver, Emma (0016-16-5) • Patrie, Lewis E. (0013-1) (0013-2) • Paul, Bobbie (0016-7-5) • Roberts, Ashley (0016-13-3)

Appendix E

Table E-2. (contd)

Comment Category	Commenter (Comment ID)
	• Sheppard, Deborah (0003-5) • Thomas, Ellen (0008-3) (0008-4) • Thomas, Russel (0003-5) • Trujillo, Dianne (0003-5) • Utley, Charles (0016-6-3) (0016-6-4) • Villarreal, Tasha (0003-5)
Ecology-Aquatic	• Hogue, Gregory (0010-2) • Mueller, Heinz J. (0019-2)
Ecology-Terrestrial	• Hogue, Gregory (0010-1) (0010-3) • Mueller, Heinz J. (0019-11)
Environmental Justice	• Arnold, Judy (0003-7) • Boatenreiter, Glenn (0003-7) • Carroll, Glenn (0002-7) • Carter, Pat (0003-7) • Cumbow, Kay (0003-7) • Dawson, Daneille (0003-7) • Dooley, Gerald (0003-7) • Falconer, Kimberly (0003-7) • Hatch, Sarah (0003-7) • Kasenow, Lisa (0003-7) • Kushner, Adele (0005-7) • Lewis, Marvin (0003-7) • Lomas, Judith (0003-7) • Lusk, Phil (0003-7) • McConnell, Joy (0003-7) • McNulty, Joy (0003-7) • Mills, Nancy (0003-7) • Ogley-Oliver, Emma (0016-16-1) • Sheppard, Deborah (0003-7) • Thomas, Russel (0003-7) • Trujillo, Dianne (0003-7) • Utley, Charles (0016-6-7) • Valentin, Dianne (0016-19-1) (0016-19-8) (0016-19-9) (0016-19-11) • Villarreal, Tasha (0003-7) • Zeller, Lou (0009-2) (0009-3) (0009-6) (0009-7)
Health-Nonradiological	• Mueller, Heinz J. (0019-10)
Health-Radiological	• Carroll, Glenn (0016-11-1) • Henson, Courtney (0016-15-4) • Michetti, Susan (0007-9) • Ogley-Oliver, Emma (0016-16-3)

Appendix E

Table E-2. (contd)

Comment Category	Commenter (Comment ID)
	• Paul, Bobbie (0016-7-4) • Rivard, Betsy (0016-20-5) (0016-20-6) (0016-20-8) (0016-20-10) (0016-20-11) • Stephens, Annie Laura (0016-10-3) (0016-10-4) • Taylor, F (0004-5) • Utley, Charles (0016-6-6) (0016-6-8) (0016-6-9) (0016-6-10) • Valentin, Dianne (0016-19-3) (0016-19-5) (0016-19-6) (0016-19-7) • Vincent, Patricia (0016-18-2)
Historic and Cultural Resources	• Celestine, Bryant (0012-1) • Mueller, Heinz J. (0019-12) • Valentin, Dianne (0016-19-12)
Hydrology-Groundwater	• Rivard, Betsy (0016-20-9) • Utley, Charles (0016-6-2) • Valentin, Dianne (0016-19-2) (0016-19-4)
Hydrology-Surface Water	• Arnold, Judy (0003-4) • Barczak, Sara (00160-5-4) (0016-5-9) (0017-1-4) (0017-1-8) • Boatenreiter, Glenn (0003-4) • Booher, Sam (0016-21-1) (0016-21-2) (0016-21-3) (0016-21-4) (0016-21-5) (0016-21-6) (0016-21-7) • Carroll, Glenn (0002-4) • Carter, Pat (0003-4) • Cumbow, Kay (0003-4) • Dawson, Daneille (0003-4) • Dooley, Gerald (0003-4) • Falconer, Kimberly (0003-4) • Hatch, Sarah (0003-4) • Kasenow, Lisa (0003-4) • Kushner, Adele (0005-4) • Lewis, Marvin (0003-4) • Lomas, Judith (0003-4) • Lusk, Phil (0003-4) • McConnell, Joy (0003-4) • McNulty, Joy (0003-4) • Michetti, Susan (0007-6) • Mills, Nancy (0003-4) • Mueller, Heinz J. (0019-1) (0019-8) • Pierce, Charles (0014-1) • Sardi, David (0016-22-3) (0016-22-10) (0016-22-13) • Sheppard, Deborah (0003-4) • Taylor, F (0004-4) • Thomas, Russel (0003-4)

Appendix E

Table E-2. (contd)

Comment Category	Commenter (Comment ID)
	• Trujillo, Dianne (0003-4) • Utley, Charles (0016-6-1) • Villarreal, Tasha (0003-4)
Land Use-Transmission Lines	• Mueller, Heinz J. (0019-7)
Meteorology and Air Quality	• Arnold, Judy (0003-9) • Boatenreiter, Glenn (0003-9) • Carroll, Glenn (0002-9) • Carter, Pat (0003-9) • Cumbow, Kay (0003-9) • Dawson, Daneille (0003-9) • Dooley, Gerald (0003-9) • Falconer, Kimberly (0003-9) • Hatch, Sarah (0003-9) • Howard, Claude (0016-17-1) • Kasenow, Lisa (0003-9) • Kushner, Adele (0005-9) • Lewis, Marvin (0003-9) • Lomas, Judith (0003-9) • Lusk, Phil (0003-9) • McConnell, Joy (0003-9) • McNulty, Joy (0003-9) • Michetti, Susan (0007-16) • Mills, Nancy (0003-9) • Mueller, Heinz J. (0019-9) (0019-14) (0019-15) • Sheppard, Deborah (0003-9) • Thomas, Russel (0003-9) • Trujillo, Dianne (0003-9) • Villarreal, Tasha (0003-9)
Need for Power	• Parr, Sue (0016-12-3)
Opposition-Licensing Action	• Arnold, Judy (0003-1) • Barczak, Sara (0016-5-1) (0016-5-10) (0016-5-11) (0017-1-1) (0017-1-9) (0017-1-10) • Baxter, Farouk (0001-1) • Boatenreiter, Glenn (0003-1) • Carroll, Glenn (0002-1) • Carter, Pat (0003-1) • Cumbow, Kay (0003-1) • Dawson, Daneille (0003-1) • Dooley, Gerald (0003-1) • Falconer, Kimberly (0003-1) • Hatch, Sarah (0003-1)

Appendix E

Table E-2. (contd)

Comment Category	Commenter (Comment ID)
	• Henson, Courtney (0016-15-1) (0016-15-7) • Kasenow, Lisa (0003-1) • Lewis, Marvin (0003-1) • Lomas, Judith (0003-1) • Lusk, Phil (0003-1) • McConnell, Joy (0003-1) • McNulty, Joy (0003-1) • Mills, Nancy (0003-1) • Ogley-Oliver, Emma (0016-16-4) • Patrie, Lewis E. (0013-4) • Paul, Bobbie (0016-7-2) • Sardi, David (0016-22-1) • Sheppard, Deborah (0003-1) • Stephens, Annie Laura (0016-10-1) • Taylor, F (0004-1) • Thomas, Ellen (0008-6) • Thomas, Russel (0003-1) • Trujillo, Dianne (0003-1) • Villarreal, Tasha (0003-1) • Zeller, Lou (0009-4)
Opposition-Licensing Process	• Carroll, Glenn (0016-11-7) • Kushner, Adele (0005-1) • Sardi, David (0016-22-15)
Opposition-Nuclear Power	• Barczak, Sara (0016-5-8) (0017-1-7) • Carroll, Glenn (0016-11-8) • Howard, Claude (0016-17-3) • Michetti, Susan (0007-1) (0007-7) (0007-15) • Ogley-Oliver, Emma (0016-16-6) • Sardi, David (0016-22-5) (0016-22-12) • Smith, Nathan (0006-1) • Vincent, Patricia (0016-18-1)
Outside Scope-Emergency Preparedness	• Paul, Bobbie (0016-7-6)
Outside Scope-Miscellaneous	• Howard, Claude (0016-17-2) • Paul, Bobbie (0016-7-1) • Rivard, Betsy (0016-20-2) • Sardi, David (0016-22-9) • Utley, Charles (0016-6-5)
Outside Scope-NRC Oversight	• Byne, Dick (0016-2-1) • DeLoach, George (0016-1-4)

Appendix E

Table E-2. (contd)

Comment Category	Commenter (Comment ID)
Outside Scope-Safety	• Arnold, Judy (0003-3) • Barczak, Sara (0016-5-3) (0017-1-3) • Baxter, Farouk (0001-2) (0001-3) • Boatenreiter, Glenn (0003-3) • Carroll, Glenn (0002-3) • Carter, Pat (0003-3) • Cumbow, Kay (0003-3) • Dawson, Daneille (0003-3) • Dooley, Gerald (0003-3) • Falconer, Kimberly (0003-3) • Hatch, Sarah (0003-3) • Henson, Courtney (0016-15-2) (0016-15-6) • Kasenow, Lisa (0003-3) • Kushner, Adele (0005-3) • Lewis, Marvin (0003-3) • Lomas, Judith (0003-3) • Lusk, Phil (0003-3) • McConnell, Joy (0003-3) • McNulty, Joy (0003-3) • Michetti, Susan (0007-11) • Mills, Nancy (0003-3) • Mueller, Heinz J. (0019-5) • Rivard, Betsy (0016-20-1) • Sheppard, Deborah (0003-3) • Stone, Jesse (0016-14-2) • Taylor, F (0004-3) • Thomas, Russel (0003-3) • Trujillo, Dianne (0003-3) • Villarreal, Tasha (0003-3)
Outside Scope-Security and Terrorism	• Sardi, David (0016-22-2) (0016-22-6)
Process-ESP-COL	• Parr, Sue (0016-12-5) • Paul, Bobbie (0016-7-3) • Rivard, Betsy (0016-20-3) (0016-20-4) • Valentin, Dianne (0016-19-10) (0016-19-13)
Process-NEPA	• Mueller, Heinz J. (0019-6) • Zeller, Lou (0009-1) • Vejdani, Vivianne (0011-1)
Socioeconomics	• Baxley, Robin (0016-9-1) (0016-9-4) • Carroll, Glenn (0016-11-3)

Appendix E

Table E-2. (contd)

Comment Category	Commenter (Comment ID)
	• Elam, Terry (0016-4-2) (0016-4-4) (0016-4-5) (0016-4-7) (0018-1-2) (0018-1-4) (0018-1-6) • Mitchell, Tommy (0016-8-2) • Parr, Sue (0016-12-6) • Rivard, Betsy (0016-20-7) • Roberts, Ashley (0016-13-4) (0016-13-5) (0016-13-6) • Sardi, David (0016-22-11) • Stone, Jesse (0016-14-1) (0016-14-7)
Support-Licensing Action	• Byne, Dick (0016-2-2) (0016-2-4) • DeLoach, George (0016-1-2) • Elam, Terry (0016-4-1) (0016-4-3) (0016-4-6) (0018-1-1) (0018-1-3) (0018-1-5) • Parr, Sue (0016-12-1) • Roberts, Ashley (0016-13-1) (0016-13-2) (0016-13-8)
Support-Licensing Process	• Abrams, Lucious (0016-3-2) • DeLoach, George (0016-1-3) • Parr, Sue (0016-12-2) (0016-12-4) • Stone, Jesse (0016-14-6) (0016-14-8)
Support-Nuclear Power	• Baxley, Robin (0016-9-3) • Stone, Jesse (0016-14-4)
Support-Plant	• Abrams, Lucious (0016-3-1) • Baxley, Robin (0016-9-2) • Byne, Dick (0016-2-3) • DeLoach, George (0016-1-1) • Mitchell, Tommy (0016-8-1) • Roberts, Ashley (0016-13-7) • Stephens, Annie Laura (0016-10-2) • Stone, Jesse (0016-14-3) (0016-14-5)
Transportation	• Mueller, Heinz J. (0019-4)
Uranium Fuel Cycle	• Barczak, Sara (0016-5-5) • Michetti, Susan (0007-8) (0007-12) (0007-13) (0007-14) • Mueller, Heinz J. (0019-3) • Ogley-Oliver, Emma (0016-16-2) • Utley, Charles (0016-6-11)

E.2.1 Comments Concerning Process – COL

Comment: So I'm looking at the notes that you gave in your PowerPoints and it's smart for you to put something like new and significant because nine times out of ten, the people in the

Appendix E

community who are already really oppressed both financially and socially are not going to be able to provide you with new and significant information. (**0016-19-10** [Valentin, Dianne])

Comment: I turned the page and it says how impacts are quantified and you have small, moderate and large. And having worked in that community and met with a lot of people, watched people die from painful cancers in that community, I wondered how people feel -- you know, they're watching their friends and their families in hospital beds, if they consider the tubes and the death a moderate effect, a large effect or a small effect? I'm not understanding how you're making these determinations when you haven't come into the community and talked to the people. (**0016-19-13** [Valentin, Dianne])

Comment: The Supplemental EIS is difficult to comment on, as I said, because it mostly says the staff is not aware of any new and significant site-specific or reactor-specific information, blah, blah, blah. And therefore, our conclusion remains valid. The problem is the NRC is dependent on Southern Company to provide that information, which I think is a little strange. They're the ones to provide the new and significant information. And so it just seems odd that they would be the ones to provide it. Would the NRC be talking to people in Finland that are still waiting for the AP1000? (**0016-20-4** [Rivard, Betsy])

Response: The environmental review conducted by NRC at the COL stage is informed by the EIS prepared at the ESP stage, and that previous review is incorporated by reference in the COL SEIS.

Pursuant to 10 CFR 52.39, matters resolved in the ESP proceedings are considered to be resolved in any subsequent proceedings, absent identification of new and significant information. Consequently, the focus in the environmental review of a COL application referencing an ESP is on identifying developments since the ESP review that have significance for the conclusions previously reached. Accordingly, as required by NRC regulations, the COL SEIS for the proposed VEGP Unit 3 and 4 focuses on new and significant information identified after issuance of the final ESP EIS (NRC 2008a). Furthermore, as explained in Section 1.6 of the SEIS, while Southern submitted its own assessment of whether new and significant information had been identified, the NRC staff independently considered Southern's process and other available information, and determined for itself whether there was new and significant information that warranted further analysis.

The SMALL, MODERATE and LARGE significance levels are used by the review team after completing its analyses to communicate the results of its assessment of the environmental impacts of the proposed action and alternative to the action. The structure for the significance levels was based on Council on Environmental Quality (CEQ) guidance (40 CFR 1508.27) and on discussions with the CEQ and the EPA when it was first implemented for licensing actions. Definitions of the three significance levels are provided in Table B-1 of 10 CFR Part 51, Subpart A, Appendix B, and are provided in Section 1.1.1.1 of this SEIS. These comments did not result in changes to the SEIS.

Appendix E

Comment: And so I would suggest that the NRC in all its deliberations and all the things before you from the intervenors and companies and whatever, get to know the people, not just what we say here. (**0016-7-3** [Paul, Bobbie])

Comment: While the construction and operation of the new units is certain to impact the environment and people amongst whom it is built, the Draft Supplemental EIS provides a thorough consideration of those impacts and recommends that the positive impacts justify continued construction and licensure. (**0016-12-5** [Parr, Sue])

Comment: But it's just a little unnerving to hear about early site permits, combined operating licenses, et cetera. It implies that everything is kind of in flux, there's no set design and I don't really know too many people that would build a house without a design. And I have heard that things are being built. I mean it's not just a flat level piece of dirt -- I don't think so anyway. (**0016-20-3** [Rivard, Betsy])

Response: These comments express general views regarding the COL licensing process. These comments provide no new and significant information. Therefore, no changes were made to the SEIS.

E.2.2 Comments Concerning Process – NEPA

Comment: However, the NRC's definitions of new and significant are either outside the meaning of the statutory definition or wholly absent from NEPA and are, therefore, artificial and improper limitations on the extant NEPA proceeding. (**0009-1** [Zeller, Lou])

Response: This comment expresses opposition to NRC's process for the environmental review of a COL application referencing an ESP. 10 CFR 51.92 reflects NRC's obligation under NEPA to address new and significant information for a COL that references an ESP. As outlined in the Federal Register notice of August 28, 2007 (72 FR 49429), the NRC rules indicate that issuance of an ESP and a COL are major Federal actions significantly affecting the quality of the human environment and that each action would require the preparation of an EIS. However, 10 CFR Part 52 does provide finality for issues previously resolved in an ESP proceeding. Thus, the environmental review conducted by NRC at the COL stage is informed by the EIS that was prepared at the ESP stage, and information from the ESP review can be incorporated by reference in the COL SEIS. The COL SEIS for the proposed Units 3 and 4 focuses on new and significant information identified after issuance of the final ESP EIS (NRC 2008a) and ensures that all environmental terms and conditions included in the ESP relevant to the COL will be satisfied by the date of issuance of the COL SEIS. No change was made to the SEIS as a result of this comment.

Comment: We note that the NRC considers transmission lines to be "preconstruction" activities (discussed in the EIS for the ESP), and that preconstruction activities are considered in the context of cumulative impacts. EPA is concerned about the impacts of transmission lines and supporting infrastructure for the project and, in accordance with NEPA, considers these activities as part of the project, and not a separate action. (**0019-6** [Mueller, Heinz J.])

Appendix E

Response: Because the Vogtle COL application references the Vogtle ESP, the NRC staff's environmental review focuses primarily on whether any new and significant information has been identified with respect to the impacts previously discussed in the ESP EIS (see, 10 CFR 51.50(c)(1), 51.92(e)(7)). The proposed extent of the potential new transmission line right-of-way (ROW) is discussed in section 4.1.2 of the ESP EIS, and the impacts of the new transmission ROW are discussed in Chapters 4 and 5 of the ESP EIS (NRC 2008a). With respect to impacts associated with the transmission line ROW, no significant new information was identified during the preparation of the SEIS. Under NRC regulations, preconstruction activities such as the building of transmission lines are excluded from the definition of "construction" because they are outside the NRC's regulatory jurisdiction and are not authorized by NRC's licensing action (see 10 CFR 50.10(a); 72 FR 57416 (2007)). The Commission has therefore explained that the impacts of those activities are to be analyzed in the environmental review for a COL application, but in the context of cumulative impacts (see, 72 FR 57421). The comment did not provide new and significant information. Therefore, no change was made to the SEIS.

Comment: Personnel with the South Carolina Department of Natural Resources (DNR) have reviewed the above-referenced Supplement Environmental Impact Statement for the Vogtle Electric Generating Plant Units 3 and 4. DNR has no comment at this time. (**0011-1** [Vejdani, Vivianne])

Response: The NRC acknowledges the South Carolina DNR review of the draft SEIS through the NEPA process. There was no change to the SEIS as a result of this comment.

E.2.3   Comments Concerning Land Use – Transmission Lines

Comment: The DSEIS (pages 3-7 and 3-8) discusses the construction of a new transmission line through a "macro-right-of-way." This term should be defined in the text, with details given regarding the proposed extent and impacts of this new transmission line. The FSEIS should also clarify whether there are plans to issue a Limited Work Authorization (LWA) for these lines pursuant to the NRC's LWA process. (**0019-7** [Mueller, Heinz J.])

Response: Sections 2.7.1 and 3.3 of the SEIS have been updated to clarify the definition of the transmission line Representative Delineated Corridor that was the focus of the staff's analysis of transmission line impacts in the ESP EIS. Construction of the new transmission line ROW would not require an LWA issued by NRC because construction of new transmission facilities is not considered construction under NRC regulations (10 CFR 51.4).

Appendix E

E.2.4 Comments Concerning Meteorology and Air Quality

Comment: COMPARISONS AND ALTERNATIVES
Table 7-1 compares only three energy types: coal, oil and nuclear. The table must be revised to compare also wind, solar, conservation and efficiency. In the likelihood that NRC does not feel qualified to make the assessment on alternative energy, an outside contractor should conduct the study. The study should include the DOE and Oceana reports as well as IEER's Carbon-Free and Nuclear-Free report. As stated above, the cost savings and benefits of readily available alternatives must be contrasted with long-range, high-risk speculation on new reactor build. Georgia's excellent potential for offshore wind must be incorporated into the EIS.
(**0003-9** [Arnold, Judy] [Boatenreiter, Glenn] [Carter, Pat] [Cumbow, Kay] [Dawson, Daneille] [Dooley, Gerald] [Falconer, Kimberly] [Hatch, Sarah] [Kasenow, Lisa] [Lewis, Marvin] [Lomas, Judith] [Lusk, Phil] [McConnell, Joy] [McNulty, Joy] [Mills, Nancy] [Sheppard, Deborah] [Thomas, Russel] [Trujillo, Dianne] [Villarreal, Tasha]); (**0002-9** [Carroll, Glenn])

Comment: COMPARISONS AND ALTERNATIVES
Table 7-1 compares only three energy types: coal, oil and nuclear. The table must be revised to compare also wind, solar, conservation and efficiency. In the likelihood that NRC does not feel qualified to make the assessment on alternative energy, an outside contractor should conduct the study. The study should include the DOE and Oceana reports as well as IEER's Carbon-Free and Nuclear-Free report. As stated above, the cost savings and benefits of readily available alternatives must be contrasted with long-range, high-risk speculation on new reactor build. Georgia's excellent potential for offshore wind must be incorporated into the EIS. Rather than continue with this flawed and dangerous model, which puts at immediate risk the community which lives close to the existing reactors, the only responsible action is to study and evaluate alternatives in the studies by NC WARN, Arjun Makhijani, SACE, DOE, and OCEANA. Otherwise we would be repeating the wasteful and dangerous use of valuable resources particularly people's lives which were spent in building the first two reactors. (**0005-9** [Kushner, Adele])

Comment: Another reason not to permit new construction is that it contributes to global warming from cradle to grave with an mean of 66 grams CO_2 equivalent per kWh. The highest renewable energy has a mean of 41 grams CO_2 equivalent per kWh life cycle. We must conserve CO2 emissions everywhere possible, and new plant construction with too many other disadvantages doesn't pass the acceptable test. (**0007-16** [Michetti, Susan])

Response: Table 7-1 in this SEIS provides estimates of CO_2 emission rates for a variety of sources, including coal-fired, natural-gas-fired, and nuclear power plants. The table is intended to provide context for comparing CO_2 emission rates from base-load power sources evaluated in Section 9.2 of the ESP EIS (NRC 2008a). As discussed in Section 9.2 of the ESP EIS, the NRC staff concluded that wind, solar, and other alternatives not requiring new generating capacity were not reasonable alternatives for base-load power generation. Because these comments did not provide new and significant information, no change was made to the SEIS.

Appendix E

Comment: Sometimes I wonder why is black smoke coming out of the stacks sometimes? My question is what's going on out there polluting and poisoning people within the community? (**0016-17-1** [Howard, Claude])

Response: Air quality impacts are addressed in Section 5.2 of the ESP EIS (NRC 2008a). As stated in Section 5.2, Units 3 and 4 would have standby auxiliary diesel generators and boilers that will be used on an infrequent basis, and the pollutants discharged will be permitted in accordance with the Georgia Department of Natural Resources. This comment appears to be referring to the existing units at the Vogtle site and does not provide any specific information relating to environmental effects of the proposed action. Because this comment did not provide new and significant information, no change was made to the SEIS.

Comment: EPA also recommends a discussion of best management practices to reduce GHGs and other air emissions during construction and operation of the facility. Specifically, clean energy options such as energy efficiency and renewable energy should be a consideration in the use of construction and maintenance equipment and vehicles. For example, equipment and vehicles that use conventional petroleum (e.g., diesel) should incorporate clean diesel technologies and fuels to reduce emissions of GHGs and other pollutants, and should adhere to anti-idling policies to the extent possible. Alternate fuel vehicles (e.g., natural gas, electric) are also possibilities. (**0019-9** [Mueller, Heinz J.])

Response: Measures to be taken to reduce construction-related pollutant emissions, such as maintaining equipment in good operating condition and developing a construction management traffic plan, are discussed in Sections 4.2 and 4.8 of the ESP EIS (NRC 2008a). Such measures will generally reduce greenhouse gas emissions (GHG). Section 7.2 of this SEIS provides comparative estimates of GHG emissions associated with nuclear, coal, and natural gas base-load power generation. As discussed in that section, the greatest contribution to GHG emissions from nuclear power plants is associated with the nuclear fuel cycle; contributions from construction, operations, or decommissioning are comparatively less. The staff also determined in that section of the SEIS that the conclusion from the ESP EIS (NRC 2008a) regarding GHG impacts remain valid. Nevertheless, the staff agrees that best management practices, such as using clean diesel technologies or alternative fuel vehicles, keeping equipment in good working order, and reducing idling time, could be implemented to reduce GHG emissions associated with construction and operation of the plant. No change was made to the SEIS as a result of this comment.

Comment: CEQ Draft Guidance on GHG Analysis within NEPA: On February 18, 2010, the Council on Environmental Quality (CEQ) proposed four steps to modernize and reinvigorate NEPA. In particular, the CEQ issued draft guidance for public comment on, among other issues, when and how Federal agencies must consider greenhouse gas emissions and climate change in their proposed actions. (Reference: http://www.whitehouse.gov/administration/epa)

The draft guidance explains how Federal agencies should analyze the environmental impacts of greenhouse gas emissions and climate change when they describe the environmental impacts of a proposed action under NEPA. It provides practical tools for agency reporting, including a

Appendix E

presumptive threshold of 25,000 metric tons of carbon dioxide equivalent (CO_2e) emissions from the proposed action to trigger a quantitative analysis, and instructs Federal agencies regarding how to assess the effects of climate change on the proposed action and their design. The draft guidance does not apply to land and resource management actions and does not propose to regulate greenhouse gases.

While this guidance is not yet final (and thus, not required), we recommend that the FSEIS explicitly reference the draft guidance, describe the elements of the draft guidance, and to the relevant extent, provide the assessments suggested by the guidance. (**0019-14** [Mueller, Heinz J.])

Response: Discussion of GHG emissions and climate change in this SEIS was guided by the focus on new and significant information subsequent to the issuance of the ESP EIS (NRC 2008a). As noted in Section 7.2 of the SEIS, although the staff considered GHG emissions in the ESP EIS and the issue is therefore not new, the staff re-examined its previous analysis to demonstrate conformance with the November 3, 2009, Commission Order CLI-09-21 (NRC 2009). This order is consistent with the objectives of the CEQ guidance by directing NRC staff to include consideration of GHG emissions from operation, construction, and the uranium fuel cycle in environmental reviews under NEPA. Because this comment did not provide new and significant information, no change was made to the SEIS.

Comment: (Note that the discussion in Section 7.2 and referencing the Sovacool paper (see footnote 1 below) regarding the derivation of 447,000 metric tons/year of CO2 emissions from a 1000 MW nuclear power plant is difficult to follow. For example, we could not find the "1 percent to 5 percent" citation noted as being in the Sovacool paper. It would be helpful to show a detailed derivation of the amount of direct and indirect CO2-equivalent emissions expected specifically from this project.) (**0019-15** [Mueller, Heinz J.])

Response: Section 7.2 of the SEIS has been modified to clarify the discussion regarding estimates of CO_2 emissions from nuclear power plants. With respect to derivation of project-specific emissions, the level of detail in the staff's discussion of GHG emissions and climate change in this SEIS was guided by the focus on new and significant information subsequent to the issuance of the ESP EIS (NRC 2008a). Moreover, as discussed in Section 7.2, the vast majority of GHG emissions associated with nuclear power generation as reported in Table 7-1 are from the uranium fuel cycle.

E.2.5 Comments Concerning Hydrology – Surface Water

Comment: The draft EIS fails to analyze the continuance of an historic 10-year drought already impacting the Savannah River Basin. The Savannah River is currently the fourth most polluted body of water in the U.S. The two reactors at Vogtle already withdraw over 68 million gallons of water each day, more than the combined daily usage of Atlanta, Savannah and Augusta.

Not all of the water withdrawn is returned to the river, much of it is lost as water vapor, a greenhouse gas, a point which also is not analyzed in the draft EIS. The water that the reactors return to the river will be hotter than the water which was withdrawn. Two-thirds of the heat

Appendix E

generated in the nuclear reactor will not be turned into electricity, but will be vented to the local environment as waste heat contributing to drought conditions. This environmental impact has not been analyzed in the draft EIS. (**0002-4** [Carroll, Glenn])

Comment: The draft EIS fails to analyze the continuance of an historic 10-year drought already impacting the Savannah River Basin. The Savannah River is currently the fourth most polluted body of water in the U.S. The two reactors at Vogtle already withdraw over 68 million gallons of water each day, more than the combined daily usage of Atlanta, Savannah and Augusta.

- Not all of the water withdrawn is returned to the river, much of it is lost as water vapor, a greenhouse gas, a point which is not analyzed in the draft EIS.

- The water that the reactors return to the river will be hotter than the water which was withdrawn.

- Two-thirds of the heat generated in the nuclear reactor will not be turned into electricity, but will be vented to the local environment as waste heat contributing to drought conditions. This environmental impact has not been analyzed in the draft EIS. (**0004-4** [Taylor, F])

Comment: ENVIRONMENTAL IMPACTS The draft EIS fails to analyze the continuance of an historic 10-year drought already impacting the Savannah River Basin. The Savannah River is currently the fourth most polluted body of water in the U.S. The two reactors at Vogtle already withdraw over 68 million gallons of water each day, more than the combined daily usage of Atlanta, Savannah and Augusta. (**0003-4** [Arnold, Judy] [Boatenreiter, Glenn] [Carter, Pat] [Cumbow, Kay] [Dawson, Daneille] [Dooley, Gerald] [Falconer, Kimberly] [Hatch, Sarah] [Kasenow, Lisa] [Lewis, Marvin] [Lomas, Judith] [Lusk, Phil] [McConnell, Joy] [McNulty, Joy] [Mills, Nancy] [Sheppard, Deborah] [Thomas, Russel] [Trujillo, Dianne] [Villarreal, Tasha]); (**0005-4** [Kushner, Adele])

Comment: Citizens were not allowed to use water to water organic gardens in Georgia during recent droughts, while the nuclear plants were not shut down during the same droughts. It is unacceptable the nuclear plants in Georgia are already diverting water use away from citizens in order to not overheat and jeopardize public health. (**0007-6** [Michetti, Susan])

Comment: Additionally, since four years ago, this region suffered through a severe drought and the reliability of existing nuclear plants were tested, and there were failures then that have continued even through this year. The powering back or shutting down of TVA's Browns Ferry reactors along the Tennessee River in Alabama, for example. And yet somehow the NRC is able to recommend approving the combined operating license for Vogtle even though the reactor design that Southern Company intends to build here has yet to be approved and water concerns remain and other issues are yet to be resolved. (**0016-5-4** [Barczak, Sara]); (**0017-1-4** [Barczak, Sara])

Comment: Does the NRC even care that if Plant Vogtle is expanded less water will be available in the Savannah River for other users both upstream and downstream? People have heard me state this statistic before, but I'm going to do it again tonight. To put the consumptive water loss in perspective from Plant Vogtle -- that is the water that does not go back into the

Appendix E

river -- with average per capita daily water use in Georgia at 75 gallons from surface and groundwater sources, this means the two existing and two proposed reactors could use enough water to supply 1.4 to 2.3 million Georgians. Somehow the NRC thinks that is a small impact. Read the EIS, they consider it a small impact. We disagree and we believe that the future communities upstream and downstream of the plant will vehemently disagree as climate change impacts are observed and droughts get longer and more severe and everyone is fighting over water. But it'll be too late by then. (**0016-5-9** [Barczak, Sara])

Comment: ...the high water that's being consumed by the plant is just astronomical. (**0016-6-1** [Utley, Charles])

Comment: The operating permit must address drought conditions, when are they critical and what are the limits. (**0016-21-7** [Booher, Sam])

Comment: It's my belief that granting this permit will ...endanger our water supply, especially in the face of global warming. (**0016-22-3** [Sardi, David])

Comment: ...I ask you to consider the power plant's impact on our water resources. Power plants require both certain temperature and enormous quantities as was just said, and the new plants will make a large impact on our water supply in the future, which is expected to be a lot more limited. Electricity supplies threaten the water resources that's an important aspect of the region -- tourism, agriculture, fishing industries and sensitive biodiversity. (**0016-22-10** [Sardi, David])

Comment: France, as has already been mentioned today, generates the majority of their energy from nuclear energy, has already been forced to shut down power plants days at a time for these reasons [water temperature increases and droughts]... (**0016-22-13** [Sardi, David])

Comment: Does the NRC even care that if Plant Vogtle is expanded less water will be available in the Savannah River for other users both upstream and downstream? To put this consumptive water loss in perspective from Plant Vogtle, with average per capita daily water use in Georgia at 75 gallons from surface and ground water sources, this means the two existing and two proposed reactors could use enough water to supply 1.4 to 2.3 million Georgians. Somehow the NRC thinks that is a "small" impact. We disagree and we believe that the future communities upstream and downstream of the plant will vehemently disagree as climate change impacts are observed and droughts get longer and more severe and everyone is fighting over water. But it'll be too late by then. (**0017-1-8** [Barczak, Sara])

Response: *The impacts of Savannah River water use during drought flows were evaluated in ESP EIS Section 5.3.2.1 (NRC 2008a). The cumulative impacts associated with existing water use by VEGP Units 1 and 2, including water use during drought conditions, were evaluated in ESP EIS Section 7.3. The thermal effects of plant discharge to the Savannah River were evaluated in ESP EIS Section 5.3.3.1 as well as in Section 5.3 of this SEIS. Cooling tower impacts were evaluated in ESP EIS Section 5.2.1 (NRC 2008a). Because these comments did not provide new and significant information, no change was made to the SEIS.*

Appendix E

Comment: The problem is that in the summer with low flows and the river water a lot warmer, Plant Vogtle needs a lot more water just to cool the current two reactors efficiently. I would offer that during low flow and drought conditions, 83 million gallons will not be sufficient to cool the water for all four reactors. (**0016-21-1** [Booher, Sam])

Comment: The second problem is that I believe federal law requires Georgia Power to keep track of the temperature and quantity of the river water they remove and record the temperature and quantity of the water being discharged back into the river. Equally important is the need to keep the NRC and the public informed of this information. (**0016-21-2** [Booher, Sam])

Comment: When the Tennessee Valley Authority, TVA, finds the Tennessee River temperature is too warm, TVA is required to reduce the energy production. Why, with the current two and soon to be four reactors will Georgia Power not be required to monitor the Savannah River water temperature they remove from the river? I can see 160 million gallons of very warm river water needed to cool all four reactors. (**0016-21-3** [Booher, Sam])

Comment: The Savannah River needs to be allowed to retain some dissolved oxygen for Savannah. During drought conditions four unconstrained reactors will not allow sufficient dissolved oxygen downstream.

The problem is that the current Georgia Power operating permit from EPD, the current Vogtle permit, does not have a requirement for anyone to keep track of how much Savannah River water and its daily temperature of that water is removed from the river. Nor is there any daily record of the amount of water and its temperature being discharged back into the Savannah River provided EPD, NRC or the public. (**0016-21-4** [Booher, Sam])

Response: Pursuant to the Clean Water Act, the EPA and the U.S. Army Corps of Engineers have authority to require water quality monitoring for nonradiological material in the waters of the United States. The NRC has no authority to place water monitoring requirements on any facility, except for radiological monitoring. Withdrawals from and discharge to the Savannah River are governed by state permits as described in ESP EIS Section 5.3 (NRC 2008a). Because these comments did not provide new and significant information, no change was made to the SEIS.

Comment: Last, I read paragraph 5.3 in the Draft Supplemental EIS. My understanding is that your staff's result is from modeling and not real water withdrawal and discharge or actual on-site data. Also, your modeling does averaging and it's not based on low flow and drought conditions, the water temperature, which is my only concern. The law says the returning water cannot be more than five degrees greater than the temperature of the original water as withdrawn from the river.

We will be having more drought conditions before reactors 3 and 4 go back on line. You need to check your data under these conditions, not averaging. (**0016-21-5** [Booher, Sam])

Appendix E

Comment: Since federal law requires this information of TVA, why is EPD allowed to issue permits to Georgia Power without following the same legal daily water temperature requirements? I do not believe the law talks about diluting plumes out in the river. It is my understanding that EPD allows periodic testing of water temperature downstream from the discharge point. I offer these diluted plumes in the river as nothing more than a way to get around federal law and it should be reviewed by your office. (0016-21-6 [Booher, Sam])

Response: Pursuant to the Clean Water Act, the EPA and the U.S. Army Corps of Engineers have authority to require water quality monitoring for nonradiological material in the waters of the United States. The NRC has no authority to place water quality monitoring requirements on any facility, except for radiological monitoring. Withdrawals from and discharge to the Savannah River are governed by state permits as described in ESP EIS Section 5.3 (NRC 2008a). Water quality impacts, including the thermal effects of plant discharge to the Savannah River under normal and low flow conditions, were evaluated in ESP EIS Section 5.3.3.1 as well as in Section 5.3 of this SEIS. Because these comments did not provide new and significant information, no change was made to the SEIS.

Comment: Title IV, Permits and Licenses section of the Federal Water Pollution Control Act requires "any applicant for a Federal license or permit to conduct any activity including, but not limited to, the construction or operation of facilities, which may result in any discharge into the navigable waters, [to] provide the licensing or permitting agency a certification from the State" indicating that the discharges will comply with applicable provisions in the Federal Water Pollution Control Act. This requirement was included in Sections 1.5 and 4.3, and Table H-1 of the VEGP Draft SEIS and states that a 401 Water Quality Certification be received from the Georgia Department of Natural Resources (GDNR) to support the VEGP Units 3 and 4 COL process. On June 1, 2010, the GDNR issued a 401 Water Quality Certification (JPN 200701837) to SNC for the VEGP Units 3 and 4 site. Please find enclosed as Attachment 1 a copy of the 401 Water Quality Certification.

Furthermore, as described in Condition #3 of the Certification, the 401 Water Quality Certification will be effective once the GDNR has issued a Stream Buffer Variance for the project. On April 29, 2010, the GDNR issued a Stream Buffer Variance to SNC for the VEGP Units 3 and 4 site. Please find enclosed as Attachment 2 a copy of the Stream Buffer Variance.

SNC requests that Sections 1.5 and 4.3, and Table H-1 be revised to indicate that a 401 Water Quality Certification has been issued by the GDNR for VEGP Units 3 and 4. (0014-1 [Pierce, Charles])

Response: Sections 1.5 and 4.3 and Table H-1 were modified to reflect receipt of the 401 Water Quality Certification (Southern 2010).

Comment: The Final Supplemental Environmental Impact Statement (FSEIS) should include a graph of the plume showing the temperature profile, and a discussion of how the increase will (or will not) cause a violation of Georgia's water quality standard for temperature at the point of discharge.

Appendix E

In addition, the design and location of the proposed new cooling water intake structure has changed. The NRC determined that this new location would not alter conclusions presented in the previous ESP FEIS. (**0019-1** [Mueller, Heinz J.])

Comment: Southern indicated that there would be an operations-related three percent increase in the thermal discharge flow. The NRC determined that the thermal plume would remain small compared to the width of the Savannah River at this location, and that it would not impede fish passage in the river (Section 5.4.2). In addition, the design and location of the proposed new cooling water intake structure has changed. The NRC determined that this new location would not alter conclusions in the previous ESP FEIS. Pursuant to our review, the following areas need clarification:

- Temperature: The discussion of the 3% increase in the thermal discharge should include a graph of the plume showing the temperature profile, and a discussion of how the increase will (or will not) cause a violation of Georgia's water quality standard for temperature at the point of discharge.

- Cooling Water Intake: For clarity, the FSEIS should restate the requirements for the cooling water intake structure. (**0019-8** [Mueller, Heinz J.])

Response: Requirements of the cooling water intake structure were described in ESP EIS Section 5.4.2.2 (NRC 2008a). These requirements under the Clean Water Act include a through-screen velocity of 0.2 m/s (0.5 ft/s) and withdrawal of less than 5 percent of the source water body mean annual flow (66 FR 65256). As described in Section 5.4.2 of this SEIS, there was no change in these requirements and the changes to the design and location of the cooling water intake structure, described in Section 3.2.2, did not result in a change to the environmental impacts.

The NRC staff believes the comment may reflect a misconception with respect to the discharge. A 3 percent increase in the discharge flow was evaluated in Section 5.3. As noted in Table I-1 of Appendix I, however, the cooling water system cooling tower blowdown temperature was unchanged from the value reported in the ESP EIS. The maximum temperature of the blowdown used in the evaluation of the thermal plume in Section 5.3 was therefore identical to that used in the ESP EIS (NRC 2008a). Section 5.4.2 was modified to clarify that the aquatic impacts evaluated were from the 3 percent increase in discharge flow, not an increase in discharge temperature.

An illustration of the size of the thermal plume resulting from the conservative analysis described in ESP EIS 5.3.3.1 was provided in ESP EIS Figure 5-1 (NRC 2008a). The increase in the size of this plume resulting from the 3 percent increase in discharge flow was quantified in Section 5.3 of this SEIS.

Clarifying changes were made to Section 5.4.2 and a figure (Figure 5-1) was added as a result of these comments.

Appendix E

E.2.6 Comments Concerning Hydrology – Groundwater

Comment: And when you think about it, you know, there are farmers who are going around now putting wells down and they're going deeper and deeper and deeper because as Sara said earlier, there is a drought. ...And yet we don't care about ours [our water], so we'll let them just suck and suck and suck all the way to the aquifer, that beautiful water that's underneath the earth and nobody should be even bothering with it. (**0016-6-2** [Utley, Charles])

Comment: You have considered birds, you have considered fishes, you considered a lot of things, but nobody came to the communities that live in the shadow of these reactors and watched the water pressure change as Plant Vogtle does its flushing systems. (**0016-19-2** [Valentin, Dianne])

Comment: But those who are not adversely affected should not disrespect the people who are. And should now not consider the people that are adversely impacted by the groundwater contamination... (**0016-19-4** [Valentin, Dianne])

Response: Groundwater impacts were evaluated in the ESP EIS (NRC 2008a), and no new or significant information was identified by the NRC staff in its review of the COL. These comments do not provide any specific information relating to the environmental effects of Units 3 and 4 construction or operation. Because these comments did not provide new and significant information, no change was made to the SEIS.

Comment: Does the NRC monitor groundwater or is it Southern Company that does the monitoring? I would think it should be NRC. Is information public? What about rainwater, offsite groundwater? NRC should require that Southern Company provide the information to the public if they're doing the monitoring, but I really think NRC should be doing the monitoring. (**0016-20-9** [Rivard, Betsy])

Response: Radiological monitoring required by the NRC was described in Section 5.9.6 of the ESP EIS (NRC 2008a). Any requirement for nonradiological groundwater monitoring near site facilities would be implemented through the applicable state permitting process. Maximum groundwater withdrawals are specified by state-issued permits, which were considered by the NRC staff. Because this comment did not provide new and significant information, no change was made to the SEIS.

E.2.7 Comments Concerning Ecology – Terrestrial

Comment: By letter dated September 19, 2008, we concurred with the findings of NRC's Biological Assessment for the effects of early site preparation and preliminary construction activities at the VEGP site. The list of species protected under the Endangered Species Act (ESA) that occur in the project area has not changed since September 2008, and includes the wood stork, red-cockaded woodpecker, indigo snake, and Canby's dropwort. The DEIS indicates that the NRC is preparing a second Biological Assessment for construction and operations effects. As transmission line corridors and other pertinent construction details are

Appendix E

more precisely defined, please coordinate directly with the US Fish and Wild Life Service's Coastal Georgia Sub-office supervisor, Strant Colwell, at (912) 832-8739, to conclude the ESA consultation process for the project. (**0010-1** [Hogue, Gregory])

Comment: The DSEIS states that a biological assessment documenting potential impact on the federally listed threatened or endangered terrestrial special as a result of operation of the proposed new units and proposed transmission line is in development. The FSEIS should provided updated information on this assessment. (**0019-11** [Mueller, Heinz J.])

Response: As part of the NRC's responsibilities under Section 7 of the Endangered Species Act, the staff prepared a biological assessment (BA) that documents potential impacts on the Federally listed threatened or endangered species as a result of the site preparation (including construction of the onsite portion of the new 500-kV transmission line) and construction of Units 3 and 4 on the VEGP site. The BA for the ESP EIS was submitted to the U.S. Fish and Wildlife Service (FWS) on January 25, 2008 (NRC 2008b), and FWS concurred with the findings on September 19, 2008 (FWS 2008). In a letter dated January 7, 2010, NRC requested that the FWS Field Office in Brunswick, Georgia, provide information regarding Federally listed species and critical habitat that may have changed since the 2008 consultation (NRC 2010b). On February 12, 2010, FWS provided a response letter indicating listed species under FWS purview had been adequately addressed for limited site-preparation activities on the VEGP site (FWS 2010a). On October 20, 2010, FWS provided an updated list of Federally listed threatened or endangered species that can be expected to occur in the project area (FWS 2010b). NRC submitted a BA to FWS on February 24, 2011 to document potential impacts on Federally listed threatened or endangered terrestrial species resulting from operation of Units 3 and 4 and ancillary facilities, as well as construction and operation of the proposed transmission line ROW. This BA is included in Appendix F of this SEIS.

Comment: The DEIS notes that bird collisions with tall structures and transmission lines are among the impacts of building and operating the proposed project (pages 4-6 and 5-3), but does not describe mitigation measures for these impacts. The Department recommends that the NRC and Southern coordinate with us and the Georgia Department of Natural Resources Wildlife Division in the development of an Avian Protection Plan (APP). The Migratory Bird Treaty Act (MBTA) prohibits take of migratory birds except when specifically authorized by the Department of the Interior. The regulations implementing the MBTA (50 CFR Part 21) do not provide for permits authorizing take of migratory birds that may be killed or injured by activities that are otherwise lawful, such as by the construction and operation of power transmission lines. The Bald and Golden Eagle Protection Act provides for very limited issuance of permits that authorize take of eagles when such take is associated with otherwise lawful activities, is unavoidable despite implementation of advanced conservation practices, and is compatible with the goal of stable or increasing eagle breeding populations. The overall goal of the APP would be to minimize avian mortality associated with the proposed facilities. (**0010-3** [Hogue, Gregory])

Response: Georgia Power Company (GPC) has developed an Avian Protection Program (APP) that includes guidelines for siting new transmission lines (GPC 2006). When siting new transmission lines, substations, or other GPC facilities, available information on migratory and

Appendix E

resident bird populations will be taken into account to ensure that the lines or facilities will have as little adverse impact as practicable on these bird species. GPC has implemented the APP to monitor and address the impacts of transmission lines on birds. Information on the APP was included in Section 5.4.1.6 of the ESP EIS (NRC 2008a). Additional information on the mitigation measures proposed in the APP to minimize bird collisions during construction and operation activities has been added to Sections 4.4.1 and 5.4.1 of this SEIS.

E.2.8 Comments Concerning Ecology – Aquatic

Comment: The Department had been concerned about the possible impacts of dredging the channel for barge delivery of reactors, containment vessels, and other large equipment; however, the DEIS notes (page 7-6) that Southern will instead deliver large components and materials by rail, and will not construct a barge slip or seek dredging of the Savannah River navigation channel. This change in the project plans eliminates our concerns related to ESA-protected aquatic species, such as the robust redhorse. (**0010-2** [Hogue, Gregory])

Response: As indicated in this SEIS, the shipment of large components and materials by rail rather than using barges will eliminate the need to further consider the cumulative impacts to aquatic biota resulting from the potential dredging of the Savannah River navigation channel to accommodate barge traffic to the site. Because this comment did not provide new and significant information, no change was made to the SEIS.

Comment: Continuing measures to limit bioentrainment and other impacts to aquatic species from surface water withdrawals and discharges should be referenced in the FSEIS, and should continue to be addressed as the project progresses, in compliance with the NPDES Permit. (**0019-2** [Mueller, Heinz J.])

Response: Because it incorporates the ESP-stage analysis by reference, this SEIS focuses on new information discovered since the ESP review. There were no additional measures identified to limit entrainment or impingement beyond those specified in the ESP EIS (NRC 2008a), the COL draft SEIS (NRC 2010a), and the ESP evidentiary hearing. Southern has indicated that it will comply with any requirements or restrictions in the National Pollutant Discharge Elimination System (NPDES) permit for thermal or chemical discharges (Southern 2007). The current NPDES permit for Units 1 and 2 does not have requirements or restrictions on entrainment or impingement. Until very recently, NPDES permits in the State of Georgia have not included restrictions or requirements for entrainment and impingement. A new State of Georgia regulation – "R.61-9, Water Pollution Control Permits" – that became effective on November 26, 2010 (Ga. Code Ann. 2010) provides consistency in permitting with the EPA regulations related to cooling water intake structures and will allow state rules to have specific requirements to minimize entrainment and impingement of aquatic organisms. The staff expects that the State of Georgia will include requirements to minimize entrainment and impingement of aquatic organisms as appropriate in the NPDES permits for Units 3 and 4. No change was made to the SEIS as a result of this comment.

Appendix E

E.2.9 Comments Concerning Socioeconomics

Comment: But what I heard in Shell Bluff that dropped my world and changed me very profoundly was that in that part of the county, said the community, "we don't have a grocery store. If we have a fire, it takes 45 minutes for emergency personnel to get here". That is not in this EIS. Economic benefits are not created equal in Burke County. That needs to be noted. (**0016-11-3** [Carroll, Glenn])

Response: The NRC regulates the civilian use of nuclear materials to protect the public health and safety and the environment. Issues related solely to the location of commercial establishments, distribution of economic benefits and of taxes for infrastructure and/or services, such as fire services, are outside NRC's mission and authority, and are not addressed in this SEIS. Socioeconomic impacts during construction and operation, including impacts to public services in Burke County as a result of the plant workforce during construction and operation, were addressed in Sections 4.5 and 5.5 of the ESP EIS (NRC 2008a). No change was made to the SEIS.

Comment: The expansion of Plant Vogtle is key to the growth of the region because it will provide employment opportunities to this part of the state, with steelworkers and well-paying jobs. At peak construction, over 3500 construction jobs and 800 permanent jobs at the site in a vast array of levels from administrative to technical to security. Permanent jobs will be a driver of the local economy, bringing with it small businesses and services that will benefit both the transient and permanent jobs that will be created at the site. (**0016-4-2** [Elam, Terry])

Comment: The expansion will drive students to our technical college to develop fundamentals in math, science and other technologies that would be applicable to Southern Nuclear's employment needs and help create a more educated workforce in general. (**0016-4-4** [Elam, Terry]); (**0018-1-4** [Elam, Terry])

Comment: But probably the most significant step in the process of finding a workforce to make this plant a very safe and reliable operation is that we have partnered with Southern Nuclear and have developed a two-year associate degree program in nuclear engineering technology. (**0016-4-5** [Elam, Terry])

Comment: The impact [of the expansion of Southern Nuclear Company] on creating an educated workforce and the potential for additional businesses will greatly benefit the local economy of Burke County. We will also benefit because we will receive students who will need training and taking advantage of the educational opportunities with the current and future crop of students. (**0016-4-7** [Elam, Terry])

Comment: As a business we are members of other chambers of commerce and different things like that around the area and Burke County and the local community are very lucky to have this as far as financial impact. And people are jealous and envious that we have this in our area and I think we need to embrace that. (**0016-9-4** [Baxley, Robin])

Appendix E

Comment: The Augusta Metro Chamber of Commerce is pleased to support the expansion of Plant Vogtle. We believe that the facility is a good neighbor, supplying a needed commodity in an efficient and safe fashion. (**0016-12-6** [Parr, Sue])

Comment: In addition, the thousands of short-term jobs created during the construction as well as the permanent jobs, once they are added, will provide a much needed boost to our economy. (**0016-13-4** [Roberts, Ashley])

Comment: I can tell you that this [the new units] is not just important to Burke County, this is important for our region of the state. This is the economic engine for what's moving our economy forward. These are jobs that won't be exported. (**0016-14-1** [Stone, Jesse])

Comment: Employment opportunities to a part of the State that needs skilled workers and well paying jobs. At peak construction, 3500 construction jobs will need to be filled and after completion, there will be 800 permanent jobs at the site in a vast array of levels from administrative to technical to security. Permanent jobs will be a driver of the local economy, bringing with it small businesses and services that will benefit both the transient and permanent jobs that will be created at the site. (**0018-1-2** [Elam, Terry])

Comment: The impact on creating an educated workforce and the potential for additional business creation will greatly benefit the local economy of Burke County. Augusta Technical College will also benefit by providing training and educational opportunities for current and future students. (**0018-1-6** [Elam, Terry])

Response: These comments generally express support for Southern's plans to add two new units to the VEGP site, based on the potential positive socioeconomic impacts that this expansion would be expected to bring to the region. Socioeconomic impacts including employment, tax revenue, and economic impacts during construction and operation were discussed in Sections 4.5 and 5.5 are the ESP EIS (NRC 2008a), respectively. No new and significant information was provided in these comments. Therefore, no change was made to the SEIS.

Comment: Our educational facilities are second to none in this area; and due to the taxes generated from Plant Vogtle, our school board has been able to maintain a relatively low millage rate in comparison to many other counties throughout the state.

The poverty rate in Burke County is relatively high; and due to the taxes from Plant Vogtle, it levels the playing field giving opportunities to students that would never have those opportunities otherwise.

The education today is the engine of our future economic growth and development. Due to Plant Vogtle's contributions, we are able to provide a quality education to all of our students here in Burke County.

Appendix E

A key question asked by companies and even families seeking to move to this area is about the quality of the public education system where they would be located. Because of Plant Vogtle's involvement, we are able in Burke County to answer that question with satisfaction and pride.

We are extremely proud to have Plant Vogtle in our community. (**0016-8-2** [Mitchell, Tommy])

Comment: And of course, from an economic standpoint, it has been great for us and helped us through this economy in the last year, this expansion. (**0016-9-1** [Baxley, Robin])

Comment: Also, the tax revenues that we receive from Plant Vogtle allow our local government to provide a menu of services to our residents, all of our residents. And I think our EMA Director in the back would argue the fact that it would take 45 minutes to get to Shell Bluff. (**0016-13-5** [Roberts, Ashley])

Comment: It also affords our Board of Education the opportunity to provide outstanding educational opportunities to benefit the children in our community, all of our community including the kids from Shell Bluff. (**0016-13-6** [Roberts, Ashley])

Comment: I can tell you that this is going to have a positive (sic) impact on us if for some reason it doesn't go forward; to many, many people, to the people in Jenkins County where unemployment is 21 percent. It's only that low because they're able to commute up to Burke County to work up here, all the surrounding counties. We are hoping and praying that this project will go forward. (**0016-14-7** [Stone, Jesse])

Response: These comments discuss past economic benefits to the community with regard to Units 1 and 2 and recent expansion activities related to Units 3 and 4. They provide some context for expectations regarding future behavior. Socioeconomic impacts including economic and tax impacts during construction and operation were addressed in Sections 4.5 and 5.5 of the ESP EIS (NRC 2008a). No new and significant information was provided in these comments; therefore, no change was made to the SEIS.

Comment: Let's see, how many local people are actually employed -- will be employed by building Plant Vogtle, or were actually employed by building Vogtle 1 and 2? I'm of the impression they brought in a lot of people from the outside, I don't think it really had a big impact on employment in the county and how many will be brought in for 3 and 4. Burke County has a very high unemployment rate of 11.5 percent. Is that going to be substantially decreased by building Vogtle 3 and 4? (**0016-20-7** [Rivard, Betsy])

Response: Employment impacts from construction and operation were addressed in Sections 4.5 and 5.5 of the ESP EIS (NRC 2008a), respectively. This comment provided no new and significant information. Therefore, no change was made to the SEIS.

Comment: This is an interesting fact, in 2006, over 3.7 million people spent almost $3.5 billion on ecotourism, hunting and fishing, just in the state of Georgia. And so draining the water, decreasing the water supply to produce nuclear energy is going to hurt our economy tomorrow.

Appendix E

We can't sacrifice thousands of permanent jobs tomorrow for temporary jobs today. (**0016-22-11** [Sardi, David])

Response: Socioeconomic impacts including employment and recreational impacts during construction and operation were addressed in Sections 4.5 and 5.5 of the ESP EIS (NRC 2008a). No new and significant information was provided in this comment. Therefore, no change was made to the SEIS.

E.2.10 Comments Concerning Historic and Cultural Resources

Comment: Upon review of your September 2, 2010 submission, we reiterate our January 7, 2010 electronic message to decline the opportunity to participate in this consultation. Burke County currently exists beyond our scope of interest for the state of Georgia. No known impacts to religious, cultural, or historical assets of the Alabama-Coushatta Tribe of Texas will occur in conjunction with this proposal. No further consultation with our Tribe regarding this project is anticipated at this time. (**0012-1** [Celestine, Bryant])

Response: NRC requested the participation of the State Historic Preservation Office (SHPO), the Advisory Council on Historic Preservation, and 25 Tribes in identifying new and significant information concerning historic properties that may be impacted by this licensing action. Appendixes C and F of this SEIS have been revised to include the consultation correspondence from the Alabama-Coushatta Tribe of Texas.

Comment: We appreciate the thorough discussion of cultural and historic resources in the DSEIS. Pursuant to the location of a historic cemetery on the VEGP site, Southern entered into a Memorandum of Understanding (SHPO) with the Georgia State Historic Preservation Office (SHPO). We also note SCE&G's cultural resources awareness training and inadvertent discovery procedure training for staff working at the site. The FSEIS should include an update of coordination activities with the SHPO. (**0019-12** [Mueller, Heinz J.])

Response: This comment refers to general cultural resources management and the status of consultation documented in the SEIS. Southern (as opposed to South Carolina Electric and Gas as cited in the comment) has an inadvertent discovery procedure, which is described in the ESP EIS (NRC 2008a). The SEIS includes the updated status of consultation activities with the SHPO.

Comment: I'm wondering if you talked to anybody from the Yemassee Tribe. But yet maybe it's not new or maybe it's not significant and maybe you wouldn't consider it either. (**0016-19-12** [Valentin, Dianne])

Response: The NRC requested the participation of the SHPO, the Advisory Council on Historic Preservation, and 25 Federally Recognized Tribes in identifying new and significant information concerning historic properties that may be impacted by this licensing action. As of August 5, 2010, the Yemassee Tribe was not listed as a Federally Recognized Tribe in the Federal Register by the Bureau of Indian Affairs. Consultation activities associated with this licensing action are discussed in Chapter 2 of this SEIS. Appendix C contains a complete listing of the

Appendix E

25 Federally Recognized Tribes with which NRC consulted. Because this comment did not provide new and significant information, no change was made to the SEIS.

E.2.11 Comments Concerning Environmental Justice

Comment: I really don't think that as a regulatory agency you have met your, what should be a standard when you allow the research and information that you use primarily to come from those that you regulate and you don't go into the communities that are affected adversely by the presence of these reactors, the two that are already here and the ones that are coming.

It's important that people be considered. (**0016-19-1** [Valentin, Dianne])

Comment: Now if you don't live in an area of Waynesboro that is impacted by the contaminations from Vogtle 1 and 2 and you don't have to be afraid of the environmental impacts of Vogtle 3 and 4, that's great. But at least give consideration to the people who do. You don't know them? Go get to know them, see what is actually happening in their communities, understand, talk to them because obviously you have not, because there is no way that you could sit through conversations with these people who live in these communities and not be personally impacted even if you don't think the environment is impacted. (**0016-19-8** [Valentin, Dianne])

Response: On February 11, 1994, the President issued Executive Order 12898, "Federal Actions to Address Environmental Justice in Minority Populations and Low Income Populations (Executive Order 1994)." This order requires each Federal executive branch agency to identify and address, as appropriate, disproportionately high and adverse human health or environmental effects on minority and low-income populations resulting from its actions. The memorandum accompanying the Executive Order directed Federal executive agencies to consider environmental justice, and CEQ provided guidance for addressing environmental justice. Although complying with the executive order is not mandatory for independent agencies, the Commission has voluntarily committed to undertake environmental justice reviews as part of its NEPA responsibilities. The Commission's "Policy Statement on the Treatment of Environmental Justice Matters in NRC Regulatory and Licensing Actions" contains guidance and information for addressing issues of environmental justice (69 FR 52040). To perform a review of environmental justice in the vicinity of a nuclear power plant, the NRC staff examines the geographic distribution of minority and low-income populations within 80 km (50 mi) of the site. The staff uses the most recent census data available. The staff also supplements its analysis with field inquiries to groups such as county planning departments, social service agencies, local churches, and private social service agencies. Once the locations of minority and low-income populations are identified, the staff evaluates whether any of the environmental impacts of the proposed action could affect these populations in a disproportionately high and adverse manner. The staff used this process during preparation of the ESP EIS (NRC 2008a); the environmental justice process and analysis was documented in Sections 2.10, 4.7, and 5.7. No change was made to the SEIS.

Appendix E

Comment: ENVIRONMENTAL JUSTICE Impacts on the community living directly adjacent to reactors already at the Vogtle site are LARGE. The Shell Bluff community is in the emergency planning zone of the reactors, and yet its residents do not enjoy emergency fire, police and health protection. This under-served community does not have a grocery store, yet could be permanently dislocated following an accidental radiation release from either the existing, or proposed, reactors at Vogtle. This environmental justice issue must be acknowledged and analyzed in the EIS. Health studies suggest cancer and death rates have risen in Burke County since Vogtle reactors 1 and 2 started operating. (**0002-7** [Carroll, Glenn])

Comment: ENVIRONMENTAL JUSTICE Impacts on the community living directly adjacent to reactors already at the Vogtle site are LARGE. The Shell Bluff community is in the emergency planning zone of the reactors, and yet its residents do not enjoy emergency fire, police and health protection. This under-served community does not have a grocery store, yet could be permanently dislocated following an accidental radiation release from either the existing, or proposed, reactors at Vogtle. This environmental justice issue must be acknowledged and analyzed in the EIS. (**0003-7** [Arnold, Judy] [Boatenreiter, Glenn] [Carter, Pat] [Cumbow, Kay] [Dawson, Daneille] [Dooley, Gerald] [Falconer, Kimberly] [Hatch, Sarah] [Kasenow, Lisa] [Lewis, Marvin] [Lomas, Judith] [Lusk, Phil] [McConnell, Joy] [McNulty, Joy] [Mills, Nancy] [Sheppard, Deborah] [Thomas, Russel] [Trujillo, Dianne] [Villarreal, Tasha]); (**0005-7** [Kushner, Adele])

Comment: These two reactors were brought to Shell Bluff, Waynesboro to boost an economically depressed area. It was proposed that the area would be saved by the nuclear industry. It was proposed that the people residing in Shell Bluff would be saved by the nuclear industry. Maybe we thought that these two reactors would define Shell Bluff or the larger Waynesboro area as a celebrated zone, a special zone.

However, we've come to think at Shell Bluff as a sacrifice zone. What does this mean? The local government and big businesses have taken advantage of people who are economically and politically powerless. My friends from Shell Bluff have not been saved by the nuclear industry. The wider area of Waynesboro has not been saved by the nuclear industry. Reactors 1 and 2 have brought daily radioactive releases. Reactors 1 and 2 prevent locals from eating from the river. Reactors 1 and 2 prevent locals from drinking the local tap water. It's hot. Reactors 1 and 2 produce significant amounts of waste -- not minuscule amounts of waste. (**0016-16-1** [Ogley-Oliver, Emma])

Comment: You have to be aware of the fact that it [fallout from the atmosphere] is a major thing when you have children in our impacted area, it is a disproportionate environmental injustice for one community to stand all the pollution being poured on them. (**0016-6-7** [Utley, Charles])

Response: The comments concern potential effects from construction and operation of the existing and proposed reactors at the VEGP site as a potential environmental justice issue. The environmental justice analysis was conducted in accordance with NRC guidance. Issues related to the distribution of taxes for infrastructure and/or services such as fire, police and health services are outside the NRC's mission and authority, and will not be addressed in the SEIS. To the extent the comments address the adequacy of Southern's emergency plan,

Appendix E

emergency planning and preparedness is reviewed by the NRC as a part of its safety review and therefore is outside the scope of the environmental review. Environmental justice, the potential for disproportionate and adverse environmental impact on minority and low-income communities, including from socioeconomic impact, was addressed in Sections 4.7 and 5.7 of the ESP EIS (NRC 2008a). Sections 4.8 and 4.9 of the ESP EIS include discussions of the nonradiological and radiological health impacts on the public during construction, and Sections 5.8 and 5.9 include discussions of the nonradiological and radiological impacts on the public during operation of the proposed facility. Information in Sections 7.7 and 7.8 of the ESP EIS addresses all potential cumulative nonradiological and radiological impacts on the public from operation of the proposed facility. The environmental justice analysis provided in Sections 4.7 and 5.7 of the ESP EIS addresses disproportionately adverse human health impacts on minority and low-income communities that could potentially be produced by the construction and operation of the proposed facility, and information in Section 7.6 addresses cumulative impacts in terms of environmental justice (NRC 2008a). The review team found that all environmental emissions and operation dose assessments are well within NRC and EPA regulations, and no demographic subgroup is affected differently then another subgroup. No studies were identified that indicated minority and low-income individuals would be more susceptible to nonradiological emissions or radiological doses. No change was made to the SEIS as a result of these comments.

Comment: The Final EIS for an early site permit for Plant Vogtle's Units 3 and 4 was completed in July 2008. The FEIS concluded: "[T]he impacts of plant operations on environmental justice would be SMALL because no environmental pathways, health characteristics, or other preconditions of the minority and low-income population were found that would lead to adverse and disproportionate impacts." Unbelievably, the report attributed the high percentage of minority and low-income people on the "sparseness" of the rural population. The data collection for this report consisted of interviews with just three residents. The application for a Vogtle combined operating license with environmental report was submitted to the NRC on March 31, 2008.

In 2009, subsequent to the Vogtle COLA and ESP-FEIS, a nuclear power siting study was published which suggests that there is a "reactor-related environmental injustice" at Plant Vogtle. Attachment C contains the full article (Alldred and Shrader-Frechette 2009). The study found:

"The mining, fuel enrichment-fabrication, and waste-management stages of the US commercial nuclear fuel cycle have been documented as involving environmental injustices affecting, respectively, indigenous uranium miners, nuclear workers, and minorities and poor people living near radioactive-waste storage facilities. After surveying these three environmental-injustice problems, the article asks whether US nuclear-reactor siting also involves environmental injustice. For instance, because high percentages of minorities and poor people live near the proposed Vogtle reactors in Georgia, would siting new reactors at the Vogtle facility involve environmental injustice? If so, would this case be an isolated instance of environmental injustice, or is the apparent Georgia inequity generally representative of environmental injustice associated with nuclear-reactor siting throughout the US? Providing a preliminary answer to

Appendix E

these questions, the article uses census data, paired t-tests, and z-tests to compare each state's percentages of minorities and poor people to the percentages living in zip codes and census tracts having commercial reactors. Although further studies are needed to fully evaluate apparent environmental injustices, preliminary results indicate that, while reactor-siting-related environmental injustice is not obvious at the census-tract level (perhaps because census tracts are designed to be demographically homogenous), zipcode-scale data suggest reactor-related environmental injustice may threaten poor people ($p < 0.001$), at least in the southeastern United States.

The summary conclusions of the ESP Final EIS are plainly wrong or at least premature. The NRC must include this new information in its analysis. (**0009-2** [Zeller, Lou])

Response: As a part of the environmental review required by the NEPA process during the ESP review, NRC conducted a scoping meeting that was announced in the Federal Register, on the NRC website, and in local and regional newspapers prior to the public meeting. Participants in the scoping process were provided an opportunity to submit oral and written comments to which the NRC staff responded. Consistent with its environmental review guidance, the staff also conducted interviews with local county government and social services agencies. A complete list of organizations contacted is in Appendix B of the ESP EIS (NRC 2008a). Possible environmental justice impacts occurring outside the impact region described the ESP EIS (such as those associated with mining and spent fuel storage) are beyond the scope of this environmental review and are not addressed in the ESP EIS (NRC 2008a). As stated in the NRC guidance, analyses of census data is done at the census block group level and provides information for geographic areas of approximately 1000 people each, on average, and as such provides sufficient geographic detail to assess the impact of VEGP Units 3 and 4 on minority and low-income populations. The ESP EIS concluded impacts of plant construction and operations on environmental justice would be SMALL because no environmental pathways, health characteristics, or other preconditions of the minority and low-income population were found that would lead to adverse and disproportionate impacts. This comment, including the referenced article, did not provide new information regarding the demographic composition around the VEGP site leading to the presence of additional environmental justice communities or environmental pathways, health characteristics, or other preconditions that would lead to disproportionately high and adverse impacts to minority and low-income communities. Therefore, no change was made to the SEIS.

Comment: They [the people impacted] don't live a block away from this building, they don't work a block away from this building. And unfortunately, their children die or they move away, so they don't have as many children in these schools that Southern Nuclear is building and making, you know, the community shine and polish. But I don't think you know that because you never came and you never asked. (**0016-19-9** [Valentin, Dianne])

Appendix E

Comment: You have a list of staff conclusions that did not change and you -- I thought it was kind of nervy for you to list environmental justice, especially since nobody came to talk to the people in Shell Bluff. But you talk about socioeconomics and the people of Shell Bluff are getting poorer. (**0016-19-11** [Valentin, Dianne])

Response: As a part of the environmental review required in the NEPA process during the ESP review, NRC conducted a scoping meeting that was announced in the Federal Register, on the NRC website, and in local and regional newspapers prior to the public meeting. Participants in the scoping process were provided an opportunity to submit oral and written comments to which the NRC staff responded. Consistent with its environmental review guidance, the NRC staff also conducted interviews with local county government and social services agencies. A list of organizations contacted is provided in Appendix B of the ESP EIS (NRC 2008a). As explained in the draft SEIS, the staff did not identify new and significant information regarding environmental justice during its COL review. These comments did not provide any new and significant information. Therefore, no change was made to the SEIS.

Comment: Section 3-301(b) of Executive Order 12898 states that "Environmental human health analyses, whenever practicable and appropriate, shall identify multiple and cumulative exposures." A missing factor in the assessment of Vogtle's impact is the proximity of the nuclear power station to the Department of Energy's Savannah River Site. Vogtle and SRS emissions intermingle, making independent assessment challenging. The principal contractor at the Savannah River Site publishes annual reports which contain the following data.

Tritium Transport in Streams

Year	SRS emissions	Vogtle emissions	Total curies
2003	4010	1900	5910
2004	2430	1200	3630
2005	2620	1860	4480

The discharge of Tritium in the form of radioactive water pollutes the Savannah River all the way to the ocean. Downstream drinking water wells are contaminated. Does the pollution come from SRS or Vogtle? The answer is "both." Until a few years ago, the Georgia Department of Natural Resources Environmental Protection Division published reports on its radiation monitoring program. The program tested samples of air, surface water, groundwater, rain, sediments, fish, soil, vegetation, milk and agricultural crops near facilities which are known to emit ionizing radiation and compares these data to background levels. Test results for Vogtle from 1995 to 2002 indicated that the nuclear power plant is the source of a variety of radionuclides which contaminate sediment, river water, fish and drinking water. The state's test results reveal striking elevations of harmful radionuclides. The test results range from 2 times to 50 times above background level. (**0009-6** [Zeller, Lou])

Comment: A study conducted by the University of South Carolina has shown that there is a higher than average instance of cervical cancer in black women, and a higher rate of esophageal cancer in black men, within a fifty mile radius of Plant Vogtle. Georgia EPD

Appendix E

monitoring indicates much of the radioactive pollution comes from the two nuclear reactors at Plant Vogtle. Studies of U.S. Centers for disease Control and Prevention data indicate that the death rate per 100,000 population from all cancers in Burke County increased by 24.2% and that infant deaths increased by 70.1% in Burke County after the Plant Vogtle reactors went online. (**0009-7** [Zeller, Lou])

Response: These comments concern radiological impacts and tritium releases from the existing Units 1 and 2, and potential tritium releases from the proposed Units 3 and 4. The expected radiation doses to the public from all radioactive effluents, including tritium, from the proposed Units 3 and 4 are addressed in Section 5.9 and Appendix G of the ESP EIS (NRC 2008a). Section 5.9 and Appendix G also address the expected combined radiation doses from operation of all four units; these estimates include tritium. As discussed in Section 5.9, the doses to the maximally exposed individual are estimated to be less than 3 mrem/yr to the total body and less than 10 mrem/yr to the organ with the highest dose. These estimates include tritium and the drinking water exposure pathway. These doses are considered to be small by the NRC because they are lower than the NRC and EPA dose standards and much lower than the average dose of 311 mrem/yr to the total body from natural sources of radiation. NRC accepts the theory that there is some health risk associated with any amount of radiation exposure. However, according to the International Commission on Radiological Protection (ICRP) and the National Council on Radiation Protection and Measurements (NCRP), doses at this level would most likely result in zero excess health effects (NCRP 1995; ICRP 2007). Furthermore in Section 7.8 of the ESP EIS (NRC 2008a), the staff considered the cumulative radiological health impacts from operations at both the existing and proposed reactors at the VEGP site, as well as the Savannah River Site, and other nuclear facilities, including from tritium, and concluded that these impacts would be SMALL. As the staff in the ESP EIS found no unusual resource dependencies or practices or environmental pathways through which minority and low-income populations would be disproportionately affected, the cumulative environmental justice impacts would remain small even when considering the radiological health impacts of the Savannah River Site. No change was made to the SEIS as a result of these comments.

Comment: Section 4-401 of Executive Order 12898 states: "In order to assist in identifying the need for ensuring protection of populations with differential patterns of subsistence consumption of fish and wildlife, Federal agencies, whenever practicable and appropriate, shall collect, maintain, and analyze information on the consumption patterns of populations who principally rely on fish and/or wildlife for subsistence."

Local residents depend on the Savannah River for fish to feed their families. Radiological monitoring reveals that Savanna River fish are contaminated with Cesium-137. Tests in the vicinity of Plant Vogtle routinely find Cesium-137 in the edible parts of fish.

Radioactive Cesium-137 is of particular concern because levels actually increase when fish is cooked. One study found that cesium levels increase by 32% when fish are fried with breading, and by 62% when fried without breading.

Appendix E

African American and low-income individuals are at specific heightened risk from hazardous materials in the Savannah River, and although individuals from all socioeconomic backgrounds engage in fishing in the area, African Americans in particular commonly engage in subsistence fishing along the Savannah River and have a higher than average consumption of fish, frequently surpassing allowable contaminated fish consumption levels. (**0009-3** [Zeller, Lou])

Response: The comment concerns subsistence consumption of fish contaminated with cesium-137. Sections 4.7 and 5.7 of the ESP EIS (NRC 2008) addressed subsistence fishing. Section 5.9 of the ESP EIS (NRC 2008), estimates the potential radiation doses to members of the public from liquid effluent releases from the proposed two new units at the Vogtle site, including consumption of fish caught in the Savannah River. The expected total body dose to the maximally exposed individual in the public from all liquid dose pathways for two new reactors, including fish consumption, was estimated to be 0.034 mrem/yr. The highest dose to any organ was estimated to be 0.042 mrem/yr to the liver of a child. The ESP EIS (NRC 2008) further estimated the dose to the maximally exposed individual from all four reactors (two existing reactors and two proposed reactors) to be 2.4 mrem/yr from all liquid and gaseous effluents (less than 0.1 mrem/yr is from fish consumption). These estimates are based on projected release rates for the two proposed reactors and the typical measured and reported release rates from the existing reactors. These estimates include all radionuclides released from the Vogtle reactors including cesium-137; the expected release rates all radionuclides in liquid effluents from the proposed reactors are shown in Table G-1 of the ESP EIS (NRC 2008a). Also, as shown in Table G-1, these doses are based on assumed annual fish consumption rates of 21 kg (about 46 lb) for an adult, 16 kg (35 lb) for a teen, and 6.9 kg (15 lb) for a child. Subsistence fishermen might consume more fish than these assumed rates; however, even if someone consumed a pound of fish every day, the doses would increase by a factor of ten to only about 1 mrem/yr. This is a very small dose compared to the average annual dose to an individual in the United States from natural radiation sources of 311 mrem/yr. As the comment suggested, the concentration of radionuclides such as cesium-137 may increase because of weight loss when the fish is cooked, and the person may consume a bigger portion of fish as a result. However, even increases in the dose of 30 to 60 percent would still result in a very small dose. No change was made to the SEIS as a result of this comment.

E.2.12 Comments Concerning Health – Nonradiological

Comment: In addition to the EPA's concerns regarding climate change effects and GHG emissions, the National Institute for Occupational Safety and Health (NIOSH) has determined that diesel exhaust is a potential human carcinogen, based on a combination of chemical, genotoxicity, and carcinogenicity data. In addition, acute exposures to diesel exhaust have been linked to health problems such as eye and nose irritation, headaches, nausea, and asthma.

Although every construction site is unique, common actions can reduce exposure to diesel exhaust. EPA recommends that the following actions be considered for construction equipment:

- Using low-sulphur diesel fuel (less than 0.05% sulphur).

Appendix E

- Retrofit engines with an exhaust filtration device to capture DPM before it enters the workplace.

- Position the exhaust pipe so that diesel fumes are directed away from the operator and nearby workers, thereby reducing the fume concentration to which personnel are exposed.

- A catalytic converter reduces carbon monoxide, aldehydes, and hydrocarbons in diesel fumes. These devices must be used with low sulphur fuels.

- Ventilate wherever diesel equipment operates indoors. Roof vents, open doors and windows, roof fans, or other mechanical systems help move fresh air through work areas. As buildings under construction are gradually enclosed, remember that fumes from diesel equipment operating indoors can build up to dangerous levels without adequate ventilation.

- Attach a hose to the tailpipe of a diesel vehicle running indoors and exhaust the fumes outside, where they cannot reenter the workplace. Inspect hoses regularly for defects and damage.

- Use enclosed, climate-controlled cabs pressurized and equipped with high efficiency particulate air (HEPA) filters to reduce operators' exposure to diesel fumes. Pressurization ensures that air moves from inside to outside. HEPA filters ensure that any air coming in is filtered first.

- Regular maintenance of diesel engines is essential to keep exhaust emissions low. Follow, the manufacturer's recommended maintenance schedule and procedures. Smoke color can signal the need for maintenance. For example, blue/black smoke indicates that an engine requires servicing or tuning.

- Work practices and training can help reduce exposure. For example, measures such as turning off engines when vehicles are stopped for more than a few minutes; training diesel-equipment operators to perform routine inspection and maintenance of filtration devices.

- When purchasing a new vehicle, ensure that it is equipped with the most advanced emission control systems available.

- With older vehicles, use electric starting aids such as block heaters to warm the engine, avoid difficulty starting, and thereby reduce diesel emissions.

- Respirators are only an interim measure to control exposure to diesel emissions. In most cases an N95 respirator is adequate. Respirators are for interim use only, until primary controls such as ventilation can be implemented. Workers must be trained and fit-tested before they wear respirators. Personnel familiar with the selection, care, and use of respirators must perform the fit testing. Respirators must bear a National Institute of Occupational Safety and Health (NIOSH) approval number. Never use paper masks or surgical masks without NIOSH approval numbers. (**0019-10** [Mueller, Heinz J.])

Appendix E

Response: The comment concerns known and potential health effects of exposure to diesel exhaust, and offers strategies to mitigate such exposures. Construction equipment exhaust was discussed in Sections 4.2 and 5.2 the ESP EIS (NRC 2008a). While the NRC determined that nonradiological impacts would be SMALL, it agrees that the measures identified in the comment would further reduce exposure to diesel exhaust. No changes were made to the draft SEIS.

E.2.13 Comments Concerning Health – Radiological

Comment: Health studies suggest cancer and death rates have risen in Burke County since Vogtle reactors 1 and started operating. (**0004-5** [Taylor, F])

Comment: New construction for nuclear industry is unacceptable for the following reasons:...

- routine and accidental releases of unseen radionuclides into the air and water, some of which are persistent and/or biohazardous toxins, affecting the health of downwind or downstream communities and watersheds (**0007-9** [Michetti, Susan])

Comment: At the same time, you're having fallout from the atmosphere, you have all that to breathe, coming down on you. (**0016-6-6** [Utley, Charles])

Comment: You can come and say well, they come from Atlanta, they come from Savannah, they come from Wisconsin. It doesn't affect them. But yeah, look at that one community, Shell Bluff, look at those folks. There is a definite impact on those who live near plants. Brain tumors in a year old -- think about it. Babies are susceptible... (**0016-6-8** [Utley, Charles])

Comment: They some people out of Australia who live as far as their reactor is from here almost to California. They had iodine. How many of you got it in your water? (**0016-6-9** [Utley, Charles])

Comment: We talk about these same issues but yet they're not here. We talk about FEMA, we talk about GEMA, we talk about all of these acronyms that's supposed to be helping us, but where are they when you're on your sick bed and all you're getting is radiation and fallout and you're trying to say send it over here, we're not going to take it. (**0016-6-10** [Utley, Charles])

Comment: ...investigate that the cancer rates since '87 and '89 when 1 and 2 went on line, have gone perhaps from 11 percent below the national average to 26 percent above -- look at those CDC figures and investigate for yourself. (**0016-7-4** [Paul, Bobbie])

Comment: Hearing from other persons, we realize that there is cancer and no amount of money can ease the suffering that I have encountered in this community. And it's just not only blacks, but it's whites also that are suffering from a high rate -- an increase of cancer. (**0016-10-3** [Stephens, Annie Laura])

Appendix E

Comment: And I heard you mention about moving the species, certain species -- well, what about mankind like over on the South Carolina side where SRS moved six communities from that site. Do you all plan to look into that as moving us as a people to another place? Take that into consideration and see if that will impact your decision when it comes down to humanity. That's what we are praying for, the health and welfare for all humanity in this area and all other areas where these plants are built. (**0016-10-4** [Stephens, Annie Laura])

Comment: One thing very fundamental has changed since reactors 1 and 2 and that is this county has experienced nuclear reactors in its community. It has come to know cancer and now we know a lot more going into Vogtle 3 and 4 than we did when we talked about Vogtle 1 and 2. (**0016-11-1** [Carroll, Glenn])

Comment: And finally, the addition of the two new reactors even further increases the environmental and health and safety dangers that the communities around Plant Vogtle face every day. (**0016-15-4** [Henson, Courtney])

Comment: The area is contaminated. The people are sick with cancer. Local government and big businesses profit, everyday folk suffer. We have a choice -- health or radiation; prosperity or devastation. (**0016-16-3** [Ogley-Oliver, Emma])

Comment: For example, I was sitting talking to somebody and she told me about a woman that lived in the Shell Bluff community. She knew 30 people that had cancer. This is something that really hit home to me, because my mom, she just completed her radiation therapy that she had to go through for breast cancer. So can you imagine 30 people with cancer that you know personally? To me, I picture my mom and 30 versions of my mom, you know, with cancer. (**0016-18-2** [Vincent, Patricia])

Comment: I don't think you have sent anybody into the communities and asked or investigated in any way where people who thought their dog had mange took them to the vet and found out that they had cancer from eating the foliage out of the yard and drinking from the puddles. I don't think you sent anybody into the communities where I saw a beautiful black lab that turned around and had a huge tumor hanging off of its side. (**0016-19-3** [Valentin, Dianne])

Comment: But those who are not adversely affected should not disrespect the people who are. And should now not consider the people that are adversely impacted by the ...contamination of the land and soil. (**0016-19-5** [Valentin, Dianne])

Comment: So I think it's important that the NRC give consideration to the fact that there are people living in these communities who have to deal with awful things as a result of the reactors being there, very awful things including cancers, adverse health effects. (**0016-19-6** [Valentin, Dianne])

Comment: You don't want to consider human life? Consider the lives of the pets if you don't want to consider human life, because it seems that you're very willing to consider how birds and fishes are impacted but not how humans are impacted. (**0016-19-7** [Valentin, Dianne])

Appendix E

Comment: The Report of the National Academy concluded that -- in their report -- they concluded their report with this hypothesis: Every exposure to radiation produces a corresponding cancer risk. Even if it's low, it all adds up. And tritium releases constitute the largest routine releases from nuclear power plants. And these releases have caused widespread contamination of water bodies at low levels. Tritium becomes tritiated water and that can cross the placenta, we know that. Non-cancer fetal risks are not part of the regulatory framework and I got this from an IEER publication, Institute for Energy -- Environmental Energy Research. (**0016-20-10** [Rivard, Betsy])

Comment: Vogtle 1 and 2, for 2006, the average amount of picocuries per liter in drinking water was 746 and 766. And the surface water for 2000, 307 picocuries. Well, Ontario, California has lowered their standards so that -- actually it's kind of difficult to see where they lowered them, but they have changed their limit to -- Ontario has changed their limit to 540 picocuries per liter and California has a public health goal at 400 picocuries per liter. This is for drinking water. And these figures, 746 and 766, that's the average daily amount in the drinking water that's higher than the standards for Ontario and California. And of course if you consider that the EPA says we can allow 20,000 picocuries, 700 sounds pretty good. But people are becoming more aware that tritium in your drinking water is not good for you.

So I just feel like the NRC should address this issue and I think that considering that this impact is small, you know, we don't really know if it's small or not. (**0016-20-11** [Rivard, Betsy])

Response: The comments concern the potential health effects from radiation exposure in the vicinity of the existing or proposed Vogtle reactors. Section 5.9 of the ESP EIS (NRC 2008a) estimates the potential radiation doses to a member of the public from operation of all four reactors (two currently operating and two proposed) at Vogtle. The doses to the maximally exposed individual are estimated to be less than 3 mrem/yr to the total body and less than 10 mrem/yr to the organ with the highest dose. These estimates include tritium and the drinking water exposure pathway. These doses are considered to be small by the NRC because they are lower than the NRC and EPA dose standards and much lower than the average dose of 311 mrem/yr to the total body from natural sources of radiation. The NRC accepts the theory that there is some health risk associated with any amount of radiation exposure. However, according to the International Commission on Radiological Protection (ICRP) and the National Council on Radiation Protection and Measurements (NCRP), doses at this level would most likely result in zero excess health effects (NCRP 1995, ICRP 2007). Southern conducts an environmental radiological monitoring program around Vogtle that sample air, crops, river water, well water, soil, fish, and sediment. This program monitors the level of radioactive material in the environment from all sources, including Vogtle and the Savannah River Site. The Georgia Environmental Protection Division also conducts a radiological environmental monitoring around the VEGP site. Results from these monitoring programs confirm that there is no significant buildup of radioactive material from Vogtle in the environment. Because these comments provided no new and significant information, no change was made to the SEIS.

Comment: I did notice that there is something in the Supplemental EIS about a new dairy in Gerard, Georgia, which will only be six miles south of the site. That's a concern to me, what

Appendix E

radionuclides are looked for when they do check the milk? Is the information on the monitoring of the existing dairies, which I think there are like within 50 miles -- is that open to public scrutiny? What is an acceptable amount of radiation in milk? I don't know. (**0016-20-5** [Rivard, Betsy])

Response: The comment concerns monitoring of dairy milk for radioactive material. As discussed in Section 5.9.1 of this SEIS, Southern will sample milk from the new dairy as part of the Radiological Environmental Monitoring Program. The results of this monitoring program are submitted to NRC in the Annual Radiological Environmental Operating Report; these reports are available to the public. The reports include information about when the samples were taken and what radionuclides were found. The samples would be analyzed for a number of radionuclides including iodine-131. Because these comments did not provide new and significant information, no change was made to the SEIS.

Comment: There's a new off-site dose calculation manual mentioned. Is that produced by NRC or by Southern Company? (**0016-20-6** [Rivard, Betsy])

Response: The comment concerns the offsite dose calculation manual (ODCM). The ODCM is a license requirement for VEGP Units 1 and 2; Southern would also be required to have and use an ODCM for proposed Units 3 and 4. The ODCM is produced by Southern and reviewed by NRC inspectors. No changes were made to the SEIS on the basis of this comment.

Comment: Radiological impacts are something I wanted to address but the Supplemental EIS does not provide me with much information. (**0016-20-8** [Rivard, Betsy])

Response: The comment concerns the level of information about radiological impacts in the SEIS. This SEIS addresses new and significant information. Radiological impacts during construction of proposed Units 3 and 4 are estimated in Section 4.9 of the ESP EIS (NRC 2008a). Estimates of the radiological impacts of normal operations are provided in Section 5.9 and Appendix G of the ESP EIS (NRC 2008a). No changes were made to the SEIS on the basis of this comment.

E.2.14 Comments Concerning Accidents – Design Basis

Comment: you also have to consider the potential for accidents.

Now the H.B. Robinson Nuclear Plant in Hartsville, South Carolina, not too far from where I live, has already shut down three times this year due to mechanical failures. And dealing with these mechanical failures, while minor, are somewhat common. Increasing the amount of power plants in the state and the region only allows for more chances for something catastrophic to occur. We've heard so many times tonight about, you know, the horrible things that happen to people very close to the power plant. So that's something you really have to keep in mind. (**0016-22-8** [Sardi, David])

Appendix E

Response: *The potential for accidents is discussed in Section 5.10 of both the ESP EIS and this SEIS. The comment provides no new and significant information; therefore, no changes were made to the SEIS.*

E.2.15 Comments Concerning Accidents – Severe

Comment: DESIGN BASIS ACCIDENT The draft EIS is based on a reactor design which has not been granted a license. The Westinghouse AP100 reactor was recently issued a Notice of Violation by NRC review staff for submitting an unrealistic assessment of impacts from a direct airplane strike. The AP1000 also has a basic design defect, in an accident, the so- called shield building would funnel radionuclides directly to the environment as shown in the Gundersen report. Either of these issues is sufficient to conclude that the environmental impact from an accident would be LARGE. (**0003-8** [Arnold, Judy] [Boatenreiter, Glenn] [Carter, Pat] [Cumbow, Kay] [Dawson, Daneille] [Dooley, Gerald] [Falconer, Kimberly] [Hatch, Sarah] [Kasenow, Lisa] [Lewis, Marvin] [Lomas, Judith] [Lusk, Phil] [McConnell, Joy] [McNulty, Joy] [Mills, Nancy] [Sheppard, Deborah] [Thomas, Russel] [Trujillo, Dianne] [Villarreal, Tasha]); (**0002-8** [Carroll, Glenn]); (**0005-8** [Kushner, Adele])

Response: *The issues raised in the comments are being addressed in the staff's separate review of Westinghouse's proposed amendment to the AP1000 design certification. Moreover, the impacts of severe accidents, including an accident with release of fission products to the environment, were considered in Section 5.10.2 of the ESP EIS. The comments provide no new and significant information; therefore, no changes were made to the SEIS.*

E.2.16 Comments Concerning the Uranium Fuel Cycle

Comment: New construction for nuclear industry is unacceptable for the following reasons:...dangerous wastes - some of which need isolation from the biosphere for millions of years. (**0007-12** [Michetti, Susan])

Comment: The failure to be able to solve the waste problem is the largest reason for stopping these proposals, regardless of the stage they may currently be at.

It is obvious that no solution to the waste problem can be found, and this makes new construction negligent and reckless with public health and environmental quality from many different places of concern. (**0007-13** [Michetti, Susan])

Comment: ...I expect the highest standards of regulation for public protection around the world from nuclear wastes generated in the US. (**0007-14** [Michetti, Susan])

Comment: You've got the Savannah River Site across the river, they don't want it, nobody wants it but you say you've got enough space to keep it. Where in the world are you going to put it? Nonsense. Don't fool yourself. Everybody in the world is looking at Waynesboro, they want to know what are you going to do with all that radiation when it gets here, because it's going to be a glow in the dark. The world is going to know. Think about it, it's your choice. (**0016-6-11** [Utley, Charles])

Appendix E

Comment: And I mentioned earlier, the proposed nuclear waste dump at Yucca Mountain to store the nation's radioactive waste from the existing Vogtle units, all the reactors across the country, [and] the new reactors being proposed [Yucca Mountain's funding] have been suspended, zeroed out in the budget. (**0016-5-5** [Barczak, Sara])

Comment: This area is contaminated just as the areas in France are contaminated. They have reprocessing -- reprocessing, which is a way to deal with the waste, they have it in my hometown in England, there's lots of leukemia there too. So if we think by producing more waste, we're going to have a way to deal with it, let's speak to our friends in England and in France. It's not happening. (**0016-16-2** [Ogley-Oliver, Emma])

Comment: The FSEIS should clarify the impact of this revision on the proposed project, as this new determination finds that spent nuclear fuel can be stored safely and securely without significant environmental impacts for at least 60 years after operation at any nuclear power plant. EPA recommends that the FSEIS cite any new analyses for longer-term storage regarding scientific knowledge relating to spent fuel storage and disposal. The FSEIS should also mention any developments with the Presidential Blue Ribbon Commission on alternatives for dealing with high-level radioactive waste, if there are such updates before FSEIS publication. (**0019-3** [Mueller, Heinz J.])

Response: *The comments concern interim storage and ultimate disposal of spent fuel and other high-level radioactive waste. Sections 5.9 of both the ESP EIS and this SEIS address the radiological impacts during operation of the proposed Units 3 and 4, including the storage of spent fuel in the spent fuel pool and in the independent spent fuel storage installation. Interim storage and ultimate disposal of spent fuel and high-level radioactive waste are discussed in Sections 6.1.6 of both the ESP EIS (NRC 2008a) and this SEIS. Section 6.1.6 presents Yucca Mountain, Nevada as an example of a possible of a high-level waste repository; the conclusions in Section 6.1.6 do not depend on whether Yucca Mountain, or another site, is ultimately the destination for spent fuel and high-level radioactive waste. Moreover, as indicated at 10 CFR 51.23(a), "... The Commission has made a generic determination that, if necessary, spent fuel generated in any reactor can be stored safely and without significant environmental impacts for at least 60 years beyond the licensed life for operation (which may include the term of a revised or renewed license) of that reactor in a combination of storage in its spent fuel storage basin and at either onsite or offsite independent spent fuel storage installations. Further, the Commission believes there is reasonable assurance that sufficient mined geologic repository capacity will be available to dispose of the commercial high-level radioactive waste and spent fuel generated in any reactor when necessary." In addition, 10 CFR 51.23(b) applies the generic determination in section 51.23(a) to provide that "... no discussion of any environmental impact of spent fuel storage in reactor facility storage pools or independent spent fuel storage installations for the period following the term of the....reactor combined license or amendment....is required in any....environmental impact statement....prepared in connection withthe.... issuance or amendment of a combined license for a nuclear power reactor under*

Appendix E

parts 52 or 54 of this chapter." Section 6.1.6 of the SEIS has been updated to reflect the current language of the Waste Confidence Decision.

Comment: New construction for nuclear industry is unacceptable for the following reasons:

- the need for uranium mining and costly, dirty processing, which destroys the health of both watersheds and communities for very long periods of time

(**0007-8** [Michetti, Susan])

Response: The comment concerns uranium mining and processing. As explained in Section 6.1 of the ESP EIS, impacts from the uranium fuel cycle have been tabulated in 10 CFR 51.51 Table S-3, which is used as the basis for evaluating the contribution of the environmental effects of uranium mining and milling to the environmental costs of licensing the nuclear power reactor (NRC 2008a). Associated effects also discussed in 10 CFR 51.51 include the production of uranium hexafluoride, isotopic enrichment, fuel fabrication, reprocessing of irradiated fuel, transportation of radioactive materials, and management of low-level wastes and high-level wastes related to uranium fuel-cycle activities. Impacts of the uranium fuel cycle are addressed in Section 6.1 of the ESP EIS (NRC 2008a). This comment provides no new and significant information. Therefore, no changes have been made to the SEIS.

E.2.17 Comments Concerning Transportation

Comment: We understand that shipping casks have not yet been designed for the spent fuel from advanced reactor designs such as the Westinghouse AP1000. Information in the Early Site Permit Environmental Report Sections and Supporting Documentation (INEEL 2003) indicated that advanced light water reactor (LWR) fuel designs would not be significantly different from existing LWR designs; therefore, current shipping cask designs were used for the analysis of Westinghouse AP1000 reactor spent fuel shipments. EPA recommends that when shipping casks are designed for the spent fuel for the Westinghouse AP1000, the analysis should be repeated. (**0019-4** [Mueller, Heinz J.])

Response: The comment concerns the lack of availability of a certified transportation cask design for AP1000 spent fuel. The commenter is correct; shipping casks designed specifically to transport Westinghouse AP1000 spent fuel have not been developed, and the NRC staff's analysis of transportation impacts was based on current shipping cask designs for LWR spent fuel. The key shipping cask design related parameters used in the analysis in Section 6.2 of the SEIS are the cargo capacities and external radiation dose rates. The shipping cask capacities used in the NRC staff's analysis are conservative; that is, they are substantially smaller than the cargo capacities anticipated for shipping casks designed for Westinghouse AP1000 spent fuel. The small cargo capacity assumed by the NRC staff results in substantially larger numbers of spent fuel shipments and radiological impacts than would actually be expected when this plant would begin to ship spent fuel offsite. Furthermore, radiation dose rates emitted from spent fuel shipments were set to the regulatory dose rate limit in the NRC staff's analysis.

Appendix E

Actual radiation dose rates cannot be higher and are likely to be lower than the regulatory limits. Consequently, the NRC staff concludes that the transportation impact analysis presented in the ESP EIS is bounding. The staff expects that further analysis to incorporate future shipping cask designs would result in lower impacts and would not affect the NRC staff's conclusion that transportation impacts are SMALL. No changes were made to the SEIS as a result of this comment.

E.2.18 Comments Concerning the Need for Power

Comment: In the future, our community will need the clean, dependable energy provided by the new units at Plant Vogtle (**0016-12-3** [Parr, Sue])

Response: In Section 8 of the draft SEIS (NRC 2010a), the staff concluded there was a need for the power that would be generated by the proposed nuclear units. No change to the SEIS was made as a result of this comment.

E.2.19 Comments Concerning Alternatives – Energy

Comment: For the reasons stated below, the Nuclear Regulatory Commission (NRC) should:... 2) Issue a finding of LARGE environmental impacts for the proposed nuclear reactors and the conclusion that off-shore wind is the preferred alternative to nuclear. (**0003-2** [Arnold, Judy] [Boatenreiter, Glenn] [Carter, Pat] [Cumbow, Kay] [Dawson, Daneille] [Dooley, Gerald] [Falconer, Kimberly] [Hatch, Sarah] [Kasenow, Lisa] [Lewis, Marvin] [Lomas, Judith] [Lusk, Phil] [McConnell, Joy] [McNulty, Joy] [Mills, Nancy] [Sheppard, Deborah] [Thomas, Russel] [Trujillo, Dianne] [Villarreal, Tasha]); (**0002-2** [Carroll, Glenn]); (**0004-2** [Taylor, F]); (**0005-2** [Kushner, Adele])

Comment: Section 9.2 about energy alternatives. Well, the word is out -- sorry I don't have the book, I hope everybody will read it off the newsletter and this darling little piece, you can get both of these out on the table out there -- Carbon Free and Nuclear Free by 2050.

Well, a skeptic said you've got to have coal, nuclear, one of these big baseload types of energy to keep on business as usual in this world. The name is Arjun Makhijani and he works for Institute for Energy and Environmental Research. Well, he was challenged to prove that we're stuck with these large polluting, poisonous power sources. And what he found, much to his surprise, a skeptic, was that we can, with existing technology (**0016-11-4** [Carroll, Glenn])

Comment: Now it is official, in 30 years, if we will get it together, we can be off all poison power. He even covered the transportation sector's use of oil. (**0016-11-6** [Carroll, Glenn])

Response: The staff reviewed the following report cited in the comments: Carbon-Free and Nuclear-Free: A Roadmap for U.S. Energy Policy issued in 2007 by the Institute for Energy and Environmental Research (Makhijani 2007). The principal focus of the Makhijani report is to create a roadmap for zero CO_2 emissions from energy production in the United States, the phase out of existing U.S. nuclear power plants, and no licensing of new nuclear power plants in the United States.

Appendix E

In Chapter 9 of the ESP EIS (NRC 2008a), the NRC staff considered alternatives to new nuclear units at the VEGP site, including viable base-load generation alternatives involving new generating capacity such as coal-fired and natural-gas-fired power plants. The staff considered the impacts, including air quality impacts, associated with those alternatives and determined that none of the viable alternatives would be environmentally preferable to the proposed new nuclear units. In Section 7.2 of the draft SEIS (NRC 2010a), the staff also described the relative annual CO_2 emission rates of coal-fired and natural-gas-fired power plants and determined that the CO_2 emissions associated with a new nuclear power plant (including the associated fuel cycle processes and operations) would be considerably less than emissions for alternative coal-fired or natural-gas-fired plants. The staff concluded that these emissions and their impacts were small both in isolation and cumulatively when compared to these other viable sources of base-load energy. Accordingly, the discussion of CO_2 emissions in the Makhijani report did not change the staff's determinations in the ESP EIS and this SEIS with respect to air quality impacts or the comparison of energy alternatives.

Makhijani asserts in his book that nuclear power should not be part of the energy future of the United States for the following reasons: (1) the connections between nuclear power and nuclear weapons technologies and infrastructure; (2) the risks arising from severe nuclear accidents, (3) issues associated with disposal of nuclear waste; (4) the financial risks associated with nuclear power; and (5) the issue of government-provided insurance for nuclear power. Impacts from severe accidents, issues associated with nuclear waste disposal, and the costs and benefits associated with the proposed action are discussed in this SEIS and the ESP EIS (NRC 2008a) (Sections 5.10, 6.1.6, and 11.6, respectively). As discussed in this SEIS and the ESP EIS (NRC 2008a), the staff determined that impacts associated with severe accidents and the nuclear fuel cycle were small, and the staff explained the basis for its conclusion that the accrued benefits of the proposed action would most likely outweigh the economic, environmental, and social costs associated with constructing and operating two new nuclear units at the VEGP site. General policy considerations such as nonproliferation and the appropriateness of government-provided insurance are outside the scope of this SEIS. As the relevant impacts, costs, and benefits discussed in the Makhijani report and within the scope of a NEPA review were already analyzed in the ESP EIS and this SEIS, the report did not identify new and significant information with respect to the staff's conclusions.

Makhijani also suggests in his book that wind, solar, geothermal, wave energy, biomass, and hydropower should be important parts of future U.S. electricity production. NRC does not establish national energy policy, and the staff's review of alternatives pursuant to NEPA focuses on reasonable alternatives to the proposed action. In any event, all of these energy sources, except wave energy, are discussed in Section 9.2 of the ESP EIS (NRC 2008a). Wave energy is an emerging technology with limited commercial application to date (Loew 2010). The staff concluded in the ESP EIS that (1) wind and solar could not supply base-load power comparable

Appendix E

to the output of a new nuclear unit without a substantial energy storage mechanism; (2) it would be highly unlikely that energy storage such as pumped hydropower storage and compressed air energy storage could be combined with an intermittent electricity source such as wind or solar to produce a quantity of base-load power comparable to a new nuclear generating unit; and (3) new hydroelectric, geothermal, and biomass plants could not supply base-load power in the region of interest comparable to the output of a new nuclear generating unit. Accordingly, the Makhijani report did not provide new and significant information with respect to the staff's evaluation of viable alternatives to the proposed action.

Based on its review, the staff was not persuaded that the 2007 Makhijani report had the potential to change any of the staff's conclusions in Section 9.2 of this SEIS relating to alternative energy technologies. Accordingly, no change to the SEIS was made as a result of the comments that referenced the 2007 Makhijani report.

Comment: The cost for nuclear reactors must be compared and contrasted with cheaper costs and quicker build-time of shovel-ready wind and solar projects. Recent reports by DOE and Oceana Institute highlight the potential for offshore wind and Georgia is particularly well suited as is Georgia Power which already holds offshore rights from Georgia. (**0003-6** [Arnold, Judy] [Boatenreiter, Glenn] [Carter, Pat] [Cumbow, Kay] [Dawson, Daneille] [Dooley, Gerald] [Falconer, Kimberly] [Hatch, Sarah] [Kasenow, Lisa] [Lewis, Marvin] [Lomas, Judith] [Lusk, Phil] [McConnell, Joy] [McNulty, Joy] [Mills, Nancy] [Sheppard, Deborah] [Thomas, Russel] [Trujillo, Dianne] [Villarreal, Tasha]); (**0002-6** [Carroll, Glenn]); (**0005-6** [Kushner, Adele])

Comment: Diversion of $8.3 billion in tight recession funds into long-term, high-risk nuclear projects squanders funds that would bring quick returns if invested in solar, and in Georgia, especially, offshore wind.
(**0007-3** [Michetti, Susan])

Comment: The economic impacts of the Georgia Vogtle reactor proposal are of national concern, as Southern Company has signed a deal with the U.S. Department of Energy to receive $8.3 BILLION in tax-funded loans. Diversion of tight recession funds into long-term, high-risk nuclear projects squanders funds that would bring quick returns if invested in solar, and in Georgia, especially, offshore wind. (**0008-5** [Thomas, Ellen])

Comment: I could go on about the fact that there are more cost effective, less water-intensive energy choices that would actually save money in the long run, keep money here at home and protect people's health and the environment such as energy efficiency and conservation and renewables including biopower, solar and wind. (**0016-5-7** [Barczak, Sara]); (**0017-1-6** [Barczak, Sara])

Comment: Solar -- the historic cross over happened two months ago. Solar is now equal in price to nuclear. And do you think we are giving $8.2 billion to anybody to do that? Wind power generation has surpassed nuclear on the planet. This is happening. (**0016-11-5** [Carroll, Glenn])

Appendix E

Response: Under its guidance in Environmental Standard Review Plan – Review Plans for Environmental Reviews for Nuclear Power Plants, NUREG-1555 (NRC 2000), NRC only considers the cost of energy alternatives if the alternatives are found to be environmentally preferable to the proposed nuclear alternative and if the energy alternatives satisfy the purpose and need for the proposed project. The staff concluded in Section 9.2.5 of the ESP EIS (NRC 2008a) that, from an environmental perspective, none of the viable energy alternatives are clearly preferable to construction of a new base-load nuclear power generating plant. As discussed in section 9.2 of this COL SEIS, the staff did not identify any information related to energy alternatives that was both new and significant. These comments do not provide information that is both new and significant. Accordingly, no change was made to the SEIS as a result of the comments.

Comment: The two reactors proposed by Southern Company and Georgia Power to be built at Vogtle on the Savannah River are not needed. They have large environmental impacts that can be avoided by pursuing offshore wind power. (**0007-4** [Michetti, Susan])

Comment: The two reactors proposed by Southern Company and Georgia Power to be built at Vogtle on the Savannah River are not needed and have large environmental impacts that can be avoided by pursuing offshore wind power. (**0008-1** [Thomas, Ellen])

Response: The staff concluded in Section 8 of this SEIS that there is a need for power from the proposed Units 3 and 4. The staff concluded in Section 9.2.3.2 of the ESP EIS (NRC 2008a) that a wind energy facility at or in the vicinity of the VEGP site currently would not be a reasonable alternative to construction of the proposed nuclear units. The staff's impact characterizations for the proposed nuclear units and for alternative coal, natural gas, and a combination of energy resources are provided in Table 9-4 of the ESP EIS. As stated in the Executive Summary of this SEIS, matters resolved in the ESP proceedings are considered to be resolved in any subsequent proceedings absent identification of new and significant information. The staff's discussion of what constitutes new and significant information is in Section 1.6 of this SEIS. As discussed in Section 9.2 of this SEIS, the staff did not identify any information related to energy alternatives that was both new and significant. These comments do not provide information that is both new and significant. Accordingly, no change was made to the SEIS as a result of the comments.

Comment: Low-cost, low-carbon technologies could be in place more quickly at lower costs than new nuclear reactors and would be more than ample to meet electricity needs for the future. These actions would create less risks for the health and safety of workers and the public. Considering the fossil fuel needed to mine, process, create reactor fuel, build nuclear reactors and deal with long term management of the end products of nuclear reactors, there would be a significant reduction in carbon emissions for the foreseeable future. (**0013-3** [Patrie, Lewis E.])

Response: The staff concluded in Section 9.2 of the ESP EIS (NRC 2008a) that coal-fired and natural-gas-fired power plants or a combination of alternatives to which at least one of these two sources would be a significant contibutor were the only viable alternatives at the present time to providing base-load power in the amount of the proposed nuclear units. The staff compared the

Appendix E

annual CO_2 emission rates of coal, natural gas, and nuclear power plants in Table 7-1 of this SEIS. As stated in the Executive Summary of this SEIS, matters resolved in the ESP proceedings are considered to be resolved in any subsequent proceedings absent identification of new and significant information. The NRC staff's discussion of what constitutes new and significant information is in Section 1.6 of this SEIS. As discussed in Section 9.2 of this SEIS, the staff
did not identify any information related to energy alternatives that was both new and significant. This comment does not provide information that is both new and significant. Accordingly, no change was made to the SEIS as a result of this comment.

Comment: Why don't we turn it into something that uses sustainable energy, like solar or wind? You know, we're not saying get rid of the plant entirely, but I think it's better to find a way to use energy that's less dangerous. You know, you could still bring jobs to the community, still have better schools and I think a lot of people within the Shell Bluff area would be -- could sleep better at night too. (**0016-18-3** [Vincent, Patricia])

Comment: More studies should be done and I'm in favor of using alternate sustainable sources like wind and solar. (**0016-20-12** [Rivard, Betsy])

Response: The staff concluded in Sections 9.2.3.2 and 9.2.3.3 of the ESP EIS (NRC 2008a) that a wind or solar energy facility at or in the vicinity of the VEGP site currently would not be a reasonable alternative to construction of the proposed nuclear units. The staff's impact characterizations for the proposed nuclear units and alternative coal, natural gas, and a combination of energy resources are in Table 9-4 of the ESP EIS (NRC 2008a). As stated in the Executive Summary of this SEIS, matters resolved in the ESP proceedings are considered to be resolved in any subsequent proceedings absent identification of new and significant information. The staff's discussion of what constitutes new and significant information is in Section 1.6 of this SEIS. As discussed in Section 9.2 of this SEIS, the staff did not identify any information related to energy alternatives that was both new and significant. These comments do not provide information that is both new and significant. Accordingly, no change was made to the SEIS as a result of the comments.

Comment: Now given there are other and better available energy sources, such as wind, solar, biomass, I respectfully ask that you reconsider your preliminary recommendation. (**0016-22-4** [Sardi, David])

Comment: ...given the dangers and uncertainties of nuclear energy, it would be more prudent to continue to develop renewable energy such as wind, solar and biomass. Georgia has great potential in these types of energy and its potential greatly outweighs that of nuclear energy. (**0016-22-14** [Sardi, David])

Response: The staff concluded in Sections 9.2.3.2, 9.2.3.3, 9.2.3.6, and 9.2.3.8 of the ESP EIS (NRC 2008a) that a wind, solar, wood, or other biomass-derived fuel energy facility at or in the vicinity of the VEGP site currently would not be a reasonable alternative to construction of the proposed nuclear units. The staff's impact characterizations for the proposed nuclear units and alternative coal, natural gas, and a combination of energy resources are in Table 9-4 of the

Appendix E

ESP EIS. As stated in the Executive Summary of this SEIS, matters resolved in the ESP proceedings are considered to be resolved in any subsequent proceedings absent identification of new and significant information. The staff's discussion of what constitutes new and significant information is in Section 1.6 of this SEIS. As discussed in Section 9.2 of this SEIS, the NRC staff did not identify any information related to energy alternatives that was both new and significant. These comments do not provide information that is both new and significant. Accordingly, no change was made to the SEIS as a result of the comments.

E.2.20 Comments Concerning Benefit-Cost Balance

Comment: Escalating costs for massive reactor projects which will take the better part of a decade to complete strain taxpayer resources in a time of historic recession. The financing of the project with taxpayer, and Georgia ratepayer, funds should be analyzed for socioeconomic impacts. (**0003-5** [Arnold, Judy] [Boatenreiter, Glenn] [Carter, Pat] [Cumbow, Kay] [Dawson, Daneille] [Dooley, Gerald] [Falconer, Kimberly] [Hatch, Sarah] [Kasenow, Lisa] [Lewis, Marvin] [Lomas, Judith] [Lusk, Phil] [McConnell, Joy] [McNulty, Joy] [Mills, Nancy] [Sheppard, Deborah] [Thomas, Russell] [Trujillo, Dianne] [Villarreal, Tasha]); (**0005-5** [Kushner, Adele]); (**0002-5** [Carroll, Glenn])

Comment: New construction for nuclear industry is unacceptable for the following reasons:...

- huge cost overruns and little accountability to the public, either with safety or financially (**0007-10** [Michetti, Susan])

Comment: The cost of building nuclear reactors has greatly escalated. Any assumption that the traditional 'learning by doing' observed in other industries is certainly untrue for nuclear reactors, which must be built to unique and certain specifications on-site. Half of the many reactors previously ordered decades ago were never completed. (**0013-1** [Patrie, Lewis E.])

Comment: Presently the proponents of new Vogtle reactors are willing to gamble billions of tax dollars and ratepayers fees in advance.

New nuclear reactors will cost two to three times more than renewable and efficiency technologies. (**0013-2** [Patrie, Lewis E.])

Comment: Section 5.5 on page 5-6, it's about socio-economic impact. Now I don't see it really discussing the tax giveaway in the middle of the worst recession since the Great Depression. (**0016-11-2** [Carroll, Glenn])

Response: *The staff analyzed the costs and benefits of the proposed action in Chapter 11 of the ESP EIS (NRC 2008a). Cost estimates for VEGP Units 3 and 4 relied on the best available estimate of project timing and duration, noting uncertainties associated with projections into the future. NRC does not have authority to ensure that the proposed plant is the least expensive alternative to provide energy services under any particular set of assumptions concerning future circumstances. Judgments concerning the appropriate level of public funding for energy infrastructure are most often the role of State regulatory authorities, such as public service*

Appendix E

commissions. Any additional consideration by the review team would be speculative because of the dynamic nature of the rate-setting process. These comments provide no new and significant information. Therefore, no change was made to the SEIS.

Comment: There are more than a few deficiencies and oversights in the Nuclear Regulatory Commission's draft EIS on Vogtle, for instance, wind, solar and conservation are not considered in a comparison chart of different energy costs and benefits. In considering economic benefits, the EIS fails to consider that the Shell Bluff community residing in the emergency planning zone of the proposed, and existing, reactors at Vogtle does not receive basic police, fire and health services despite the purported local economic benefits of hosting a nuclear reactor. (**0007-5** [Michetti, Susan])

Comment: There are more than a few deficiencies and oversights in the Nuclear Regulatory Commission's draft EIS on Vogtle, for instance, wind, solar and conservation are not considered in a comparison chart of different energy costs and benefits. (**0008-3** [Thomas, Ellen])

Response: *Under its guidance in NUREG-1555 (NRC 2000), NRC only considers the cost of energy alternatives if the alternatives are found to be environmentally preferable to the proposed nuclear alternative and if the energy alternatives satisfy the purpose and need for the proposed project. The staff concluded in Section 9.2.5 of the ESP EIS (NRC 2008a) that, from an environmental perspective, none of the viable energy alternatives is clearly preferable to construction of a new base-load nuclear power generating plant. These comments provide no new and significant information. Therefore, no change was made to the SEIS.*

Comment: Taxpayers are fed up with funding corporations and their unfunded externalities that harm public interests, including health, environment, and financial interests. Our taxpayer dollars are not for gambling on expensive, risky, unsafe nuclear power, no matter what state the proposal to build is in. In terms of taxpayer money, this is extremely wasteful due to the inefficiencies in the extremely high financial costs of nuclear power. The financial cost is indefensible compared to renewables without the history of cradle to grave accidents and risks to public health. (**0007-2** [Michetti, Susan])

Comment: Georgia Power ratepayers now are saddled with a bum deal that will cause their electric bills to start going up come January, because of the Georgia legislature passing anti-consumer legislation in 2009 to help finance the new reactors. This nuclear power tax is a prepayment scheme that takes money out of Georgians' pocketbooks today, instead of from the wallets of Southern Company shareholders and the big industrials who managed to get exempted from this scheme, for something that may never come to fruition tomorrow -- and there will be no rebate. You are not going to get a check in the mail if this plant doesn't get built. And this all happened as the country is stuck in the middle of an historic recession that has devastated the economy, families and our overall future. And this recession has also impacted the fact that future energy projections have fallen putting projects such as this in serious question -- but nothing in the draft NRC report touches on these realities.

If Vogtle is abandoned, Southern Company and its utility partners managed to also feed from

Appendix E

the trough of the U.S. Treasury over these last four years, which is ultimately the U.S. taxpayers' checkbook, by getting an $8.3 billion conditional loan guarantee award from the Obama Administration that was awarded in February. All of us in this room could be on the hook financially for this boondoggle. ...No wonder Georgia utilities remain doggedly set on pushing the Vogtle reactors forward -- they have very little in this game and are proposing a very risk project in a very regulatory friendly environment that is shrouded in secrecy. (**0016-5-6** [Barczak, Sara]); (**0017-1-5** [Barczak, Sara])

Response: In determining the costs and benefits of the proposed action, NRC does not have authority to ensure that the proposed plant is the least expensive alternative to provide energy services under any particular set of assumptions concerning future circumstances. Judgments concerning the appropriate level of public funding for energy infrastructure most often are the role of State regulatory authorities such as public service commissions. Any additional consideration by the review team would be speculative because of the dynamic nature of the rate-setting process. The ESP EIS (NRC 2008a) considered the potential for alternative non-nuclear technologies to provide electricity that could be generated by the proposed plant and their environmental impacts in Chapter 9. An analysis of the history of the nuclear power industry that goes beyond the proposed reactors and the alternatives is beyond the scope of this SEIS. In early 2010, President Obama and the U.S. Department of Energy announced $8.3 billion in loan guarantees for Units 3 and 4 authorized by the Energy Policy Act of 2005. The loan guarantees are contingent on Southern receiving all regulatory approvals, including a COL. NRC does not have the authority to grant or restrict loan guarantees. In its COL review, the staff did not identify new and significant information concerning the projected financial costs of the proposed units. No new and significant information was provided in these comments. Therefore, no changes were made to the SEIS.

Comment: In considering economic benefits, the EIS fails to consider that the Shell Bluff community residing in the emergency planning zone of the proposed, and existing, reactors at Vogtle does not receive basic police, fire and health services despite the purported local economic benefits of hosting a nuclear reactor. (**0008-4** [Thomas, Ellen])

Response: NRC regulates the nuclear industry to protect the public health and safety. Issues related solely to the economic benefits and distribution of taxes for infrastructure and/or services, such as fire, police and health services, are outside the NRC's mission and authority and will not be addressed in the SEIS. Socioeconomic impacts of construction and operation, including impacts to public services in Burke County as a result of the plant workforce during construction and operation, were addressed in Sections 4.5 and 5.5 of the ESP EIS (NRC 2008a). This comment does not provide new and significant information. Therefore, no changes were made to the SEIS.

Comment: [We believe this expansion will allow us to continue to receive] cost-effective and reliable energy to serve our community as well as the state. (**0016-13-3** [Roberts, Ashley])

Appendix E

Response: *The comments express general support for VEGP Units 3 and 4. No new and significant information was provided in this comment. Therefore, no changes were made to the SEIS.*

Comment: The promise of $1.30 a month for the first year starting in 2011 has now turned to $3.73 a month, almost tripling what was proposed. Too many promises have been broken, financially and spiritually and I can tell you that people are afraid. (**0016-7-5** [Paul, Bobbie])

Comment: Second, Georgia Power continues to implement rate hikes to pay for these new reactors and that's burdening myself financially and I'm sure other Georgians as well. (**0016-15-3** [Henson, Courtney])

Comment: We can choose to build the reactors and continue to burden Georgians financially (**0016-15-5** [Henson, Courtney])

Response: *The comments relate to the costs of power generation that are passed on to customers. NRC's responsibility is to regulate the nuclear industry to protect the public health and safety policy. NRC is not involved in establishing the rates paid by customers. No new and significant information was provided in this comment. Therefore, no changes were made to the SEIS.*

Comment: But it's okay [to take water out of the aquifer] because we're going to use it for the almighty dollar. (**0016-6-3** [Utley, Charles])

Comment: And with those two natural things [air and sun], why in the world am I upsetting what God has given me to live on? And then I'm going to build two more of them [reactors]. Why? It's easy because it's not out of my pocket, it's out of those people who live in Georgia. (**0016-6-4** [Utley, Charles])

Comment: Reactors 3 and 4 will represent a continuation of environmental destruction. ...more money for local government and big businesses. Who suffers? (**0016-16-5** [Ogley-Oliver, Emma])

Response: *These comments express opposition to the costs of Units 3 and 4, but do not provide any new and significant information. Therefore, no changes were made to the SEIS as a result of these comments.*

Comment: Costs [of building nuclear reactors] have gone through the roof. New reactors proposed in Florida have more than tripled in cost. In fact, in just over the course of one year, Progress' estimate for the Levy County reactors in Florida sits at $5 billion more than it did in 2009, it's now $22 billion overall for the two AP1000 reactors, and they now have a five-year delay to boot. (**0016-5-2** [Barczak, Sara]); (**0017-1-2** [Barczak, Sara])

Response: *The NRC is not involved in establishing national energy policy nor does it have the authority to ensure that the proposed plant is the least costly alternative to provide energy services. Rather, it regulates the nuclear industry to protect the public health and safety within existing policy. The purpose of the ESP EIS (NRC 2008a) and this SEIS is to disclose potential*

Appendix E

environmental impacts of building and operating the proposed nuclear power plant. Chapter 11 of the ESP EIS addressed the estimated overall costs and environmental impacts of the proposed project, relying on the best available estimate of project timing and duration, while noting possible uncertainties that may affect those estimates. In its COL review, the staff did not identify new and significant information concerning the projected financial costs of the proposed units. No new and significant information was provided in these comments. Therefore, no changes were made to the SEIS as a result of these comments.

E.2.21 General Comments in Support of the Licensing Action

Comment: ...we've got a lot of our infrastructure in place now and we expect a lot of growth and with the economic situation like it is now, you know, we're excited about the future. (**0016-1-2** [DeLoach, George])

Comment: I believe in this nuclear power plant, I believe it will be good for Waynesboro, Burke County, Georgia and this great country. I feel like this panel has been thorough up to this point, I expect them to continue and I believe in the men and women of Georgia Power and the Southern Company. (**0016-2-2** [Byne, Dick])

Comment: I believe in this plant and I know it will work and I feel very comfortable with it. (**0016-2-4** [Byne, Dick])

Comment: To the Nuclear Regulatory Commission, on behalf of Augusta Technical College, a Georgia-based two-year technical college, we offer our support regarding the expansion of Plant Vogtle in Burke County by the Southern Nuclear Company. (**0016-4-1** [Elam, Terry]); (**0018-1-1** [Elam, Terry])

Comment: The expansion of Plant Vogtle opens up opportunities for innovations in training and for the industry to continue improving on its already existing high quality standards. (**0016-4-3** [Elam, Terry]); (0018-1-3 [Elam, Terry])

Comment: Augusta Technical College endorses expansion of Southern Nuclear Company's efforts in Burke County. (**0016-4-6** [Elam, Terry]); (**0018-1-5** [Elam, Terry])

Comment: Since 2005, I've been traveling to Waynesboro and we've [the Augusta Metro Chamber of Commerce] been actively engaged in the regulatory and licensing process for Vogtle's reactors 3 and 4. Our organization is a strong advocate for diversified clean and safe solutions that will meet our growing energy needs.

The Augusta Metro Chamber of Commerce supports the construction of reactors 3 and 4 at the Vogtle Generating site. (**0016-12-1** [Parr, Sue])

Comment: On behalf of the Burke County Chamber of Commerce and the Board of Directors, I would like to state that we are in full support of Georgia Power in the expansion of Plant Vogtle. (**0016-13-1** [Roberts, Ashley])

Appendix E

Comment: We believe this expansion will allow us to continue to receive clean [energy to serve our community as well as the state] (**0016-13-2** [Roberts, Ashley])

Comment: While many may argue that the community leaders such as ourselves support this expansion and Plant Vogtle because we are blinded by the dollar signs of a project of this magnitude, I can promise you there is no amount of money that would be worth sacrificing the safety and security of my family and my community. Instead, we support the company and this project because of the relationship we have developed and the safe and reliable record that they have earned over the past 20 years in our community. (**0016-13-8** [Roberts, Ashley])

Response: These comments provide general information in support of Southern's COL application. Because these comments did not provide new and significant information, no change was made to the SEIS.

E.2.22 General Comments in Support of the Licensing Process

Comment: ...I'd like to thank the NRC for having this public meeting here in Waynesboro. No other countries have the freedom that we have of dissent and being for something. (**0016-1-3** [DeLoach, George])

Comment: But beyond that, today when I was in the meeting with NRC today, not only Plant Vogtle and Southern Nuclear, all of them, how they handle themselves professionally, they have all these agencies -- and it just blew my mind today how they have to make sure that every screw, every bolt, every grain of dirt, has to be right. So I feel comfortable, and whatever we can do from the Board of Commissioners, Burke County Board of Commissioners, we're here to assist you because you are true professionals, you're a blessing for Burke County and whatever we can do to continue this relationship, we support you. (**0016-3-2** [Abrams, Lucious])

Comment: We have confidence in the regulatory process that has occurred thus far and we believe it has provided the necessary oversight to ensure the best possible outcome for our community. The Draft Supplemental Environmental Impact Statement, the DSEIS, further supports our opinion. The staff conclusion that the DSEIS finds no reason to deny the future issuance of combined operating license and an additional Limited Work Authorization is good news for Georgians. (**0016-12-2** [Parr, Sue])

Comment: [T]he continuing regulatory process assures safe and responsible construction [of the new units]. (**0016-12-4** [Parr, Sue])

Comment: We are very grateful for all concerned, everybody in this room, but most particularly the NRC in the thoroughness that you have devoted in studying the plans for this reactor expansion. (**0016-14-6** [Stone, Jesse])

Comment: I appreciate y'all opening up this forum for public comment and look forward to listening to all the thoughtful comments that are coming ahead. (**0016-14-8** [Stone, Jesse])

Appendix E

Response: These comments provide general information in support of the NRC's COL review process. Because these comments did not provide new and significant information, no change was made to the SEIS.

E.2.23 General Comments in Support of Nuclear Power

Comment: I think there is always risk in anything that we do. My business is a couple of blocks -- a block away. We have a big railroad going behind it, you know, I mean a train accident would kill us all. But we still have to take those risks...we welcome them to our community, it has been a great asset. (**0016-9-3** [Baxley, Robin])

Comment: This is -- we're not only proud to have it in our backyard, we're proud to be on the forefront of leading our country to energy independence. And we are just sorry that it has taken so long for us to get back on track. We need to catch up with other countries like Japan and France, and lead the nation in the way we need to go. (**0016-14-4** [Stone, Jesse])

Response: These comments provide general information in support of nuclear power. Because these comments did not provide new and significant information, no change was made to the SEIS.

E.2.24 General Comments in Support of the Existing Plant

Comment: Plant Vogtle has meant a great deal to this town and county and we expect it to have a great impact on us in the next five to ten years. (**0016-1-1** [DeLoach, George])

Comment: I feel they [Georgia Power and the Southern Company] have the best workforce in the southeast. ...They have treated me with respect and have answered my questions as well as can be expected. I feel very confident in their work ethics, I trust them and I appreciate their willingness to come to Burke County. (**0016-2-3** [Byne, Dick])

Comment: They [Georgia Power and Plant Vogtle] have been true professionals. And not only with being a very true professional in whatever they do, the workers, how they handle their business, how they work in the communities, and beyond the impact, we know that it's a blessing due to the economy, the way everything is going on. (**0016-3-1** [Abrams, Lucious])

Comment: From a public school perspective, we are very proud to have Plant Vogtle in our community. (**0016-8-1** [Mitchell, Tommy])

Comment: But it has been very interesting to me to see all the things that they are doing for safety and EPD and it's been a great thing. I love to learn and they are following guidelines and welcome that accountability, from what I see. ... I know that they had to recently wait -- and this is not on the record exact figures -- four months for some bird eggs that were in an area that they had to wait to purge some land. I mean the land is changing every day and I think that they are going by those guidelines and doing those things to try do research and make it as safe as possible. (**0016-9-2** [Baxley, Robin])

Appendix E

Comment: Now Georgia Power is here, Plant Vogtle is here. We can't do anything about that, two are already here. And these two, Georgia Power and Plant Vogtle, has become bread to this community. And I say don't fight the hand that feeds you bread. (**0016-10-2** [Stephens, Annie Laura])

Comment: I would like to say that Plant Vogtle is one of the finest corporate citizens a community could ask for and we are proud to have them in ours. Whether it is through civic involvement or a charitable cause, we can always count on overwhelming support of the company and the employees. (**0016-13-7** [Roberts, Ashley])

Comment: We are proud to have Plant Vogtle in our backyard. (**0016-14-3** [Stone, Jesse])

Comment: Now we're blessed so much it's hard to describe, at having Plant Vogtle here. (**0016-14-5** [Stone, Jesse])

Response: These comments express support of the existing units at the site. Because these comments did not provide new and significant information, no change was made to the SEIS.

E.2.25 General Comments in Opposition to the Licensing Action

Comment: The NRC is urged not to issue a Combined Licenses for Vogtle Electric Generating Plant Units 3 and 4 because of flawed electrical systems inherent to the AP1000 which fail to meet AP1000 compliance documents as well as NRC safety requirements and regulations. (**0001-1** [Baxter, Farouk])

Comment: For the reasons stated below, the Nuclear Regulatory Commission (NRC) should: 1) Deny Southern Company and its subsidiaries additional limited work authorization (LWA) for further construction related to proposed, unneeded, and still-unlicensed reactors on the Vogtle site in Georgia. (**0003-1** [Arnold, Judy] [Boatenreiter, Glenn] [Carter, Pat] [Cumbow, Kay] [Dawson, Daneille] [Dooley, Gerald] [Falconer, Kimberly] [Hatch, Sarah] [Kasenow, Lisa] [Lewis, Marvin] [Lomas, Judith] [Lusk, Phil] [McConnell, Joy] [McNulty, Joy] [Mills, Nancy] [Sheppard, Deborah] [Thomas, Russel] [Trujillo, Dianne] [Villarreal, Tasha]); (**0002-1** [Carroll, Glenn]); (**0004-1** [Taylor, F])

Comment: Please halt these plans. (**0008-6** [Thomas, Ellen])

Comment: If NRC permits Georgia Power to add two more, it would double the danger of radiation exposure, double the risk of nuclear accidents, and double the impact on future generations. (**0009-4** [Zeller, Lou])

Comment: [T]he proposed construction of nuclear reactors would be counter productive, considering the proposed alternatives. (**0013-4** [Patrie, Lewis E.])

Comment: Regulators, in our opinion, continue to have blinders on. We again believe that the NRC has failed to protect the public by recommending approval of Georgia Power and its utility partners' push to build two new reactors here for an estimated $14 billion price tag.

Appendix E

I mentioned four years ago that the issue of building more nuclear reactors would affect not just this local community, but Georgia as a whole and our region overall. And I had hoped that the NRC staff understood that it was important to do something that would benefit all, not just a select few. Sadly, that has not happened. (**0016-5-1** [Barczak, Sara]); (**0017-1-1** [Barczak, Sara])

Comment: In closing, we hope that the NRC and other regulators overseeing this project will step back and rethink all of this, will step back from all the hoopla surrounding this boondoggle and do what is right for the public and our natural resources and deny the license for the proposed Vogtle reactors. ...As I said four years ago, the future of not only this community, but many, many others are at stake. (**0016-5-10** [Barczak, Sara])

Comment: It is not fair for the power companies to be given the biggest straw to pull from our precious water resources and a blank check from our wallets. (**0016-5-11** [Barczak, Sara]); (**0017-1-10** [Barczak, Sara])

Comment: ...we're all connected here in this country, whether we live in Waynesboro; Shell Bluff, four miles from the reactor as the crow flies; Atlanta; Rockville, Maryland; whatever. And we know that DOE and NRC and EPA and DNR and South Carolina DHEC and all these people talk to each other. And I've come here to plead with you that what was rejected in this or whatever the term is -- no change -- are the things that deeply impact the people that live in this community, especially around the reactors. (**0016-7-2** [Paul, Bobbie])

Comment: I can see that Georgia Power, Plant Vogtle are determined to build two more new reactors to the two existing reactors not regarding the affliction, the burden and the confusion that they are bringing to the community of Burke County and all other communities where these reactors are located. (**0016-10-1** [Stephens, Annie Laura])

Comment: I've been deeply concerned about the two new nuclear reactors that are proposed at the Vogtle site. (**0016-15-1** [Henson, Courtney])

Comment: I hope we will consider the latter [choose to stop the construction and take one step forward to a better, safer Georgia]. (**0016-15-7** [Henson, Courtney])

Comment: Reactors 3 and 4 will represent a continuation of environmental destruction. More polluted land and water,...Who suffers? (**0016-16-4** [Ogley-Oliver, Emma])

Comment: I am greatly concerned over Vogtle's proposed new reactors here in Waynesboro. (**0016-22-1** [Sardi, David])

Comment: In closing, we hope that the NRC and other regulators overseeing this project will step back and rethink all of this - will step back from all the hoopla surrounding this boondoggle and do what is best for the public and our natural resources and deny the license for the proposed Vogtle reactors. (**0017-1-9** [Barczak, Sara])

Appendix E

Response: *These comments provide general information in opposition to Southern's COL application. Because these comments did not provide new and significant information, no change was made to the SEIS.*

E.2.26 General Comments in Opposition to the Licensing Process

Comment: For the reasons stated below, the Nuclear Regulatory Commission (NRC) should: 1) Deny Southern Company and its subsidiaries additional limited work authorization (LWA) for further construction related to proposed, unneeded, and still-unlicensed reactors on the Vogtle site in Georgia. (**0005-1** [Kushner, Adele])

Comment: Well, something needs to be re-looked at in the EIS. (**0016-11-7** [Carroll, Glenn])

Comment: I do not believe the NRC should approve this permit. Thank you for your time and consideration. (**0016-22-15** [Sardi, David])

Response: *These comments provide general information in opposition to NRC's COL review process. Because these comments did not provide new and significant information, no change was made to the SEIS.*

E.2.27 General Comments in Opposition to Nuclear Power

Comment: I would like to state my displeasure with the idea of having two new nuclear reactors in Georgia. (**0006-1** [Smith, Nathan])

Comment: I wish to be on record opposing all taxpayer funding and construction of new nuclear power plants, specifically this includes the two Vogtle plants for Georgia. (**0007-1** [Michetti, Susan])

Comment: The global warming and associated climate change are introducing new hazards to the use of nuclear plants that does not exist with renewable energy. It is no longer acceptable in public opinion to built new nuclear power plants that harm public interests in many ways. (**0007-7** [Michetti, Susan])

Comment: I ask the NRC to not permit new plant construction due to this unsolved waste problem that appears to have no solution that doesn't endanger public health with long-term consequences of radioactivity. (**0007-15** [Michetti, Susan])

Comment: But why bother [talking about alternative energy sources]? As it all falls upon deaf ears in terms of the NRC and I'm afraid of other regulators overseeing this project. Let's face it, Georgia is using its natural resources, impacting its citizens' health, and allowing radioactive nuclear waste to pile up within its borders to power other states' air conditioning units and to line Southern Company's shareholders' wallets. (**0016-5-8** [Barczak, Sara])

Comment: This is not a proper way of doing business. This is a relatively new way of doing business, it can't make it, it's going out of business. This is happening. We can get ripped off

Appendix E

until the cows come home. I predict no reactor will ever come on line in this country again. We should save our money, we should give the good folks in Shell Bluff emergency services and a grocery store at a minimum. (**0016-11-8** [Carroll, Glenn])

Comment: We have a choice today, let's choose health and prosperity, not radiation and devastation. (**0016-16-6** [Ogley-Oliver, Emma])

Comment: [This is a poem written by the commenter.] There was a community that was a peaceful area, they made their living off the land. They had strong family moral values and they passed it on to the next generation, their land and their homes. But as time moved on, there was a pimp that observed the way that they lived. He disguised himself to take advantage of the community.

The pimp decided to bring two females and to take the man from his family. So if you kill the head, the body will die. Those two females were prostitutes, they had a disease that is called AIDS. So he got the man out and he began to enjoy the pleasures of life. The man did not know that these two females had AIDS. The pimp knew because he was their master, so he thought. The pimp made good profit on the two prostitutes. He had nowhere to take them after being used but to store all their venom in the land. Their scent got in the air, water and soil. The community started dying because of them.

The pimp saw how much wealth he had made. So he got him two more prostitutes to bring in the area. But this time he shared some of the wealth with some of the community, so they were blinded by their desire and did not warn the community of the lies and the sickness in the land.

For she has cast down many wounded, many strong men have been slain by her. Her house is the way to hell, going down to the chambers of death. When you allow the dollar and human lives to control your decisions, then God will handle you. (**0016-17-3** [Howard, Claude])

Comment: Now I don't know about everybody here, but personally I'm comfortable sleeping in my bed that's not near a nuclear reactor. I do not value nuclear energy because from what I've seen and heard, they bring death to communities that are near them. (**0016-18-1** [Vincent, Patricia])

Comment: Hailing nuclear energy as a replacement for fossil fuels as a solution for global warming would be dangerous and irresponsible in a post-9/11 world. First of all, the United States will lose all moral authority in trying to deny North Korea and Iran their right to pursue nuclear energy. We can't champion nuclear energy as the future and at the same time reasonably keep it from the rest of the world. (**0016-22-5** [Sardi, David])

Comment: For over five decades, nuclear power has diverted major funds away from the development of more benign but powerful forms of energy production... (**0016-22-12** [Sardi, David])

Comment: But why bother as it falls upon deaf ears in terms of the NRC and I'm afraid other regulators overseeing this project? Let's face it. Georgia is using its natural resources, impacting

Appendix E

its citizens' health, and allowing radioactive nuclear waste to pile up within in its borders to power other states' air conditioning units and to line Southern Company's shareholders' wallets. (**0017-1-7** [Barczak, Sara])

Response: *These comments provide general information in opposition to nuclear power. Because these comments did not provide new and significant information, no change was made to the SEIS.*

E.2.28 Comments Concerning Issues Outside Scope – Emergency Preparedness

Comment: ...I'd like to say that tonight I was disappointed that there were not the booklets that are handed out when you go to Plant Vogtle or when you ask for evacuation routes. There was PR on Vogtle, there was one line in there that mentioned the public or public safety about evacuation. I've looked at this book with a checklist. There are four levels of radiation releases and you're supposed to look at it and determine which one is safe to stay in your house and which one you get in your car and get the hell out of Dodge. How to put a cloth on your mouth, turn off your air conditioning, shut down your heating, shut the windows. It's a new form of terrorism for the people living around these reactor sites. And I just ask for further screening on the NRC's part. (**0016-7-6** [Paul, Bobbie])

Response: *These comments relate to the adequacy of emergency plans, which is a safety issue that is outside the scope of the staff's environmental review. As part of its site safety review, the NRC staff will determine, after consultation with the U.S. Department of Homeland Security and the Federal Emergency Management Agency, whether emergency plans submitted by Southern meet applicable requirements. No change was made to the SEIS as a result of this comment.*

E.2.29 Comments Concerning Issues Outside Scope – Miscellaneous

Comment: Walked in to visit with her [the commenter's aunt] and she said, Well, tell me one thing about it, Charles, I just can't understand why my electric bill keeps going up. I said, Because they decided that you need to pay to build something that you're going to give a blank check and when they get through, you're going to pay to use it because every time you cut it on, you're paying. I said, You know, that's a good concept. Why don't I come up with something and you pay me to build it and I in return sell it back to you and you then buy it back from me. Isn't that crazy? That's what you're doing, that's exactly what you're doing. (**0016-6-5** [Utley, Charles])

Comment: I heard that the monitoring from the Department of Energy to the state of Georgia had been cut after years. And I couldn't believe it. ... This was right when the secret energy talks were happening in Washington and no one would disclose who was in them. We know Southern Company was there. I wondered today, when that was cut if those findings that our Georgia EPD -- Environmental Protection Division of DNR, DOE, everything -- had found or had explored, sampled and tested in beer, peanuts, pears, fish, the river -- I wondered why we didn't want that information any more. Who didn't want to have information about their community,

Appendix E

about their environment? ...radiation doesn't acknowledge state boundaries.

So we've been working for about 14 months to restore that and we have a commitment from the Department of Energy and our state -- Georgia EPD --... (**0016-7-1** [Paul, Bobbie])

Comment: As was said earlier, Georgia Power has the power to do what they want to do. They have the ability to buy who they want to buy. (**0016-17-2** [Howard, Claude])

Comment: The AP1000 in Finland, I have heard about, it's not on line yet and they have made multiple design changes. I don't know if their design is design or not, but they've had many cost overruns and it's still not on line yet and it's way behind schedule. (**0016-20-2** [Rivard, Betsy])

Comment: So for these reasons, continued and increased reliance on nuclear energy does not and cannot make sense within America's national security policy. (**0016-22-9** [Sardi, David])

Response: This environmental review focuses on significant issues related to the proposed action. Having a defined scope for the environmental review allows the NRC to concentrate on the essential issues for actions under consideration rather than on issues that may have been or are being evaluated through different regulatory review processes. The issues raised in these comments are outside the scope of the environmental review process and were not addressed in the SEIS.

E.2.30 Comments Concerning Issues Outside Scope – NRC Oversight

Comment: I'd just like to say ... my relationship with the NRC at city hall and others has been very professional and I thank you for what you are doing and most of all I thank you for having the safety of the general public first in your mind. (**0016-1-4** [DeLoach, George])

Comment: If there's anything that I have learned -- the more that I learn, the less that I know. And I think that's the reason we have to ask Georgia Power, we have to ask the NRC questions and we have to continue to ask questions and you have to hold them accountable. (**0016-2-1** [Byne, Dick])

Response: These comments provide general information regarding the NRC oversight process. These comments provide no new and significant information. Therefore, no change was made to the SEIS.

E.2.31 Comments Concerning Issues Outside Scope – Safety

Comment: Section 3.2 of the Draft Supplemental Environmental Impact Statement indicates that the AP1000 design has been certified by the NRC, but is presently undergoing further review by the NRC. The flawed electrical design of the AP1000 identified herein should also be resolved by the NRC prior to issue of the COL. The AP1000 design is flawed because it has failed to comply with the requirements of IEEE Standard 603 requiring the electrical portion of the safety systems that perform safety functions be classified as Class 1E. Compliance with

Appendix E

IEEE Standard 603 would require the Ancillary Diesel Generators to be classified as Class 1E versus the present Non-Safety Related Commercial Grade classification. IEEE Standard 603 is listed by AP1000 as a compliance document with no exceptions; however, AP1000 does not comply with its requirements. IEEE Standard 603 is also endorsed by NRC Regulatory Guide 1.153, and defines the functional requirements of the Safety System, and directs that electrical portions of the Safety System be classified as Class 1E; AP1000 also indicates complete conformance with Regulatory Guide 1.153, but the design does not comply. (**0001-2** [Baxter, Farouk])

Response: The issues raised in this comment are outside the scope of the environmental review and are not addressed in the SEIS. The safety assessment for the proposed licensing action was provided as part of the application. The NRC is in the process of developing a safety evaluation report that analyzes all aspects of reactor and operational safety. The issues raised in the comment that are specific to the AP1000 design are being addressed in the staff's separate review of Westinghouse's proposed amendment to the AP1000 design certification. The NRC will issue a license or permit only if there is reasonable assurance that (1) the activities authorized by the license or permit can be conducted without endangering the health and safety of the public and (2) such activities will be conducted in compliance with the rules and regulations of the Commission. No change was made to the SEIS as a result of this comment.

Comment: The details of flawed electrical of electrical design are identified in the six attachments of detailed correspondence between Mr. Michael Johnson, NRC Director, New Reactors, and his staff. I had initially written to Mr. Johnson identifying safety flaws in the electrical design of the AP1000, and though a response was received from Mr. Johnson, as well as from Mr. Bergman, Mr. Chopra, and Mr. Jaffe; the final disposition from the Mr. Jaffe was that NRC did not have the time to review every concern that was brought to their attention; and therefore, no further action was planned to be undertaken by the NRC. (**0001-3** [Baxter, Farouk])

Comment: LIMITED WORK AUTHORIZATION Westinghouse's AP1000 reactor design has unresolved safety issues likely to impact the outcome of the licensing review as well as the final cost of the proposed reactors. Containment failure in an accident and analysis of impacts from a direct airline strike are unresolved safety issues of concern even to the NRC license review staff. These factors are likely to affect financing for, and the viability of, the proposed project. Therefore, it is premature to authorize any further work to the Vogtle site.
(**0003-3** [Arnold, Judy] [Boatenreiter, Glenn] [Carter, Pat] [Cumbow, Kay] [Dawson, Daneille] [Dooley, Gerald] [Falconer, Kimberly] [Hatch, Sarah] [Kasenow, Lisa] [Lewis, Marvin] [Lomas, Judith] [Lusk, Phil] [McConnell, Joy] [McNulty, Joy] [Mills, Nancy] [Sheppard, Deborah] [Thomas, Russel] [Trujillo, Dianne] [Villarreal, Tasha]); (**0005-3** [Kushner, Adele]); (**0002-3** [Carroll, Glenn])

Comment: Westinghouse's AP1000 reactor design has unresolved safety issues likely to impact the outcome of the licensing review as well as the final cost of the proposed reactors. (**0004-3** [Taylor, F])

Comment: New construction for nuclear industry is unacceptable for the following reasons:

Appendix E

- the threat of serious accidents or incidents that could contaminate land and waters and all who live there, locally and perhaps globally - for a very long period of time

(**0007-11** [Michetti, Susan])

Comment: Most astonishingly -- and we heard it discussed just this evening -- the AP1000 design still is not certified, Revision. I think all of us vividly remember being told that having a certified design would make this process much smoother, save money and on and on. Well, that hasn't happened and the most recent news is that Westinghouse has again missed another deadline. Yes, maybe eventually they'll get it together and the NRC will approve the design, but it has certainly been a long and bumpy road. (**0016-5-3** [Barczak, Sara]); (**0017-1-3** [Barczak, Sara])

Comment: I can understand all of your concerns about safety, and believe me, we are concerned too and we're not dumb, we read these preliminary reports and we study them and we know the experiences of our friends, our workers, our family members, our colleagues who work out there and have worked out there for years. We know the safety record that Southern Company, Southern Nuclear and Georgia Power and all the other partners in that venture have chalked up. (**0016-14-2** [Stone, Jesse])

Comment: First, the AP1000 design has gone through several revisions and it's still not safe. (**0016-15-2** [Henson, Courtney])

Comment: [We can choose to build the reactors] and put their safety at risk (**0016-15-6** [Henson, Courtney])

Comment: ...my concern is the design is not complete. (**0016-20-1** [Rivard, Betsy])

Comment: EPA understands that concerns have been raised by the NRC that certain structural components of the revised AP1000 shield building may not be suitable to withstand design loads. The shield building is designed to protect the reactor's primary containment from severe weather and other events, as well as serving as a radiation barrier and also supporting an emergency cooling water tank. It is EPA's understanding that the NRC is currently reviewing the remainder of the next- generation reactor's design certification amendment application, and that Westinghouse is expected to make design modifications and conduct safety testing to ensure the shield building design can meet its safety functions.

The FSEIS should address the status of the Westinghouse AP1000 certification review and related issues, particularly the analysis of the structural integrity of the AP1000. We understand that the Safety Evaluation Report will address these issues in even more detail, and that the certification review may be completed as soon as December 2010. EPA understands that Revision 15 of the AP1000 design is codified in 10 CFR Part 52, Appendix D. EPA concurs with NRC's plan to conduct an additional environmental review if changes result in the final design being significantly different from the design considered in the DEIS. (**0019-5** [Mueller, Heinz J.])

Appendix E

Response: The NRC's principal responsibility is to protect the health and safety of the public when authorizing the use of radioactive material. Because NEPA regulations do not include a safety review, the NRC has codified the regulations for preparing an EIS separately from the regulations for reviewing safety issues. The regulations governing the environmental review are set forth in 10 CFR Part 51, and the regulations covering the safety review are in 10 CFR Part 52. For this reason, the license process includes an environmental review that is distinct and separate from the safety review. Because the two reviews are separate, operational safety issues are considered outside the scope of the environmental review, just as environmental issues are not considered part of the safety review. However, the staff forwards safety issues that are raised during the environmental review to the appropriate NRC organization for consideration and appropriate action. At this time, the staff has identified no changes in the design being evaluated in the AP1000 design certification amendment proceeding that differ significantly from the design considered in this SEIS. These comments are related to safety and are outside the scope of the staff's environmental review. Therefore, no changes were made to the SEIS.

E.2.32 Comments Concerning Issues Outside Scope – Security and Terrorism

Comment: It's my belief that granting this permit will impact our national security... (**0016-22-2** [Sardi, David])

Comment: Second, the dangerous materials could potentially make it a prime target for terrorists attempting to harm the United States. (**0016-22-6** [Sardi, David])

Response: Comments related to security and terrorism are safety issues that are outside the scope of the environmental review. However, NRC is devoting substantial time and attention to terrorism-related matters, including coordination with the U.S. Department of Homeland Security. As part of its mission to protect public health and safety and the common defense and security pursuant to the Atomic Energy Act, the NRC staff is conducting vulnerability assessments for the domestic utilization of radioactive material. Since September 2001, NRC has identified the need for license holders to implement compensatory measures and has issued several orders to license holders imposing enhanced security requirements. Finally, NRC has taken actions to ensure that applicants and license holders maintain vigilance and a high degree of security awareness. Consequently, NRC will continue to consider measures to prevent and mitigate the consequences of acts of terrorism in fulfilling its safety mission. Additional information about the NRC staff's actions regarding physical security since September 11, 2001, can be found on NRC's public web site (www.nrc.gov). No change was made to the SEIS as a result of these comments.

E.3 References

10 CFR Part 20. Code of Federal Regulations, Title 10, *Energy*, Part 20, "Standards for Protection Against Radiation."

Appendix E

10 CFR Part 50. Code of Federal Regulations, Title 10, *Energy*, Part 50, "Domestic Licensing of Production and Utilization Facilities."

10 CFR Part 51. Code of Federal Regulations, Title 10, *Energy*, Part 51, "Environmental Protection Regulations for Domestic Licensing and Related Regulatory Functions."

10 CFR Part 52. Code of Federal Regulations, Title 10, *Energy*, Part 52, "Licenses, Certifications, and Approvals for Nuclear Power Plants."

40 CFR Part 1508. Code of Federal Regulations, Title 40, *Protection of Environment*, Part 1508, "Terminology and Index."

64 FR 68005. December 6, 1999. "Waste Confidence Decision Review: Status." *Federal Register*, U.S. Nuclear Regulatory Commission.

66 FR 65256. December 18, 2001. "National Pollutant Discharge Elimination System: Regulations Addressing Cooling Water Intake Structures for New Facilities." *Federal Register*, Environmental Protection Agency.

69 FR 52040. August 24, 2004. "Policy Statement on the Treatment of Environmental Justice Matters in NRC Regulatory and Licensing Actions." *Federal Register*, U.S. Nuclear Regulatory Commission.

72 FR 49429. August 28, 2007. "Licenses, Certification, and Approvals for Nuclear Power Plants; Final Rule." *Federal Register*, U.S. Nuclear Regulatory Commission.

72 FR 57416. October 9, 2007. "Limited Work Authorizations for Nuclear Power Plant; Final Rule." *Federal Register*, U.S. Nuclear Regulatory Commission.

73 FR 59551. October 9, 2008. "Waste Confidence Decision Update." *Federal Register*, U.S. Nuclear Regulatory Commission.

75 FR 54190. September 3, 2010. "Southern Nuclear Operating Company; Notice of Availability of the Draft Supplemental Environmental Impact Statement for Combined Licenses (COLs) for Vogtle Electric Generating Plant Units 3 and 4 and Associated Public Meeting." *Federal Register*, U.S. Nuclear Regulatory Commission.

Atomic Energy Act of 1954. 42 U.S.C. 2011, et seq.

Alldred, M. and K. Shrader-Frechette. 2009. "Environmental Injustice in Siting Nuclear Plants." *Environmental Justice*. 2(2):85-96.

Appendix E

Baker, P.J. and D.G. Hoel. 2007. "Meta-analysis of standardized incidence and mortality rates of childhood leukemia in proximity to nuclear facilities." *European Journal of Cancer Care* 16(4):355-363.

Clean Water Act (CWA). 33 USC 1251, et seq. (also called the Federal Water Pollution Control Act [FWPCA])

Connecticut Academy of Science and Engineering. 2000. *Study of Radiation Exposure From the Connecticut Yankee Nuclear Power Plant.* Available at http://www.ctcase.org/reports/index.html.

Endangered Species Act of 1973. 16 USC 1531, et seq.

Executive Order 12898. 1994. "Federal Actions to Address Environmental Justice in Minority Populations and Low-Income Populations." Office of the President, Washington, D.C.

Ga. Code Ann. 61-9. 2010. "Water Pollution Control Permits." *Georgia Code Annotated.*

Georgia Power Corporation (GPC). 2006. *Avian Protection Program for Georgia Power Company, Rev..1.* March 14, 2006. Accession No. ML063000228.

Health Physics Society (HPS). 2004. *Radiation Risk in Perspective: Position Statement of the Health Physics Society, Rev.* Health Physics Society, McLean, Virginia.

Hoffmann, W., C.Terschueren, and D.B. Richardson. 2007. "Childhood leukemia in the vicinity of the Geesthacht nuclear establishments near Hamburg, Germany." *Environmental Health Perspectives* 115(6):947-952.

International Commission on Radiological Protection (ICRP). 1971. "1971 Recommendations of the ICRP." ICRP Publication 26, *Annals of the ICRP* 1(3).

International Commission on Radiological Protection (ICRP). 1978. "Limits for Intakes of Radionuclides by Workers." ICRP Publication 30, *Annals of the ICRP* 8(4).

Jablon, S.Z. Hrubec, J. D. Boice, Jr., and B. J. Stone. 1990. *Cancer in Populations Living Near Nuclear Facilities.* NIH Pub. No. 90-874, National Institutes of Health, Washington, D.C.

Loew, T. 2010. "Oregon is first U.S. site for a wave-power farm." *USA TODAY.* February 17, 2010.

Makhijani, A. 2007. *Carbon-Free and Nuclear-Free: A Roadmap for U.S. Energy Policy.* Joint Project of the Nuclear Policy Research Institute and the Institute for Energy and Environmental

Appendix E

Research, IEER Press and RDR Books, Takoma Park, Maryland. Free PDF download available at http://www.ieer.org/carbonfree/.

Mangano, J.J., J. Sherman, C. Chang, A. Dave, E. Feinberg, and M. Frimer. 2003. "Elevated childhood cancer incidence proximate to U.S. nuclear power plants." *Archives of Environmental Health* 58(2):74-82.

Meinke, W.W. and T.H. Essig. 1991a. *Offsite Dose Calculation Manual Guidance: Standard Radiological Effluent Controls for Boiling Water Reactors.* NUREG-1302, U.S. Nuclear Regulatory Commission, Washington, D.C.

Meinke, W.W. and T.H. Essig. 1991b. *Offsite Dose Calculation Manual Guidance: Standard Radiological Effluent Controls for Pressurized Water Reactors.* NUREG-1301, U.S. Nuclear Regulatory Commission, Washington, D.C.

National Research Council. 2006. *Health Risks for Exposure to Low Levels of Ionizing Radiation: BEIR VII - Phase 2.* National Academies Press, Washington, D.C.

Silva-Mato, A., D. Viana, M.I. Fernandez-SanMartin, J. Cobos, and M. Viana. 2003. "Cancer risk around the nuclear power plants of Trillo and Zorita (Spain)." *Occupational and Environmental Medicine* 60(7):521-527.

Southern Nuclear Operating Company, Inc. (Southern). 2007. *Vogtle Early Site Permit Application.* Rev. 2, Southern Company, Birmingham, Alabama.

Southern Nuclear Operating Company, Inc. (Southern). 2010c. Letter from C.R. Pierce, Southern, to NRC. "Southern Nuclear Operating Company, Vogtle Electric Generating Plant Units 3 and 4 Combined License Application Comments on Draft Supplemental Environmental Impact Statement." Letter ND-10-2199. November 23,1010. Southern Company, Birmingham, Alabama. Accession No. ML103300035.

Talbott, E.O., A.O. Youk, K.P. Hugh-Pemu, and J.V. Zborowski. 2003. "Long-term follow-up of the residents of the Three Mile Island accident area: 1979-1998." *Environmental Health Perspective* 111(3):341-348.

U.S. Atomic Energy Commission (AEC). 1974. *Measuring, Evaluating, and Reporting Radioactivity in Solid Wastes and Releases of Radioactive Materials in Liquid and Gaseous Effluents form Light-Water-Cooled Nuclear Power Plants.* Regulatory Guide 1.21, Washington, D.C.

U.S. Atomic Energy Commission (AEC). 1975. *Programs for Monitoring of Nuclear Power Plants.* Regulatory Guide 4.1, Washington, D.C.

Appendix E

U.S. Fish and Wildlife Service (FWS). 2008. Letter from FWS to NRC regarding USFWS Log# 08-FA-0473. September 19, 2008. Accession No. ML082760694.

U.S. Fish and Wildlife Service (FWS). 2010a. Letter from FWS to NRC regarding USFWS Log# 2009-1387. February 12, 2010. Accession No. ML100500426.

U.S. Fish and Wildlife Service (FWS). 2010b. Letter from FWS to NRC regarding USFWS Log# 2010-1254. October 20, 2010. Accession No. ML103010076.

U.S. Nuclear Regulatory Commission (NRC). 2000. *Environmental Standard Review Plan—Review Plans for Environmental Reviews for Nuclear Power Plants.* NUREG-1555, NRC, Washington, D.C. Includes 2007 updates.

U.S. Nuclear Regulatory Commission (NRC). 2006. *Liquid Radioactive Release Lessons Learned Task Force Final Report.* Accessed at www.nrc.gov/reactors/operating/ops experience/tritium/lr-release-lessons-learned.pdf. Accession No. ML083220312.

U.S. Nuclear Regulatory Commission (NRC). 2007. *Quality Assurance for Radiological Monitoring Programs (Inception Through Normal Operations to License Termination) – Effluent Streams and the Environment.* Regulatory Guide 4.15, Washington, D.C.

U.S. Nuclear Regulatory Commission (NRC). 2008a. *Final Environmental Impact Statement for an Early Site Permit (ESP) at the Vogtle Electric Generating Plant Site.* NUREG-1872, Vols. 1 and 2, and Errata, Washington, D.C. Accession Nos. ML082240145; ML082240165, ML082260203; ML082550040.

U.S. Nuclear Regulatory Commission (NRC). 2008b. Letter from NRC to FWS dated January 25, 2008. Subject: Biological Assessment for Threatened and Endangered Species and Designated Critical Habitat for the Vogtle Electric Generating Plant Early Site Permit Application. Accession No. ML080100512.

U.S. Nuclear Regulatory Commission (NRC). 2009. *Duke Energy Carolinas, LLC and Tennessee Valley Authority* CLI-09-21, Docket No. 52-018-COL, 52-019-COL, 52-014-COL, 52-015-COL.

U.S. Nuclear Regulatory Commission (NRC). 2010a. *Draft Supplemental Impact Statement for a Combined Licenses (COLs) for Vogtle Electric Generating Plant Units 3 and 4 –NUREG-1947,* Washington, D.C. Accession No. ML102000138.

U.S. Nuclear Regulatory Commission (NRC). 2010b. Letter from G.P. Hatchett, NRC, to S. Tucker, FWS, "Request for a List of Protected Species Within the Area Under Evaluation for the Vogtle Electric Generating Plant, Units 3 and 4 Combined License Applications," January 7, 2010. Washington, D.C. Accession No. ML092600684.

Appendix E

U.S. Nuclear Regulatory Commission (NRC). 2010c. *Summary of the Public Meeting for the Draft Supplemental Environmental Impact Statement for the Combined Licenses for Vogtle Electric Generating Plant, Units 3 and 4.* Washington, D.C. Accession No. ML103130579.

U.S. Nuclear Regulatory Commission (NRC), Biological Assessment for Threatened and Endangered Species and Designed Critical Habitat for the Vogtle Electrical Generating Plant, Units 3 and 4 Combined License Application

Appendix F

Key Consultation Correspondence

Appendix F

Key Consultation Correspondence

Key consultation correspondence during the evaluation process of the application for combined licenses for Units 3 and 4 at the Vogtle Electric Generating Plant (VEGP) is identified in Table F-1. A list of pertinent correspondence generated during the preparation of this supplemental environmental impact statement is located in Appendix C. Copies of the correspondence listed in Table F-1 are included at the end of this appendix. Correspondence information relative to the early site permit (ESP) review of Units 3 and 4 can be found in Appendix F of ESP environmental impact statement, dated August 2008.

Table F-1. Key Consultation Correspondence Regarding the Combined Operating License Application for Units 3 and 4 at the VEGP Site

Source	Recipient	Date of Letter
U.S. Nuclear Regulatory Commission (NRC) (Mr. Gregory P. Hatchett)	Georgia Department of National Resources (Dr. Dave Crass)	December 9, 2009
NRC (Mr. Gregory P. Hatchett)	Poarch Band of Creek Indians (Ms. Stephanie Rolin)	December 10, 2009
NRC (Mr. Gregory P. Hatchett)	United Keetoowah Band of Cherokee Indians (Ms. Emma Sue Holland)	December 10, 2009
NRC (Mr. Gregory P. Hatchett)	Poarch Band of Creek Indians (Mr. Eddie Tullis)	December 10, 2009
NRC (Mr. Gregory P. Hatchett)	Eastern Band of Cherokee Indians (Ms. Kathy McCoy)	December 10, 2009
NRC (Mr. Gregory P. Hatchett)	Coushatta Tribe Louisiana (Mr. John Zachary)	December 10, 2009
NRC (Mr. Gregory P. Hatchett)	Kialegee Tribal Town (Ms Evelyn Bucktrot)	December 10, 2009
NRC (Mr. Gregory P. Hatchett)	Miccosukee Tribe of Indians of Florida (Mr. Steven Terry)	December 10, 2009
NRC (Mr. Gregory P. Hatchett)	Poarch Band of Creek Indians (Ms. Gale Thrower)	December 10, 2009
NRC (Mr. Gregory P. Hatchett)	Thlopthlocco Tribal Town (Mr. Louis McGertt)	December 10, 2009
NRC (Mr. Gregory P. Hatchett)	Muscogee National (Mr. A. D. Ellis)	December 10, 2009
NRC (Mr. Gregory P. Hatchett)	Cherokee Nation of Oklahoma (Mr. Richard L. Allen)	December 10, 2009

Appendix F

Table F-1. (contd)

Source	Recipient	Date of Letter
NRC (Mr. Gregory P. Hatchett)	Chickasaw Nation (Ms. Gingy Nail)	December 10, 2009
NRC (Mr. Gregory P. Hatchett)	Chickasaw Nation of Oklahoma (Mr. Bill Anoatubby)	December 10, 2009
NRC (Mr. Gregory P. Hatchett)	Georgia Tribe of Easter Cherokee (Mr. Charles Thurmond)	December 10, 2009
NRC (Mr. Gregory P. Hatchett)	Alabama-Quassarte Tribal Town (Mr Tarpie Yargee)	December 10, 2009
NRC (Mr. Gregory P. Hatchett)	Seminole Nation of Oklahoma (Mr. Pare Bowlegs)	December 10, 2009
NRC (Mr. Gregory P. Hatchett)	Eastern Band of Cherokee Indians (Mr. Michell Hicks)	December 10, 2009
NRC (Mr. Gregory P. Hatchett)	United Keetoowah Band of Cherokee Indians (Mr. Dallas Proctor)	December 10, 2009
NRC (Mr. Gregory P. Hatchett)	Cultural/Historic Preservation Department (Ms. Karen Kaniatobe)	December 10, 2009
NRC (Mr. Gregory P. Hatchett)	Alabama-Coushatta Tribe of Texas (Ms. Debbie Thomas)	December 10, 2009
NRC (Mr. Gregory P. Hatchett)	Muscogee (Creek) Nation of Oklahoma (Ms. Joyce Bear)	December 10, 2009
NRC (Mr. Gregory P. Hatchett)	Cherokee Nation of Oklahoma (Mr. Chadwick Smith)	December 10, 2009
NRC (Mr. Gregory P. Hatchett)	Catawba Indian Tribe (Mr. Gilbert Blue)	December 10, 2009
NRC (Mr. Gregory P. Hatchett)	Seminole Tribe of Florida (Mr. Willard Steele)	December 10, 2009
NRC (Mr. Gregory P. Hatchett)	Mississippi Band of Choctaw Indians (Mr. Kenneth Carleton)	December 10, 2009
NRC (Ms. Mallecia Sutton)	South Carolina Department of Natural Resources (Ms. Julie Holling)	December 15, 2009
South Carolina Department of Natural Resources (Ms. Julie Holling)	NRC (Ms. Mallecia Sutton)	December 15, 2009
NRC (Ms. Mallecia Sutton)	Georgia Department of Natural Resources (Mr. Matt Elliot)	December 15, 2009
Georgia Department of Natural Resources (Mr. Matt Elliot)	NRC (Ms. Mallecia Sutton)	December 16, 2009
Georgia Department of Natural Resources (Ms. Katrina Morris)	NRC (Ms. Mallecia Sutton)	December 17, 2009
NRC (Mr. Gregory P. Hatchett)	U.S. Advisory Council on Historic	December 23, 2009

Appendix F

Table F-1. (contd)

Source	Recipient	Date of Letter
	Preservation (Mr. Don Klima)	
NRC (Mr. Gregory P. Hatchett)	U.S. Fish and Wildlife Service Coastal Sub Office (Ms. Sandra S. Tucker)	January 7, 2010
NRC (Mr. Gregory P. Hatchett)	Catawba Indian Nation (Mr. Donald Rodgers)	January 7, 2010
U.S. Fish and Wildlife Service (Ms. Sandra Tucker)	NRC (Ms. Mallecia Sutton)	February 12, 2010
Georgia Department of Natural Resources (Ms. Elizabeth Shirk)	NRC (Ms. Mallecia Sutton)	June 17, 2010
NRC (Mr. Gregory P. Hatchett)	Muscogee (Creek) Nation (Mr. A.D. Ellis)	September 2, 2010
NRC (Mr. Gregory P. Hatchett)	United Keetoowah Band of Cherokee Indians (Mr. Dallas Proctor)	September 2, 2010
NRC (Mr. Gregory P. Hatchett)	Poarch Band of Creek Indians (Mr. Eddie Tullis)	September 2, 2010
NRC (Mr. Gregory P. Hatchett)	United Keetoowah Band of Cherokee Indians (Ms. Emma Sue Holland)	September 2, 2010
NRC (Mr. Gregory P. Hatchett)	Eastern Band of Cherokee Indians (Ms. Kathy McCoy)	September 2, 2010
NRC (Mr. Gregory P. Hatchett)	Coushatta Tribe of Louisiana (Mr. John Zachary)	September 2, 2010
NRC (Mr. Gregory P. Hatchett)	Kialegee Tribe, (Ms. Evelyn Bucktrot)	September 2, 2010
NRC (Mr. Gregory P. Hatchett)	Miccosukee Tribe of Indians of Florida (Mr. Steven Terry)	September 2, 2010
NRC (Mr. Gregory P. Hatchett)	Poarch Band of Creek Indians (Ms. Gale Thrower)	September 2, 2010
NRC (Mr. Gregory P. Hatchett)	Thlopthlocco Tribe (Mr. Louis McGertt)	September 2, 2010
NRC (Mr. Gregory P. Hatchett)	Cherokee Nation of Oklahoma (Mr. Richard Allen)	September 2, 2010
NRC (Mr. Gregory P. Hatchett)	Chickasaw Nation (Ms. Gingy [Virginia] Hail)	September 2, 2010
NRC (Mr. Gregory P. Hatchett)	Chickasaw Nation (Mr. Bill Anoatubby)	September 2, 2010
NRC (Mr. Gregory P. Hatchett)	Georgia Tribe of Eastern Cherokee (Mr. Charles Thurmond)	September 2, 2010
NRC (Mr. Gregory P. Hatchett)	Alabama-Quassarte Tribe (Mr. Tarpie Yargee)	September 2, 2010
NRC (Mr. Gregory P. Hatchett)	Seminole Nation of Oklahoma (Mr. Pare Bowlegs)	September 2, 2010
NRC (Mr. Gregory P. Hatchett)	Eastern Band of Cherokee Indians (Mr.	September 2, 2010

Appendix F

Table F-1. (contd)

Source	Recipient	Date of Letter
	Michell Hicks)	
NRC (Mr. Gregory P. Hatchett)	Absentee-Shawnee Tribe of Oklahoma (Ms. Karen Kaniatobe)	September 2, 2010
NRC (Mr. Gregory P. Hatchett)	Alabama-Coushatta Tribe of Texas (Ms. Debbie Thomas)	September 2, 2010
NRC (Mr. Gregory P. Hatchett)	Muscogee (Creek) Nation of Oklahoma (Ms. Joyce A. Bear)	September 2, 2010
NRC (Mr. Gregory P. Hatchett)	Cherokee Nation of Oklahoma (Mr. Chadwick Smith)	September 2, 2010
NRC (Mr. Gregory P. Hatchett)	Seminole Tribe of Florida (Mr. Willard Steele)	September 2, 2010
NRC (Mr. Gregory P. Hatchett)	Mississippi Band of Choctaw Indians (Mr. Kenneth H. Carleton)	September 2, 2010
NRC (Mr. Gregory P. Hatchett)	Poarch Band of Creek Indians (Ms. Stephanie Rolin)	September 2, 2010
NRC (Mr. Gregory P. Hatchett)	U.S. Army Corp of Engineers, Savannah District (Ms. Carol Bernstein)	September 3, 2010
NRC (Mr. Gregory P. Hatchett)	National Marine Fisheries Service (Mr. David Bernhart)	September 3, 2010
NRC (Mr. Gregory P. Hatchett)	South Carolina Department of Natural Resources (Mr. Robert Perry)	September 3, 2010
NRC (Mr. Gregory P. Hatchett)	U.S. Fish and Wildlife Service (Ms. Sandra Tucker)	September 3, 2010
Alabama-Coushatta Tribe of Texas (Mr. Bryant J. Celestine)	NRC	October 6, 2010
U.S. Environmental Protection Agency (Mr. Heinz Mueller)	NRC	November 15, 2010
U.S. Department of the Interior (Mr. Gregory Hogue)	NRC	November 21, 2010
NRC (Mr. Gregory P. Hatchett)	U.S. Fish and Wildlife Service (Ms. Sandra Tucker)	February 24, 2011
NRC (Mr. Gregory P. Hatchett)	U.S. National Marine Fisheries Service (Mr. David Bernhart)	February 24, 2011

Appendix F

December 09, 2009

Dr. Dave Crass, Acting Division Director and Deputy SHPO
State of Georgia Historic Preservation Officer
Historic Preservation Division
Department of Natural Resources
254 Washington Street, NW (Ground-level)
Atlanta, GA 30334

SUBJECT: VOGTLE ELECTRIC GENERATING PLANT, UNITS 3 AND 4 COMBINED LICENSE APPLICATION REVIEW

Dear Dr. Crass:

The U.S. Nuclear Regulatory Commission (NRC) staff is reviewing an application submitted by Southern Nuclear Operating Company, Inc. (Southern or SNC), on behalf of itself and several co-applicants, for a combined license (COL) for construction and operation of two new nuclear units at its Vogtle Electric Generating Plant (Vogtle or VEGP) site in Burke County, Georgia. The purpose of this letter is to invite the State of Georgia Historic Preservation Officer (SHPO) to consult with the NRC regarding the proposed action, pursuant to NRC regulations at 10 CFR Part 51 and Section 106 of the National Historic Preservation Act (NHPA) of 1966, as amended. The NRC plans to coordinate compliance with Section 106 of the NHPA using the National Environmental Policy Act (NEPA) of 1969, as amended process identified in 36 CFR 800.8(c) in lieu of the procedures set forth in §§ 800.3 through 800.6. Additionally, the NRC staff will rely on procedures in its regulations at 10 CFR 51.92 for supplementing a final environmental impact statement (FEIS).

Section 51.26(d) of the NRC regulations describes the processes for determining the need to supplement an FEIS, issue a notice of intent, and determine whether a formal scoping process will be conducted. As required by 10 CFR 51.92, the NRC staff will prepare a supplement to the FEIS for the early site permit (ESP) issued on August 26, 2009, to SNC and the same co-applicants. In this case, the NRC staff has determined that it will not conduct a formal scoping process for the development of the supplemental environmental impact statement (SEIS).

An ESP is a Commission approval of a site suitable for construction and operation of one or more new nuclear units. Under 10 CFR 51.50(c), a COL applicant referencing an ESP need not submit information or analyses regarding environmental issues that were resolved in the ESP EIS, except to the extent the COL applicant has identified any new and potentially significant information.

In accordance with the provisions in 36 CFR § 800.8, the NRC wishes to ensure that if you have an interest in any potential historic properties in the area of potential effect (APE), you will be afforded the opportunity to identify your concerns, provide advice on the identification and evaluation of historic properties, including those of traditional religious and cultural importance, and, if necessary, participate in the resolution of any adverse effects to such properties.

Appendix F

D. Crass   -2-

Thus, to support the NRC staff's review of any new and potentially significant circumstances or information relevant to the environmental concerns related to the proposed action or its impacts, we request you submit written comments you may have to offer on the environmental review of the COL by January 15, 2010.

The NRC intends to describe and propose measures in the SEIS for the COL analyses for potential impacts to historical and cultural resources, including developing alternatives and proposing measures that might avoid, minimize, or mitigate any adverse effects of the propose action on historic properties. To complete consultation under § 800.8(c), the NRC staff will forward the SEIS on the COL to you for your review and comment, and will address your comments in the final SEIS on the COL.

The proposed new reactors, Vogtle Units 3 and 4, would be located approximately 26 miles southeast of Augusta, Georgia, near the town of Waynesboro, Georgia. SNC currently operates two reactors, Vogtle Units 1 and 2, on the site, and plans to construct the new units adjacent to the existing reactors. The application for the COL was accepted for docketing on May 30, 2008 and is available through the web-based version of the NRC Agency-wide Documents Access Management System (ADAMS), which can be found at http://www.nrc.gov/readingrm/adams.html. The Environmental Report for the application is listed under ADAMS accession number ML081050181. The VEGP COL application is also available on the Internet at http://www/nrc.gov/reactors/new-licensing/col/vogtle.html. During the ESP environmental review, the NRC consulted with your office. The detailed review by the NRC of the environmental impacts of constructing and operating proposed Units 3 and 4 is documented in NUREG-1872, "Final Environmental Impact Statement for an Early Site Permit (ESP) at the Vogtle Electric Generating Plant Site." (ESP FEIS Volumes 1 and 2 are available in ADAMS under ADAMS accession numbers ML082240145 and ML082240165, respectively.)

Please submit comments either by mail to Mallecia Sutton, Environmental Project Manager, Division of Site and Environmental Reviews, Mail Stop T-7E18, U.S. Nuclear Regulatory Commission, Washington, D.C.20555-0001 or via e-mail to Vogtle.COLAEIS@nrc.gov by January 15, 2010.

If you have any questions or require additional information, please contact Mrs. Sutton at 301 415-0673 or via e-mail to Mallecia.Sutton@nrc.gov.

                                    Sincerely,

                                    /RA/

                                    Gregory P. Hatchett, Chief
                                    Environmental Projects Branch 1
                                    Division of Site and Environmental Reviews
                                    Office of New Reactors

Docket Nos. 52-025 and 52-026

cc: See next page

Appendix F

December 10, 2009

Ms. Stephanie Rolin
NAGPRA Contact
Poarch Band of Creek Indians
5811 Jack Springs Road
Atmore, AL 36502

SUBJECT: U.S. NUCLEAR REGULATORY COMMISSION'S SUPPLEMENTAL ENVIRONMENTAL IMPACT STATEMENT FOR SOUTHERN NUCLEAR OPERATING COMPANY'S COMBINED LICENSE APPLICATION FOR THE PROPOSED CONSTRUCTION AND OPERATION OF UNITS 3 AND 4 AT THE VOGTLE ELECTRIC GENERATING PLANT IN WAYNESBORO, GEORGIA

Dear Ms. Rolin:

The U.S. Nuclear Regulatory Commission (NRC) staff is reviewing an application submitted by Southern Nuclear Operating Company, Inc. (Southern or SNC), on behalf of itself and several co-applicants, for a combined license (COL) for construction and operation of two new nuclear units at its Vogtle Electric Generating Plant (Vogtle or VEGP) site in Burke County, Georgia. The purpose of this letter is to invite the Poarch Band of Creek Indians to consult with the NRC regarding the proposed action, pursuant to NRC regulations at 10 CFR Part 51 and Section 106 of the National Historic Preservation Act (NHPA) of 1966, as amended. The NRC plans to coordinate compliance with Section 106 of the NHPA using the National Environmental Policy Act (NEPA) process identified in 36 CFR 800.8(c) in lieu of the procedures set forth in §§ 800.3 through 800.6. Additionally, the NRC staff will rely on procedures in its regulations at 10 CFR 51.92 for supplementing a final environmental impact statement (FEIS).

Section 51.26(d) of the NRC regulations describes the processes for determining the need to supplement a FEIS, issue a notice of intent, and determine whether or not a formal scoping process will be conducted. As required by 10 CFR 51.92, the NRC staff will prepare a supplement to the FEIS for the early site permit (ESP) issued on August 26, 2009, to SNC and the same co-applicants. In this case, the NRC staff has determined that it will not conduct a formal scoping process for the development of the supplemental environmental impact statement (SEIS).

An ESP is a Commission approval of a site suitable for construction and operation of one or more new nuclear units. Under 10 CFR 51.50(c), a COL applicant referencing an ESP need not submit information or analyses regarding environmental issues that were resolved in the ESP EIS, except to the extent the COL applicant has identified any new and potentially significant information.

In accordance with the provisions in 36 CFR § 800.8, the NRC wishes to ensure that Indian tribes that might have an interest in any potential historic properties in the area of potential effect (APE) are afforded the opportunity to identify their concerns, provide advice on the identification and evaluation of historic properties, including those of traditional religious and cultural importance and, if necessary, participate in the resolution of any adverse effects to such properties.

Appendix F

S. Rolin   -2-

Thus, to support the NRC staff's review of any new and potentially significant circumstances or information relevant to the environmental concerns related to the proposed action or its impacts, we request you submit written comments, if any; your tribe may have to offer on the environmental review of the COL by January 15, 2010.

The NRC intends to propose measures in the SEIS for the COL analyses for potential impacts to historical and cultural resources including developing alternatives and proposing measures that might avoid, minimize, or mitigate any adverse effects of the propose action on historic properties. To complete consultation under § 800.8(c), the NRC staff will forward the SEIS on the COL to you for your review and comment, and will address your comments in the final SEIS on the COL.

The proposed new reactors, Vogtle Units 3 and 4 would be located approximately 26 miles southeast of Augusta, Georgia, near the town of Waynesboro, Georgia. SNC currently operates two reactors, Vogtle Units 1 and 2 on the site and plans to construct the new units adjacent to the existing reactors. The application for the COL was accepted for docketing on May 30, 2008, and is available through the web-based version of the NRC Agency-wide Documents Access Management System (ADAMS), which can be found at http://www.nrc.gov/readingrm/adams.html. The Environmental Report for the application is listed under ADAMS accession number ML081050181. The Vogtle Plant COL application is also available on the Internet at http://www.nrc.gov/reactors/new-licensing/col/vogtle.html. A detailed review by NRC of the environmental impacts of constructing and operating proposed Units 3 and 4 is documented in NUREG-1872, "Final Environmental Impact Statement for an Early Site Permit (ESP) at the Vogtle Electric Generating Plant Site." (ESP FEIS Volumes 1 and 2 are available in ADAMS under ADAMS accession numbers ML082240145 and ML082240165, respectively.)

Please submit comments either by mail to Mrs. Mallecia Sutton, Environmental Project Manager, Division of Site and Environmental Reviews, Mail Stop T-7E18, U.S. Nuclear Regulatory Commission, Washington, D.C. 20555-0001 or via e-mail to VogtleCOLASEIS@nrc.gov by January 15, 2010.

If you have any questions or require additional information, please contact Mrs. Sutton at (301) 415-0673 or via e-mail to Mallecia.Sutton@nrc.gov.

Sincerely,

/RA/

Gregory P. Hatchett, Branch Chief
Environmental Projects Branch 1
Division of Site and Environmental Reviews
Office of New Reactors

Docket Nos. 52-025 and 52-026

cc: See next page

Appendix F

December 10, 2009

Ms. Emma Sue Holland, NAGPRA Contact
United Keetoowah Band of Cherokee Indians
P.O. Box 746
Tahlequah, OK 74465

SUBJECT: U.S. NUCLEAR REGULATORY COMMISSION'S SUPPLEMENTAL ENVIRONMENTAL IMPACT STATEMENT FOR SOUTHERN NUCLEAR OPERATING COMPANY'S COMBINED LICENSE APPLICATION FOR THE PROPOSED CONSTRUCTION AND OPERATION OF UNITS 3 AND 4 AT THE VOGTLE ELECTRIC GENERATING PLANT IN WAYNESBORO, GEORGIA

Dear Ms. Holland:

The U.S. Nuclear Regulatory Commission (NRC) staff is reviewing an application submitted by Southern Nuclear Operating Company, Inc. (Southern or SNC), on behalf of itself and several co-applicants, for a combined license (COL) for construction and operation of two new nuclear units at its Vogtle Electric Generating Plant (Vogtle or VEGP) site in Burke County, Georgia. The purpose of this letter is to invite the United Keetoowah Band of Cherokee Indians to consult with the NRC regarding the proposed action, pursuant to NRC regulations at 10 CFR Part 51 and Section 106 of the National Historic Preservation Act (NHPA) of 1966, as amended. The NRC plans to coordinate compliance with Section 106 of the NHPA using the National Environmental Policy Act (NEPA) process identified in 36 CFR 800.8(c) in lieu of the procedures set forth in §§ 800.3 through 800.6. Additionally, the NRC staff will rely on procedures in its regulations at 10 CFR 51.92 for supplementing a final environmental impact statement (FEIS).

Section 51.26(d) of the NRC regulations describes the processes for determining the need to supplement a FEIS, issue a notice of intent, and determine whether or not a formal scoping process will be conducted. As required by 10 CFR 51.92, the NRC staff will prepare a supplement to the FEIS for the early site permit (ESP) issued on August 26, 2009, to SNC and the same co-applicants. In this case, the NRC staff has determined that it will not conduct a formal scoping process for the development of the supplemental environmental impact statement (SEIS).

An ESP is a Commission approval of a site suitable for construction and operation of one or more new nuclear units. Under 10 CFR 51.50(c), a COL applicant referencing an ESP need not submit information or analyses regarding environmental issues that were resolved in the ESP EIS, except to the extent the COL applicant has identified any new and potentially significant information.

In accordance with the provisions in 36 CFR § 800.8, the NRC wishes to ensure that Indian tribes that might have an interest in any potential historic properties in the area of potential effect (APE) are afforded the opportunity to identify their concerns, provide advice on the identification and evaluation of historic properties, including those of traditional religious and cultural importance and, if necessary, participate in the resolution of any adverse effects to such properties.

Appendix F

Thus, to support the NRC staff's review of any new and potentially significant circumstances or information relevant to the environmental concerns related to the proposed action or its impacts, we request you submit written comments, if any, your tribe may have to offer on the environmental review of the COL by January 15, 2010.

The NRC intends to propose measures in the SEIS for the COL analyses for potential impacts to historical and cultural resources including developing alternatives and proposing measures that might avoid, minimize, or mitigate any adverse effects of the propose action on historic properties. To complete consultation under § 800.8(c), the NRC staff will forward the SEIS on the COL to you for your review and comment, and will address your comments in the final SEIS on the COL.

The proposed new reactors, Vogtle Units 3 and 4 would be located approximately 26 miles southeast of Augusta, Georgia, near the town of Waynesboro, Georgia. SNC currently operates two reactors, Vogtle Units 1 and 2 on the site and plans to construct the new units adjacent to the existing reactors. The application for the COL was accepted for docketing on May 30, 2008, and is available through the web-based version of the NRC Agency-wide Documents Access Management System (ADAMS), which can be found at http://www.nrc.gov/readingrm/adams.html. The Environmental Report for the application is listed under ADAMS accession number ML081050181. The Vogtle Plant COL application is also available on the Internet at http://www.nrc.gov/reactors/new-licensing/col/vogtle.html. A detailed review by NRC of the environmental impacts of constructing and operating proposed Units 3 and 4 is documented in NUREG-1872, "Final Environmental Impact Statement for an Early Site Permit (ESP) at the Vogtle Electric Generating Plant Site." (ESP FEIS Volumes 1 and 2 are available in ADAMS under ADAMS accession numbers ML082240145 and ML082240165, respectively.)

Please submit comments either by mail to Mrs. Mallecia Sutton, Environmental Project Manager, Division of Site and Environmental Reviews, Mail Stop T-7E18, U.S. Nuclear Regulatory Commission, Washington, D.C. 20555-0001 or via e-mail to Vogtle.COLAEIS@nrc.gov by January 15, 2010.

If you have any questions or require additional information, please contact Mrs. Sutton at (301) 415-0673 or via e-mail to Mallecia.Sutton@nrc.gov.

Sincerely,

/RA/

Gregory P. Hatchett, Branch Chief
Environmental Projects Branch 1
Division of Site and Environmental Reviews
Office of New Reactors

Docket Nos. 52-025 and 52-026

cc: See next page

Appendix F

December 10, 2009

Mr. Eddie Tullis, Chairperson
Poarch Band of Creek Indians
5811 Jack Springs Rd
Atmore, AL 36502

SUBJECT: U.S. NUCLEAR REGULATORY COMMISSION'S SUPPLEMENTAL ENVIRONMENTAL IMPACT STATEMENT FOR SOUTHERN NUCLEAR OPERATING COMPANY'S COMBINED LICENSE APPLICATION FOR THE PROPOSED CONSTRUCTION AND OPERATION OF UNITS 3 AND 4 AT THE VOGTLE ELECTRIC GENERATING PLANT IN WAYNESBORO, GEORGIA

Dear Mr. Tullis:

The U.S. Nuclear Regulatory Commission (NRC) staff is reviewing an application submitted by Southern Nuclear Operating Company, Inc. (Southern or SNC), on behalf of itself and several co-applicants, for a combined license (COL) for construction and operation of two new nuclear units at its Vogtle Electric Generating Plant (Vogtle or VEGP) site in Burke County, Georgia. The purpose of this letter is to invite the Poarch Band of Creek Indians to consult with the NRC regarding the proposed action, pursuant to NRC regulations at 10 CFR Part 51 and Section 106 of the National Historic Preservation Act (NHPA) of 1966, as amended. The NRC plans to coordinate compliance with Section 106 of the NHPA using the National Environmental Policy Act (NEPA) process identified in 36 CFR 800.8(c) in lieu of the procedures set forth in §§ 800.3 through 800.6. Additionally, the NRC staff will rely on procedures in its regulations at 10 CFR 51.92 for supplementing a final environmental impact statement (FEIS).

Section 51.26(d) of the NRC regulations describes the processes for determining the need to supplement a FEIS, issue a notice of intent, and determine whether or not a formal scoping process will be conducted. As required by 10 CFR 51.92, the NRC staff will prepare a supplement to the FEIS for the early site permit (ESP) issued on August 26, 2009, to SNC and the same co-applicants. In this case, the NRC staff has determined that it will not conduct a formal scoping process for the development of the supplemental environmental impact statement (SEIS).

An ESP is a Commission approval of a site suitable for construction and operation of one or more new nuclear units. Under 10 CFR 51.50(c), a COL applicant referencing an ESP need not submit information or analyses regarding environmental issues that were resolved in the ESP EIS, except to the extent the COL applicant has identified any new and potentially significant information.

In accordance with the provisions in 36 CFR § 800.8, the NRC wishes to ensure that Indian tribes that might have an interest in any potential historic properties in the area of potential effect (APE) are afforded the opportunity to identify their concerns, provide advice on the identification and evaluation of historic properties, including those of traditional religious and cultural importance and, if necessary, participate in the resolution of any adverse effects to such properties.

Appendix F

E. Tullis   -2-

Thus, to support the NRC staff's review of any new and potentially significant circumstances or information relevant to the environmental concerns related to the proposed action or its impacts, we request you submit written comments, if any, your tribe may have to offer on the environmental review of the COL by January 15, 2010.

The NRC intends to propose measures in the SEIS for the COL analyses for potential impacts to historical and cultural resources including developing alternatives and proposing measures that might avoid, minimize, or mitigate any adverse effects of the propose action on historic properties. To complete consultation under § 800.8(c), the NRC staff will forward the SEIS on the COL to you for your review and comment, and will address your comments in the final SEIS on the COL.

The proposed new reactors, Vogtle Units 3 and 4 would be located approximately 26 miles southeast of Augusta, Georgia, near the town of Waynesboro, Georgia. SNC currently operates two reactors, Vogtle Units 1 and 2 on the site and plans to construct the new units adjacent to the existing reactors. The application for the COL was accepted for docketing on May 30, 2008, and is available through the web-based version of the NRC Agency-wide Documents Access Management System (ADAMS), which can be found at http://www.nrc.gov/readingrm/adams.html. The Environmental Report for the application is listed under ADAMS accession number ML081050181. The Vogtle Plant COL application is also available on the Internet at http://www.nrc.gov/reactors/new-licensing/col/vogtle.html. A detailed review by NRC of the environmental impacts of constructing and operating proposed Units 3 and 4 is documented in NUREG-1872, "Final Environmental Impact Statement for an Early Site Permit (ESP) at the Vogtle Electric Generating Plant Site." (ESP FEIS Volumes 1 and 2 are available in ADAMS under ADAMS accession numbers ML082240145 and ML082240165, respectively.)

Please submit comments either by mail to Mrs. Mallecia Sutton, Environmental Project Manager, Division of Site and Environmental Reviews, Mail Stop T-7E18, U.S. Nuclear Regulatory Commission, Washington, D.C. 20555-0001 or via e-mail to Vogtle.COLAEIS@nrc.gov by January 15, 2010.

If you have any questions or require additional information, please contact Mrs. Sutton at (301) 415-0673 or via e-mail to Mallecia.Sutton@nrc.gov.

Sincerely,

/RA/

Gregory P. Hatchett, Branch Chief
Environmental Projects Branch 1
Division of Site and Environmental Reviews
Office of New Reactors

Docket Nos. 52-025 and 52-026

cc: See next page

Appendix F

December 10, 2009

Ms. Kathy McCoy, NAGPRA Contact
Eastern Band of Cherokee Indians
P.O. Box 455
Cherokee, NC 28719

SUBJECT: U.S. NUCLEAR REGULATORY COMMISSION'S SUPPLEMENTAL ENVIRONMENTAL IMPACT STATEMENT FOR SOUTHERN NUCLEAR OPERATING COMPANY'S COMBINED LICENSE APPLICATION FOR THE PROPOSED CONSTRUCTION AND OPERATION OF UNITS 3 AND 4 AT THE VOGTLE ELECTRIC GENERATING PLANT IN WAYNESBORO, GEORGIA

Dear Ms. McCoy:

The U.S. Nuclear Regulatory Commission (NRC) staff is reviewing an application submitted by Southern Nuclear Operating Company, Inc. (Southern or SNC), on behalf of itself and several co-applicants, for a combined license (COL) for construction and operation of two new nuclear units at its Vogtle Electric Generating Plant (Vogtle or VEGP) site in Burke County, Georgia. The purpose of this letter is to invite the Eastern Band of Cherokee Indians to consult with the NRC regarding the proposed action, pursuant to NRC regulations at 10 CFR Part 51 and Section 106 of the National Historic Preservation Act (NHPA) of 1966, as amended. The NRC plans to coordinate compliance with Section 106 of the NHPA using the National Environmental Policy Act (NEPA) process identified in 36 CFR 800.8(c) in lieu of the procedures set forth in §§ 800.3 through 800.6. Additionally, the NRC staff will rely on procedures in its regulations at 10 CFR 51.92 for supplementing a final environmental impact statement (FEIS).

Section 51.26(d) of the NRC regulations describes the processes for determining the need to supplement a FEIS, issue a notice of intent, and determine whether or not a formal scoping process will be conducted. As required by 10 CFR 51.92, the NRC staff will prepare a supplement to the FEIS for the early site permit (ESP) issued on August 26, 2009, to SNC and the same co-applicants. In this case, the NRC staff has determined that it will not conduct a formal scoping process for the development of the supplemental environmental impact statement (SEIS).

An ESP is a Commission approval of a site suitable for construction and operation of one or more new nuclear units. Under 10 CFR 51.50(c), a COL applicant referencing an ESP need not submit information or analyses regarding environmental issues that were resolved in the ESP EIS, except to the extent the COL applicant has identified any new and potentially significant information.

In accordance with the provisions in 36 CFR § 800.8, the NRC wishes to ensure that Indian tribes that might have an interest in any potential historic properties in the area of potential effect (APE) are afforded the opportunity to identify their concerns, provide advice on the identification and evaluation of historic properties, including those of traditional religious and cultural importance and, if necessary, participate in the resolution of any adverse effects to such properties.

Appendix F

K. McCoy   -2-

Thus, to support the NRC staff's review of any new and potentially significant circumstances or information relevant to the environmental concerns related to the proposed action or its impacts, we request you submit written comments, if any, your tribe may have to offer on the environmental review of the COL by January 15, 2010.

The NRC intends to propose measures in the SEIS for the COL analyses for potential impacts to historical and cultural resources including developing alternatives and proposing measures that might avoid, minimize, or mitigate any adverse effects of the propose action on historic properties. To complete consultation under § 800.8(c), the NRC staff will forward the SEIS on the COL to you for your review and comment, and will address your comments in the final SEIS on the COL.

The proposed new reactors, Vogtle Units 3 and 4 would be located approximately 26 miles southeast of Augusta, Georgia, near the town of Waynesboro, Georgia. SNC currently operates two reactors, Vogtle Units 1 and 2 on the site and plans to construct the new units adjacent to the existing reactors. The application for the COL was accepted for docketing on May 30, 2008, and is available through the web-based version of the NRC Agency-wide Documents Access Management System (ADAMS), which can be found at http://www.nrc.gov/readingrm/adams.html. The Environmental Report for the application is listed under ADAMS accession number ML081050181. The Vogtle Plant COL application is also available on the Internet at http://www.nrc.gov/reactors/new-licensing/col/vogtle.html. A detailed review by NRC of the environmental impacts of constructing and operating proposed Units 3 and 4 is documented in NUREG-1872, "Final Environmental Impact Statement for an Early Site Permit (ESP) at the Vogtle Electric Generating Plant Site." (ESP FEIS Volumes 1 and 2 are available in ADAMS under ADAMS accession numbers ML082240145 and ML082240165, respectively.)

Please submit comments either by mail to Mrs. Mallecia Sutton, Environmental Project Manager, Division of Site and Environmental Reviews, Mail Stop T-7E18, U.S. Nuclear Regulatory Commission, Washington, D.C. 20555-0001 or via e-mail to Vogtle.COLAEIS@nrc.gov by January 15, 2010.

If you have any questions or require additional information, please contact Mrs. Sutton at (301) 415-0673 or via e-mail to Mallecia.Sutton@nrc.gov.

Sincerely,

/RA/

Gregory P. Hatchett, Branch Chief
Environmental Projects Branch 1
Division of Site and Environmental Reviews
Office of New Reactors

Docket Nos. 52-025 and 52-026

cc: See next page

Appendix F

December 10, 2009

Mr. John Zachary, Attorney at Law
c/o Coushatta Tribe of Louisiana
P.O. Box 12730
Alexandria, LA 71315-2730

SUBJECT: U.S. NUCLEAR REGULATORY COMMISSION'S SUPPLEMENTAL ENVIRONMENTAL IMPACT STATEMENT FOR SOUTHERN NUCLEAR OPERATING COMPANY'S COMBINED LICENSE APPLICATION FOR THE PROPOSED CONSTRUCTION AND OPERATION OF UNITS 3 AND 4 AT THE VOGTLE ELECTRIC GENERATING PLANT IN WAYNESBORO, GEORGIA

Dear Mr. Zachary:

The U.S. Nuclear Regulatory Commission (NRC) staff is reviewing an application submitted by Southern Nuclear Operating Company, Inc. (Southern or SNC), on behalf of itself and several co-applicants, for a combined license (COL) for construction and operation of two new nuclear units at its Vogtle Electric Generating Plant (Vogtle or VEGP) site in Burke County, Georgia. The purpose of this letter is to invite the Coushatta Tribe of Louisiana to consult with the NRC regarding the proposed action, pursuant to NRC regulations at 10 CFR Part 51 and Section 106 of the National Historic Preservation Act (NHPA) of 1966, as amended. The NRC plans to coordinate compliance with Section 106 of the NHPA using the National Environmental Policy Act (NEPA) process identified in 36 CFR 800.8(c) in lieu of the procedures set forth in §§ 800.3 through 800.6. Additionally, the NRC staff will rely on procedures in its regulations at 10 CFR 51.92 for supplementing a final environmental impact statement (FEIS).

Section 51.26(d) of the NRC regulations describes the processes for determining the need to supplement a FEIS, issue a notice of intent, and determine whether or not a formal scoping process will be conducted. As required by 10 CFR 51.92, the NRC staff will prepare a supplement to the FEIS for the early site permit (ESP) issued on August 26, 2009, to SNC and the same co-applicants. In this case, the NRC staff has determined that it will not conduct a formal scoping process for the development of the supplemental environmental impact statement (SEIS).

An ESP is a Commission approval of a site suitable for construction and operation of one or more new nuclear units. Under 10 CFR 51.50(c), a COL applicant referencing an ESP need not submit information or analyses regarding environmental issues that were resolved in the ESP EIS, except to the extent the COL applicant has identified any new and potentially significant information.

In accordance with the provisions in 36 CFR § 800.8, the NRC wishes to ensure that Indian tribes that might have an interest in any potential historic properties in the area of potential effect (APE) are afforded the opportunity to identify their concerns, provide advice on the identification and evaluation of historic properties, including those of traditional religious and cultural importance and, if necessary, participate in the resolution of any adverse effects to such properties.

Appendix F

Thus, to support the NRC staff's review of any new and potentially significant circumstances or information relevant to the environmental concerns related to the proposed action or its impacts, we request you submit written comments, if any, your tribe may have to offer on the environmental review of the COL by January 15, 2010.

The NRC intends to propose measures in the SEIS for the COL analyses for potential impacts to historical and cultural resources including developing alternatives and proposing measures that might avoid, minimize, or mitigate any adverse effects of the propose action on historic properties. To complete consultation under § 800.8(c), the NRC staff will forward the SEIS on the COL to you for your review and comment, and will address your comments in the final SEIS on the COL.

The proposed new reactors, Vogtle Units 3 and 4 would be located approximately 26 miles southeast of Augusta, Georgia, near the town of Waynesboro, Georgia. SNC currently operates two reactors, Vogtle Units 1 and 2 on the site and plans to construct the new units adjacent to the existing reactors. The application for the COL was accepted for docketing on May 30, 2008, and is available through the web-based version of the NRC Agency-wide Documents Access Management System (ADAMS), which can be found at http://www.nrc.gov/readingrm/adams.html. The Environmental Report for the application is listed under ADAMS accession number ML081050181. The Vogtle Plant COL application is also available on the Internet at http://www.nrc.gov/reactors/new-licensing/col/vogtle.html. A detailed review by NRC of the environmental impacts of constructing and operating proposed Units 3 and 4 is documented in NUREG-1872, "Final Environmental Impact Statement for an Early Site Permit (ESP) at the Vogtle Electric Generating Plant Site." (ESP FEIS Volumes 1 and 2 are available in ADAMS under ADAMS accession numbers ML082240145 and ML082240165, respectively.)

Please submit comments either by mail to Mrs. Mallecia Sutton, Environmental Project Manager, Division of Site and Environmental Reviews, Mail Stop T-7E18, U.S. Nuclear Regulatory Commission, Washington, D.C. 20555-0001 or via e-mail to Vogtle.COLAEIS@nrc.gov by January 15, 2010.

If you have any questions or require additional information, please contact Mrs. Sutton at (301) 415-0673 or via e-mail to Mallecia.Sutton@nrc.gov.

Sincerely,

/RA/

Gregory P. Hatchett, Branch Chief
Environmental Projects Branch 1
Division of Site and Environmental Reviews
Office of New Reactors

Docket Nos. 52-025 and 52-026

cc: See next page

Appendix F

December 10, 2009

Ms. Evelyn Bucktrot, Town King
Kialegee Tribal Town
P.O. Box 332
Wetumka, OK 74883

SUBJECT: U.S. NUCLEAR REGULATORY COMMISSION'S SUPPLEMENTAL ENVIRONMENTAL IMPACT STATEMENT FOR SOUTHERN NUCLEAR OPERATING COMPANY'S COMBINED LICENSE APPLICATION FOR THE PROPOSED CONSTRUCTION AND OPERATION OF UNITS 3 AND 4 AT THE VOGTLE ELECTRIC GENERATING PLANT IN WAYNESBORO, GEORGIA

Dear Ms. Bucktrot:

The U.S. Nuclear Regulatory Commission (NRC) staff is reviewing an application submitted by Southern Nuclear Operating Company, Inc. (Southern or SNC), on behalf of itself and several co-applicants, for a combined license (COL) for construction and operation of two new nuclear units at its Vogtle Electric Generating Plant (Vogtle or VEGP) site in Burke County, Georgia. The purpose of this letter is to invite the Kialegee Tribal Town to consult with the NRC regarding the proposed action, pursuant to NRC regulations at 10 CFR Part 51 and Section 106 of the National Historic Preservation Act (NHPA) of 1966, as amended. The NRC plans to coordinate compliance with Section 106 of the NHPA using the National Environmental Policy Act (NEPA) process identified in 36 CFR 800.8(c) in lieu of the procedures set forth in §§ 800.3 through 800.6. Additionally, the NRC staff will rely on procedures in its regulations at 10 CFR 51.92 for supplementing a final environmental impact statement (FEIS).

Section 51.26(d) of the NRC regulations describes the processes for determining the need to supplement a FEIS, issue a notice of intent, and determine whether or not a formal scoping process will be conducted. As required by 10 CFR 51.92, the NRC staff will prepare a supplement to the FEIS for the early site permit (ESP) issued on August 26, 2009, to SNC and the same co-applicants. In this case, the NRC staff has determined that it will not conduct a formal scoping process for the development of the supplemental environmental impact statement (SEIS).

An ESP is a Commission approval of a site suitable for construction and operation of one or more new nuclear units. Under 10 CFR 51.50(c), a COL applicant referencing an ESP need not submit information or analyses regarding environmental issues that were resolved in the ESP EIS, except to the extent the COL applicant has identified any new and potentially significant information.

In accordance with the provisions in 36 CFR § 800.8, the NRC wishes to ensure that Indian tribes that might have an interest in any potential historic properties in the area of potential effect (APE) are afforded the opportunity to identify their concerns, provide advice on the identification and evaluation of historic properties, including those of traditional religious and cultural importance and, if necessary, participate in the resolution of any adverse effects to such properties.

Appendix F

E. Bucktrot   -2-

Thus, to support the NRC staff's review of any new and potentially significant circumstances or information relevant to the environmental concerns related to the proposed action or its impacts, we request you submit written comments, if any, your tribe may have to offer on the environmental review of the COL by January 15, 2010.

The NRC intends to propose measures in the SEIS for the COL analyses for potential impacts to historical and cultural resources including developing alternatives and proposing measures that might avoid, minimize, or mitigate any adverse effects of the propose action on historic properties. To complete consultation under § 800.8(c), the NRC staff will forward the SEIS on the COL to you for your review and comment, and will address your comments in the final SEIS on the COL.

The proposed new reactors, Vogtle Units 3 and 4 would be located approximately 26 miles southeast of Augusta, Georgia, near the town of Waynesboro, Georgia. SNC currently operates two reactors, Vogtle Units 1 and 2 on the site and plans to construct the new units adjacent to the existing reactors. The application for the COL was accepted for docketing on May 30, 2008, and is available through the web-based version of the NRC Agency-wide Documents Access Management System (ADAMS), which can be found at http://www.nrc.gov/readingrm/adams.html. The Environmental Report for the application is listed under ADAMS accession number ML081050181. The Vogtle Plant COL application is also available on the Internet at http://www.nrc.gov/reactors/new-licensing/col/vogtle.html. A detailed review by NRC of the environmental impacts of constructing and operating proposed Units 3 and 4 is documented in NUREG-1872, "Final Environmental Impact Statement for an Early Site Permit (ESP) at the Vogtle Electric Generating Plant Site." (ESP FEIS Volumes 1 and 2 are available in ADAMS under ADAMS accession numbers ML082240145 and ML082240165, respectively.)

Please submit comments either by mail to Mrs. Mallecia Sutton, Environmental Project Manager, Division of Site and Environmental Reviews, Mail Stop T-7E18, U.S. Nuclear Regulatory Commission, Washington, D.C. 20555-0001 or via e-mail to Vogtle.COLAEIS@nrc.gov by January 15, 2010.

If you have any questions or require additional information, please contact Mrs. Sutton at (301) 415-0673 or via e-mail to Mallecia.Sutton@nrc.gov.

Sincerely,

/RA/

Gregory P. Hatchett, Branch Chief
Environmental Projects Branch 1
Division of Site and Environmental Reviews
Office of New Reactors

Docket Nos. 52-025 and 52-026

cc: See next page

Appendix F

December 10, 2009

Mr. Steven Terry
Land Resource Manager
Miccosukee Tribe of Indians of Florida
Real Estate Services, Mile Marker 70
US 41 at Admin. Bldg.
Miami, FL 33194

SUBJECT: U.S. NUCLEAR REGULATORY COMMISSION'S SUPPLEMENTAL ENVIRONMENTAL IMPACT STATEMENT FOR SOUTHERN NUCLEAR OPERATING COMPANY'S COMBINED LICENSE APPLICATION FOR THE PROPOSED CONSTRUCTION AND OPERATION OF UNITS 3 AND 4 AT THE VOGTLE ELECTRIC GENERATING PLANT IN WAYNESBORO, GEORGIA

Dear Mr. Terry:

The U.S. Nuclear Regulatory Commission (NRC) staff is reviewing an application submitted by Southern Nuclear Operating Company, Inc. (Southern or SNC), on behalf of itself and several co-applicants, for a combined license (COL) for construction and operation of two new nuclear units at its Vogtle Electric Generating Plant (Vogtle or VEGP) site in Burke County, Georgia. The purpose of this letter is to invite the Miccosulkee Tribe of Indians to consult with the NRC regarding the proposed action, pursuant to NRC regulations at 10 CFR Part 51 and Section 106 of the National Historic Preservation Act (NHPA) of 1966, as amended. The NRC plans to coordinate compliance with Section 106 of the NHPA using the National Environmental Policy Act (NEPA) process identified in 36 CFR 800.8(c) in lieu of the procedures set forth in §§ 800.3 through 800.6. Additionally, the NRC staff will rely on procedures in its regulations at 10 CFR 51.92 for supplementing a final environmental impact statement (FEIS).

Section 51.26(d) of the NRC regulations describes the processes for determining the need to supplement a FEIS, issue a notice of intent, and determine whether or not a formal scoping process will be conducted. As required by 10 CFR 51.92, the NRC staff will prepare a supplement to the FEIS for the early site permit (ESP) issued on August 26, 2009, to SNC and the same co-applicants. In this case, the NRC staff has determined that it will not conduct a formal scoping process for the development of the supplemental environmental impact statement (SEIS).

An ESP is a Commission approval of a site suitable for construction and operation of one or more new nuclear units. Under 10 CFR 51.50(c), a COL applicant referencing an ESP need not submit information or analyses regarding environmental issues that were resolved in the ESP EIS, except to the extent the COL applicant has identified any new and potentially significant information.

In accordance with the provisions in 36 CFR § 800.8, the NRC wishes to ensure that Indian tribes that might have an interest in any potential historic properties in the area of potential effect (APE) are afforded the opportunity to identify their concerns, provide advice on the identification and evaluation of historic properties, including those of traditional religious and cultural importance and, if necessary, participate in the resolution of any adverse effects to such properties.

Appendix F

Thus, to support the NRC staff's review of any new and potentially significant circumstances or information relevant to the environmental concerns related to the proposed action or its impacts, we request you submit written comments, if any, your tribe may have to offer on the environmental review of the COL by January 15, 2010.

The NRC intends to propose measures in the SEIS for the COL analyses for potential impacts to historical and cultural resources including developing alternatives and proposing measures that might avoid, minimize, or mitigate any adverse effects of the propose action on historic properties. To complete consultation under § 800.8(c), the NRC staff will forward the SEIS on the COL to you for your review and comment, and will address your comments in the final SEIS on the COL.

The proposed new reactors, Vogtle Units 3 and 4 would be located approximately 26 miles southeast of Augusta, Georgia, near the town of Waynesboro, Georgia. SNC currently operates two reactors, Vogtle Units 1 and 2 on the site and plans to construct the new units adjacent to the existing reactors. The application for the COL was accepted for docketing on May 30, 2008, and is available through the web-based version of the NRC Agency-wide Documents Access Management System (ADAMS), which can be found at http://www.nrc.gov/readingrm/adams.html. The Environmental Report for the application is listed under ADAMS accession number ML081050181. The Vogtle Plant COL application is also available on the Internet at http://www.nrc.gov/reactors/new-licensing/col/vogtle.html. A detailed review by NRC of the environmental impacts of constructing and operating proposed Units 3 and 4 is documented in NUREG-1872, "Final Environmental Impact Statement for an Early Site Permit (ESP) at the Vogtle Electric Generating Plant Site." (ESP FEIS Volumes 1 and 2 are available in ADAMS under ADAMS accession numbers ML082240145 and ML082240165, respectively.)

Please submit comments either by mail to Mrs. Mallecia Sutton, Environmental Project Manager, Division of Site and Environmental Reviews, Mail Stop T-7E18, U.S. Nuclear Regulatory Commission, Washington, D.C. 20555-0001 or via e-mail to Vogtle.COLAEIS@nrc.gov by January 15, 2010.

If you have any questions or require additional information, please contact Mrs. Sutton at (301) 415-0673 or via e-mail to Mallecia.Sutton@nrc.gov.

Sincerely,

/RA/

Gregory P. Hatchett, Branch Chief
Environmental Projects Branch 1
Division of Site and Environmental Reviews
Office of New Reactors

Docket Nos. 52-025 and 52-026

cc: See next page

Appendix F

December 10, 2009

Ms. Gale Thrower, NAGPRA Contact
Poarch Band of Creek Indians
5811 Jack Springs Road
Atmore, AL 36502

SUBJECT: U.S. NUCLEAR REGULATORY COMMISSION'S SUPPLEMENTAL ENVIRONMENTAL IMPACT STATEMENT FOR SOUTHERN NUCLEAR OPERATING COMPANY'S COMBINED LICENSE APPLICATION FOR THE PROPOSED CONSTRUCTION AND OPERATION OF UNITS 3 AND 4 AT THE VOGTLE ELECTRIC GENERATING PLANT IN WAYNESBORO, GEORGIA

Dear Ms. Thrower:

The U.S. Nuclear Regulatory Commission (NRC) staff is reviewing an application submitted by Southern Nuclear Operating Company, Inc. (Southern or SNC), on behalf of itself and several co-applicants, for a combined license (COL) for construction and operation of two new nuclear units at its Vogtle Electric Generating Plant (Vogtle or VEGP) site in Burke County, Georgia. The purpose of this letter is to invite the Poarch Band of Creek Indians to consult with the NRC regarding the proposed action, pursuant to NRC regulations at 10 CFR Part 51 and Section 106 of the National Historic Preservation Act (NHPA) of 1966, as amended. The NRC plans to coordinate compliance with Section 106 of the NHPA using the National Environmental Policy Act (NEPA) process identified in 36 CFR 800.8(c) in lieu of the procedures set forth in §§ 800.3 through 800.6. Additionally, the NRC staff will rely on procedures in its regulations at 10 CFR 51.92 for supplementing a final environmental impact statement (FEIS).

Section 51.26(d) of the NRC regulations describes the processes for determining the need to supplement a FEIS, issue a notice of intent, and determine whether or not a formal scoping process will be conducted. As required by 10 CFR 51.92, the NRC staff will prepare a supplement to the FEIS for the early site permit (ESP) issued on August 26, 2009, to SNC and the same co-applicants. In this case, the NRC staff has determined that it will not conduct a formal scoping process for the development of the supplemental environmental impact statement (SEIS).

An ESP is a Commission approval of a site suitable for construction and operation of one or more new nuclear units. Under 10 CFR 51.50(c), a COL applicant referencing an ESP need not submit information or analyses regarding environmental issues that were resolved in the ESP EIS, except to the extent the COL applicant has identified any new and potentially significant information.

In accordance with the provisions in 36 CFR § 800.8, the NRC wishes to ensure that Indian tribes that might have an interest in any potential historic properties in the area of potential effect (APE) are afforded the opportunity to identify their concerns, provide advice on the identification and evaluation of historic properties, including those of traditional religious and cultural importance and, if necessary, participate in the resolution of any adverse effects to such properties.

Appendix F

Thus, to support the NRC staff's review of any new and potentially significant circumstances or information relevant to the environmental concerns related to the proposed action or its impacts, we request you submit written comments, if any, your tribe may have to offer on the environmental review of the COL by January 15, 2010.

The NRC intends to propose measures in the SEIS for the COL analyses for potential impacts to historical and cultural resources including developing alternatives and proposing measures that might avoid, minimize, or mitigate any adverse effects of the propose action on historic properties. To complete consultation under § 800.8(c), the NRC staff will forward the SEIS on the COL to you for your review and comment, and will address your comments in the final SEIS on the COL.

The proposed new reactors, Vogtle Units 3 and 4 would be located approximately 26 miles southeast of Augusta, Georgia, near the town of Waynesboro, Georgia. SNC currently operates two reactors, Vogtle Units 1 and 2 on the site and plans to construct the new units adjacent to the existing reactors. The application for the COL was accepted for docketing on May 30, 2008, and is available through the web-based version of the NRC Agency-wide Documents Access Management System (ADAMS), which can be found at http://www.nrc.gov/readingrm/adams.html. The Environmental Report for the application is listed under ADAMS accession number ML081050181. The Vogtle Plant COL application is also available on the Internet at http://www.nrc.gov/reactors/new-licensing/col/vogtle.html. A detailed review by NRC of the environmental impacts of constructing and operating proposed Units 3 and 4 is documented in NUREG-1872, "Final Environmental Impact Statement for an Early Site Permit (ESP) at the Vogtle Electric Generating Plant Site." (ESP FEIS Volumes 1 and 2 are available in ADAMS under ADAMS accession numbers ML082240145 and ML082240165, respectively.)

Please submit comments either by mail to Mrs. Mallecia Sutton, Environmental Project Manager, Division of Site and Environmental Reviews, Mail Stop T-7E18, U.S. Nuclear Regulatory Commission, Washington, D.C. 20555-0001 or via e-mail to Vogtle.COLAEIS@nrc.gov by January 15, 2010.

If you have any questions or require additional information, please contact Mrs. Sutton at (301) 415-0673 or via e-mail to Mallecia.Sutton@nrc.gov.

Sincerely,

/RA/

Gregory P. Hatchett, Branch Chief
Environmental Projects Branch 1
Division of Site and Environmental Reviews
Office of New Reactors

Docket Nos. 52-025 and 52-026

cc: See next page

Appendix F

December 10, 2009

Mr. Louis McGertt, Town King
Thlopthlocco Tribal Town
P.O. Box 188
Okema, OK 74859

SUBJECT: U.S. NUCLEAR REGULATORY COMMISSION'S SUPPLEMENTAL ENVIRONMENTAL IMPACT STATEMENT FOR SOUTHERN NUCLEAR OPERATING COMPANY'S COMBINED LICENSE APPLICATION FOR THE PROPOSED CONSTRUCTION AND OPERATION OF UNITS 3 AND 4 AT THE VOGTLE ELECTRIC GENERATING PLANT IN WAYNESBORO, GEORGIA

Dear Mr. McGertt:

The U.S. Nuclear Regulatory Commission (NRC) staff is reviewing an application submitted by Southern Nuclear Operating Company, Inc. (Southern or SNC), on behalf of itself and several co-applicants, for a combined license (COL) for construction and operation of two new nuclear units at its Vogtle Electric Generating Plant (Vogtle or VEGP) site in Burke County, Georgia. The purpose of this letter is to invite the Thlopthlocco Tribal Town to consult with the NRC regarding the proposed action, pursuant to NRC regulations at 10 CFR Part 51 and Section 106 of the National Historic Preservation Act (NHPA) of 1966, as amended. The NRC plans to coordinate compliance with Section 106 of the NHPA using the National Environmental Policy Act (NEPA) process identified in 36 CFR 800.8(c) in lieu of the procedures set forth in §§ 800.3 through 800.6. Additionally, the NRC staff will rely on procedures in its regulations at 10 CFR 51.92 for supplementing a final environmental impact statement (FEIS).

Section 51.26(d) of the NRC regulations describes the processes for determining the need to supplement a FEIS, issue a notice of intent, and determine whether or not a formal scoping process will be conducted. As required by 10 CFR 51.92, the NRC staff will prepare a supplement to the FEIS for the early site permit (ESP) issued on August 26, 2009, to SNC and the same co-applicants. In this case, the NRC staff has determined that it will not conduct a formal scoping process for the development of the supplemental environmental impact statement (SEIS).

An ESP is a Commission approval of a site suitable for construction and operation of one or more new nuclear units. Under 10 CFR 51.50(c), a COL applicant referencing an ESP need not submit information or analyses regarding environmental issues that were resolved in the ESP EIS, except to the extent the COL applicant has identified any new and potentially significant information.

In accordance with the provisions in 36 CFR § 800.8, the NRC wishes to ensure that Indian tribes that might have an interest in any potential historic properties in the area of potential effect (APE) are afforded the opportunity to identify their concerns, provide advice on the identification and evaluation of historic properties, including those of traditional religious and cultural importance and, if necessary, participate in the resolution of any adverse effects to such properties.

Appendix F

L. McGertt   -2-

Thus, to support the NRC staff's review of any new and potentially significant circumstances or information relevant to the environmental concerns related to the proposed action or its impacts, we request you submit written comments, if any, your tribe may have to offer on the environmental review of the COL by January 15, 2010.

The NRC intends to propose measures in the SEIS for the COL analyses for potential impacts to historical and cultural resources including developing alternatives and proposing measures that might avoid, minimize, or mitigate any adverse effects of the propose action on historic properties. To complete consultation under § 800.8(c), the NRC staff will forward the SEIS on the COL to you for your review and comment, and will address your comments in the final SEIS on the COL.

The proposed new reactors, Vogtle Units 3 and 4 would be located approximately 26 miles southeast of Augusta, Georgia, near the town of Waynesboro, Georgia. SNC currently operates two reactors, Vogtle Units 1 and 2 on the site and plans to construct the new units adjacent to the existing reactors. The application for the COL was accepted for docketing on May 30, 2008, and is available through the web-based version of the NRC Agency-wide Documents Access Management System (ADAMS), which can be found at http://www.nrc.gov/readingrm/adams.html. The Environmental Report for the application is listed under ADAMS accession number ML081050181. The Vogtle Plant COL application is also available on the Internet at http://www.nrc.gov/reactors/new-licensing/col/vogtle.html. A detailed review by NRC of the environmental impacts of constructing and operating proposed Units 3 and 4 is documented in NUREG-1872, "Final Environmental Impact Statement for an Early Site Permit (ESP) at the Vogtle Electric Generating Plant Site." (ESP FEIS Volumes 1 and 2 are available in ADAMS under ADAMS accession numbers ML082240145 and ML082240165, respectively.)

Please submit comments either by mail to Mrs. Mallecia Sutton, Environmental Project Manager, Division of Site and Environmental Reviews, Mail Stop T-7E18, U.S. Nuclear Regulatory Commission, Washington, D.C. 20555-0001 or via e-mail to Vogtle.COLAEIS@nrc.gov by January 15, 2010.

If you have any questions or require additional information, please contact Mrs. Sutton at (301) 415-0673 or via e-mail to Mallecia.Sutton@nrc.gov.

Sincerely,

/RA/

Gregory P. Hatchett, Branch Chief
Environmental Projects Branch 1
Division of Site and Environmental Reviews
Office of New Reactors

Docket Nos. 52-025 and 52-026

cc: See next page

Appendix F

December 10, 2009

Mr. A.D. Ellis, Principal Chief
Muscogee (Creek) Nation
P.O. Box 580
Okmulgee, OK 74447

SUBJECT: U.S. NUCLEAR REGULATORY COMMISSION'S SUPPLEMENTAL ENVIRONMENTAL IMPACT STATEMENT FOR SOUTHERN NUCLEAR OPERATING COMPANY'S COMBINED LICENSE APPLICATION FOR THE PROPOSED CONSTRUCTION AND OPERATION OF UNITS 3 AND 4 AT THE VOGTLE ELECTRIC GENERATING PLANT IN WAYNESBORO, GEORGIA

Dear Mr. Ellis:

The U.S. Nuclear Regulatory Commission (NRC) staff is reviewing an application submitted by Southern Nuclear Operating Company, Inc. (Southern or SNC), on behalf of itself and several co-applicants, for a combined license (COL) for construction and operation of two new nuclear units at its Vogtle Electric Generating Plant (Vogtle or VEGP) site in Burke County, Georgia. The purpose of this letter is to invite the Muscogee (Creek Nation) to consult with the NRC regarding the proposed action, pursuant to NRC regulations at 10 CFR Part 51 and Section 106 of the National Historic Preservation Act (NHPA) of 1966, as amended. The NRC plans to coordinate compliance with Section 106 of the NHPA using the National Environmental Policy Act (NEPA) process identified in 36 CFR 800.8(c) in lieu of the procedures set forth in §§ 800.3 through 800.6. Additionally, the NRC staff will rely on procedures in its regulations at 10 CFR 51.92 for supplementing a final environmental impact statement (FEIS).

Section 51.26(d) of the NRC regulations describes the processes for determining the need to supplement a FEIS, issue a notice of intent, and determine whether or not a formal scoping process will be conducted. As required by 10 CFR 51.92, the NRC staff will prepare a supplement to the FEIS for the early site permit (ESP) issued on August 26, 2009, to SNC and the same co-applicants. In this case, the NRC staff has determined that it will not conduct a formal scoping process for the development of the supplemental environmental impact statement (SEIS).

An ESP is a Commission approval of a site suitable for construction and operation of one or more new nuclear units. Under 10 CFR 51.50(c), a COL applicant referencing an ESP need not submit information or analyses regarding environmental issues that were resolved in the ESP EIS, except to the extent the COL applicant has identified any new and potentially significant information.

In accordance with the provisions in 36 CFR § 800.8, the NRC wishes to ensure that Indian tribes that might have an interest in any potential historic properties in the area of potential effect (APE) are afforded the opportunity to identify their concerns, provide advice on the identification and evaluation of historic properties, including those of traditional religious and cultural importance and, if necessary, participate in the resolution of any adverse effects to such properties.

Appendix F

A. D. Ellis   -2-

Thus, to support the NRC staff's review of any new and potentially significant circumstances or information relevant to the environmental concerns related to the proposed action or its impacts, we request you submit written comments, if any, your tribe may have to offer on the environmental review of the COL by January 15, 2010.

The NRC intends to propose measures in the SEIS for the COL analyses for potential impacts to historical and cultural resources including developing alternatives and proposing measures that might avoid, minimize, or mitigate any adverse effects of the propose action on historic properties. To complete consultation under § 800.8(c), the NRC staff will forward the SEIS on the COL to you for your review and comment, and will address your comments in the final SEIS on the COL.

The proposed new reactors, Vogtle Units 3 and 4 would be located approximately 26 miles southeast of Augusta, Georgia, near the town of Waynesboro, Georgia. SNC currently operates two reactors, Vogtle Units 1 and 2 on the site and plans to construct the new units adjacent to the existing reactors. The application for the COL was accepted for docketing on May 30, 2008, and is available through the web-based version of the NRC Agency-wide Documents Access Management System (ADAMS), which can be found at http://www.nrc.gov/readingrm/adams.html. The Environmental Report for the application is listed under ADAMS accession number ML081050181. The Vogtle Plant COL application is also available on the Internet at http://www.nrc.gov/reactors/new-licensing/col/vogtle.html. A detailed review by NRC of the environmental impacts of constructing and operating proposed Units 3 and 4 is documented in NUREG-1872, "Final Environmental Impact Statement for an Early Site Permit (ESP) at the Vogtle Electric Generating Plant Site." (ESP FEIS Volumes 1 and 2 are available in ADAMS under ADAMS accession numbers ML082240145 and ML082240165, respectively.)

Please submit comments either by mail to Mrs. Mallecia Sutton, Environmental Project Manager, Division of Site and Environmental Reviews, Mail Stop T-7E18, U.S. Nuclear Regulatory Commission, Washington, D.C. 20555-0001 or via e-mail to Vogtle.COLAEIS@nrc.gov by January 15, 2010.

If you have any questions or require additional information, please contact Mrs. Sutton at (301) 415-0673 or via e-mail to Mallecia.Sutton@nrc.gov.

Sincerely,

/RA/

Gregory P. Hatchett, Branch Chief
Environmental Projects Branch 1
Division of Site and Environmental Reviews
Office of New Reactors

Docket Nos. 52-025 and 52-026

cc: See next page

Appendix F

December 10, 2009

Mr. Richard L. Allen, NAGPRA Contact
Cherokee Nation of Oklahoma
P.O. Box 948
Tahleque, OK 74465-0948

SUBJECT: U.S. NUCLEAR REGULATORY COMMISSION'S SUPPLEMENTAL ENVIRONMENTAL IMPACT STATEMENT FOR SOUTHERN NUCLEAR OPERATING COMPANY'S COMBINED LICENSE APPLICATION FOR THE PROPOSED CONSTRUCTION AND OPERATION OF UNITS 3 AND 4 AT THE VOGTLE ELECTRIC GENERATING PLANT IN WAYNESBORO, GEORGIA

Dear Mr. Allen:

The U.S. Nuclear Regulatory Commission (NRC) staff is reviewing an application submitted by Southern Nuclear Operating Company, Inc. (Southern or SNC), on behalf of itself and several co-applicants, for a combined license (COL) for construction and operation of two new nuclear units at its Vogtle Electric Generating Plant (Vogtle or VEGP) site in Burke County, Georgia. The purpose of this letter is to invite the Cherokee Nation of Oklahoma to consult with the NRC regarding the proposed action, pursuant to NRC regulations at 10 CFR Part 51 and Section 106 of the National Historic Preservation Act (NHPA) of 1966, as amended. The NRC plans to coordinate compliance with Section 106 of the NHPA using the National Environmental Policy Act (NEPA) process identified in 36 CFR 800.8(c) in lieu of the procedures set forth in §§ 800.3 through 800.6. Additionally, the NRC staff will rely on procedures in its regulations at 10 CFR 51.92 for supplementing a final environmental impact statement (FEIS).

Section 51.26(d) of the NRC regulations describes the processes for determining the need to supplement a FEIS, issue a notice of intent, and determine whether or not a formal scoping process will be conducted. As required by 10 CFR 51.92, the NRC staff will prepare a supplement to the FEIS for the early site permit (ESP) issued on August 26, 2009, to SNC and the same co-applicants. In this case, the NRC staff has determined that it will not conduct a formal scoping process for the development of the supplemental environmental impact statement (SEIS).

An ESP is a Commission approval of a site suitable for construction and operation of one or more new nuclear units. Under 10 CFR 51.50(c), a COL applicant referencing an ESP need not submit information or analyses regarding environmental issues that were resolved in the ESP EIS, except to the extent the COL applicant has identified any new and potentially significant information.

In accordance with the provisions in 36 CFR § 800.8, the NRC wishes to ensure that Indian tribes that might have an interest in any potential historic properties in the area of potential effect (APE) are afforded the opportunity to identify their concerns, provide advice on the identification and evaluation of historic properties, including those of traditional religious and cultural importance and, if necessary, participate in the resolution of any adverse effects to such properties.

Appendix F

Thus, to support the NRC staff's review of any new and potentially significant circumstances or information relevant to the environmental concerns related to the proposed action or its impacts, we request you submit written comments, if any, your tribe may have to offer on the environmental review of the COL by January 15, 2010.

The NRC intends to propose measures in the SEIS for the COL analyses for potential impacts to historical and cultural resources including developing alternatives and proposing measures that might avoid, minimize, or mitigate any adverse effects of the propose action on historic properties. To complete consultation under § 800.8(c), the NRC staff will forward the SEIS on the COL to you for your review and comment, and will address your comments in the final SEIS on the COL.

The proposed new reactors, Vogtle Units 3 and 4 would be located approximately 26 miles southeast of Augusta, Georgia, near the town of Waynesboro, Georgia. SNC currently operates two reactors, Vogtle Units 1 and 2 on the site and plans to construct the new units adjacent to the existing reactors. The application for the COL was accepted for docketing on May 30, 2008, and is available through the web-based version of the NRC Agency-wide Documents Access Management System (ADAMS), which can be found at http://www.nrc.gov/readingrm/adams.html. The Environmental Report for the application is listed under ADAMS accession number ML081050181. The Vogtle Plant COL application is also available on the Internet at http://www.nrc.gov/reactors/new-licensing/col/vogtle.html. A detailed review by NRC of the environmental impacts of constructing and operating proposed Units 3 and 4 is documented in NUREG-1872, "Final Environmental Impact Statement for an Early Site Permit (ESP) at the Vogtle Electric Generating Plant Site." (ESP FEIS Volumes 1 and 2 are available in ADAMS under ADAMS accession numbers ML082240145 and ML082240165, respectively.)

Please submit comments either by mail to Mrs. Mallecia Sutton, Environmental Project Manager, Division of Site and Environmental Reviews, Mail Stop T-7E18, U.S. Nuclear Regulatory Commission, Washington, D.C. 20555-0001 or via e-mail to Vogtle.COLAEIS@nrc.gov by January 15, 2010.

If you have any questions or require additional information, please contact Mrs. Sutton at (301) 415-0673 or via e-mail to Mallecia.Sutton@nrc.gov.

Sincerely,

/RA/

Gregory P. Hatchett, Branch Chief
Environmental Projects Branch 1
Division of Site and Environmental Reviews
Office of New Reactors

Docket Nos. 52-025 and 52-026

cc: See next page

Appendix F

December 10, 2009

Ms. Gingy (Virginia) Nail
NAGPRA Contact
Chickasaw Nation
P.O. Box 1548
Ada, OK 74883

SUBJECT: U.S. NUCLEAR REGULATORY COMMISSION'S SUPPLEMENTAL ENVIRONMENTAL IMPACT STATEMENT FOR SOUTHERN NUCLEAR OPERATING COMPANY'S COMBINED LICENSE APPLICATION FOR THE PROPOSED CONSTRUCTION AND OPERATION OF UNITS 3 AND 4 AT THE VOGTLE ELECTRIC GENERATING PLANT IN WAYNESBORO, GEORGIA

Dear Ms. Nail:

The U.S. Nuclear Regulatory Commission (NRC) staff is reviewing an application submitted by Southern Nuclear Operating Company, Inc. (Southern or SNC), on behalf of itself and several co-applicants, for a combined license (COL) for construction and operation of two new nuclear units at its Vogtle Electric Generating Plant (Vogtle or VEGP) site in Burke County, Georgia. The purpose of this letter is to invite the Chickasaw Nation to consult with the NRC regarding the proposed action, pursuant to NRC regulations at 10 CFR Part 51 and Section 106 of the National Historic Preservation Act (NHPA) of 1966, as amended. The NRC plans to coordinate compliance with Section 106 of the NHPA using the National Environmental Policy Act (NEPA) process identified in 36 CFR 800.8(c) in lieu of the procedures set forth in §§ 800.3 through 800.6. Additionally, the NRC staff will rely on procedures in its regulations at 10 CFR 51.92 for supplementing a final environmental impact statement (FEIS).

Section 51.26(d) of the NRC regulations describes the processes for determining the need to supplement a FEIS, issue a notice of intent, and determine whether or not a formal scoping process will be conducted. As required by 10 CFR 51.92, the NRC staff will prepare a supplement to the FEIS for the early site permit (ESP) issued on August 26, 2009, to SNC and the same co-applicants. In this case, the NRC staff has determined that it will not conduct a formal scoping process for the development of the supplemental environmental impact statement (SEIS).

An ESP is a Commission approval of a site suitable for construction and operation of one or more new nuclear units. Under 10 CFR 51.50(c), a COL applicant referencing an ESP need not submit information or analyses regarding environmental issues that were resolved in the ESP EIS, except to the extent the COL applicant has identified any new and potentially significant information.

In accordance with the provisions in 36 CFR § 800.8, the NRC wishes to ensure that Indian tribes that might have an interest in any potential historic properties in the area of potential effect (APE) are afforded the opportunity to identify their concerns, provide advice on the identification and evaluation of historic properties, including those of traditional religious and cultural importance and, if necessary, participate in the resolution of any adverse effects to such properties.

Appendix F

G. Nail   -2-

Thus, to support the NRC staff's review of any new and potentially significant circumstances or information relevant to the environmental concerns related to the proposed action or its impacts, we request you submit written comments, if any, your tribe may have to offer on the environmental review of the COL by January 15, 2010.

The NRC intends to propose measures in the SEIS for the COL analyses for potential impacts to historical and cultural resources including developing alternatives and proposing measures that might avoid, minimize, or mitigate any adverse effects of the propose action on historic properties. To complete consultation under § 800.8(c), the NRC staff will forward the SEIS on the COL to you for your review and comment, and will address your comments in the final SEIS on the COL.

The proposed new reactors, Vogtle Units 3 and 4 would be located approximately 26 miles southeast of Augusta, Georgia, near the town of Waynesboro, Georgia. SNC currently operates two reactors, Vogtle Units 1 and 2 on the site and plans to construct the new units adjacent to the existing reactors. The application for the COL was accepted for docketing on May 30, 2008, and is available through the web-based version of the NRC Agency-wide Documents Access Management System (ADAMS), which can be found at http://www.nrc.gov/readingrm/adams.html. The Environmental Report for the application is listed under ADAMS accession number ML081050181. The Vogtle Plant COL application is also available on the Internet at http://www.nrc.gov/reactors/new-licensing/col/vogtle.html. A detailed review by NRC of the environmental impacts of constructing and operating proposed Units 3 and 4 is documented in NUREG-1872, "Final Environmental Impact Statement for an Early Site Permit (ESP) at the Vogtle Electric Generating Plant Site." (ESP FEIS Volumes 1 and 2 are available in ADAMS under ADAMS accession numbers ML082240145 and ML082240165, respectively.)

Please submit comments either by mail to Mrs. Mallecia Sutton, Environmental Project Manager, Division of Site and Environmental Reviews, Mail Stop T-7E18, U.S. Nuclear Regulatory Commission, Washington, D.C. 20555-0001 or via e-mail to VogtleCOLAEIS@nrc.gov by January 15, 2010.

If you have any questions or require additional information, please contact Mrs. Sutton at (301) 415-0673 or via e-mail to Mallecia.Sutton@nrc.gov.

                                        Sincerely,

                                        /RA/

                                        Gregory P. Hatchett, Branch Chief
                                        Environmental Projects Branch 1
                                        Division of Site and Environmental Reviews
                                        Office of New Reactors

Docket Nos. 52-025 and 52-026

cc: See next page

Appendix F

December 10, 2009

Mr. Bill Anoatubby, Governor
Chickasaw Nation of Oklahoma
P.O. Box 1548
Ada, OK 74821-1548

SUBJECT: U.S. NUCLEAR REGULATORY COMMISSION'S SUPPLEMENTAL ENVIRONMENTAL IMPACT STATEMENT FOR SOUTHERN NUCLEAR OPERATING COMPANY'S COMBINED LICENSE APPLICATION FOR THE PROPOSED CONSTRUCTION AND OPERATION OF UNITS 3 AND 4 AT THE VOGTLE ELECTRIC GENERATING PLANT IN WAYNESBORO, GEORGIA

Dear Mr. Anoatubby:

The U.S. Nuclear Regulatory Commission (NRC) staff is reviewing an application submitted by Southern Nuclear Operating Company, Inc. (Southern or SNC), on behalf of itself and several co-applicants, for a combined license (COL) for construction and operation of two new nuclear units at its Vogtle Electric Generating Plant (Vogtle or VEGP) site in Burke County, Georgia. The purpose of this letter is to invite the Chickasaw Nation of Oklahoma to consult with the NRC regarding the proposed action, pursuant to NRC regulations at 10 CFR Part 51 and Section 106 of the National Historic Preservation Act (NHPA) of 1966, as amended. The NRC plans to coordinate compliance with Section 106 of the NHPA using the National Environmental Policy Act (NEPA) process identified in 36 CFR 800.8(c) in lieu of the procedures set forth in §§ 800.3 through 800.6. Additionally, the NRC staff will rely on procedures in its regulations at 10 CFR 51.92 for supplementing a final environmental impact statement (FEIS).

Section 51.26(d) of the NRC regulations describes the processes for determining the need to supplement a FEIS, issue a notice of intent, and determine whether or not a formal scoping process will be conducted. As required by 10 CFR 51.92, the NRC staff will prepare a supplement to the FEIS for the early site permit (ESP) issued on August 26, 2009, to SNC and the same co-applicants. In this case, the NRC staff has determined that it will not conduct a formal scoping process for the development of the supplemental environmental impact statement (SEIS).

An ESP is a Commission approval of a site suitable for construction and operation of one or more new nuclear units. Under 10 CFR 51.50(c), a COL applicant referencing an ESP need not submit information or analyses regarding environmental issues that were resolved in the ESP EIS, except to the extent the COL applicant has identified any new and potentially significant information.

In accordance with the provisions in 36 CFR § 800.8, the NRC wishes to ensure that Indian tribes that might have an interest in any potential historic properties in the area of potential effect (APE) are afforded the opportunity to identify their concerns, provide advice on the identification and evaluation of historic properties, including those of traditional religious and cultural importance and, if necessary, participate in the resolution of any adverse effects to such properties.

Appendix F

Thus, to support the NRC staff's review of any new and potentially significant circumstances or information relevant to the environmental concerns related to the proposed action or its impacts, we request you submit written comments, if any, your tribe may have to offer on the environmental review of the COL by January 15, 2010.

The NRC intends to propose measures in the SEIS for the COL analyses for potential impacts to historical and cultural resources including developing alternatives and proposing measures that might avoid, minimize, or mitigate any adverse effects of the propose action on historic properties. To complete consultation under § 800.8(c), the NRC staff will forward the SEIS on the COL to you for your review and comment, and will address your comments in the final SEIS on the COL.

The proposed new reactors, Vogtle Units 3 and 4 would be located approximately 26 miles southeast of Augusta, Georgia, near the town of Waynesboro, Georgia. SNC currently operates two reactors, Vogtle Units 1 and 2 on the site and plans to construct the new units adjacent to the existing reactors. The application for the COL was accepted for docketing on May 30, 2008, and is available through the web-based version of the NRC Agency-wide Documents Access Management System (ADAMS), which can be found at http://www.nrc.gov/readingrm/adams.html. The Environmental Report for the application is listed under ADAMS accession number ML081050181. The Vogtle Plant COL application is also available on the Internet at http://www.nrc.gov/reactors/new-licensing/col/vogtle.html. A detailed review by NRC of the environmental impacts of constructing and operating proposed Units 3 and 4 is documented in NUREG-1872, "Final Environmental Impact Statement for an Early Site Permit (ESP) at the Vogtle Electric Generating Plant Site." (ESP FEIS Volumes 1 and 2 are available in ADAMS under ADAMS accession numbers ML082240145 and ML082240165, respectively.)

Please submit comments either by mail to Mrs. Mallecia Sutton, Environmental Project Manager, Division of Site and Environmental Reviews, Mail Stop T-7E18, U.S. Nuclear Regulatory Commission, Washington, D.C. 20555-0001 or via e-mail to Vogtle.COLAEIS@nrc.gov by January 15, 2010.

If you have any questions or require additional information, please contact Mrs. Sutton at (301) 415-0673 or via e-mail to Mallecia.Sutton@nrc.gov.

Sincerely,

/RA/

Gregory P. Hatchett, Branch Chief
Environmental Projects Branch 1
Division of Site and Environmental Reviews
Office of New Reactors

Docket Nos. 52-025 and 52-026

cc: See next page

Appendix F

December 10, 2009

Mr. Charles Thurmond, NAGPRA Contact
Georgia Tribe of Eastern Cherokee
P.O. Box 1324
Clayton, GA 30525

SUBJECT: U.S. NUCLEAR REGULATORY COMMISSION'S SUPPLEMENTAL ENVIRONMENTAL IMPACT STATEMENT FOR SOUTHERN NUCLEAR OPERATING COMPANY'S COMBINED LICENSE APPLICATION FOR THE PROPOSED CONSTRUCTION AND OPERATION OF UNITS 3 AND 4 AT THE VOGTLE ELECTRIC GENERATING PLANT IN WAYNESBORO, GEORGIA

Dear Mr. Thurmond:

The U.S. Nuclear Regulatory Commission (NRC) staff is reviewing an application submitted by Southern Nuclear Operating Company, Inc. (Southern or SNC), on behalf of itself and several co-applicants, for a combined license (COL) for construction and operation of two new nuclear units at its Vogtle Electric Generating Plant (Vogtle or VEGP) site in Burke County, Georgia. The purpose of this letter is to invite the Georgia Tribe of Eastern Cherokee to consult with the NRC regarding the proposed action, pursuant to NRC regulations at 10 CFR Part 51 and Section 106 of the National Historic Preservation Act (NHPA) of 1966, as amended. The NRC plans to coordinate compliance with Section 106 of the NHPA using the National Environmental Policy Act (NEPA) process identified in 36 CFR 800.8(c) in lieu of the procedures set forth in §§ 800.3 through 800.6. Additionally, the NRC staff will rely on procedures in its regulations at 10 CFR 51.92 for supplementing a final environmental impact statement (FEIS).

Section 51.26(d) of the NRC regulations describes the processes for determining the need to supplement a FEIS, issue a notice of intent, and determine whether or not a formal scoping process will be conducted. As required by 10 CFR 51.92, the NRC staff will prepare a supplement to the FEIS for the early site permit (ESP) issued on August 26, 2009, to SNC and the same co-applicants. In this case, the NRC staff has determined that it will not conduct a formal scoping process for the development of the supplemental environmental impact statement (SEIS).

An ESP is a Commission approval of a site suitable for construction and operation of one or more new nuclear units. Under 10 CFR 51.50(c), a COL applicant referencing an ESP need not submit information or analyses regarding environmental issues that were resolved in the ESP EIS, except to the extent the COL applicant has identified any new and potentially significant information.

In accordance with the provisions in 36 CFR § 800.8, the NRC wishes to ensure that Indian tribes that might have an interest in any potential historic properties in the area of potential effect (APE) are afforded the opportunity to identify their concerns, provide advice on the identification and evaluation of historic properties, including those of traditional religious and cultural importance and, if necessary, participate in the resolution of any adverse effects to such properties.

Appendix F

Thus, to support the NRC staff's review of any new and potentially significant circumstances or information relevant to the environmental concerns related to the proposed action or its impacts, we request you submit written comments, if any, your tribe may have to offer on the environmental review of the COL by January 15, 2010.

The NRC intends to propose measures in the SEIS for the COL analyses for potential impacts to historical and cultural resources including developing alternatives and proposing measures that might avoid, minimize, or mitigate any adverse effects of the propose action on historic properties. To complete consultation under § 800.8(c), the NRC staff will forward the SEIS on the COL to you for your review and comment, and will address your comments in the final SEIS on the COL.

The proposed new reactors, Vogtle Units 3 and 4 would be located approximately 26 miles southeast of Augusta, Georgia, near the town of Waynesboro, Georgia. SNC currently operates two reactors, Vogtle Units 1 and 2 on the site and plans to construct the new units adjacent to the existing reactors. The application for the COL was accepted for docketing on May 30, 2008, and is available through the web-based version of the NRC Agency-wide Documents Access Management System (ADAMS), which can be found at http://www.nrc.gov/readingrm/adams.html. The Environmental Report for the application is listed under ADAMS accession number ML081050181. The Vogtle Plant COL application is also available on the Internet at http://www.nrc.gov/reactors/new-licensing/col/vogtle.html. A detailed review by NRC of the environmental impacts of constructing and operating proposed Units 3 and 4 is documented in NUREG-1872, "Final Environmental Impact Statement for an Early Site Permit (ESP) at the Vogtle Electric Generating Plant Site." (ESP FEIS Volumes 1 and 2 are available in ADAMS under ADAMS accession numbers ML082240145 and ML082240165, respectively.)

Please submit comments either by mail to Mrs. Mallecia Sutton, Environmental Project Manager, Division of Site and Environmental Reviews, Mail Stop T-7E18, U.S. Nuclear Regulatory Commission, Washington, D.C. 20555-0001 or via e-mail to Vogtle.COLAEIS@nrc.gov by January 15, 2010.

If you have any questions or require additional information, please contact Mrs. Sutton at (301) 415-0673 or via e-mail to Mallecia.Sutton@nrc.gov.

Sincerely,

/RA/

Gregory P. Hatchett, Branch Chief
Environmental Projects Branch 1
Division of Site and Environmental Reviews
Office of New Reactors

Docket Nos. 52-025 and 52-026

cc: See next page

Appendix F

December 10, 2009

Mr. Tarpie Yargee
Alabama-Quassarte Tribal Town
P.O. Box 187
Wetumka, OK 74883

SUBJECT: U.S. NUCLEAR REGULATORY COMMISSION'S SUPPLEMENTAL ENVIRONMENTAL IMPACT STATEMENT FOR SOUTHERN NUCLEAR OPERATING COMPANY'S COMBINED LICENSE APPLICATION FOR THE PROPOSED CONSTRUCTION AND OPERATION OF UNITS 3 AND 4 AT THE VOGTLE ELECTRIC GENERATING PLANT IN WAYNESBORO, GEORGIA

Dear Mr. Yargee:

The U.S. Nuclear Regulatory Commission (NRC) staff is reviewing an application submitted by Southern Nuclear Operating Company, Inc. (Southern or SNC), on behalf of itself and several co-applicants, for a combined license (COL) for construction and operation of two new nuclear units at its Vogtle Electric Generating Plant (Vogtle or VEGP) site in Burke County, Georgia. The purpose of this letter is to invite the Alabama-Quassarte Tribal Town to consult with the NRC regarding the proposed action, pursuant to NRC regulations at 10 CFR Part 51 and Section 106 of the National Historic Preservation Act (NHPA) of 1966, as amended. The NRC plans to coordinate compliance with Section 106 of the NHPA using the National Environmental Policy Act (NEPA) process identified in 36 CFR 800.8(c) in lieu of the procedures set forth in §§ 800.3 through 800.6. Additionally, the NRC staff will rely on procedures in its regulations at 10 CFR 51.92 for supplementing a final environmental impact statement (FEIS).

Section 51.26(d) of the NRC regulations describes the processes for determining the need to supplement a FEIS, issue a notice of intent, and determine whether or not a formal scoping process will be conducted. As required by 10 CFR 51.92, the NRC staff will prepare a supplement to the FEIS for the early site permit (ESP) issued on August 26, 2009, to SNC and the same co-applicants. In this case, the NRC staff has determined that it will not conduct a formal scoping process for the development of the supplemental environmental impact statement (SEIS).

An ESP is a Commission approval of a site suitable for construction and operation of one or more new nuclear units. Under 10 CFR 51.50(c), a COL applicant referencing an ESP need not submit information or analyses regarding environmental issues that were resolved in the ESP EIS, except to the extent the COL applicant has identified any new and potentially significant information.

In accordance with the provisions in 36 CFR § 800.8, the NRC wishes to ensure that Indian tribes that might have an interest in any potential historic properties in the area of potential effect (APE) are afforded the opportunity to identify their concerns, provide advice on the identification and evaluation of historic properties, including those of traditional religious and cultural importance and, if necessary, participate in the resolution of any adverse effects to such properties.

Appendix F

T. Yargee   -2-

Thus, to support the NRC staff's review of any new and potentially significant circumstances or information relevant to the environmental concerns related to the proposed action or its impacts, we request you submit written comments, if any, your tribe may have to offer on the environmental review of the COL by January 15, 2010.

The NRC intends to propose measures in the SEIS for the COL analyses for potential impacts to historical and cultural resources including developing alternatives and proposing measures that might avoid, minimize, or mitigate any adverse effects of the propose action on historic properties. To complete consultation under § 800.8(c), the NRC staff will forward the SEIS on the COL to you for your review and comment, and will address your comments in the final SEIS on the COL.

The proposed new reactors, Vogtle Units 3 and 4 would be located approximately 26 miles southeast of Augusta, Georgia, near the town of Waynesboro, Georgia. SNC currently operates two reactors, Vogtle Units 1 and 2 on the site and plans to construct the new units adjacent to the existing reactors. The application for the COL was accepted for docketing on May 30, 2008, and is available through the web-based version of the NRC Agency-wide Documents Access Management System (ADAMS), which can be found at http://www.nrc.gov/readingrm/adams.html. The Environmental Report for the application is listed under ADAMS accession number ML081050181. The Vogtle Plant COL application is also available on the Internet at http://www.nrc.gov/reactors/new-licensing/col/vogtle.html. A detailed review by NRC of the environmental impacts of constructing and operating proposed Units 3 and 4 is documented in NUREG-1872, "Final Environmental Impact Statement for an Early Site Permit (ESP) at the Vogtle Electric Generating Plant Site." (ESP FEIS Volumes 1 and 2 are available in ADAMS under ADAMS accession numbers ML082240145 and ML082240165, respectively.)

Please submit comments either by mail to Mrs. Mallecia Sutton, Environmental Project Manager, Division of Site and Environmental Reviews, Mail Stop T-7E18, U.S. Nuclear Regulatory Commission, Washington, D.C. 20555-0001 or via e-mail to VogtleCOLASEIS@nrc.gov by January 15, 2010.

If you have any questions or require additional information, please contact Mrs. Sutton at (301) 415-0673 or via e-mail to Mallecia.Sutton@nrc.gov.

Sincerely,

/RA/

Gregory P. Hatchett, Branch Chief
Environmental Projects Branch 1
Division of Site and Environmental Reviews
Office of New Reactors

Docket Nos. 52-025 and 52-026

cc: See next page

Appendix F

December 10, 2009

Mr. Pare Bowlegs
Seminole Nation of Oklahoma
P.O. Box 1498
Wewoka, OK 74884

SUBJECT: U.S. NUCLEAR REGULATORY COMMISSION'S SUPPLEMENTAL ENVIRONMENTAL IMPACT STATEMENT FOR SOUTHERN NUCLEAR OPERATING COMPANY'S COMBINED LICENSE APPLICATION FOR THE PROPOSED CONSTRUCTION AND OPERATION OF UNITS 3 AND 4 AT THE VOGTLE ELECTRIC GENERATING PLANT IN WAYNESBORO, GEORGIA

Dear Mr. Bowlegs:

The U.S. Nuclear Regulatory Commission (NRC) staff is reviewing an application submitted by Southern Nuclear Operating Company, Inc. (Southern or SNC), on behalf of itself and several co-applicants, for a combined license (COL) for construction and operation of two new nuclear units at its Vogtle Electric Generating Plant (Vogtle or VEGP) site in Burke County, Georgia. The purpose of this letter is to invite the Seminole Nation of Oklahoma to consult with the NRC regarding the proposed action, pursuant to NRC regulations at 10 CFR Part 51 and Section 106 of the National Historic Preservation Act (NHPA) of 1966, as amended. The NRC plans to coordinate compliance with Section 106 of the NHPA using the National Environmental Policy Act (NEPA) process identified in 36 CFR 800.8(c) in lieu of the procedures set forth in §§ 800.3 through 800.6. Additionally, the NRC staff will rely on procedures in its regulations at 10 CFR 51.92 for supplementing a final environmental impact statement (FEIS).

Section 51.26(d) of the NRC regulations describes the processes for determining the need to supplement a FEIS, issue a notice of intent, and determine whether or not a formal scoping process will be conducted. As required by 10 CFR 51.92, the NRC staff will prepare a supplement to the FEIS for the early site permit (ESP) issued on August 26, 2009, to SNC and the same co-applicants. In this case, the NRC staff has determined that it will not conduct a formal scoping process for the development of the supplemental environmental impact statement (SEIS).

An ESP is a Commission approval of a site suitable for construction and operation of one or more new nuclear units. Under 10 CFR 51.50(c), a COL applicant referencing an ESP need not submit information or analyses regarding environmental issues that were resolved in the ESP EIS, except to the extent the COL applicant has identified any new and potentially significant information.

In accordance with the provisions in 36 CFR § 800.8, the NRC wishes to ensure that Indian tribes that might have an interest in any potential historic properties in the area of potential effect (APE) are afforded the opportunity to identify their concerns, provide advice on the identification and evaluation of historic properties, including those of traditional religious and cultural importance and, if necessary, participate in the resolution of any adverse effects to such properties.

Appendix F

P. Bowlegs   -2-

Thus, to support the NRC staff's review of any new and potentially significant circumstances or information relevant to the environmental concerns related to the proposed action or its impacts, we request you submit written comments, if any, your tribe may have to offer on the environmental review of the COL by January 8, 2010.

The NRC intends to propose measures in the SEIS for the COL analyses for potential impacts to historical and cultural resources including developing alternatives and proposing measures that might avoid, minimize, or mitigate any adverse effects of the propose action on historic properties. To complete consultation under § 800.8(c), the NRC staff will forward the SEIS on the COL to you for your review and comment, and will address your comments in the final SEIS on the COL.

The proposed new reactors, Vogtle Units 3 and 4 would be located approximately 26 miles southeast of Augusta, Georgia, near the town of Waynesboro, Georgia. SNC currently operates two reactors, Vogtle Units 1 and 2 on the site and plans to construct the new units adjacent to the existing reactors. The application for the COL was accepted for docketing on May 30, 2008, and is available through the web-based version of the NRC Agency-wide Documents Access Management System (ADAMS), which can be found at http://www.nrc.gov/readingrm/adams.html. The Environmental Report for the application is listed under ADAMS accession number ML081050181. The Vogtle Plant COL application is also available on the Internet at http://www.nrc.gov/reactors/new-licensing/col/vogtle.html. A detailed review by NRC of the environmental impacts of constructing and operating proposed Units 3 and 4 is documented in NUREG-1872, "Final Environmental Impact Statement for an Early Site Permit (ESP) at the Vogtle Electric Generating Plant Site." (ESP FEIS Volumes 1 and 2 are available in ADAMS under ADAMS accession numbers ML082240145 and ML082240165, respectively.)

Please submit comments either by mail to Mrs. Mallecia Sutton, Environmental Project Manager, Division of Site and Environmental Reviews, Mail Stop T-7E18, U.S. Nuclear Regulatory Commission, Washington, D.C. 20555-0001 or via e-mail to VogtleCOLASEIS@nrc.gov by January 8, 2010.

If you have any questions or require additional information, please contact Mrs. Sutton at (301) 415-0673 or via e-mail to Mallecia.Sutton@nrc.gov.

Sincerely,

/RA/

Gregory P. Hatchett, Branch Chief
Environmental Projects Branch 1
Division of Site and Environmental Reviews
Office of New Reactors

Docket Nos. 52-025 and 52-026

cc: See next page

Appendix F

December 10, 2009

Mr. Michell Hicks, Principal Chief
Eastern Band of Cherokee Indians
P.O. Box 455
Qualla Boundary
Cherokee, NC 28719

SUBJECT: U.S. NUCLEAR REGULATORY COMMISSION'S SUPPLEMENTAL ENVIRONMENTAL IMPACT STATEMENT FOR SOUTHERN NUCLEAR OPERATING COMPANY'S COMBINED LICENSE APPLICATION FOR THE PROPOSED CONSTRUCTION AND OPERATION OF UNITS 3 AND 4 AT THE VOGTLE ELECTRIC GENERATING PLANT IN WAYNESBORO, GEORGIA

Dear Mr. Hicks:

The U.S. Nuclear Regulatory Commission (NRC) staff is reviewing an application submitted by Southern Nuclear Operating Company, Inc. (Southern or SNC), on behalf of itself and several co-applicants, for a combined license (COL) for construction and operation of two new nuclear units at its Vogtle Electric Generating Plant (Vogtle or VEGP) site in Burke County, Georgia. The purpose of this letter is to invite the Eastern Band of Cherokee Indians to consult with the NRC regarding the proposed action, pursuant to NRC regulations at 10 CFR Part 51 and Section 106 of the National Historic Preservation Act (NHPA) of 1966, as amended. The NRC plans to coordinate compliance with Section 106 of the NHPA using the National Environmental Policy Act (NEPA) process identified in 36 CFR 800.8(c) in lieu of the procedures set forth in §§ 800.3 through 800.6. Additionally, the NRC staff will rely on procedures in its regulations at 10 CFR 51.92 for supplementing a final environmental impact statement (FEIS).

Section 51.26(d) of the NRC regulations describes the processes for determining the need to supplement a FEIS, issue a notice of intent, and determine whether or not a formal scoping process will be conducted. As required by 10 CFR 51.92, the NRC staff will prepare a supplement to the FEIS for the early site permit (ESP) issued on August 26, 2009, to SNC and the same co-applicants. In this case, the NRC staff has determined that it will not conduct a formal scoping process for the development of the supplemental environmental impact statement (SEIS).

An ESP is a Commission approval of a site suitable for construction and operation of one or more new nuclear units. Under 10 CFR 51.50(c), a COL applicant referencing an ESP need not submit information or analyses regarding environmental issues that were resolved in the ESP EIS, except to the extent the COL applicant has identified any new and potentially significant information.

In accordance with the provisions in 36 CFR § 800.8, the NRC wishes to ensure that Indian tribes that might have an interest in any potential historic properties in the area of potential effect (APE) are afforded the opportunity to identify their concerns, provide advice on the identification and evaluation of historic properties, including those of traditional religious and cultural importance and, if necessary, participate in the resolution of any adverse effects to such properties.

Appendix F

M. Hicks   -2-

Thus, to support the NRC staff's review of any new and potentially significant circumstances or information relevant to the environmental concerns related to the proposed action or its impacts, we request you submit written comments, if any, your tribe may have to offer on the environmental review of the COL by January 15, 2010.

The NRC intends to propose measures in the SEIS for the COL analyses for potential impacts to historical and cultural resources including developing alternatives and proposing measures that might avoid, minimize, or mitigate any adverse effects of the propose action on historic properties. To complete consultation under § 800.8(c), the NRC staff will forward the SEIS on the COL to you for your review and comment, and will address your comments in the final SEIS on the COL.

The proposed new reactors, Vogtle Units 3 and 4 would be located approximately 26 miles southeast of Augusta, Georgia, near the town of Waynesboro, Georgia. SNC currently operates two reactors, Vogtle Units 1 and 2 on the site and plans to construct the new units adjacent to the existing reactors. The application for the COL was accepted for docketing on May 30, 2008, and is available through the web-based version of the NRC Agency-wide Documents Access Management System (ADAMS), which can be found at http://www.nrc.gov/readingrm/adams.html. The Environmental Report for the application is listed under ADAMS accession number ML081050181. The Vogtle Plant COL application is also available on the Internet at http://www.nrc.gov/reactors/new-licensing/col/vogtle.html. A detailed review by NRC of the environmental impacts of constructing and operating proposed Units 3 and 4 is documented in NUREG-1872, "Final Environmental Impact Statement for an Early Site Permit (ESP) at the Vogtle Electric Generating Plant Site." (ESP FEIS Volumes 1 and 2 are available in ADAMS under ADAMS accession numbers ML082240145 and ML082240165, respectively.)

Please submit comments either by mail to Mrs. Mallecia Sutton, Environmental Project Manager, Division of Site and Environmental Reviews, Mail Stop T-7E18, U.S. Nuclear Regulatory Commission, Washington, D.C. 20555-0001 or via e-mail to Vogtle.COLAEIS@nrc.gov by January 15, 2010.

If you have any questions or require additional information, please contact Mrs. Sutton at (301) 415-0673 or via e-mail to Mallecia.Sutton@nrc.gov.

                                        Sincerely,

                                        /RA/

                                        Gregory P. Hatchett, Branch Chief
                                        Environmental Projects Branch 1
                                        Division of Site and Environmental Reviews
                                        Office of New Reactors

Docket Nos. 52-025 and 52-026

cc: See next page

Appendix F

December 10, 2009

Mr. Dallas Proctor, Chief
United Keetoowah Band of Cherokee Indians
P.O. Box 746
Tahlequah, OK 74465

SUBJECT: U.S. NUCLEAR REGULATORY COMMISSION'S SUPPLEMENTAL ENVIRONMENTAL IMPACT STATEMENT FOR SOUTHERN NUCLEAR OPERATING COMPANY'S COMBINED LICENSE APPLICATION FOR THE PROPOSED CONSTRUCTION AND OPERATION OF UNITS 3 AND 4 AT THE VOGTLE ELECTRIC GENERATING PLANT IN WAYNESBORO, GEORGIA

Dear Chief Proctor:

The U.S. Nuclear Regulatory Commission (NRC) staff is reviewing an application submitted by Southern Nuclear Operating Company, Inc. (Southern or SNC), on behalf of itself and several co-applicants, for a combined license (COL) for construction and operation of two new nuclear units at its Vogtle Electric Generating Plant (Vogtle or VEGP) site in Burke County, Georgia. The purpose of this letter is to invite the United Keetoowah Band of Cherokee Indians to consult with the NRC regarding the proposed action, pursuant to NRC regulations at 10 CFR Part 51 and Section 106 of the National Historic Preservation Act (NHPA) of 1966, as amended. The NRC plans to coordinate compliance with Section 106 of the NHPA using the National Environmental Policy Act (NEPA) process identified in 36 CFR 800.8(c) in lieu of the procedures set forth in §§ 800.3 through 800.6. Additionally, the NRC staff will rely on procedures in its regulations at 10 CFR 51.92 for supplementing a final environmental impact statement (FEIS).

Section 51.26(d) of the NRC regulations describes the processes for determining the need to supplement a FEIS, issue a notice of intent, and determine whether or not a formal scoping process will be conducted. As required by 10 CFR 51.92, the NRC staff will prepare a supplement to the FEIS for the early site permit (ESP) issued on August 26, 2009, to SNC and the same co-applicants. In this case, the NRC staff has determined that it will not conduct a formal scoping process for the development of the supplemental environmental impact statement (SEIS).

An ESP is a Commission approval of a site suitable for construction and operation of one or more new nuclear units. Under 10 CFR 51.50(c), a COL applicant referencing an ESP need not submit information or analyses regarding environmental issues that were resolved in the ESP EIS, except to the extent the COL applicant has identified any new and potentially significant information.

In accordance with the provisions in 36 CFR § 800.8, the NRC wishes to ensure that Indian tribes that might have an interest in any potential historic properties in the area of potential effect (APE) are afforded the opportunity to identify their concerns, provide advice on the identification and evaluation of historic properties, including those of traditional religious and cultural importance and, if necessary, participate in the resolution of any adverse effects to such properties.

Appendix F

D. Proctor   -2-

Thus, to support the NRC staff's review of any new and potentially significant circumstances or information relevant to the environmental concerns related to the proposed action or its impacts, we request you submit written comments, if any, your tribe may have to offer on the environmental review of the COL by January 8, 2010.

The NRC intends to propose measures in the SEIS for the COL analyses for potential impacts to historical and cultural resources including developing alternatives and proposing measures that might avoid, minimize, or mitigate any adverse effects of the propose action on historic properties. To complete consultation under § 800.8(c), the NRC staff will forward the SEIS on the COL to you for your review and comment, and will address your comments in the final SEIS on the COL.

The proposed new reactors, Vogtle Units 3 and 4 would be located approximately 26 miles southeast of Augusta, Georgia, near the town of Waynesboro, Georgia. SNC currently operates two reactors, Vogtle Units 1 and 2 on the site and plans to construct the new units adjacent to the existing reactors. The application for the COL was accepted for docketing on May 30, 2008, and is available through the web-based version of the NRC Agency-wide Documents Access Management System (ADAMS), which can be found at http://www.nrc.gov/readingrm/adams.html. The Environmental Report for the application is listed under ADAMS accession number ML081050181. The Vogtle Plant COL application is also available on the Internet at http://www.nrc.gov/reactors/new-licensing/col/vogtle.html. A detailed review by NRC of the environmental impacts of constructing and operating proposed Units 3 and 4 is documented in NUREG-1872, "Final Environmental Impact Statement for an Early Site Permit (ESP) at the Vogtle Electric Generating Plant Site." (ESP FEIS Volumes 1 and 2 are available in ADAMS under ADAMS accession numbers ML082240145 and ML082240165, respectively.)

Please submit comments either by mail to Mrs. Mallecia Sutton, Environmental Project Manager, Division of Site and Environmental Reviews, Mail Stop T-7E18, U.S. Nuclear Regulatory Commission, Washington, D.C. 20555-0001 or via e-mail to Vogtle.COLAEIS@nrc.gov by January 8, 2010.

If you have any questions or require additional information, please contact Mrs. Sutton at (301) 415-0673 or via e-mail to Mallecia.Sutton@nrc.gov.

Sincerely,

/RA/

Gregory P. Hatchett, Branch Chief
Environmental Projects Branch 1
Division of Site and Environmental Reviews
Office of New Reactors

Docket Nos. 52-025 and 52-026

cc: See next page

Appendix F

December 10, 2009

Ms. Karen Kaniatobe
Director of the Cultural/Historical Preservation Department
Absentee-Shawnee Tribe of Oklahoma
2025 S. Gordon Cooper Drive
Shawnee, OK 74801

SUBJECT: U.S. NUCLEAR REGULATORY COMMISSION'S SUPPLEMENTAL ENVIRONMENTAL IMPACT STATEMENT FOR SOUTHERN NUCLEAR OPERATING COMPANY'S COMBINED LICENSE APPLICATION FOR THE PROPOSED CONSTRUCTION AND OPERATION OF UNITS 3 AND 4 AT THE VOGTLE ELECTRIC GENERATING PLANT IN WAYNESBORO, GEORGIA

Dear Ms Kaniatobe:

The U.S. Nuclear Regulatory Commission (NRC) staff is reviewing an application submitted by Southern Nuclear Operating Company, Inc. (Southern or SNC), on behalf of itself and several co-applicants, for a combined license (COL) for construction and operation of two new nuclear units at its Vogtle Electric Generating Plant (Vogtle or VEGP) site in Burke County, Georgia The purpose of this letter is to invite the Absentee-Shawnee Tribe of Oklahoma to consult with the NRC regarding the proposed action, pursuant to NRC regulations at 10 CFR Part 51 and Section 106 of the National Historic Preservation Act (NHPA) of 1966, as amended. The NRC plans to coordinate compliance with Section 106 of the NHPA using the National Environmental Policy Act (NEPA) process identified in 36 CFR 800.8(c) in lieu of the procedures set forth in §§ 800.3 through 800.6. Additionally, the NRC staff will rely on procedures in its regulations at 10 CFR 51.92 for supplementing a final environmental impact statement (FEIS).

Section 51.26(d) of the NRC regulations describes the processes for determining the need to supplement a FEIS, issue a notice of intent, and determine whether or not a formal scoping process will be conducted. As required by 10 CFR 51.92, the NRC staff will prepare a supplement to the FEIS for the early site permit (ESP) issued on August 26, 2009, to SNC and the same co-applicants. In this case, the NRC staff has determined that it will not conduct a formal scoping process for the development of the supplemental environmental impact statement (SEIS).

An ESP is a Commission approval of a site suitable for construction and operation of one or more new nuclear units. Under 10 CFR 51.50(c), a COL applicant referencing an ESP need not submit information or analyses regarding environmental issues that were resolved in the ESP EIS, except to the extent the COL applicant has identified any new and potentially significant information.

In accordance with the provisions in 36 CFR § 800.8, the NRC wishes to ensure that Indian tribes that might have an interest in any potential historic properties in the area of potential effect (APE) are afforded the opportunity to identify their concerns, provide advice on the identification and evaluation of historic properties, including those of traditional religious and cultural importance and, if necessary, participate in the resolution of any adverse effects to such properties.

Appendix F

Thus, to support the NRC staff's review of any new and potentially significant circumstances or information relevant to the environmental concerns related to the proposed action or its impacts, we request you submit written comments, if any, your tribe may have to offer on the environmental review of the COL by January 15, 2010.

The NRC intends to propose measures in the SEIS for the COL analyses for potential impacts to historical and cultural resources including developing alternatives and proposing measures that might avoid, minimize, or mitigate any adverse effects of the propose action on historic properties. To complete consultation under § 800.8(c), the NRC staff will forward the SEIS on the COL to you for your review and comment, and will address your comments in the final SEIS on the COL.

The proposed new reactors, Vogtle Units 3 and 4 would be located approximately 26 miles southeast of Augusta, Georgia, near the town of Waynesboro, Georgia. SNC currently operates two reactors, Vogtle Units 1 and 2 on the site and plans to construct the new units adjacent to the existing reactors. The application for the COL was accepted for docketing on May 30, 2008, and is available through the web-based version of the NRC Agency-wide Documents Access Management System (ADAMS), which can be found at http://www.nrc.gov/readingrm/adams.html. The Environmental Report for the application is listed under ADAMS accession number ML081050181. The Vogtle Plant COL application is also available on the Internet at http://www.nrc.gov/reactors/new-licensing/col/vogtle.html. A detailed review by NRC of the environmental impacts of constructing and operating proposed Units 3 and 4 is documented in NUREG-1872, "Final Environmental Impact Statement for an Early Site Permit (ESP) at the Vogtle Electric Generating Plant Site." (ESP FEIS Volumes 1 and 2 are available in ADAMS under ADAMS accession numbers ML082240145 and ML082240165, respectively.)

Please submit comments either by mail to Mrs. Mallecia Sutton, Environmental Project Manager, Division of Site and Environmental Reviews, Mail Stop T-7E18, U.S. Nuclear Regulatory Commission, Washington, D.C. 20555-0001 or via e-mail to Vogtle.COLAEIS@nrc.gov by January 15, 2010.

If you have any questions or require additional information, please contact Mrs. Sutton at (301) 415-0673 or via e-mail to Mallecia.Sutton@nrc.gov.

Sincerely,

/RA/

Gregory P. Hatchett, Branch Chief
Environmental Projects Branch 1
Division of Site and Environmental Reviews
Office of New Reactors

Docket Nos. 52-025 and 52-026

cc: See next page

Appendix F

December 10, 2009

Ms. Debbie Thomas
Tribal Historic Preservation Officer
NAGPRA Coordinator
Alabama-Coushatta Tribe of Texas
571 State Park Road 56
Livingston, TX 77351

SUBJECT: U.S. NUCLEAR REGULATORY COMMISSION'S SUPPLEMENTAL ENVIRONMENTAL IMPACT STATEMENT FOR SOUTHERN NUCLEAR OPERATING COMPANY'S COMBINED LICENSE APPLICATION FOR THE PROPOSED CONSTRUCTION AND OPERATION OF UNITS 3 AND 4 AT THE VOGTLE ELECTRIC GENERATING PLANT IN WAYNESBORO, GEORGIA

Dear Ms. Thomas:

The U.S. Nuclear Regulatory Commission (NRC) staff is reviewing an application submitted by Southern Nuclear Operating Company, Inc. (Southern or SNC), on behalf of itself and several co-applicants, for a combined license (COL) for construction and operation of two new nuclear units at its Vogtle Electric Generating Plant (Vogtle or VEGP) site in Burke County, Georgia. The purpose of this letter is to invite the Alabama-Coushatta Tribe of Texas to consult with the NRC regarding the proposed action, pursuant to NRC regulations at 10 CFR Part 51 and Section 106 of the National Historic Preservation Act (NHPA) of 1966, as amended. The NRC plans to coordinate compliance with Section 106 of the NHPA using the National Environmental Policy Act (NEPA) process identified in 36 CFR 800.8(c) in lieu of the procedures set forth in §§ 800.3 through 800.6. Additionally, the NRC staff will rely on procedures in its regulations at 10 CFR 51.92 for supplementing a final environmental impact statement (FEIS).

Section 51.26(d) of the NRC regulations describes the processes for determining the need to supplement a FEIS, issue a notice of intent, and determine whether or not a formal scoping process will be conducted. As required by 10 CFR 51.92, the NRC staff will prepare a supplement to the FEIS for the early site permit (ESP) issued on August 26, 2009, to SNC and the same co-applicants. In this case, the NRC staff has determined that it will not conduct a formal scoping process for the development of the supplemental environmental impact statement (SEIS).

An ESP is a Commission approval of a site suitable for construction and operation of one or more new nuclear units. Under 10 CFR 51.50(c), a COL applicant referencing an ESP need not submit information or analyses regarding environmental issues that were resolved in the ESP EIS, except to the extent the COL applicant has identified any new and potentially significant information.

In accordance with the provisions in 36 CFR § 800.8, the NRC wishes to ensure that Indian tribes that might have an interest in any potential historic properties in the area of potential effect (APE) are afforded the opportunity to identify their concerns, provide advice on the identification and evaluation of historic properties, including those of traditional religious and cultural importance and, if necessary, participate in the resolution of any adverse effects to such properties.

Appendix F

D. Thomas   -2-

Thus, to support the NRC staff's review of any new and potentially significant circumstances or information relevant to the environmental concerns related to the proposed action or its impacts, we request you submit written comments, if any, your tribe may have to offer on the environmental review of the COL by January 15, 2010.

The NRC intends to propose measures in the SEIS for the COL analyses for potential impacts to historical and cultural resources including developing alternatives and proposing measures that might avoid, minimize, or mitigate any adverse effects of the propose action on historic properties. To complete consultation under § 800.8(c), the NRC staff will forward the SEIS on the COL to you for your review and comment, and will address your comments in the final SEIS on the COL.

The proposed new reactors, Vogtle Units 3 and 4 would be located approximately 26 miles southeast of Augusta, Georgia, near the town of Waynesboro, Georgia. SNC currently operates two reactors, Vogtle Units 1 and 2 on the site and plans to construct the new units adjacent to the existing reactors. The application for the COL was accepted for docketing on May 30, 2008, and is available through the web-based version of the NRC Agency-wide Documents Access Management System (ADAMS), which can be found at http://www.nrc.gov/readingrm/adams.html. The Environmental Report for the application is listed under ADAMS accession number ML081050181. The Vogtle Plant COL application is also available on the Internet at http://www.nrc.gov/reactors/new-licensing/col/vogtle.html. A detailed review by NRC of the environmental impacts of constructing and operating proposed Units 3 and 4 is documented in NUREG-1872, "Final Environmental Impact Statement for an Early Site Permit (ESP) at the Vogtle Electric Generating Plant Site." (ESP FEIS Volumes 1 and 2 are available in ADAMS under ADAMS accession numbers ML082240145 and ML082240165, respectively.)

Please submit comments either by mail to Mrs. Mallecia Sutton, Environmental Project Manager, Division of Site and Environmental Reviews, Mail Stop T-7E18, U.S. Nuclear Regulatory Commission, Washington, D.C. 20555-0001 or via e-mail to Vogtle.COLAEIS@nrc.gov by January 15, 2010.

If you have any questions or require additional information, please contact Mrs. Sutton at (301) 415-0673 or via e-mail to Mallecia.Sutton@nrc.gov.

Sincerely,

/RA/

Gregory P. Hatchett, Branch Chief
Environmental Projects Branch 1
Division of Site and Environmental Reviews
Office of New Reactors

Docket Nos. 52-025 and 52-026

cc: See next page

Appendix F

December 10, 2009

Mrs. Joyce A. Bear, NAGPRA Contact
Muscogee (Creek) Nation of Oklahoma
P.O. Box 580
Okmulgee, OK 74447

SUBJECT: U.S. NUCLEAR REGULATORY COMMISSION'S SUPPLEMENTAL ENVIRONMENTAL IMPACT STATEMENT FOR SOUTHERN NUCLEAR OPERATING COMPANY'S COMBINED LICENSE APPLICATION FOR THE PROPOSED CONSTRUCTION AND OPERATION OF UNITS 3 AND 4 AT THE VOGTLE ELECTRIC GENERATING PLANT IN WAYNESBORO, GEORGIA

Dear Mrs. Bear:

The U.S. Nuclear Regulatory Commission (NRC) staff is reviewing an application submitted by Southern Nuclear Operating Company, Inc. (Southern or SNC), on behalf of itself and several co-applicants, for a combined license (COL) for construction and operation of two new nuclear units at its Vogtle Electric Generating Plant (Vogtle or VEGP) site in Burke County, Georgia. The purpose of this letter is to invite the Muscogee (Creek) Nation of Oklahoma to consult with the NRC regarding the proposed action, pursuant to NRC regulations at 10 CFR Part 51 and Section 106 of the National Historic Preservation Act (NHPA) of 1966, as amended. The NRC plans to coordinate compliance with Section 106 of the NHPA using the National Environmental Policy Act (NEPA) process identified in 36 CFR 800.8(c) in lieu of the procedures set forth in §§ 800.3 through 800.6. Additionally, the NRC staff will rely on procedures in its regulations at 10 CFR 51.92 for supplementing a final environmental impact statement (FEIS).

Section 51.26(d) of the NRC regulations describes the processes for determining the need to supplement a FEIS, issue a notice of intent, and determine whether or not a formal scoping process will be conducted. As required by 10 CFR 51.92, the NRC staff will prepare a supplement to the FEIS for the early site permit (ESP) issued on August 26, 2009, to SNC and the same co-applicants. In this case, the NRC staff has determined that it will not conduct a formal scoping process for the development of the supplemental environmental impact statement (SEIS).

An ESP is a Commission approval of a site suitable for construction and operation of one or more new nuclear units. Under 10 CFR 51.50(c), a COL applicant referencing an ESP need not submit information or analyses regarding environmental issues that were resolved in the ESP EIS, except to the extent the COL applicant has identified any new and potentially significant information.

In accordance with the provisions in 36 CFR § 800.8, the NRC wishes to ensure that Indian tribes that might have an interest in any potential historic properties in the area of potential effect (APE) are afforded the opportunity to identify their concerns, provide advice on the identification and evaluation of historic properties, including those of traditional religious and cultural importance and, if necessary, participate in the resolution of any adverse effects to such properties.

Appendix F

J. A. Bear   -2-

Thus, to support the NRC staff's review of any new and potentially significant circumstances or information relevant to the environmental concerns related to the proposed action or its impacts, we request you submit written comments, if any, your tribe may have to offer on the environmental review of the COL by January 15, 2010.

The NRC intends to propose measures in the SEIS for the COL analyses for potential impacts to historical and cultural resources including developing alternatives and proposing measures that might avoid, minimize, or mitigate any adverse effects of the propose action on historic properties. To complete consultation under § 800.8(c), the NRC staff will forward the SEIS on the COL to you for your review and comment, and will address your comments in the final SEIS on the COL.

The proposed new reactors, Vogtle Units 3 and 4 would be located approximately 26 miles southeast of Augusta, Georgia, near the town of Waynesboro, Georgia. SNC currently operates two reactors, Vogtle Units 1 and 2 on the site and plans to construct the new units adjacent to the existing reactors. The application for the COL was accepted for docketing on May 30, 2008, and is available through the web-based version of the NRC Agency-wide Documents Access Management System (ADAMS), which can be found at http://www.nrc.gov/readingrm/adams.html. The Environmental Report for the application is listed under ADAMS accession number ML081050181. The Vogtle Plant COL application is also available on the Internet at http://www.nrc.gov/reactors/new-licensing/col/vogtle.html. A detailed review by NRC of the environmental impacts of constructing and operating proposed Units 3 and 4 is documented in NUREG-1872, "Final Environmental Impact Statement for an Early Site Permit (ESP) at the Vogtle Electric Generating Plant Site." (ESP FEIS Volumes 1 and 2 are available in ADAMS under ADAMS accession numbers ML082240145 and ML082240165, respectively.)

Please submit comments either by mail to Mrs. Mallecia Sutton, Environmental Project Manager, Division of Site and Environmental Reviews, Mail Stop T-7E18, U.S. Nuclear Regulatory Commission, Washington, D.C. 20555-0001 or via e-mail to Vogtle.COLAEIS@nrc.gov by January 15, 2010.

If you have any questions or require additional information, please contact Mrs. Sutton at (301) 415-0673 or via e-mail to Mallecia.Sutton@nrc.gov.

Sincerely,

/RA/

Gregory P. Hatchett, Branch Chief
Environmental Projects Branch 1
Division of Site and Environmental Reviews
Office of New Reactors

Docket Nos. 52-025 and 52-026

cc: See next page

Appendix F

December 10, 2009

Mr. Chadwick Smith, Principal Chief
Cherokee Nation of Oklahoma
P.O. Box 948
Tahlequa, OK 74465

SUBJECT: U.S. NUCLEAR REGULATORY COMMISSION'S SUPPLEMENTAL ENVIRONMENTAL IMPACT STATEMENT FOR SOUTHERN NUCLEAR OPERATING COMPANY'S COMBINED LICENSE APPLICATION FOR THE PROPOSED CONSTRUCTION AND OPERATION OF UNITS 3 AND 4 AT THE VOGTLE ELECTRIC GENERATING PLANT IN WAYNESBORO, GEORGIA

Dear Mr. Smith:

The U.S. Nuclear Regulatory Commission (NRC) staff is reviewing an application submitted by Southern Nuclear Operating Company, Inc. (Southern or SNC), on behalf of itself and several co-applicants, for a combined license (COL) for construction and operation of two new nuclear units at its Vogtle Electric Generating Plant (Vogtle or VEGP) site in Burke County, Georgia. The purpose of this letter is to invite the Cherokee Nation of Oklahoma to consult with the NRC regarding the proposed action, pursuant to NRC regulations at 10 CFR Part 51 and Section 106 of the National Historic Preservation Act (NHPA) of 1966, as amended. The NRC plans to coordinate compliance with Section 106 of the NHPA using the National Environmental Policy Act (NEPA) process identified in 36 CFR 800.8(c) in lieu of the procedures set forth in §§ 800.3 through 800.6. Additionally, the NRC staff will rely on procedures in its regulations at 10 CFR 51.92 for supplementing a final environmental impact statement (FEIS).

Section 51.26(d) of the NRC regulations describes the processes for determining the need to supplement a FEIS, issue a notice of intent, and determine whether or not a formal scoping process will be conducted. As required by 10 CFR 51.92, the NRC staff will prepare a supplement to the FEIS for the early site permit (ESP) issued on August 26, 2009, to SNC and the same co-applicants. In this case, the NRC staff has determined that it will not conduct a formal scoping process for the development of the supplemental environmental impact statement (SEIS).

An ESP is a Commission approval of a site suitable for construction and operation of one or more new nuclear units. Under 10 CFR 51.50(c), a COL applicant referencing an ESP need not submit information or analyses regarding environmental issues that were resolved in the ESP EIS, except to the extent the COL applicant has identified any new and potentially significant information.

In accordance with the provisions in 36 CFR § 800.8, the NRC wishes to ensure that Indian tribes that might have an interest in any potential historic properties in the area of potential effect (APE) are afforded the opportunity to identify their concerns, provide advice on the identification and evaluation of historic properties, including those of traditional religious and cultural importance and, if necessary, participate in the resolution of any adverse effects to such properties.

Appendix F

Thus, to support the NRC staff's review of any new and potentially significant circumstances or information relevant to the environmental concerns related to the proposed action or its impacts, we request you submit written comments, if any, your tribe may have to offer on the environmental review of the COL by January 15, 2010.

The NRC intends to propose measures in the SEIS for the COL analyses for potential impacts to historical and cultural resources including developing alternatives and proposing measures that might avoid, minimize, or mitigate any adverse effects of the propose action on historic properties. To complete consultation under § 800.8(c), the NRC staff will forward the SEIS on the COL to you for your review and comment, and will address your comments in the final SEIS on the COL.

The proposed new reactors, Vogtle Units 3 and 4 would be located approximately 26 miles southeast of Augusta, Georgia, near the town of Waynesboro, Georgia. SNC currently operates two reactors, Vogtle Units 1 and 2 on the site and plans to construct the new units adjacent to the existing reactors. The application for the COL was accepted for docketing on May 30, 2008, and is available through the web-based version of the NRC Agency-wide Documents Access Management System (ADAMS), which can be found at http://www.nrc.gov/readingrm/adams.html. The Environmental Report for the application is listed under ADAMS accession number ML081050181. The Vogtle Plant COL application is also available on the Internet at http://www.nrc.gov/reactors/new-licensing/col/vogtle.html. A detailed review by NRC of the environmental impacts of constructing and operating proposed Units 3 and 4 is documented in NUREG-1872, "Final Environmental Impact Statement for an Early Site Permit (ESP) at the Vogtle Electric Generating Plant Site." (ESP FEIS Volumes 1 and 2 are available in ADAMS under ADAMS accession numbers ML082240145 and ML082240165, respectively.)

Please submit comments either by mail to Mrs. Mallecia Sutton, Environmental Project Manager, Division of Site and Environmental Reviews, Mail Stop T-7E18, U.S. Nuclear Regulatory Commission, Washington, D.C. 20555-0001 or via e-mail to Vogtle.COLAEIS@nrc.gov by January 15, 2010.

If you have any questions or require additional information, please contact Mrs. Sutton at (301) 415-0673 or via e-mail to Mallecia.Sutton@nrc.gov.

Sincerely,

/RA/

Gregory P. Hatchett, Branch Chief
Environmental Projects Branch 1
Division of Site and Environmental Reviews
Office of New Reactors

Docket Nos. 52-025 and 52-026

cc: See next page

Appendix F

December 10, 2009

Mr. Gilbert Blue, Chairperson
Catawba Indian Tribe
P.O. box 188
Catawaba, SC 29704

SUBJECT: U.S. NUCLEAR REGULATORY COMMISSION'S SUPPLEMENTAL ENVIRONMENTAL IMPACT STATEMENT FOR SOUTHERN NUCLEAR OPERATING COMPANY'S COMBINED LICENSE APPLICATION FOR THE PROPOSED CONSTRUCTION AND OPERATION OF UNITS 3 AND 4 AT THE VOGTLE ELECTRIC GENERATING PLANT IN WAYNESBORO, GEORGIA

Dear Mr. Blue:

The U.S. Nuclear Regulatory Commission (NRC) staff is reviewing an application submitted by Southern Nuclear Operating Company, Inc. (Southern or SNC), on behalf of itself and several co-applicants, for a combined license (COL) for construction and operation of two new nuclear units at its Vogtle Electric Generating Plant (Vogtle or VEGP) site in Burke County, Georgia. The purpose of this letter is to invite the Catawba Indian Tribe to consult with the NRC regarding the proposed action, pursuant to NRC regulations at 10 CFR Part 51 and Section 106 of the National Historic Preservation Act (NHPA) of 1966, as amended. The NRC plans to coordinate compliance with Section 106 of the NHPA using the National Environmental Policy Act (NEPA) process identified in 36 CFR 800.8(c) in lieu of the procedures set forth in §§ 800.3 through 800.6. Additionally, the NRC staff will rely on procedures in its regulations at 10 CFR 51.92 for supplementing a final environmental impact statement (FEIS).

Section 51.26(d) of the NRC regulations describes the processes for determining the need to supplement a FEIS, issue a notice of intent, and determine whether or not a formal scoping process will be conducted. As required by 10 CFR 51.92, the NRC staff will prepare a supplement to the FEIS for the early site permit (ESP) issued on August 26, 2009, to SNC and the same co-applicants. In this case, the NRC staff has determined that it will not conduct a formal scoping process for the development of the supplemental environmental impact statement (SEIS).

An ESP is a Commission approval of a site suitable for construction and operation of one or more new nuclear units. Under 10 CFR 51.50(c), a COL applicant referencing an ESP need not submit information or analyses regarding environmental issues that were resolved in the ESP EIS, except to the extent the COL applicant has identified any new and potentially significant information.

In accordance with the provisions in 36 CFR § 800.8, the NRC wishes to ensure that Indian tribes that might have an interest in any potential historic properties in the area of potential effect (APE) are afforded the opportunity to identify their concerns, provide advice on the identification and evaluation of historic properties, including those of traditional religious and cultural importance and, if necessary, participate in the resolution of any adverse effects to such properties.

Appendix F

G. Blue   -2-

Thus, to support the NRC staff's review of any new and potentially significant circumstances or information relevant to the environmental concerns related to the proposed action or its impacts, we request you submit written comments, if any; your tribe may have to offer on the environmental review of the COL by January 15, 2010.

The NRC intends to propose measures in the SEIS for the COL analyses for potential impacts to historical and cultural resources including developing alternatives and proposing measures that might avoid, minimize, or mitigate any adverse effects of the propose action on historic properties. To complete consultation under § 800.8(c), the NRC staff will forward the SEIS on the COL to you for your review and comment, and will address your comments in the final SEIS on the COL.

The proposed new reactors, Vogtle Units 3 and 4 would be located approximately 26 miles southeast of Augusta, Georgia, near the town of Waynesboro, Georgia. SNC currently operates two reactors, Vogtle Units 1 and 2 on the site and plans to construct the new units adjacent to the existing reactors. The application for the COL was accepted for docketing on May 30, 2008, and is available through the web-based version of the NRC Agency-wide Documents Access Management System (ADAMS), which can be found at http://www.nrc.gov/readingrm/adams.html. The Environmental Report for the application is listed under ADAMS accession number ML081050181. The Vogtle Plant COL application is also available on the Internet at http://www.nrc.gov/reactors/new-licensing/col/vogtle.html. A detailed review by NRC of the environmental impacts of constructing and operating proposed Units 3 and 4 is documented in NUREG-1872, "Final Environmental Impact Statement for an Early Site Permit (ESP) at the Vogtle Electric Generating Plant Site." (ESP FEIS Volumes 1 and 2 are available in ADAMS under ADAMS accession numbers ML082240145 and ML082240165, respectively.)

Please submit comments either by mail to Mrs. Mallecia Sutton, Environmental Project Manager, Division of Site and Environmental Reviews, Mail Stop T-7E18, U.S. Nuclear Regulatory Commission, Washington, D.C. 20555-0001 or via e-mail to Vogtle.COLAEIS@nrc.gov by January 15, 2010.

If you have any questions or require additional information, please contact Mrs. Sutton at (301) 415-0673 or via e-mail to Mallecia.Sutton@nrc.gov.

Sincerely,

/RA/

Gregory P. Hatchett, Branch Chief
Environmental Projects Branch 1
Division of Site and Environmental Reviews
Office of New Reactors

Docket Nos. 52-025 and 52-026

cc: See next page

Appendix F

December 10, 2009

Mr. Willard Steele, Deputy THPO
Seminole Tribe of Florida
Ah-Tah-Thi-Ki Museum
HC 61, Box 21A
Clewiston, FL 33440

SUBJECT: U.S. NUCLEAR REGULATORY COMMISSION'S SUPPLEMENTAL ENVIRONMENTAL IMPACT STATEMENT FOR SOUTHERN NUCLEAR OPERATING COMPANY'S COMBINED LICENSE APPLICATION FOR THE PROPOSED CONSTRUCTION AND OPERATION OF UNITS 3 AND 4 AT THE VOGTLE ELECTRIC GENERATING PLANT IN WAYNESBORO, GEORGIA

Dear Mr. Steele:

The U.S. Nuclear Regulatory Commission (NRC) staff is reviewing an application submitted by Southern Nuclear Operating Company, Inc. (Southern or SNC), on behalf of itself and several co-applicants, for a combined license (COL) for construction and operation of two new nuclear units at its Vogtle Electric Generating Plant (Vogtle or VEGP) site in Burke County, Georgia. The purpose of this letter is to invite the Seminole Tribe of Florida to consult with the NRC regarding the proposed action, pursuant to NRC regulations at 10 CFR Part 51 and Section 106 of the National Historic Preservation Act (NHPA) of 1966, as amended. The NRC plans to coordinate compliance with Section 106 of the NHPA using the National Environmental Policy Act (NEPA) process identified in 36 CFR 800.8(c) in lieu of the procedures set forth in §§ 800.3 through 800.6. Additionally, the NRC staff will rely on procedures in its regulations at 10 CFR 51.92 for supplementing a final environmental impact statement (FEIS).

Section 51.26(d) of the NRC regulations describes the processes for determining the need to supplement a FEIS, issue a notice of intent, and determine whether or not a formal scoping process will be conducted. As required by 10 CFR 51.92, the NRC staff will prepare a supplement to the FEIS for the early site permit (ESP) issued on August 26, 2009, to SNC and the same co-applicants. In this case, the NRC staff has determined that it will not conduct a formal scoping process for the development of the supplemental environmental impact statement (SEIS).

An ESP is a Commission approval of a site suitable for construction and operation of one or more new nuclear units. Under 10 CFR 51.50(c), a COL applicant referencing an ESP need not submit information or analyses regarding environmental issues that were resolved in the ESP EIS, except to the extent the COL applicant has identified any new and potentially significant information.

In accordance with the provisions in 36 CFR § 800.8, the NRC wishes to ensure that Indian tribes that might have an interest in any potential historic properties in the area of potential effect (APE) are afforded the opportunity to identify their concerns, provide advice on the identification and evaluation of historic properties, including those of traditional religious and

Appendix F

W. Steele   -2-

cultural importance and, if necessary, participate in the resolution of any adverse effects to such properties. Thus, to support the NRC staff's review of any new and potentially significant circumstances or information relevant to the environmental concerns related to the proposed action or its impacts, we request you submit written comments, if any; your tribe may have to offer on the environmental review of the COL by January 15, 2010.

The NRC intends to propose measures in the SEIS for the COL analyses for potential impacts to historical and cultural resources including developing alternatives and proposing measures that might avoid, minimize, or mitigate any adverse effects of the propose action on historic properties. To complete consultation under § 800.8(c), the NRC staff will forward the SEIS on the COL to you for your review and comment, and will address your comments in the final SEIS on the COL.

The proposed new reactors, Vogtle Units 3 and 4 would be located approximately 26 miles southeast of Augusta, Georgia, near the town of Waynesboro, Georgia. SNC currently operates two reactors, Vogtle Units 1 and 2 on the site and plans to construct the new units adjacent to the existing reactors. The application for the COL was accepted for docketing on May 30, 2008, and is available through the web-based version of the NRC Agency-wide Documents Access Management System (ADAMS), which can be found at http://www.nrc.gov/readingrm/adams.html. The Environmental Report for the application is listed under ADAMS accession number ML081050181. The Vogtle Plant COL application is also available on the Internet at http://www.nrc.gov/reactors/new-licensing/col/vogtle.html. A detailed review by NRC of the environmental impacts of constructing and operating proposed Units 3 and 4 is documented in NUREG-1872, "Final Environmental Impact Statement for an Early Site Permit (ESP) at the Vogtle Electric Generating Plant Site." (ESP FEIS Volumes 1 and 2 are available in ADAMS under ADAMS accession numbers ML082240145 and ML082240165, respectively.)

Please submit comments either by mail to Mrs. Mallecia Sutton, Environmental Project Manager, Division of Site and Environmental Reviews, Mail Stop T-7E18, U.S. Nuclear Regulatory Commission, Washington, D.C. 20555-0001 or via e-mail to Vogtle.COLAEIS@nrc.gov by January 15, 2010.

If you have any questions or require additional information, please contact Mrs. Sutton at (301) 415-0673 or via e-mail to Mallecia.Sutton@nrc.gov.

Sincerely,

/RA/

Gregory P. Hatchett, Branch Chief
Environmental Projects Branch 1
Division of Site and Environmental Reviews
Office of New Reactors

Docket Nos. 52-025 and 52-026

cc: See next page

Appendix F

December 10, 2009

Mr. Kenneth H. Carleton
THPO/Tribal Archaeologist
Mississippi Band of Choctaw Indians
P.O. Box 6257/ 101 Industrial Road
Choctaw, MS 39350

SUBJECT: U.S. NUCLEAR REGULATORY COMMISSION'S SUPPLEMENTAL ENVIRONMENTAL IMPACT STATEMENT FOR SOUTHERN NUCLEAR OPERATING COMPANY'S COMBINED LICENSE APPLICATION FOR THE PROPOSED CONSTRUCTION AND OPERATION OF UNITS 3 AND 4 AT THE VOGTLE ELECTRIC GENERATING PLANT IN WAYNESBORO, GEORGIA

Dear Mr. Carleton:

The U.S. Nuclear Regulatory Commission (NRC) staff is reviewing an application submitted by Southern Nuclear Operating Company, Inc. (Southern or SNC), on behalf of itself and several co-applicants, for a combined license (COL) for construction and operation of two new nuclear units at its Vogtle Electric Generating Plant (Vogtle or VEGP) site in Burke County, Georgia. The purpose of this letter is to invite the Mississippi Band of Choctaw Indians to consult with the NRC regarding the proposed action, pursuant to NRC regulations at 10 CFR Part 51 and Section 106 of the National Historic Preservation Act (NHPA) of 1966, as amended. The NRC plans to coordinate compliance with Section 106 of the NHPA using the National Environmental Policy Act (NEPA) process identified in 36 CFR 800.8(c) in lieu of the procedures set forth in §§ 800.3 through 800.6. Additionally, the NRC staff will rely on procedures in its regulations at 10 CFR 51.92 for supplementing a final environmental impact statement (FEIS).

Section 51.26(d) of the NRC regulations describes the processes for determining the need to supplement a FEIS, issue a notice of intent, and determine whether or not a formal scoping process will be conducted. As required by 10 CFR 51.92, the NRC staff will prepare a supplement to the FEIS for the early site permit (ESP) issued on August 26, 2009, to SNC and the same co-applicants. In this case, the NRC staff has determined that it will not conduct a formal scoping process for the development of the supplemental environmental impact statement (SEIS).

An ESP is a Commission approval of a site suitable for construction and operation of one or more new nuclear units. Under 10 CFR 51.50(c), a COL applicant referencing an ESP need not submit information or analyses regarding environmental issues that were resolved in the ESP EIS, except to the extent the COL applicant has identified any new and potentially significant information.

In accordance with the provisions in 36 CFR § 800.8, the NRC wishes to ensure that Indian tribes that might have an interest in any potential historic properties in the area of potential effect (APE) are afforded the opportunity to identify their concerns, provide advice on the identification and evaluation of historic properties, including those of traditional religious and cultural importance and, if necessary, participate in the resolution of any adverse effects to such properties.

Appendix F

Thus, to support the NRC staff's review of any new and potentially significant circumstances or information relevant to the environmental concerns related to the proposed action or its impacts, we request you submit written comments, if any, your tribe may have to offer on the environmental review of the COL by January 15, 2010.

The NRC intends to propose measures in the SEIS for the COL analyses for potential impacts to historical and cultural resources including developing alternatives and proposing measures that might avoid, minimize, or mitigate any adverse effects of the propose action on historic properties. To complete consultation under § 800.8(c), the NRC staff will forward the SEIS on the COL to you for your review and comment, and will address your comments in the final SEIS on the COL.

The proposed new reactors, Vogtle Units 3 and 4 would be located approximately 26 miles southeast of Augusta, Georgia, near the town of Waynesboro, Georgia. SNC currently operates two reactors, Vogtle Units 1 and 2 on the site and plans to construct the new units adjacent to the existing reactors. The application for the COL was accepted for docketing on May 30, 2008, and is available through the web-based version of the NRC Agency-wide Documents Access Management System (ADAMS), which can be found at http://www.nrc.gov/readingrm/adams.html. The Environmental Report for the application is listed under ADAMS accession number ML081050181. The Vogtle Plant COL application is also available on the Internet at http://www.nrc.gov/reactors/new-licensing/col/vogtle.html. A detailed review by NRC of the environmental impacts of constructing and operating proposed Units 3 and 4 is documented in NUREG-1872, "Final Environmental Impact Statement for an Early Site Permit (ESP) at the Vogtle Electric Generating Plant Site." (ESP FEIS Volumes 1 and 2 are available in ADAMS under ADAMS accession numbers ML082240145 and ML082240165, respectively.)

Please submit comments either by mail to Mrs. Mallecia Sutton, Environmental Project Manager, Division of Site and Environmental Reviews, Mail Stop T-7E18, U.S. Nuclear Regulatory Commission, Washington, D.C. 20555-0001 or via e-mail to Vogtle.COLAEIS@nrc.gov by January 15, 2010.

If you have any questions or require additional information, please contact Mrs. Sutton at (301) 415-0673 or via e-mail to Mallecia.Sutton@nrc.gov.

Sincerely,

/RA/

Gregory P. Hatchett, Branch Chief
Environmental Projects Branch 1
Division of Site and Environmental Reviews
Office of New Reactors

Docket Nos. 52-025 and 52-026

cc: See next page

Appendix F

PMVogtleCOLNPEm Resource

From:	Sutton, Mallecia
Sent:	Tuesday, December 15, 2009 4:24 PM
To:	Kuntzleman, Nancy
Cc:	PMVogtleCOLNPEm Resource
Subject:	FW: SC State Threatened and Endangered Species in the Vicinity of Vogtle Electric Generating Plant

FYI

From: Julie Holling [mailto:HollingJ@dnr.sc.gov]
Sent: Tuesday, December 15, 2009 2:49 PM
To: Sutton, Mallecia
Subject: RE: SC State Threatened and Endangered Species in the Vicinity of Vogtle Electric Generating Plant

Ms. Sutton,

The species listed in your attachment are still accurate for the 10-mile radius from VEGP. Please let me know if you need additional information.

Julie

Julie Holling - Data Manager
SC Dept. of Natural Resources
Heritage Trust Program
P. O. Box 167, Columbia, SC 29202
1000 Assembly St., Columbia, SC 29201
office: 803-734-3917   fax: 803-734-3931
HollingJ@dnr.sc.gov

DNR protects and manages South Carolina's natural resources by making wise and balanced decisions for the benefit of the state's natural resources and its people. Find out more about DNR at www.dnr.sc.gov.

 Please consider the environment before printing this e-mail.

From: Sutton, Mallecia [mailto:Mallecia.Sutton@nrc.gov]
Sent: Tuesday, December 15, 2009 1:13 PM
To: Julie Holling
Cc: PMVogtleCOLNPEm Resource
Subject: SC State Threatened and Endangered Species in the Vicinity of Vogtle Electric Generating Plant

Dear Ms. Holling:

The NRC staff is currently reviewing an application submitted by Southern Nuclear Operating Company, Inc., for a combined license (COL) for construction and operation of a two new nuclear power plants at the Vogtle Electric Generating Plant (VEGP) site in Burke County, Georgia. We originally consulted with you in 2007 when we were preparing the Environmental Impact Statement (EIS) for the Early Site Permit (ESP) for VEGP Units 3 and 4. The EIS for the COL will be a supplement to the ESP EIS. In 2007, we compiled a list of state threatened and endangered species in South Carolina within 10 miles of the VEGP site using the quads on the SCDNR website. I have attached the tables we included in the ESP EIS that contained this information. Could

Appendix F

you please let us know if these lists are still accurate or if there is updated or new species information? I have included the shapefile containing the centroid for the VEGP site. Coordsys is geographic: NAD27 degrees and is provided within the .prj file. Please let me know if you can provide information to verify or update this list with new information and if anything else is required to enable this data exchange. If I need to contact someone else for this information, please advise.

I can be reached by phone or by email.

Thanks for your assistance.

Mallecia Sutton
Environmental Project Manager
U.S. Nuclear Regulatory Commission
Two White Flint North
11545 Rockville Pike
Rockville, MD 20852-27388
Mailstop:T7E18
301-415-0673

Best regards,

Appendix F

PMVogtleCOLNPEm Resource

From:	Sutton, Mallecia
Sent:	Wednesday, December 16, 2009 10:03 AM
To:	Matt Elliott
Cc:	Brett Albanese; Katrina Morris; PMVogtleCOLNPEm Resource
Subject:	RE: FW: GDNR email Vogtle COL

Thanks

Mallecia Sutton
U.S.Nuclear Regulatory Commission
Two White Flint North
11545 Rockville Pike
Rockville, MD 20852-27388
Mailstop:T7E18
301-415-0673

-----Original Message-----
From: Matt Elliott [mailto:Matt.Elliott@dnr.state.ga.us]
Sent: Wednesday, December 16, 2009 9:49 AM
To: Sutton, Mallecia
Cc: Brett Albanese; Katrina Morris
Subject: Re: FW: GDNR email Vogtle COL

Mallecia
Attached is our current GA protected species list. It has not changed since 2006 (when the last changes took effect). The attachment shows the changes that took place in 2006.

Trina Morris and Brett Albanese will work on the rest of your request.
Thanks
Matt

Matt Elliott
Program Manager
Georgia Department of Natural Resources
Wildlife Resources Division
Nongame Conservation Section
2065 US Hwy 278, SE
Social Circle, GA 30025
(770)918-6411 or (706)557-3032 - office
(404)291-8156 - cell

>>> "Sutton, Mallecia" <Mallecia.Sutton@nrc.gov> 12/15/09 1:07 PM >>>

Dear Matt:

1

Appendix F

Thanks for taking time out of your busy schedule to participate on the phone call held Tuesday, December 8 with the environmental staff working on the Vogtle COL application. As mentioned on the phone, the NRC staff is currently reviewing an application submitted by Southern Nuclear Operating Company, Inc. (SNC) for a combined license (COL) for construction and operation of two new nuclear power plants at the Vogtle Electric Generating Plant (VEGP) site in Burke County, Georgia. NRC is preparing a supplement to their 2008 Final Environmental Impact Statement (FEIS) that was prepared to support the decision to grant an Early Site Permit (ESP) to Southern for the VEGP site. NRC is particularly interested in any new Georgia state-listed species information in the vicinity of the VEGP site and the associated proposed transmission line macrocorridor.

Attached are:

1. Shapefiles for the boundary of the VEGP site and the transmission line macrocorridor.

2. Tables 2-3 through 2-6 and Tables 2-9 and 2-10 from the ESP EIS. These tables provide lists of terrestrial and aquatic Federally and State-listed species within 10 miles of the VEGP site in Burke County and listed species in the counties crossed by the proposed transmission corridor (Burke, McDuffie, Jefferson, and Warren) as of 2007.

We would appreciate it if you would please provide an updated list of the Georgia state-listed species or verify that there have been no changes. In addition, we would appreciate any new information you have on the occurrences of federally-listed species in the vicinity of Vogtle and the proposed transmission corridor.

If you have any questions, I can be reached by email or phone.

Thanks for your assistance

Best Regards,

Mallecia Sutton
U.S. Nuclear Regulatory Commission
Two White Flint North
11545 Rockville Pike
Rockville, MD 20852-27388
Mailstop: T7E18
301-415-0673

Appendix F

PMVogtleCOLPEm Resource

From:	Katrina Morris [Katrina.Morris@dnr.state.ga.us]
Sent:	Monday, December 21, 2009 11:22 AM
To:	Sutton, Mallecia
Cc:	Matt Elliott
Subject:	Re: FW: GDNR email Vogtle COL
Attachments:	ir_12784.pdf

Hi Mallecia,
Please see attached letter regarding the Vogtle COL. Let me know if you have any questions.
Thanks,
Trina

Trina Morris, Wildlife Biologist
Environmental Review Coordinator
Georgia Dept. of Natural Resources
Nongame Conservation Section
2065 U.S. Hwy. 278 S.E.
Social Circle, GA 30025-4743
Ph: 770-918-6411 or 706-557-3032
Fax: 706-557-3033
katrina.morris@dnr.state.ga.us
http://georgiawildlife.dnr.state.ga.us/

Wild about wildlife? Sign up for Georgia Wild, DNR's free e-newsletter about all things nongame, from animals to habitats. Click here to subscribe (or paste this link into your browser):
http://www.georgiawildlife.com/enewsletters.aspx

>>> "Sutton, Mallecia" <Mallecia.Sutton@nrc.gov> 12/15/09 1:07 PM >>>

Dear Matt:

Thanks for taking time out of your busy schedule to participate on the phone call held Tuesday, December 8 with the environmental staff working on the Vogtle COL application. As mentioned on the phone, the NRC staff is currently reviewing an application submitted by Southern Nuclear Operating Company, Inc. (SNC) for a combined license (COL) for construction and operation of two new nuclear power plants at the Vogtle Electric Generating Plant (VEGP) site in Burke County, Georgia. NRC is preparing a supplement to their 2008 Final Environmental Impact Statement (FEIS) that was prepared to support the decision to grant an Early Site Permit (ESP) to Southern for the VEGP site. NRC is particularly interested in any new Georgia state-listed species information in the vicinity of the VEGP site and the associated proposed transmission line macrocorridor.

Attached are:

1. Shapefiles for the boundary of the VEGP site and the transmission line macrocorridor.

2. Tables 2-3 through 2-6 and Tables 2-9 and 2-10 from the ESP EIS. These tables provide lists of terrestrial and aquatic Federally and State-listed species within 10 miles of the VEGP site in Burke County and listed species in the counties crossed by the proposed transmission corridor (Burke, McDuffie, Jefferson, and Warren) as of 2007.

We would appreciate it if you would please provide an updated list of the Georgia state-listed species or verify that there have been no changes. In addition, we would appreciate any new information you have on the occurrences of federally-listed species in the vicinity of Vogtle and the proposed transmission corridor.

Appendix F

If you have any questions, I can be reached by email or phone.

Thanks for your assistance

Best Regards,

Mallecia Sutton
U.S. Nuclear Regulatory Commission
Two White Flint North
11545 Rockville Pike
Rockville, MD 20852-27388
Mailstop:T7E18
301-415-0673

Appendix F

CHRIS CLARK
COMMISSIONER

WILDLIFE RESOURCES DIVISION

DAN FORSTER
DIRECTOR

December 17, 2009

Mallecia Sutton
US Nuclear Regulatory Commission
Two White Flint North
111545 Rockville Pike
Rockville, MD 20852-27388
Mailstop: T7E18

Subject: Known occurrences of natural communities, plants and animals of highest priority conservation status on or near Vogtle COL, Burke County, Georgia

Dear Ms. Sutton:

This is in response to your request of December 15, 2009. According to our records, **within the VEGP Boundary and the Transmission Line Macrocorridor** there are the following Natural Heritage Database occurrences:

	Scientific Name	Common Name	Status	Counties
	Ambystoma tigrinum tigrinum	Eastern Tiger Salamander		Jefferson, Mcduffie
GA	Ceratiola ericoides	Sandhill Rosemary	ST	Burke
	Desmognathus auriculatus	Southern Dusky Salamander		Burke
GA	Geomys pinetis	Southeastern Pocket Gopher	ST	Burke
GA	Haliaeetus leucocephalus	Bald Eagle	ST	Mcduffie
	Passerina ciris	Painted Bunting		Burke
	Passerina ciris	Painted Bunting		Burke
	Passerina ciris	Painted Bunting		Burke
	Passerina ciris	Painted Bunting		Burke
GA	Stewartia malacodendron	Silky Camellia	SR	Burke

NONGAME CONSERVATION SECTION
2065 U.S. HIGHWAY 278 S.E. | SOCIAL CIRCLE, GEORGIA 30025-4743
770.918.6411 or 706.557.3032 | FAX 706.557.3033 | WWW.GEORGIAWILDLIFE.COM

Appendix F

Boggy Gut Creek [High Priority Stream]
Brier Creek [High Priority Stream]
Brushy Creek [High Priority Stream]
McBean Creek [High Priority Stream]
Reedy Creek [High Priority Stream]
Sandy Run Creek [High Priority Stream]
Savannah River [High Priority Stream]

According to our records, **within 10 miles of the VEGP Boundary** there are the following Natural Heritage Database occurrences:

- US *Acipenser brevirostrum* (Shortnose Sturgeon) approx. 1.0 mi. NE of site in the Savannah River
- US *Acipenser brevirostrum* (Shortnose Sturgeon) approx. 10 mi. NW of site in the Savannah River
- US *Acipenser brevirostrum* (Shortnose Sturgeon) approx. 4.0 mi. N of site in the Savannah River
- US *Acipenser brevirostrum* (Shortnose Sturgeon) approx. 5 mi. N of site in the Savannah River
- US *Acipenser brevirostrum* (Shortnose Sturgeon) approx. 6 mi. N of site in the Savannah River
- *Acipenser oxyrinchus oxyrinchus* (Atlantic Sturgeon) [HISTORIC?] approx. 4.0 mi. N of site in the Savannah River
- *Ambystoma tigrinum tigrinum* (Eastern Tiger Salamander) [HISTORIC] approx. 8 mi. S of site
- *Cp mesic broadleaf decid.-broadleaf ever. forest* (Coastal Plain Mesic Ravine Forest) approx. 4.0 mi. S of site
- *Cp mesic broadleaf decid.-broadleaf ever. forest* (Coastal Plain Mesic Ravine Forest) approx. 3.5 mi. S of site
- *Cp mesic broadleaf decid.-broadleaf ever. forest* (Coastal Plain Mesic Ravine Forest) approx. 4.0 mi. S of site
- *Desmognathus auriculatus* (Southern Dusky Salamander) approx. 8 mi. W of site
- *Dryopteris celsa* (Log Fern) approx. 7 mi. S of site
- GA *Enneacanthus chaetodon* (Blackbanded Sunfish) approx. 10 mi. E of site in unnamed tributary #3
- GA *Enneacanthus chaetodon* (Blackbanded Sunfish) approx. 5 mi. E of site in Pen Branch
- *Etheostoma fricksium* (Savannah Darter) approx. 3.5 mi. S of site in High Head Branch
- *Etheostoma fricksium* (Savannah Darter) approx. 10 mi. E of site in Meyers Branch
- *Etheostoma fricksium* (Savannah Darter) approx. 8 mi. E of site in Steel Creek
- *Etheostoma fricksium* (Savannah Darter) approx. 8 mi. NE of site in Pen Branch
- *Etheostoma fricksium* (Savannah Darter) approx. 9 mi. E of site in Meyers Branch
- *Etheostoma fricksium* (Savannah Darter) approx. 9 mi. N of site in Upper Three Runs
- *Etheostoma serrifer* (Sawcheek Darter) approx. 1.5 mi. E of site in the Savannah River
- *Etheostoma serrifer* (Sawcheek Darter) approx. 4.0 mi. E of site in the Savannah River
- *Etheostoma serrifer* (Sawcheek Darter) approx. 4.0 mi. SE of site in the Savannah River

Appendix F

Etheostoma serrifer (Sawcheek Darter) approx. 5 mi. E of site in the Savannah River
Etheostoma serrifer (Sawcheek Darter) approx. 6 mi. E of site in the Savannah River
Etheostoma serrifer (Sawcheek Darter) approx. 7 mi. SE of site in the Savannah River
Etheostoma serrifer (Sawcheek Darter) approx. 8 mi. E of site in the Savannah River
Etheostoma serrifer (Sawcheek Darter) approx. 8 mi. NE of site in Pen Branch
Etheostoma serrifer (Sawcheek Darter) approx. 9 mi. E of site in the Savannah River
GA *Fusconaia masoni* (Atlantic Pigtoe) [EXTIRPATED?] approx. 8 mi. S of site in Brier Creek in Brier Creek
Fundulus chrysotus (Golden Topminnow) approx. 1.5 mi. E of site in the Savannah River
Fundulus chrysotus (Golden Topminnow) approx. 4.0 mi. E of site in the Savannah River
Fundulus chrysotus (Golden Topminnow) approx. 4.0 mi. SE of site in the Savannah River
Fundulus chrysotus (Golden Topminnow) approx. 7 mi. NW of site in the Savannah River
Fundulus chrysotus (Golden Topminnow) approx. 7 mi. SE of site in the Savannah River
Fundulus chrysotus (Golden Topminnow) approx. 8 mi. E of site in the Savannah River
GA *Geomys pinetis* (Southeastern Pocket Gopher) approx. 1.5 mi. NW of site
GA *Geomys pinetis* (Southeastern Pocket Gopher) approx. 2.0 mi. SW of site
GA *Geomys pinetis* (Southeastern Pocket Gopher) approx. 2.5 mi. SW of site
GA *Geomys pinetis* (Southeastern Pocket Gopher) approx. 3.0 mi. NW of site
GA *Geomys pinetis* (Southeastern Pocket Gopher) approx. 3.0 mi. W of site
GA *Geomys pinetis* (Southeastern Pocket Gopher) approx. 3.5 mi. W of site
GA *Geomys pinetis* (Southeastern Pocket Gopher) approx. 4.0 mi. NW of site
GA *Geomys pinetis* (Southeastern Pocket Gopher) approx. 5 mi. NW of site
GA *Geomys pinetis* (Southeastern Pocket Gopher) approx. 5 mi. SW of site
GA *Geomys pinetis* (Southeastern Pocket Gopher) approx. 7 mi. NW of site
Lindera subcoriacea (Bog Spicebush) approx. 4.0 mi. S of site
GA *Moxostoma robustum* (Robust Redhorse) approx. 10 mi. NW of site in the Savannah River in the Savannah River
GA *Moxostoma robustum* (Robust Redhorse) approx. 2.0 mi. E of site in the Savannah River in the Savannah River
GA *Moxostoma robustum* (Robust Redhorse) approx. 5 mi. NW of site in the Savannah River in the Savannah River
GA *Moxostoma robustum* (Robust Redhorse) approx. 7 mi. SE of site in the Savannah River in the Savannah River
Nerodia floridana (Florida Green Water Snake) approx. 9 mi. SE of site
GA *Nestronia umbellula* (Indian Olive) approx. 3.5 mi. S of site
Notropis chalybaeus (Ironcolor Shiner) approx. 1.0 mi. NE of site in the Savannah River
Notropis chalybaeus (Ironcolor Shiner) approx. 1.5 mi. E of site in the Savannah River
Notropis chalybaeus (Ironcolor Shiner) approx. 1.5 mi. N of site in the Savannah River
Notropis chalybaeus (Ironcolor Shiner) approx. 3.5 mi. NE of site in Fourmile Branch
Notropis chalybaeus (Ironcolor Shiner) approx. 4.0 mi. E of site in the Savannah River
Notropis chalybaeus (Ironcolor Shiner) approx. 4.0 mi. N of site in the Savannah River
Notropis chalybaeus (Ironcolor Shiner) approx. 4.0 mi. SE of site in the Savannah River
Notropis chalybaeus (Ironcolor Shiner) approx. 4.5 mi. E of site in the Savannah River
Notropis chalybaeus (Ironcolor Shiner) approx. 5 mi. E of site in the Savannah River
Notropis chalybaeus (Ironcolor Shiner) approx. 5 mi. N of site in the Savannah River

Appendix F

    Notropis chalybaeus (Ironcolor Shiner) approx. 6 mi. E of site in the Savannah River
    Notropis chalybaeus (Ironcolor Shiner) approx. 6 mi. NE of site in Pen Branch
    Notropis chalybaeus (Ironcolor Shiner) approx. 7 mi. NE of site in Indian Grave Branch
    Notropis chalybaeus (Ironcolor Shiner) approx. 7 mi. NW of site in the Savannah River
    Notropis chalybaeus (Ironcolor Shiner) approx. 7 mi. SE of site in the Savannah River
    Notropis chalybaeus (Ironcolor Shiner) approx. 8 mi. E of site in the Savannah River
    Notropis chalybaeus (Ironcolor Shiner) approx. 9 mi. E of site in the Savannah River
    Passerina ciris (Painted Bunting) approx. 1.0 mi. SW of site
    Passerina ciris (Painted Bunting) approx. 1.5 mi. NW of site
    Passerina ciris (Painted Bunting) approx. 2.5 mi. NW of site
    Passerina ciris (Painted Bunting) approx. 7 mi. SE of site
    Passerina ciris (Painted Bunting) approx. 8 mi. S of site
    Pituophis melanoleucus mugitus (Florida Pine Snake) approx. 6 mi. SW of site
    Pseudacris brimleyi (Brimley's Chorus Frog) [HISTORIC] approx. 5 mi. N of site
    Quercus austrina (Bluff White Oak) approx. 7 mi. NW of site
GA *Sarracenia rubra* (Sweet Pitcherplant) approx. 8 mi. SE of site
GA *Scutellaria ocmulgee* (Ocmulgee Skullcap) approx. 4.0 mi. SE of site
    Silene caroliniana (Carolina Pink) approx. 3.5 mi. S of site
    Silene caroliniana (Carolina Pink) approx. 4.0 mi. SE of site
    Silene caroliniana (Carolina Pink) approx. 7 mi. NW of site
    Umbra pygmaea (Eastern Mudminnow) approx. 10 mi. E of site in Meyers Branch
    Umbra pygmaea (Eastern Mudminnow) approx. 4.0 mi. E of site in the Savannah River
    Umbra pygmaea (Eastern Mudminnow) approx. 4.0 mi. NE of site
    Umbra pygmaea (Eastern Mudminnow) approx. 4.5 mi. E of site in the Savannah River
    Umbra pygmaea (Eastern Mudminnow) approx. 5 mi. E of site in the Savannah River
    Umbra pygmaea (Eastern Mudminnow) approx. 5 mi. N of site in the Savannah River
    Umbra pygmaea (Eastern Mudminnow) approx. 5 mi. NE of site
    Umbra pygmaea (Eastern Mudminnow) approx. 6 mi. E of site in the Savannah River
    Umbra pygmaea (Eastern Mudminnow) approx. 6 mi. NE of site in Pen Branch
    Umbra pygmaea (Eastern Mudminnow) approx. 7 mi. E of site
    Umbra pygmaea (Eastern Mudminnow) approx. 7 mi. N of site in Island Creek
    Umbra pygmaea (Eastern Mudminnow) approx. 7 mi. NW of site in the Savannah River
    Umbra pygmaea (Eastern Mudminnow) approx. 7 mi. SE of site in the Savannah River
    Umbra pygmaea (Eastern Mudminnow) approx. 8 mi. E of site in the Savannah River
    Umbra pygmaea (Eastern Mudminnow) approx. 9 mi. E of site in the Savannah River
    Umbra pygmaea (Eastern Mudminnow) approx. 9 mi. NW of site in McBean Creek

* Entries above proceeded by "US" indicates species with federal status in Georgia (Protected or Candidate). Species that are federally protected in Georgia are also state protected; "GA" indicates Georgia protected species.

Recommendations:

We have records of several high priority species within the project area (see Table). We also a large number of records of species of concern within 10 miles of the VEGP site. This includes a federally listed species, *Acipenser brevirostrum* (Shortnose Sturgeon). Section 9 of the

Appendix F

Endangered Species Act states that taking or harming of a listed species is prohibited. We recommend all requestors with projects located near federally protected species consult with the United States Fish and Wildlife Service. For southeast Georgia, please contact Strant Colwell (912-265-9336, ext.30 or Strant_Colwell@fws.gov). In southwest Georgia, please contact John Doresky (706-544-6999 or John_Doresky@fws.gov). In north Georgia, please contact Robin Goodloe (706-613-9493, ext.221 or Robin_Goodloe@fws.gov).

A record of a nesting Bald Eagle (*Haliaeetus leucocephalus*) is also within the transmission line macrocorridor area. Although Bald Eagles are no longer considered an endangered species, they are still protected by the Migratory Bird Treaty Act, the Bald and Golden Eagle Protection Act and the Georgia Endangered Species Act. These Acts continue to protect bald eagles from potentially harmful human activities. For more information on how to prevent impacts to bald eagles that could violate the Eagle Act, download the National Bald Eagle Management Guidelines:
http://www.fws.gov/migratorybirds/issues/BaldEagle/NationalBaldEagleManagementGuidelines.pdf

Though we don't have any records within the project area, there may be appropriate habitat for gopher tortoises (*Gopherus polyphemus*) within the corridor. We recommend identifying any burrows before construction and avoiding disturbance of burrows and tortoises in those areas.

In order to protect aquatic habitats and water quality, we recommend that all machinery be kept out of creeks during construction. Streams should not be culverted/forded to allow equipment access during construction or for future ROW maintenance. Further, we strongly advocate retaining at least a 25-foot vegetative buffer between each stream bank and the closest power pole, and allow this buffer to regenerate to shrub-scrub growth after the line is installed (if the landowner is willing). We realize that some trees may have to be removed, but recommend that shrubs and ground vegetation be left in place. Wider buffers may be needed for projects where land slopes sharply toward the stream being crossed. We also recommend that stringent erosion control practices be used during construction activities and that vegetation is re-established on disturbed areas as quickly as possible. Silt fences and other erosion control devices should be inspected and maintained until soil is stabilized by vegetation. Please use natural vegetation and grading techniques (e.g. vegetated swales, turn-offs, vegetated buffer strips) that will ensure that the project area does not serve as a conduit for storm water or pollutants into the water during or after construction. These measures will help protect water quality in the vicinity of the project as well as in downstream areas.

Please be aware that this project occurs near several high priority streams. As part of an effort to develop a comprehensive wildlife conservation strategy for the state of Georgia, the Wildlife Resources division has developed and mapped a list of streams that are important to the protection or restoration of rare aquatic species and aquatic communities. High priority waters and their surrounding watersheds are a high priority for a broad array of conservation activities, but do not receive any additional legal protections. We now have GIS ESRI shapefiles of GA high priority waters available on our website
(http://www.georgiawildlife.com/content/displaycontent.asp?txtDocument=89&txtPage=13).

Appendix F

Please contact the Georgia Natural Heritage Program if you would like additional information on high priority waters.

Data Available on the Nongame Conservation Section Website

By visiting the Nongame Conservation Section Website you can view the highest priority species and natural community information by Quarter Quad, County and HUC8 Watershed. To access this information, please visit our GA Rare Species and Natural Community Information page at: http://georgiawildlife.dnr.state.ga.us/content/displaycontent.asp?txtDocument=89

An ESRI shape file of our highest priority species and natural community data by quarter quad and county is also available. It can be downloaded from: http://georgiawildlife.dnr.state.ga.us/assets/documents/gnhp/gnhpds.zip

Disclaimer:

Please keep in mind the limitations of our database. The data collected by the Nongame Conservation Section comes from a variety of sources, including museum and herbarium records, literature, and reports from individuals and organizations, as well as field surveys by our staff biologists. In most cases the information is not the result of a recent on-site survey by our staff. Many areas of Georgia have never been surveyed thoroughly. Therefore, the Nongame Conservation Section can only occasionally provide definitive information on the presence or absence of rare species on a given site. Our files are updated constantly as new information is received. **Thus, information provided by our program represents the existing data in our files at the time of the request and should not be considered a final statement on the species or area under consideration.**

If you know of populations of highest priority species that are not in our database, please fill out the appropriate data collection form and send it to our office. Forms can be obtained through our web site (http://www.georgiawildlife.com) or by contacting our office. If I can be of further assistance, please let me know.

Sincerely,

Katrina Morris
Environmental Review Coordinator

Appendix F

December 23, 2009

Mr. Don Klima, Director
Office of Federal Agency Programs
Advisory Council on Historic Preservation
Old Post Office Building
1100 Pennsylvania Avenue, NW, Suite 809
Washington, DC 20004

SUBJECT: REQUEST FOR INFORMATION ON HISTORIC PROPERTIES WITHIN THE AREA UNDER EVALUATION FOR THE VOGTLE ELECTRIC GENERATING PLANT, UNITS 3 AND 4 COMBINED LICENSE APPLICATION REVIEW

Dear Mr. Klima:

The U.S. Nuclear Regulatory Commission (NRC) staff is reviewing an application submitted by Southern Nuclear Operating Company, Inc. (Southern or SNC), on behalf of itself and several co-applicants, for a combined license (COL) for construction and operation of two new nuclear units at its Vogtle Electric Generating Plant (Vogtle or VEGP) site in Burke County, Georgia. The purpose of this letter is to initiate consultation for the subject project with the Advisory Council on Historic Preservation (ACHP) pursuant to U.S. Nuclear Regulatory Commission (NRC) regulations at 10 CFR 51 and Section 106 of the National Historic Preservation Act (NHPA) of 1966, as amended. The NRC plans to coordinate compliance with Section 106 of the NHPA using the National Environmental Policy Act (NEPA) of 1969, as amended, process identified in 36 CFR 800.8(c) in lieu of the procedures set forth in §§ 800.3 through 800.6. Additionally, the NRC staff will rely on procedures in its regulations at 10 CFR 51.92 for supplementing a final environmental impact statement (FEIS).

Section 51.26(d) of the NRC regulations describes the processes for determining the need to supplement an FEIS, issue a notice of intent and determine whether a formal scoping process will be conducted. NRC staff will prepare a supplement to the FEIS for the early site permit (ESP) issued on August 26, 2009, as required by 10 CFR 51.92. In this case, the NRC staff has determined that it will not conduct a formal scoping process for the development of the SEIS.

An ESP is a Commission approval of a site suitable for construction and operation of one or more new nuclear units. Therefore, a COL applicant referencing an ESP need not submit information or analyses regarding environmental issues that were resolved in the ESP EIS, except to the extent the COL applicant has identified any new and potentially significant information.

In accordance with the provisions in 36 CFR § 800.8, the NRC wishes to ensure that if your agency has an interest in any potential historic properties in the area of potential effect (APE), it will be afforded the opportunity to identify its concerns, provide advice on the identification and evaluation of historic properties, including those of traditional religious and cultural importance, and, if necessary, participate in the resolution of any adverse effects to such properties.

Appendix F

D. Klima   2

Thus, to support the NRC staff's review of any new and potentially significant circumstances or information relevant to the environmental concerns related to the proposed action or its impacts, we request that you submit written comments, if any, you may have to offer on the environmental review of the COL by January 19, 2010.

The NRC intends to propose measures in the SEIS for the COL analyses for potential impacts to historical and cultural resources including developing alternatives and proposing measures that might avoid, minimize, or mitigate any adverse effects of the proposed action on historic properties. To complete consultation under § 800.8, the NRC staff will forward the SEIS on the COL to you for your review and comment, and will address your comments in the final SEIS on the COL.

The proposed new reactors, Vogtle Units 3 and 4, would be located approximately 26 miles southeast of Augusta, Georgia, near the town of Waynesboro, Georgia. SNC currently operates two reactors, Vogtle Units 1 and 2, on the site and plans to construct the new units adjacent to the existing reactors. The application for the COL was accepted for docketing on May 30, 2008, and is available through the web-based version of the NRC Agency-wide Documents Access Management System (ADAMS), which can be found at http://www.nrc.gov/readingrm/adams.html. The Environmental Report for the application is listed under ADAMS accession number ML081050181. The Vogtle Plant COL application is also available on the Internet at http://www.nrc.gov/reactors/new-licensing/col/vogtle.html. A detailed review by NRC of the environmental impacts of constructing and operating proposed Units 3 and 4 is documented in NUREG-1872, "Final Environmental Impact Statement for an Early Site Permit (ESP) at the Vogtle Electric Generating Plant Site." (ESP FEIS Volumes 1 and 2 are available in ADAMS under ADAMS accession numbers ML082240145 and ML082240165, respectively.)

Please submit comments either by mail to Mrs. Mallecia Sutton, Environmental Project Manager, Division of Site and Environmental Reviews, Mail Stop T-7E18, U.S. Nuclear Regulatory Commission, Washington, D.C. 20555-0001 or via e-mail to VogtleCOLAEIS@nrc.gov by December 30, 2009.

If you have any questions or require additional information, please contact Mrs. Sutton by telephone at 301- 415-0673 or via e-mail to Mallecia.Sutton@nrc.gov.

Sincerely,

/RA/ A Fetter for

Gregory P. Hatchett, Branch Chief
Environmental Projects Branch 1
Division of Site and Environmental Reviews
Office of New Reactors

Docket Nos. 52-025 and 52-026

cc: See next page

Appendix F

January 7, 2010

Sandra S. Tucker, Field Supervisor
U.S. Fish and Wildlife Service
Coastal Sub Office
4270 Norwich Street
Brunswick, GA 31520

SUBJECT: REQUEST FOR LIST OF PROTECTED SPECIES WITHIN THE AREA UNDER EVALUATION FOR THE VOGTLE ELECTRIC GENERATING PLANT, UNITS 3 AND 4 COMBINED LICENSE APPLICATIONS

Dear Ms. Tucker:

The U.S. Nuclear Regulatory Commission (NRC) staff is reviewing an application submitted by Southern Nuclear Operating Company (SNC), on behalf of itself and four co-applicants, for a combined license (COL) for construction and operation of two new nuclear power plants at the Vogtle Electric Generating Plant (VEGP) site in Burke County, Georgia. As part of the review of this COL application, the NRC is conducting an environmental review as required by Title 10, of the *Code of Federal Regulations* (10 CFR) Part 51, the NRC regulation that implements the National Environmental Policy Act of 1969, as amended (NEPA). This letter is being submitted under provisions of the Endangered Species Act of 1973 (ESA), and the Fish and Wildlife Coordination Act of 1934, as amended (FWCA).

In accordance with the procedures set forth in NRC regulations at 10 CFR 51.92, the NRC is preparing a supplement (an SEIS) to the Final Environmental Impact Statement (FEIS) that was issued in connection with the NRC's review of an early site permit (ESP) application submitted to the NRC in 2006 by SNC and the same co-applicants. The ESP was issued on August 26, 2009. Because the COL application references the Vogtle ESP, the COL SEIS will supplement the NRC staff's analysis in the ESP FEIS with an analysis of any new and significant information regarding the environmental effects of construction and operation of a new nuclear power plant at the VEGP site. Accordingly, the COL EIS will address new and significant information pertinent to the environmental issues resolved in the ESP FEIS, such as impacts to fish and wildlife, including threatened or endangered species. To support the process for preparing the SEIS on the COL application and to ensure compliance with Section 7 of the ESA, the NRC requests current information on Federally-listed, proposed and candidate species, and critical habitat that may be in the vicinity of the VEGP site. In addition, to fulfill consultation requirements of the FWCA, please provide any information you consider appropriate under the provisions of that statute.

The proposed new reactors, Plant Vogtle Units 3 and 4, would be located on a 3169 acre site in Burke County, approximately 26 miles southeast of Augusta, Georgia. SNC currently operates two reactors, Units 1 and 2, on the site, and it plans to construct Units 3 and 4 adjacent to the existing units, wholly within the existing boundaries of the VEGP site. SNC submitted the application for COL by letter dated March 28, 2008, pursuant to NRC requirements in 10 CFR

Appendix F

S. Tucker   -2-

Part 52. The application was accepted for docketing on May 30, 2008, and is available through the web-based version of the NRC Agency-wide Documents Access Management System (ADAMS), which can be found at http://www.nrc.gov/readingrm/adams.html. The Environmental Report for the application is listed under ADAMS accession number ML081050181. The VEGP COL application is also available on the Internet at http://www.nrc.gov/reactors/new-licensing/col/vogtle.html.

On August 26, 2009, the NRC issued SNC and its co-applicants an early site permit (ESP) for the VEGP site, which is the site proposed for Units 3 and 4. An ESP is a Commission approval of a site as suitable for construction and operation of one or more new nuclear units. During the ESP environmental review, the NRC consulted with the Coastal Sub Office in Brunswick, GA, and by letter dated September 19, 2008, (Enclosure 1) received concurrence on a biological assessment evaluating the impacts of limited site-preparation activities for two new reactors at the VEGP site on potentially occurring federally-listed threatened or endangered species. The NRC's detailed review of the environmental impacts of constructing and operating proposed Units 3 and 4 is documented in NUREG-1872, "Final Environmental Impact Statement for an Early site Permit (ESP) at the Vogtle Electric Generating Plant Site." (ESP FEIS Volumes 1 and 2 are available in ADAMS under ADAMS accession numbers ML082240145 and ML082240165, respectively.) Pursuant to NRC regulations in 10 CFR 51.50(c), a COL applicant referencing an ESP need not submit information or analyses regarding environmental issues that were resolved in the ESP EIS, except to the extent the COL applicant has identified new and significant information regarding such issues. Pursuant to 10 CFR 52.39, matters resolved in the ESP proceedings are considered to be resolved in any subsequent proceedings, absent identification of new and significant information.

Consequently, in this consultation, the NRC is particularly interested in any information related to Federally-listed species, critical habitat, and our interactions under the FWCA that may have changed since our last consultation. As set forth in the COL application, SNC intends to use a closed-cycle, wet cooling tower system to remove waste heat during power operation for Plant Vogtle Units 3 and 4. Make-up water for the cooling tower system would be withdrawn from the Savannah River through a new intake structure. Blow-down from the closed-cycle cooling system would be discharged to the Savannah River through a new discharge structure. As noted above, the NRC SEIS on the COL application will include, among other things, analyses of new and significant information relating to threatened or endangered species, if any.

As part of your office's participation in the consultation process, please submit by January 29, 2010, any written comments you have to offer regarding the environmental review. Comments should be submitted either by mail to Mallecia Sutton, Environmental Project Manager, Division of Site and Environmental Reviews, Mail Stop T-7E18, U.S. Nuclear Regulatory Commission, Washington, D.C. 20555-0001 or via e-mail to Vogtle.COLAEIS@nrc.gov.

Appendix F

S. Tucker   -3-

If you have any questions or require additional information, please contact Mrs. Mallecia Sutton, Environmental Project Manager by telephone at 301 415-0673 or via e-mail to Mallecia.Sutton@nrc.gov.

                              Sincerely,

                              /RA/

                              Gregory P. Hatchett, Chief
                              Environmental Projects Branch 1
                              Division of Site and Environmental Reviews
                              Office of New Reactors

Docket Nos. 52-025 and 52-026

cc: See next page

Enclosure: As stated

Appendix F

January 7, 2010

Mr. Donald Rodgers, Chief
Catawba Indian Nation
996 Avenue of the Nations
Catawba, SC 29730

SUBJECT: U.S. NUCLEAR REGULATORY COMMISSION'S SUPPLEMENTAL ENVIRONMENTAL IMPACT STATEMENT FOR SOUTHERN NUCLEAR OPERATING COMPANY'S COMBINED LICENSE APPLICATION FOR THE PROPOSED CONSTRUCTION AND OPERATION OF UNITS 3 AND 4 AT THE VOGTLE ELECTRIC GENERATING PLANT IN WAYNESBORO, GEORGIA

Dear Chief Rodgers:

The U.S. Nuclear Regulatory Commission (NRC) staff is reviewing an application submitted by Southern Nuclear Operating Company, Inc. (Southern or SNC) on behalf of itself and several co-applicants for a combined license (COL) for construction and operation of two new nuclear units at its Vogtle Electric Generating Plant (Vogtle or VEGP) site in Burke County, Georgia. The purpose of this letter is to invite the Catawba Indian Nation to consult with the NRC regarding the proposed action, pursuant to NRC regulations at 10 CFR Part 51 and Section 106 of the National Historic Preservation Act (NHPA) of 1966, as amended. The NRC plans to coordinate compliance with Section 106 of the NHPA using the National Environmental Policy Act (NEPA) process identified in 36 CFR 800.8(c) in lieu of the procedures set forth in §§ 800.3 through 800.6. Additionally, the NRC staff will rely on procedures in its regulations at 10 CFR 51.92 for supplementing a final environmental impact statement (FEIS).

Section 51.26(d) of the NRC regulations describes the processes for determining the need to supplement a FEIS, issue a notice of intent, and determine whether or not a formal scoping process will be conducted. As required by 10 CFR 51.92, the NRC staff will prepare a supplement to the FEIS for the early site permit (ESP) issued on August 26, 2009, to SNC and the same co-applicants. In this case, the NRC staff has determined that it will not conduct a formal scoping process for the development of the supplemental environmental impact statement (SEIS).

An ESP is a Commission approval of a site suitable for construction and operation of one or more new nuclear units. Under 10 CFR 51.50(c), a COL applicant referencing an ESP need not submit information or analyses regarding environmental issues that were resolved in the ESP EIS, except to the extent the COL applicant has identified any new and potentially significant information.

In accordance with the provisions in 36 CFR § 800.8, the NRC wishes to ensure that Indian tribes that might have an interest in any potential historic properties in the area of potential effect (APE) are afforded the opportunity to identify their concerns, provide advice on the identification and evaluation of historic properties, including those of traditional religious and cultural importance and, if necessary, participate in the resolution of any adverse effects to such properties.

Appendix F

Chief Rodgers   -2-

Thus, to support the NRC staff's review of any new and potentially significant circumstances or information relevant to the environmental concerns related to the proposed action or its impacts, we request you submit written comments, if any; your tribe may have to offer on the environmental review of the COL by January 29, 2010.

The NRC intends to propose measures in the SEIS for the COL analyses for potential impacts to historical and cultural resources including developing alternatives and proposing measures that might avoid, minimize, or mitigate any adverse effects of the propose action on historic properties. To complete consultation under § 800.8(c), the NRC staff will forward the SEIS on the COL to you for your review and comment, and will address your comments in the final SEIS on the COL.

The proposed new reactors, Vogtle Units 3 and 4 would be located approximately 26 miles southeast of Augusta, Georgia, near the town of Waynesboro, Georgia. SNC currently operates two reactors, Vogtle Units 1 and 2 on the site and plans to construct the new units adjacent to the existing reactors. The application for the COL was accepted for docketing on May 30, 2008, and is available through the web-based version of the NRC Agency-wide Documents Access Management System (ADAMS), which can be found at http://www.nrc.gov/readingrm/adams.html. The Environmental Report for the application is listed under ADAMS accession number ML081050181. The Vogtle Electric Generating Plant COL application is also available on the Internet at http://www.nrc.gov/reactors/new-licensing/col/vogtle.html. A detailed review by NRC of the environmental impacts of constructing and operating proposed Units 3 and 4 is documented in NUREG-1872, "Final Environmental Impact Statement for an Early Site Permit (ESP) at the Vogtle Electric Generating Plant Site." (ESP FEIS Volumes 1 and 2 are available in ADAMS under ADAMS accession numbers ML082240145 and ML082240165, respectively.)

Please submit comments either by mail to Mrs. Mallecia Sutton, Environmental Project Manager, Division of Site and Environmental Reviews, Mail Stop T-7E18, U.S. Nuclear Regulatory Commission, Washington, D.C. 20555-0001 or via e-mail to Vogtle.COLAEIS@nrc.gov by January 29, 2010.

If you have any questions or require additional information, please contact Mrs. Sutton at (301) 415-0673 or via e-mail to Mallecia.Sutton@nrc.gov.

Sincerely,

/RA/

Gregory P. Hatchett, Branch Chief
Environmental Projects Branch 1
Division of Site and Environmental Reviews
Office of New Reactors

Docket Nos. 52-025 and 52-026

cc: See next page

Appendix F

Mar 01 10 04:15p   USFWS   9128328744   p.2

United States Department of the Interior

Fish and Wildlife Service
105 West Park Drive, Suite D
Athens, Georgia 30606
Phone: (706) 613-9493
Fax: (706) 613-6059

West Georgia Sub-Office
Post Office Box 52560
Fort Benning, Georgia 31995-2560
Phone: (706) 544-6428
Fax: (706) 544-6419

FEB 1 2 2010

Coastal Sub-Office
4980 Wildlife Drive
Townsend, Georgia 31331
Phone: (912) 832-8739
Fax: (912) 832-8744

Ms. Mallecia Sutton
Environmental Project Manager
U. S. Nuclear Regulatory Commission
Division of Site and Environmental Reviews
Mail Stop T-7E18
Washington, DC 20555

Re: USFWS Log Number 2009-1387

Dear Ms. Sutton:

Thank you for your letter dated January 7, 2010, regarding your preparation for a combined license application for Plant Vogtle Units 3 and 4 in Burke County on the Savannah River near Waynesboro, Georgia. Plant Vogtle is a nuclear power, electric generating plant in the Southern Company system. The U. S. Fish and Wildlife Service provides the following comments in accordance with provisions of the Endangered Species Act (ESA) of 1973, as amended; (16 U.S.C. 1531 et seq.) and the Fish and Wildlife Coordination Act (FWCA) (48 Stat. 401, as amended; 16 U.S.C. 661 et seq.) to further the conservation of fish and wildlife resources and their habitat, including federally listed threatened and endangered species.

Our September 19, 2008, consultation letter stated we believed that listed species under our purview had been adequately addressed for limited site-preparation activities at the Vogtle site. Based on our knowledge, there is no additional information related to federally-listed species, critical habitat or other interactions under the FWCA.

We appreciate your willingness to protect all natural resources. Regarding this project, we will provide comments to the U. S. Army Corps of Engineers pursuant to Section 404 of the Clean Water Act (33 U.S.C. 1344) And Section 10 of the River and Harbors Act of 1899 (33 U.S.C. 403). If you have any questions regarding this matter, please contact our Coastal Georgia Sub Office supervisor, Strant Colwell, at 912-832-8739 extension 1.

Sincerely,

Sandra S. Tucker
Field Supervisor

Appendix F

```
-----Original Message-----
From: Elizabeth Shirk [mailto:Elizabeth.Shirk@dnr.state.ga.us]
Sent: Thursday, June 17, 2010 4:02 PM
To: Sutton, Mallecia
Subject: Vogtle Electric Generating Plant, Burke County, Georgia, Units
3 and 4 Supplement
```

Ms. Sutton:

The Historic Preservation Division (HPD) has reviewed the additional information concerning the
above referenced undertaking in Burke County, Georgia. Our comments are offered to assist
the U.S. Nuclear Regulatory Commission (NRC) and its applicants in complying with the
provisions of Section 106 of the National Historic Preservation Act of 1966, as amended.

Based on the information provided, HPD agrees with NRC that the backfill operations will have
no effect to properties listed on or eligible for listing on the National Register of Historic Places.

If we may be of further assistance, please feel free to contact me.

Sincerely,

Elizabeth (Betsy) Shirk
Environmental Review Coordinator
Historic Preservation Division
254 Washington Street, SW
Ground Floor
Atlanta, GA 30334
404-651-6624

Please Note Our New Address

Appendix F

September 2, 2010

Mr. A. D. Ellis, Principal Chief
Muscogee (Creek) Nation
P.O. Box 580
Okmulgee, OK 74447

SUBJECT: SECTION 106 CONSULTATION AND NOTIFICATION OF THE ISSUANCE AND REQUEST FOR COMMENTS ON THE DRAFT SUPPLEMENTAL ENVIRONMENTAL IMPACT STATEMENT FOR THE VOGTLE ELECTRIC GENERATING PLANT, UNITS 3 AND 4 COMBINED LICENSE APPLICATION

Dear Chief Ellis:

On behalf of the Nuclear Regulatory Commission (NRC) staff, I am forwarding a copy of the "Draft Supplemental Environmental Impact Statement for Combined Licenses (COLs) for Vogtle Electric Generating Plant Units 3 and 4," for your review and comments. The NRC is reviewing the application submitted by Southern Nuclear Operating Company, Inc. (SNC) and several co-applicants for two combined licenses (COLs) to construct and operate two new nuclear units at the Vogtle Electric Generating Plant site in Burke County, GA. As part of its review of the proposed action, the NRC staff has prepared the draft supplemental environmental impact statement (DSEIS) to include an analysis of relevant environmental issues, including potential impacts to historic properties. The DSEIS documents the NRC determination regarding the environmental impacts at the proposed site from the construction and operation of two new nuclear units.

This DSEIS is a supplement to the Final Environmental Impact Statement (FEIS) for the early site permit (ESP) issued on August 26, 2009, to SNC and the same co-applicants. An ESP is a Commission approval of a site suitable for construction and operation of one or more new nuclear units. Under Title 10 of the *Code of Federal Regulations* (10 CFR) CFR 51.50(c), a COL applicant referencing an ESP need not submit information or analyses regarding environmental issues that were resolved in the ESP EIS, except to the extent the COL applicant has identified any new and potentially significant information. Accordingly, in preparing the DSEIS, the NRC staff considered whether new and significant information has been identified, including with respect to potential impacts to historic properties. The NRC staff conducted an environmental audit at the site and reviewed historic and archaeological records. The NRC staff also contacted Indian Tribes identified as having potential interest in the proposed action.

By letter dated December 10, 2009, the NRC staff notified you that it will comply with its obligations under Section 106 of the National Historic Preservation Act of 1966, as amended, (NHPA) using the process set forth in 36 CFR 800.8(c) in lieu of the procedures set forth in 36 CFR 800.3 through 36 CFR 800.6. Pursuant to 36 CFR 800.8(c), the NRC staff is using the preparation of the DSEIS required by the National Environmental Policy Act of 1969, as amended, (NEPA), to comply with its obligations under Section 106 of the NHPA.

Chief Ellis   - 2 -

In the context of NEPA, under which the DSEIS was prepared, the NRC preliminary determination is that the impact of the two new proposed nuclear units on historical and archaeological resources remains moderate, as concluded in the ESP FEIS. In addition, SNC has entered into a Memorandum of Understanding with the Georgia State Historic Preservation Officer (SHPO). Under the provisions of the National Historic Preservation Act, the NRC preliminary determination is that, consistent with the determination in the ESP FEIS, the proposed project will affect, but not adversely affect, historic properties. Note that in Chapter 2 of the DSEIS you will find a discussion of the areas of potential effect, and impacts to historic properties from construction and operation are discussed in Chapters 4 and 5.

The NRC plans to hold a public meeting to go over the analysis and results in the DSEIS on October 7, 2010, at Augusta Technical College, Waynesboro Campus, 216 Highway 24 South, Waynesboro, Georgia 30830. The meeting will convene at 7:00 p.m. and will continue until 10 p.m., as necessary. In addition, the meeting will be preceded by an open house session from 6:00 p.m. to 7:00 p.m., during which members of the public may meet and talk with NRC staff members on an informal basis. You and your staff are invited to attend.

Pursuant to 10 CFR 51.92 and 36 CFR 800.2(c), the NRC wishes to ensure that Indian Tribes that might have an interest in any potential historic properties in the areas of potential effect are afforded the opportunity to identify their concerns, provide advice on the identification and evaluation of historic properties including those of traditional, religious, and cultural importance; and if necessary, participate in the resolution of any adverse effects to such properties.

In accordance with our December 10, 2009, letter, the NRC staff is forwarding the DSEIS for your review and comments. Pursuant to 36 CFR 800.8(c), we are requesting your comments on the DSEIS, specifically, on our preliminary conclusions regarding historic properties. Please provide any information or comments you may have on the DSEIS during the comment period, which ends on November 24, 2010. The NRC may consider additional comments after the comment period, to the extent practicable. Comments should be submitted either by mail to the Chief, Rules, Announcements, and Directives Branch, Division of Administrative Services, Office of Administration, Mailstop TWB-05-B01M, Washington, D.C. 20555-0001 or via e-mail to Vogtle.COLAEIS@nrc.gov. Your comments will be addressed in the final SEIS.

Appendix F

Chief Ellis   - 3 -

If you have any questions or require additional information, please contact Ms. Mallecia Sutton, NRC Environmental Project Manager at (301) 415-0673 or via e-mail to Mallecia.Sutton@nrc.gov.

                              Sincerely,

                              /RA/

                              Gregory P. Hatchett, Chief
                              Environmental Projects Branch 1
                              Division of Site and Environmental Reviews
                              Office of New Reactors

Docket Nos. 52-025
            52-026
Enclosure:
As stated

cc: See next page

Appendix F

September 2, 2010

Mr. Dallas Proctor, Chief
United Keetoowah Band of Cherokee Indians
P.O. Box 746
Tahlequah, OK 74465

SUBJECT: SECTION 106 CONSULTATION AND NOTIFICATION OF THE ISSUANCE AND REQUEST FOR COMMENTS ON THE DRAFT SUPPLEMENTAL ENVIRONMENTAL IMPACT STATEMENT FOR THE VOGTLE ELECTRIC GENERATING PLANT, UNITS 3 AND 4 COMBINED LICENSE APPLICATION

Dear Chief Proctor:

On behalf of the Nuclear Regulatory Commission (NRC) staff, I am forwarding a copy of the "Draft Supplemental Environmental Impact Statement for Combined Licenses (COLs) for Vogtle Electric Generating Plant Units 3 and 4," for your review and comments. The NRC is reviewing the application submitted by Southern Nuclear Operating Company, Inc. (SNC) and several co-applicants for two combined licenses (COLs) to construct and operate two new nuclear units at the Vogtle Electric Generating Plant site in Burke County, GA. As part of its review of the proposed action, the NRC staff has prepared the draft supplemental environmental impact statement (DSEIS) to include an analysis of relevant environmental issues, including potential impacts to historic properties. The DSEIS documents the NRC determination regarding the environmental impacts at the proposed site from the construction and operation of two new nuclear units.

This DSEIS is a supplement to the Final Environmental Impact Statement (FEIS) for the early site permit (ESP) issued on August 26, 2009, to SNC and the same co-applicants. An ESP is a Commission approval of a site suitable for construction and operation of one or more new nuclear units. Under Title 10 of the *Code of Federal Regulations* (10 CFR) CFR 51.50(c), a COL applicant referencing an ESP need not submit information or analyses regarding environmental issues that were resolved in the ESP EIS, except to the extent the COL applicant has identified any new and potentially significant information. Accordingly, in preparing the DSEIS, the NRC staff considered whether new and significant information has been identified, including with respect to potential impacts to historic properties. The NRC staff conducted an environmental audit at the site and reviewed historic and archaeological records. The NRC staff also contacted Indian Tribes identified as having potential interest in the proposed action.

By letter dated December 10, 2009, the NRC staff notified you that it will comply with its obligations under Section 106 of the National Historic Preservation Act of 1966, as amended, (NHPA) using the process set forth in 36 CFR 800.8(c) in lieu of the procedures set forth in 36 CFR 800.3 through 36 CFR 800.6. Pursuant to 36 CFR 800.8(c), the NRC staff is using the preparation of the DSEIS required by the National Environmental Policy Act of 1969, as amended, (NEPA), to comply with its obligations under Section 106 of the NHPA.

Appendix F

Chief Proctor   - 2 -

In the context of NEPA, under which the DSEIS was prepared, the NRC preliminary determination is that the impact of the two new proposed nuclear units on historical and archaeological resources remains moderate, as concluded in the ESP FEIS. In addition, SNC has entered into a Memorandum of Understanding with the Georgia State Historic Preservation Officer (SHPO). Under the provisions of the National Historic Preservation Act, the NRC preliminary determination is that, consistent with the determination in the ESP FEIS, the proposed project will affect, but not adversely affect, historic properties. Note that in Chapter 2 of the DSEIS you will find a discussion of the areas of potential effect, and impacts to historic properties from construction and operation are discussed in Chapters 4 and 5.

The NRC plans to hold a public meeting to go over the analysis and results in the DSEIS on October 7, 2010, at Augusta Technical College, Waynesboro Campus, 216 Highway 24 South, Waynesboro, Georgia 30830. The meeting will convene at 7:00 p.m. and will continue until 10 p.m., as necessary. In addition, the meeting will be preceded by an open house session from 5:00 p.m. to 6:00 p.m., during which members of the public may meet and talk with NRC staff members on an informal basis. You and your staff are invited to attend.

Pursuant to 10 CFR 51.92 and 36 CFR 800.2(c), the NRC wishes to ensure that Indian Tribes that might have an interest in any potential historic properties in the areas of potential effect are afforded the opportunity to identify their concerns, provide advice on the identification and evaluation of historic properties including those of traditional, religious, and cultural importance; and if necessary, participate in the resolution of any adverse effects to such properties.

In accordance with our December 10, 2009, letter, the NRC staff is forwarding the DSEIS for your review and comments. Pursuant to 36 CFR 800.8(c), we are requesting your comments on the DSEIS, specifically, on our preliminary conclusions regarding historic properties. Please provide any information or comments you may have on the DSEIS during the comment period, which ends on November 24, 2010. The NRC may consider additional comments after the comment period, to the extent practicable. Comments should be submitted either by mail to the Chief, Rules, Announcements, and Directives Branch, Division of Administrative Services, Office of Administration, Mailstop TWB-05-B01M, Washington, D.C. 20555-0001 or via e-mail to Vogtle.COLAEIS@nrc.gov. Your comments will be addressed in the final SEIS.

Appendix F

Chief Proctor   - 3 -

If you have any questions or require additional information, please contact Ms. Mallecia Sutton, NRC Environmental Project Manager at (301) 415-0673 or via e-mail to Mallecia.Sutton@nrc.gov.

           Sincerely,

           /RA/

           Gregory P. Hatchett, Chief
           Environmental Projects Branch 1
           Division of Site and Environmental Reviews
           Office of New Reactors

Docket Nos. 52-025
      52-026

Enclosure:
As stated

cc: See next page

Appendix F

September 02, 2010

Mr. Eddie Tullis, Chairperson
Poarch Band of Creek Indians
5811 Jack Springs Road
Atmore, AL 36502

SUBJECT: SECTION 106 CONSULTATION AND NOTIFICATION OF THE ISSUANCE AND REQUEST FOR COMMENTS ON THE DRAFT SUPPLEMENTAL ENVIRONMENTAL IMPACT STATEMENT FOR THE VOGTLE ELECTRIC GENERATING PLANT, UNITS 3 AND 4 COMBINED LICENSE APPLICATION

Dear Mr. Chairman:

On behalf of the Nuclear Regulatory Commission (NRC) staff, I am forwarding a copy of the "Draft Supplemental Environmental Impact Statement for Combined Licenses (COLs) for Vogtle Electric Generating Plant Units 3 and 4," for your review and comments. The NRC is reviewing the application submitted by Southern Nuclear Operating Company, Inc. (SNC) and several co-applicants for two combined licenses (COLs) to construct and operate two new nuclear units at the Vogtle Electric Generating Plant site in Burke County, GA. As part of its review of the proposed action, the NRC staff has prepared the draft supplemental environmental impact statement (DSEIS) to include an analysis of relevant environmental issues, including potential impacts to historic properties. The DSEIS documents the NRC determination regarding the environmental impacts at the proposed site from the construction and operation of two new nuclear units.

This DSEIS is a supplement to the Final Environmental Impact Statement (FEIS) for the early site permit (ESP) issued on August 26, 2009, to SNC and the same co-applicants. An ESP is a Commission approval of a site suitable for construction and operation of one or more new nuclear units. Under Title 10 of the *Code of Federal Regulations* (10 CFR) CFR 51.50(c), a COL applicant referencing an ESP need not submit information or analyses regarding environmental issues that were resolved in the ESP EIS, except to the extent the COL applicant has identified any new and potentially significant information. Accordingly, in preparing the DSEIS, the NRC staff considered whether new and significant information has been identified, including with respect to potential impacts to historic properties. The NRC staff conducted an environmental audit at the site and reviewed historic and archaeological records. The NRC staff also contacted Indian Tribes identified as having potential interest in the proposed action.

By letter dated December 10, 2009, the NRC staff notified you that it will comply with its obligations under Section 106 of the National Historic Preservation Act of 1966, as amended, (NHPA) using the process set forth in 36 CFR 800.8(c) in lieu of the procedures set forth in 36 CFR 800.3 through 36 CFR 800.6. Pursuant to 36 CFR 800.8(c), the NRC staff is using the preparation of the DSEIS required by the National Environmental Policy Act of 1969, as amended, (NEPA), to comply with its obligations under Section 106 of the NHPA.

Appendix F

E. Tullis   - 2 -

In the context of NEPA, under which the DSEIS was prepared, the NRC preliminary determination is that the impact of the two new proposed nuclear units on historical and archaeological resources remains moderate, as concluded in the ESP FEIS. In addition, SNC has entered into a Memorandum of Understanding with the Georgia State Historic Preservation Officer (SHPO). Under the provisions of the National Historic Preservation Act, the NRC preliminary determination is that, consistent with the determination in the ESP FEIS, the proposed project will affect, but not adversely affect, historic properties. Note that in Chapter 2 of the DSEIS you will find a discussion of the areas of potential effect, and impacts to historic properties from construction and operation are discussed in Chapters 4 and 5.

The NRC plans to hold a public meeting to go over the analysis and results in the DSEIS on October 7, 2010, at Augusta Technical College, Waynesboro Campus, 216 Highway 24 South, Waynesboro, Georgia 30830. The meeting will convene at 7:00 p.m. and will continue until 10 p.m., as necessary. In addition, the meeting will be preceded by an open house session from 6:00 p.m. to 7:00 p.m., during which members of the public may meet and talk with NRC staff members on an informal basis. You and your staff are invited to attend.

Pursuant to 10 CFR 51.92 and 36 CFR 800.2(c), the NRC wishes to ensure that Indian Tribes that might have an interest in any potential historic properties in the areas of potential effect are afforded the opportunity to identify their concerns, provide advice on the identification and evaluation of historic properties including those of traditional, religious, and cultural importance; and if necessary, participate in the resolution of any adverse effects to such properties.

In accordance with our December 10, 2009, letter, the NRC staff is forwarding the DSEIS for your review and comments. Pursuant to 36 CFR 800.8(c), we are requesting your comments on the DSEIS, specifically, on our preliminary conclusions regarding historic properties. Please provide any information or comments you may have on the DSEIS during the comment period, which ends on November 24, 2010. The NRC may consider additional comments after the comment period, to the extent practicable. Comments should be submitted either by mail to the Chief, Rules, Announcements, and Directives Branch, Division of Administrative Services, Office of Administration, Mailstop TWB-05-B01M, Washington, D.C. 20555-0001 or via e-mail to Vogtle.COLAEIS@nrc.gov. Your comments will be addressed in the final SEIS.

Appendix F

E. Tullis   - 3 -

If you have any questions or require additional information, please contact Ms. Mallecia Sutton, NRC Environmental Project Manager at (301) 415-0673 or via e-mail to Mallecia.Sutton@nrc.gov.

Sincerely,

/RA/

Gregory P. Hatchett, Chief
Environmental Projects Branch 1
Division of Site and Environmental Reviews
Office of New Reactors

Docket Nos. 52-025
             52-026
Enclosure:
As stated

cc: See next page

Appendix F

September 2, 2010

Ms. Emma Sue Holland
NAGPRA Contact
United Keetoowah Band
  of Cherokee Indians
P.O. Box 746
Tahlequah, OK 74465

SUBJECT: SECTION 106 CONSULTATION AND NOTIFICATION OF THE ISSUANCE AND REQUEST FOR COMMENTS ON THE DRAFT SUPPLEMENTAL ENVIRONMENTAL IMPACT STATEMENT FOR THE VOGTLE ELECTRIC GENERATING PLANT, UNITS 3 AND 4 COMBINED LICENSE APPLICATION

Dear Ms. Holland:

On behalf of the Nuclear Regulatory Commission (NRC) staff, I am forwarding a copy of the "Draft Supplemental Environmental Impact Statement for Combined Licenses (COLs) for Vogtle Electric Generating Plant Units 3 and 4," for your review and comments. The NRC is reviewing the application submitted by Southern Nuclear Operating Company, Inc. (SNC) and several co-applicants for two combined licenses (COLs) to construct and operate two new nuclear units at the Vogtle Electric Generating Plant site in Burke County, GA. As part of its review of the proposed action, the NRC staff has prepared the draft supplemental environmental impact statement (DSEIS) to include an analysis of relevant environmental issues, including potential impacts to historic properties. The DSEIS documents the NRC determination regarding the environmental impacts at the proposed site from the construction and operation of two new nuclear units.

This DSEIS is a supplement to the Final Environmental Impact Statement (FEIS) for the early site permit (ESP) issued on August 26, 2009, to SNC and the same co-applicants. An ESP is a Commission approval of a site suitable for construction and operation of one or more new nuclear units. Under Title 10 of the *Code of Federal Regulations* (10 CFR) CFR 51.50(c), a COL applicant referencing an ESP need not submit information or analyses regarding environmental issues that were resolved in the ESP EIS, except to the extent the COL applicant has identified any new and potentially significant information. Accordingly, in preparing the DSEIS, the NRC staff considered whether new and significant information has been identified, including with respect to potential impacts to historic properties. The NRC staff conducted an environmental audit at the site and reviewed historic and archaeological records. The NRC staff also contacted Indian Tribes identified as having potential interest in the proposed action.

By letter dated December 10, 2009, the NRC staff notified you that it will comply with its obligations under Section 106 of the National Historic Preservation Act of 1966, as amended, (NHPA) using the process set forth in 36 CFR 800.8(c) in lieu of the procedures set forth in 36 CFR 800.3 through 36 CFR 800.6. Pursuant to 36 CFR 800.8(c), the NRC staff is using the preparation of the DSEIS required by the National Environmental Policy Act of 1969, as amended, (NEPA), to comply with its obligations under Section 106 of the NHPA.

Appendix F

E. Holland   - 2 -

In the context of NEPA, under which the DSEIS was prepared, the NRC preliminary determination is that the impact of the two new proposed nuclear units on historical and archaeological resources remains moderate, as concluded in the ESP FEIS. In addition, SNC has entered into a Memorandum of Understanding with the Georgia State Historic Preservation Officer (SHPO). Under the provisions of the National Historic Preservation Act, the NRC preliminary determination is that, consistent with the determination in the ESP FEIS, the proposed project will affect, but not adversely affect, historic properties. Note that in Chapter 2 of the DSEIS you will find a discussion of the areas of potential effect, and impacts to historic properties from construction and operation are discussed in Chapters 4 and 5.

The NRC plans to hold a public meeting to go over the analysis and results in the DSEIS on October 7, 2010, at Augusta Technical College, Waynesboro Campus, 216 Highway 24 South, Waynesboro, Georgia 30830. The meeting will convene at 7:00 p.m. and will continue until 10 p.m., as necessary. In addition, the meeting will be preceded by an open house session from 6:00 p.m. to 7:00 p.m., during which members of the public may meet and talk with NRC staff members on an informal basis. You and your staff are invited to attend.

Pursuant to 10 CFR 51.92 and 36 CFR 800.2(c), the NRC wishes to ensure that Indian Tribes that might have an interest in any potential historic properties in the areas of potential effect are afforded the opportunity to identify their concerns, provide advice on the identification and evaluation of historic properties including those of traditional, religious, and cultural importance; and if necessary, participate in the resolution of any adverse effects to such properties.

In accordance with our December 10, 2009, letter, the NRC staff is forwarding the DSEIS for your review and comments. Pursuant to 36 CFR 800.8(c), we are requesting your comments on the DSEIS, specifically, on our preliminary conclusions regarding historic properties. Please provide any information or comments you may have on the DSEIS during the comment period, which ends on November 24, 2010. The NRC may consider additional comments after the comment period, to the extent practicable. Comments should be submitted either by mail to the Chief, Rules, Announcements, and Directives Branch, Division of Administrative Services, Office of Administration, Mailstop TWB-05-B01M, Washington, D.C. 20555-0001 or via e-mail to Vogtle.COLAEIS@nrc.gov. Your comments will be addressed in the final SEIS.

Appendix F

E. Holland   - 3 -

If you have any questions or require additional information, please contact Ms. Mallecia Sutton, NRC Environmental Project Manager at (301) 415-0673 or via e-mail to Mallecia.Sutton@nrc.gov.

        Sincerely,

        /RA/

        Gregory P. Hatchett, Chief
        Environmental Projects Branch 1
        Division of Site and Environmental Reviews
        Office of New Reactors

Docket Nos. 52-025
     52-026
Enclosure:
As stated

cc: See next page

Appendix F

September 2, 2010

Ms. Kathy McCoy, NAGPRA Contact
Eastern Band of Cherokee Indians
P.O. Box 455
Cherokee, NC 28719

SUBJECT: SECTION 106 CONSULTATION AND NOTIFICATION OF THE ISSUANCE AND REQUEST FOR COMMENTS ON THE DRAFT SUPPLEMENTAL ENVIRONMENTAL IMPACT STATEMENT FOR THE VOGTLE ELECTRIC GENERATING PLANT, UNITS 3 AND 4 COMBINED LICENSE APPLICATION

Dear Ms. McCoy:

On behalf of the Nuclear Regulatory Commission (NRC) staff, I am forwarding a copy of the "Draft Supplemental Environmental Impact Statement for Combined Licenses (COLs) for Vogtle Electric Generating Plant Units 3 and 4," for your review and comments. The NRC is reviewing the application submitted by Southern Nuclear Operating Company, Inc. (SNC) and several co-applicants for two combined licenses (COLs) to construct and operate two new nuclear units at the Vogtle Electric Generating Plant site in Burke County, GA. As part of its review of the proposed action, the NRC staff has prepared the draft supplemental environmental impact statement (DSEIS) to include an analysis of relevant environmental issues, including potential impacts to historic properties. The DSEIS documents the NRC determination regarding the environmental impacts at the proposed site from the construction and operation of two new nuclear units.

This DSEIS is a supplement to the Final Environmental Impact Statement (FEIS) for the early site permit (ESP) issued on August 26, 2009, to SNC and the same co-applicants. An ESP is a Commission approval of a site suitable for construction and operation of one or more new nuclear units. Under Title 10 of the *Code of Federal Regulations* (10 CFR) CFR 51.50(c), a COL applicant referencing an ESP need not submit information or analyses regarding environmental issues that were resolved in the ESP EIS, except to the extent the COL applicant has identified any new and potentially significant information. Accordingly, in preparing the DSEIS, the NRC staff considered whether new and significant information has been identified, including with respect to potential impacts to historic properties. The NRC staff conducted an environmental audit at the site and reviewed historic and archaeological records. The NRC staff also contacted Indian Tribes identified as having potential interest in the proposed action.

By letter dated December 10, 2009, the NRC staff notified you that it will comply with its obligations under Section 106 of the National Historic Preservation Act of 1966, as amended, (NHPA) using the process set forth in 36 CFR 800.8(c) in lieu of the procedures set forth in 36 CFR 800.3 through 36 CFR 800.6. Pursuant to 36 CFR 800.8(c), the NRC staff is using the preparation of the DSEIS required by the National Environmental Policy Act of 1969, as amended, (NEPA), to comply with its obligations under Section 106 of the NHPA.

K. McCoy   - 2 -

In the context of NEPA, under which the DSEIS was prepared, the NRC preliminary determination is that the impact of the two new proposed nuclear units on historical and archaeological resources remains moderate, as concluded in the ESP FEIS. In addition, SNC has entered into a Memorandum of Understanding with the Georgia State Historic Preservation Officer (SHPO). Under the provisions of the National Historic Preservation Act, the NRC preliminary determination is that, consistent with the determination in the ESP FEIS, the proposed project will affect, but not adversely affect, historic properties. Note that in Chapter 2 of the DSEIS you will find a discussion of the areas of potential effect, and impacts to historic properties from construction and operation are discussed in Chapters 4 and 5.

The NRC plans to hold a public meeting to go over the analysis and results in the DSEIS on October 7, 2010, at Augusta Technical College, Waynesboro Campus, 216 Highway 24 South, Waynesboro, Georgia 30830. The meeting will convene at 7:00 p.m. and will continue until 10 p.m., as necessary. In addition, the meeting will be preceded by an open house session from 6:00 p.m. to 7:00 p.m., during which members of the public may meet and talk with NRC staff members on an informal basis. You and your staff are invited to attend.

Pursuant to 10 CFR 51.92 and 36 CFR 800.2(c), the NRC wishes to ensure that Indian Tribes that might have an interest in any potential historic properties in the areas of potential effect are afforded the opportunity to identify their concerns, provide advice on the identification and evaluation of historic properties including those of traditional, religious, and cultural importance; and if necessary, participate in the resolution of any adverse effects to such properties.

In accordance with our December 10, 2009, letter, the NRC staff is forwarding the DSEIS for your review and comments. Pursuant to 36 CFR 800.8(c), we are requesting your comments on the DSEIS, specifically, on our preliminary conclusions regarding historic properties. Please provide any information or comments you may have on the DSEIS during the comment period, which ends on November 24, 2010. The NRC may consider additional comments after the comment period, to the extent practicable. Comments should be submitted either by mail to the Chief, Rules, Announcements, and Directives Branch, Division of Administrative Services, Office of Administration, Mailstop TWB-05-B01M, Washington, D.C. 20555-0001 or via e-mail to Vogtle.COLAEIS@nrc.gov. Your comments will be addressed in the final SEIS.

Appendix F

K. McCoy   - 3 -

If you have any questions or require additional information, please contact Ms. Mallecia Sutton, NRC Environmental Project Manager at (301) 415-0673 or via e-mail to Mallecia.Sutton@nrc.gov.

                                              Sincerely,

                                              /RA/

                                        Gregory P. Hatchett, Chief
                                        Environmental Projects Branch 1
                                        Division of Site and Environmental Reviews
                                        Office of New Reactors

Docket Nos. 52-025
              52-026
Enclosure:
As stated

cc: See next page

Appendix F

September 2, 2010

Mr. John Zachary, Attorney at Law
c/o Coushatta Tribe of Louisiana
P.O. Box 12730
Alexandria, LA   71315-2730

SUBJECT: SECTION 106 CONSULTATION AND NOTIFICATION OF THE ISSUANCE AND
REQUEST FOR COMMENTS ON THE DRAFT SUPPLEMENTAL
ENVIRONMENTAL IMPACT STATEMENT FOR THE VOGTLE ELECTRIC
GENERATING PLANT, UNITS 3 AND 4 COMBINED LICENSE APPLICATION

Dear Mr. Zachary:

On behalf of the Nuclear Regulatory Commission (NRC) staff, I am forwarding a copy of the "Draft Supplemental Environmental Impact Statement for Combined Licenses (COLs) for Vogtle Electric Generating Plant Units 3 and 4," for your review and comments. The NRC is reviewing the application submitted by Southern Nuclear Operating Company, Inc. (SNC) and several co-applicants for two combined licenses (COLs) to construct and operate two new nuclear units at the Vogtle Electric Generating Plant site in Burke County, GA. As part of its review of the proposed action, the NRC staff has prepared the draft supplemental environmental impact statement (DSEIS) to include an analysis of relevant environmental issues, including potential impacts to historic properties. The DSEIS documents the NRC determination regarding the environmental impacts at the proposed site from the construction and operation of two new nuclear units.

This DSEIS is a supplement to the Final Environmental Impact Statement (FEIS) for the early site permit (ESP) issued on August 26, 2009, to SNC and the same co-applicants. An ESP is a Commission approval of a site suitable for construction and operation of one or more new nuclear units. Under Title 10 of the *Code of Federal Regulations* (10 CFR) CFR 51.50(c), a COL applicant referencing an ESP need not submit information or analyses regarding environmental issues that were resolved in the ESP EIS, except to the extent the COL applicant has identified any new and potentially significant information. Accordingly, in preparing the DSEIS, the NRC staff considered whether new and significant information has been identified, including with respect to potential impacts to historic properties. The NRC staff conducted an environmental audit at the site and reviewed historic and archaeological records. The NRC staff also contacted Indian Tribes identified as having potential interest in the proposed action.

By letter dated December 10, 2009, the NRC staff notified you that it will comply with its obligations under Section 106 of the National Historic Preservation Act of 1966, as amended, (NHPA) using the process set forth in 36 CFR 800.8(c) in lieu of the procedures set forth in 36 CFR 800.3 through 36 CFR 800.6. Pursuant to 36 CFR 800.8(c), the NRC staff is using the preparation of the DSEIS required by the National Environmental Policy Act of 1969, as amended, (NEPA), to comply with its obligations under Section 106 of the NHPA.

Appendix F

J. Zachary   - 2 -

In the context of NEPA, under which the DSEIS was prepared, the NRC preliminary determination is that the impact of the two new proposed nuclear units on historical and archaeological resources remains moderate, as concluded in the ESP FEIS. In addition, SNC has entered into a Memorandum of Understanding with the Georgia State Historic Preservation Officer (SHPO). Under the provisions of the National Historic Preservation Act, the NRC preliminary determination is that, consistent with the determination in the ESP FEIS, the proposed project will affect, but not adversely affect, historic properties. Note that in Chapter 2 of the DSEIS you will find a discussion of the areas of potential effect, and impacts to historic properties from construction and operation are discussed in Chapters 4 and 5.

The NRC plans to hold a public meeting to go over the analysis and results in the DSEIS on October 7, 2010, at Augusta Technical College, Waynesboro Campus, 216 Highway 24 South, Waynesboro, Georgia 30830. The meeting will convene at 7:00 p.m. and will continue until 10 p.m., as necessary. In addition, the meeting will be preceded by an open house session from 6:00 p.m. to 7:00 p.m., during which members of the public may meet and talk with NRC staff members on an informal basis. You and your staff are invited to attend.

Pursuant to 10 CFR 51.92 and 36 CFR 800.2(c), the NRC wishes to ensure that Indian Tribes that might have an interest in any potential historic properties in the areas of potential effect are afforded the opportunity to identify their concerns, provide advice on the identification and evaluation of historic properties including those of traditional, religious, and cultural importance; and if necessary, participate in the resolution of any adverse effects to such properties.

In accordance with our December 10, 2009, letter, the NRC staff is forwarding the DSEIS for your review and comments. Pursuant to 36 CFR 800.8(c), we are requesting your comments on the DSEIS, specifically, on our preliminary conclusions regarding historic properties. Please provide any information or comments you may have on the DSEIS during the comment period, which ends on November 24, 2010. The NRC may consider additional comments after the comment period, to the extent practicable. Comments should be submitted either by mail to the Chief, Rules, Announcements, and Directives Branch, Division of Administrative Services, Office of Administration, Mailstop TWB-05-B01M, Washington, D.C. 20555-0001 or via e-mail to Vogtle.COLAEIS@nrc.gov. Your comments will be addressed in the final SEIS.

Appendix F

J. Zachary   - 3 -

If you have any questions or require additional information, please contact Ms. Mallecia Sutton,
NRC Environmental Project Manager at (301) 415-0673 or via e-mail to
Mallecia.Sutton@nrc.gov.

                              Sincerely,

                              /RA/

                              Gregory P. Hatchett, Chief
                              Environmental Projects Branch 1
                              Division of Site and Environmental Reviews
                              Office of New Reactors

Docket Nos. 52-025
            52-026
Enclosure:
As stated

cc: See next page

Appendix F

September 2, 2010

Ms. Evelyn Bucktrot, Town King
Kialegee Tribal Town
P.O. Box 332
Wetumka, OK 74883

SUBJECT: SECTION 106 CONSULTATION AND NOTIFICATION OF THE ISSUANCE AND REQUEST FOR COMMENTS ON THE DRAFT SUPPLEMENTAL ENVIRONMENTAL IMPACT STATEMENT FOR THE VOGTLE ELECTRIC GENERATING PLANT, UNITS 3 AND 4 COMBINED LICENSE APPLICATION

Dear Ms. Bucktrot:

On behalf of the Nuclear Regulatory Commission (NRC) staff, I am forwarding a copy of the "Draft Supplemental Environmental Impact Statement for Combined Licenses (COLs) for Vogtle Electric Generating Plant Units 3 and 4," for your review and comments. The NRC is reviewing the application submitted by Southern Nuclear Operating Company, Inc. (SNC) and several co-applicants for two combined licenses (COLs) to construct and operate two new nuclear units at the Vogtle Electric Generating Plant site in Burke County, GA. As part of its review of the proposed action, the NRC staff has prepared the draft supplemental environmental impact statement (DSEIS) to include an analysis of relevant environmental issues, including potential impacts to historic properties. The DSEIS documents the NRC determination regarding the environmental impacts at the proposed site from the construction and operation of two new nuclear units.

This DSEIS is a supplement to the Final Environmental Impact Statement (FEIS) for the early site permit (ESP) issued on August 26, 2009, to SNC and the same co-applicants. An ESP is a Commission approval of a site suitable for construction and operation of one or more new nuclear units. Under Title 10 of the *Code of Federal Regulations* (10 CFR) CFR 51.50(c), a COL applicant referencing an ESP need not submit information or analyses regarding environmental issues that were resolved in the ESP EIS, except to the extent the COL applicant has identified any new and potentially significant information. Accordingly, in preparing the DSEIS, the NRC staff considered whether new and significant information has been identified, including with respect to potential impacts to historic properties. The NRC staff conducted an environmental audit at the site and reviewed historic and archaeological records. The NRC staff also contacted Indian Tribes identified as having potential interest in the proposed action.

By letter dated December 10, 2009, the NRC staff notified you that it will comply with its obligations under Section 106 of the National Historic Preservation Act of 1966, as amended, (NHPA) using the process set forth in 36 CFR 800.8(c) in lieu of the procedures set forth in 36 CFR 800.3 through 36 CFR 800.6. Pursuant to 36 CFR 800.8(c), the NRC staff is using the preparation of the DSEIS required by the National Environmental Policy Act of 1969, as amended, (NEPA), to comply with its obligations under Section 106 of the NHPA.

E. Bucktrot   -2-

In the context of NEPA, under which the DSEIS was prepared, the NRC preliminary determination is that the impact of the two new proposed nuclear units on historical and archaeological resources remains moderate, as concluded in the ESP FEIS. In addition, SNC has entered into a Memorandum of Understanding with the Georgia State Historic Preservation Officer (SHPO). Under the provisions of the National Historic Preservation Act, the NRC preliminary determination is that, consistent with the determination in the ESP FEIS, the proposed project will affect, but not adversely affect, historic properties. Note that in Chapter 2 of the DSEIS you will find a discussion of the areas of potential effect, and impacts to historic properties from construction and operation are discussed in Chapters 4 and 5.

The NRC plans to hold a public meeting to go over the analysis and results in the DSEIS on October 7, 2010, at Augusta Technical College, Waynesboro Campus, 216 Highway 24 South, Waynesboro, Georgia 30830. The meeting will convene at 7:00 p.m. and will continue until 10 p.m., as necessary. In addition, the meeting will be preceded by an open house session from 6:00 p.m. to 7:00 p.m., during which members of the public may meet and talk with NRC staff members on an informal basis. You and your staff are invited to attend.

Pursuant to 10 CFR 51.92 and 36 CFR 800.2(c), the NRC wishes to ensure that Indian Tribes that might have an interest in any potential historic properties in the areas of potential effect are afforded the opportunity to identify their concerns, provide advice on the identification and evaluation of historic properties including those of traditional, religious, and cultural importance; and if necessary, participate in the resolution of any adverse effects to such properties.

In accordance with our December 10, 2009, letter, the NRC staff is forwarding the DSEIS for your review and comments. Pursuant to 36 CFR 800.8(c), we are requesting your comments on the DSEIS, specifically, on our preliminary conclusions regarding historic properties. Please provide any information or comments you may have on the DSEIS during the comment period, which ends on November 24, 2010. The NRC may consider additional comments after the comment period, to the extent practicable. Comments should be submitted either by mail to the Chief, Rules, Announcements, and Directives Branch, Division of Administrative Services, Office of Administration, Mailstop TWB-05-B01M, Washington, D.C. 20555-0001 or via e-mail to Vogtle.COLAEIS@nrc.gov. Your comments will be addressed in the final SEIS.

Appendix F

E. Bucktrot   - 3 -

If you have any questions or require additional information, please contact Ms. Mallecia Sutton, NRC Environmental Project Manager at (301) 415-0673 or via e-mail to Mallecia.Sutton@nrc.gov.

                                            Sincerely,

                                            /RA/

                                            Gregory P. Hatchett, Chief
                                            Environmental Projects Branch 1
                                            Division of Site and Environmental Reviews
                                            Office of New Reactors

Docket Nos. 52-025
                52-026
Enclosure:
As stated

cc: See next page

Appendix F

September 2, 2010

Mr. Steven Terry
Land Resource Manager
Miccosukee Tribe of Indians of Florida
Real Estate Services, Mile Marker 70
US 41 at Admin. Bldg.
Miami, FL 33194

SUBJECT: SECTION 106 CONSULTATION AND NOTIFICATION OF THE ISSUANCE AND REQUEST FOR COMMENTS ON THE DRAFT SUPPLEMENTAL ENVIRONMENTAL IMPACT STATEMENT FOR THE VOGTLE ELECTRIC GENERATING PLANT, UNITS 3 AND 4 COMBINED LICENSE APPLICATION

Dear Mr. Terry:

On behalf of the Nuclear Regulatory Commission (NRC) staff, I am forwarding a copy of the "Draft Supplemental Environmental Impact Statement for Combined Licenses (COLs) for Vogtle Electric Generating Plant Units 3 and 4," for your review and comments. The NRC is reviewing the application submitted by Southern Nuclear Operating Company, Inc. (SNC) and several co-applicants for two combined licenses (COLs) to construct and operate two new nuclear units at the Vogtle Electric Generating Plant site in Burke County, GA. As part of its review of the proposed action, the NRC staff has prepared the draft supplemental environmental impact statement (DSEIS) to include an analysis of relevant environmental issues, including potential impacts to historic properties. The DSEIS documents the NRC determination regarding the environmental impacts at the proposed site from the construction and operation of two new nuclear units.

This DSEIS is a supplement to the Final Environmental Impact Statement (FEIS) for the early site permit (ESP) issued on August 26, 2009, to SNC and the same co-applicants. An ESP is a Commission approval of a site suitable for construction and operation of one or more new nuclear units. Under Title 10 of the *Code of Federal Regulations* (10 CFR) CFR 51.50(c), a COL applicant referencing an ESP need not submit information or analyses regarding environmental issues that were resolved in the ESP EIS, except to the extent the COL applicant has identified any new and potentially significant information. Accordingly, in preparing the DSEIS, the NRC staff considered whether new and significant information has been identified, including with respect to potential impacts to historic properties. The NRC staff conducted an environmental audit at the site and reviewed historic and archaeological records. The NRC staff also contacted Indian Tribes identified as having potential interest in the proposed action.

By letter dated December 10, 2009, the NRC staff notified you that it will comply with its obligations under Section 106 of the National Historic Preservation Act of 1966, as amended, (NHPA) using the process set forth in 36 CFR 800.8(c) in lieu of the procedures set forth in 36 CFR 800.3 through 36 CFR 800.6. Pursuant to 36 CFR 800.8(c), the NRC staff is using the preparation of the DSEIS required by the National Environmental Policy Act of 1969, as amended, (NEPA), to comply with its obligations under Section 106 of the NHPA.

Appendix F

S. Terry   - 2 -

In the context of NEPA, under which the DSEIS was prepared, the NRC preliminary determination is that the impact of the two new proposed nuclear units on historical and archaeological resources remains moderate, as concluded in the ESP FEIS. In addition, SNC has entered into a Memorandum of Understanding with the Georgia State Historic Preservation Officer (SHPO). Under the provisions of the National Historic Preservation Act, the NRC preliminary determination is that, consistent with the determination in the ESP FEIS, the proposed project will affect, but not adversely affect, historic properties. Note that in Chapter 2 of the DSEIS you will find a discussion of the areas of potential effect, and impacts to historic properties from construction and operation are discussed in Chapters 4 and 5.

The NRC plans to hold a public meeting to go over the analysis and results in the DSEIS on October 7, 2010, at Augusta Technical College, Waynesboro Campus, 216 Highway 24 South, Waynesboro, Georgia 30830. The meeting will convene at 7:00 p.m. and will continue until 10 p.m., as necessary. In addition, the meeting will be preceded by an open house session from 6:00 p.m. to 7:00 p.m., during which members of the public may meet and talk with NRC staff members on an informal basis. You and your staff are invited to attend.

Pursuant to 10 CFR 51.92 and 36 CFR 800.2(c), the NRC wishes to ensure that Indian Tribes that might have an interest in any potential historic properties in the areas of potential effect are afforded the opportunity to identify their concerns, provide advice on the identification and evaluation of historic properties including those of traditional, religious, and cultural importance; and if necessary, participate in the resolution of any adverse effects to such properties.

In accordance with our December 10, 2009, letter, the NRC staff is forwarding the DSEIS for your review and comments. Pursuant to 36 CFR 800.8(c), we are requesting your comments on the DSEIS, specifically, on our preliminary conclusions regarding historic properties. Please provide any information or comments you may have on the DSEIS during the comment period, which ends on November 24, 2010. The NRC may consider additional comments after the comment period, to the extent practicable. Comments should be submitted either by mail to the Chief, Rules, Announcements, and Directives Branch, Division of Administrative Services, Office of Administration Mailstop TWB-05-B01M, Washington, D.C. 20555-0001 or via e-mail to Vogtle.COLAEIS@nrc.gov. Your comments will be addressed in the final SEIS.

S. Terry   - 3 -

If you have any questions or require additional information, please contact Ms. Mallecia Sutton, NRC Environmental Project Manager at (301) 415-0673 or via e-mail to Mallecia.Sutton@nrc.gov.

					Sincerely,

					/RA/

					Gregory P. Hatchett, Chief
					Environmental Projects Branch 1
					Division of Site and Environmental Reviews
					Office of New Reactors

Docket Nos. 52-025
            52-026

Enclosure:
As stated

cc: See next page

Appendix F

September 2, 2010

Ms. Gale Thrower, NAGPRA Contact
Poarch Band of Creek Indians
5811 Jack Springs Road
Atmore, AL 36502

SUBJECT: SECTION 106 CONSULTATION AND NOTIFICATION OF THE ISSUANCE AND
REQUEST FOR COMMENTS ON THE DRAFT SUPPLEMENTAL
ENVIRONMENTAL IMPACT STATEMENT FOR THE VOGTLE ELECTRIC
GENERATING PLANT, UNITS 3 AND 4 COMBINED LICENSE APPLICATION

Dear Ms. Thrower:

On behalf of the Nuclear Regulatory Commission (NRC) staff, I am forwarding a copy of the "Draft Supplemental Environmental Impact Statement for Combined Licenses (COLs) for Vogtle Electric Generating Plant Units 3 and 4," for your review and comments. The NRC is reviewing the application submitted by Southern Nuclear Operating Company, Inc. (SNC) and several co-applicants for two combined licenses (COLs) to construct and operate two new nuclear units at the Vogtle Electric Generating Plant site in Burke County, GA. As part of its review of the proposed action, the NRC staff has prepared the draft supplemental environmental impact statement (DSEIS) to include an analysis of relevant environmental issues, including potential impacts to historic properties. The DSEIS documents the NRC determination regarding the environmental impacts at the proposed site from the construction and operation of two new nuclear units.

This DSEIS is a supplement to the Final Environmental Impact Statement (FEIS) for the early site permit (ESP) issued on August 26, 2009, to SNC and the same co-applicants. An ESP is a Commission approval of a site suitable for construction and operation of one or more new nuclear units. Under Title 10 of the *Code of Federal Regulations* (10 CFR) CFR 51.50(c), a COL applicant referencing an ESP need not submit information or analyses regarding environmental issues that were resolved in the ESP EIS, except to the extent the COL applicant has identified any new and potentially significant information. Accordingly, in preparing the DSEIS, the NRC staff considered whether new and significant information has been identified, including with respect to potential impacts to historic properties. The NRC staff conducted an environmental audit at the site and reviewed historic and archaeological records. The NRC staff also contacted Indian Tribes identified as having potential interest in the proposed action.

By letter dated December 10, 2009, the NRC staff notified you that it will comply with its obligations under Section 106 of the National Historic Preservation Act of 1966, as amended, (NHPA) using the process set forth in 36 CFR 800.8(c) in lieu of the procedures set forth in 36 CFR 800.3 through 36 CFR 800.6. Pursuant to 36 CFR 800.8(c), the NRC staff is using the preparation of the DSEIS required by the National Environmental Policy Act of 1969, as amended, (NEPA), to comply with its obligations under Section 106 of the NHPA

G. Thrower   - 2 -

In the context of NEPA, under which the DSEIS was prepared, the NRC preliminary determination is that the impact of the two new proposed nuclear units on historical and archaeological resources remains moderate, as concluded in the ESP FEIS. In addition, SNC has entered into a Memorandum of Understanding with the Georgia State Historic Preservation Officer (SHPO). Under the provisions of the National Historic Preservation Act, the NRC preliminary determination is that, consistent with the determination in the ESP FEIS, the proposed project will affect, but not adversely affect, historic properties. Note that in Chapter 2 of the DSEIS you will find a discussion of the areas of potential effect, and impacts to historic properties from construction and operation are discussed in Chapters 4 and 5.

The NRC plans to hold a public meeting to go over the analysis and results in the DSEIS on October 7, 2010, at Augusta Technical College, Waynesboro Campus, 216 Highway 24 South, Waynesboro, Georgia 30830. The meeting will convene at 7:00 p.m. and will continue until 10 p.m., as necessary. In addition, the meeting will be preceded by an open house session from 6:00 p.m. to 7:00 p.m., during which members of the public may meet and talk with NRC staff members on an informal basis. You and your staff are invited to attend.

Pursuant to 10 CFR 51.92 and 36 CFR 800.2(c), the NRC wishes to ensure that Indian Tribes that might have an interest in any potential historic properties in the areas of potential effect are afforded the opportunity to identify their concerns, provide advice on the identification and evaluation of historic properties including those of traditional, religious, and cultural importance; and if necessary, participate in the resolution of any adverse effects to such properties.

In accordance with our December 10, 2009, letter, the NRC staff is forwarding the DSEIS for your review and comments. Pursuant to 36 CFR 800.8(c), we are requesting your comments on the DSEIS, specifically, on our preliminary conclusions regarding historic properties. Please provide any information or comments you may have on the DSEIS during the comment period, which ends on November 24, 2010. The NRC may consider additional comments after the comment period, to the extent practicable. Comments should be submitted either by mail to the Chief, Rules, Announcements, and Directives Branch, Division of Administrative Services, Office of Administration, Mailstop TWB-05-B01M, Washington, D.C. 20555-0001 or via e-mail to Vogtle.COLAEIS@nrc.gov. Your comments will be addressed in the final SEIS.

Appendix F

G. Thrower - 3 -

If you have any questions or require additional information, please contact Ms. Mallecia Sutton, NRC Environmental Project Manager at (301) 415-0673 or via e-mail to Mallecia.Sutton@nrc.gov.

Sincerely,

/RA/

Gregory P. Hatchett, Chief
Environmental Projects Branch 1
Division of Site and Environmental Reviews
Office of New Reactors

Docket Nos. 52-025
52-026
Enclosure:
As stated

cc: See next page

Appendix F

September 2, 2010

Mr. Louis McGertt, Town King
Thlopthlocco Tribal Town
P.O. Box 188
Okema, OK 74859

SUBJECT: SECTION 106 CONSULTATION AND NOTIFICATION OF THE ISSUANCE AND REQUEST FOR COMMENTS ON THE DRAFT SUPPLEMENTAL ENVIRONMENTAL IMPACT STATEMENT FOR THE VOGTLE ELECTRIC GENERATING PLANT, UNITS 3 AND 4 COMBINED LICENSE APPLICATION

Dear Mr. McGertt:

On behalf of the Nuclear Regulatory Commission (NRC) staff, I am forwarding a copy of the "Draft Supplemental Environmental Impact Statement for Combined Licenses (COLs) for Vogtle Electric Generating Plant Units 3 and 4," for your review and comments. The NRC is reviewing the application submitted by Southern Nuclear Operating Company, Inc. (SNC) and several co-applicants for two combined licenses (COLs) to construct and operate two new nuclear units at the Vogtle Electric Generating Plant site in Burke County, GA. As part of its review of the proposed action, the NRC staff has prepared the draft supplemental environmental impact statement (DSEIS) to include an analysis of relevant environmental issues, including potential impacts to historic properties. The DSEIS documents the NRC determination regarding the environmental impacts at the proposed site from the construction and operation of two new nuclear units.

This DSEIS is a supplement to the Final Environmental Impact Statement (FEIS) for the early site permit (ESP) issued on August 26, 2009, to SNC and the same co-applicants. An ESP is a Commission approval of a site suitable for construction and operation of one or more new nuclear units. Under Title 10 of the *Code of Federal Regulations* (10 CFR) CFR 51.50(c), a COL applicant referencing an ESP need not submit information or analyses regarding environmental issues that were resolved in the ESP EIS, except to the extent the COL applicant has identified any new and potentially significant information. Accordingly, in preparing the DSEIS, the NRC staff considered whether new and significant information has been identified, including with respect to potential impacts to historic properties. The NRC staff conducted an environmental audit at the site and reviewed historic and archaeological records. The NRC staff also contacted Indian Tribes identified as having potential interest in the proposed action.

By letter dated December 10, 2009, the NRC staff notified you that it will comply with its obligations under Section 106 of the National Historic Preservation Act of 1966, as amended, (NHPA) using the process set forth in 36 CFR 800.8(c) in lieu of the procedures set forth in 36 CFR 800.3 through 36 CFR 800.6. Pursuant to 36 CFR 800.8(c), the NRC staff is using the preparation of the DSEIS required by the National Environmental Policy Act of 1969, as amended, (NEPA), to comply with its obligations under Section 106 of the NHPA.

Appendix F

L. McGertt   - 2 -

In the context of NEPA, under which the DSEIS was prepared, the NRC preliminary determination is that the impact of the two new proposed nuclear units on historical and archaeological resources remains moderate, as concluded in the ESP FEIS. In addition, SNC has entered into a Memorandum of Understanding with the Georgia State Historic Preservation Officer (SHPO). Under the provisions of the National Historic Preservation Act, the NRC preliminary determination is that, consistent with the determination in the ESP FEIS, the proposed project will affect, but not adversely affect, historic properties. Note that in Chapter 2 of the DSEIS you will find a discussion of the areas of potential effect, and impacts to historic properties from construction and operation are discussed in Chapters 4 and 5.

The NRC plans to hold a public meeting to go over the analysis and results in the DSEIS on October 7, 2010, at Augusta Technical College, Waynesboro Campus, 216 Highway 24 South, Waynesboro, Georgia 30830. The meeting will convene at 7:00 p.m. and will continue until 10 p.m., as necessary. In addition, the meeting will be preceded by an open house session from 6:00 p.m. to 7:00 p.m., during which members of the public may meet and talk with NRC staff members on an informal basis. You and your staff are invited to attend.

Pursuant to 10 CFR 51.92 and 36 CFR 800.2(c), the NRC wishes to ensure that Indian Tribes that might have an interest in any potential historic properties in the areas of potential effect are afforded the opportunity to identify their concerns, provide advice on the identification and evaluation of historic properties including those of traditional, religious, and cultural importance; and if necessary, participate in the resolution of any adverse effects to such properties.

In accordance with our December 10, 2009, letter, the NRC staff is forwarding the DSEIS for your review and comments. Pursuant to 36 CFR 800.8(c), we are requesting your comments on the DSEIS, specifically, on our preliminary conclusions regarding historic properties. Please provide any information or comments you may have on the DSEIS during the comment period, which ends on November 24, 2010. The NRC may consider additional comments after the comment period, to the extent practicable. Comments should be submitted either by mail to the Chief, Rules, Announcements, and Directives Branch, Division of Administrative Services, Office of Administration Mailstop TWB-05-B01M, Washington, D.C. 20555-0001 or via e-mail to Vogtle.COLAEIS@nrc.gov. Your comments will be addressed in the final SEIS.

Appendix F

L. McGertt   - 3 -

If you have any questions or require additional information, please contact Ms. Mallecia Sutton, NRC Environmental Project Manager at (301) 415-0673 or via e-mail to Mallecia.Sutton@nrc.gov.

                                      Sincerely,

                                      /RA/

                                    Gregory P. Hatchett, Chief
                                    Environmental Projects Branch 1
                                    Division of Site and Environmental Reviews
                                    Office of New Reactors

Docket Nos. 52-025
              52-026

Enclosure:
As stated

cc: See next page

Appendix F

September 2, 2010

Mr. Richard L. Allen, NAGPRA Contact
Cherokee Nation of Oklahoma
P.O. Box 948
Tahleque, OK 74465-0948

SUBJECT: SECTION 106 CONSULTATION AND NOTIFICATION OF THE ISSUANCE AND REQUEST FOR COMMENTS ON THE DRAFT SUPPLEMENTAL ENVIRONMENTAL IMPACT STATEMENT FOR THE VOGTLE ELECTRIC GENERATING PLANT, UNITS 3 AND 4 COMBINED LICENSE APPLICATION

Dear Mr. Allen:

On behalf of the Nuclear Regulatory Commission (NRC) staff, I am forwarding a copy of the "Draft Supplemental Environmental Impact Statement for Combined Licenses (COLs) for Vogtle Electric Generating Plant Units 3 and 4," for your review and comments. The NRC is reviewing the application submitted by Southern Nuclear Operating Company, Inc. (SNC) and several co-applicants for two combined licenses (COLs) to construct and operate two new nuclear units at the Vogtle Electric Generating Plant site in Burke County, GA. As part of its review of the proposed action, the NRC staff has prepared the draft supplemental environmental impact statement (DSEIS) to include an analysis of relevant environmental issues, including potential impacts to historic properties. The DSEIS documents the NRC determination regarding the environmental impacts at the proposed site from the construction and operation of two new nuclear units.

This DSEIS is a supplement to the Final Environmental Impact Statement (FEIS) for the early site permit (ESP) issued on August 26, 2009, to SNC and the same co-applicants. An ESP is a Commission approval of a site suitable for construction and operation of one or more new nuclear units. Under Title 10 of the *Code of Federal Regulations* (10 CFR) CFR 51.50(c), a COL applicant referencing an ESP need not submit information or analyses regarding environmental issues that were resolved in the ESP EIS, except to the extent the COL applicant has identified any new and potentially significant information. Accordingly, in preparing the DSEIS, the NRC staff considered whether new and significant information has been identified, including with respect to potential impacts to historic properties. The NRC staff conducted an environmental audit at the site and reviewed historic and archaeological records. The NRC staff also contacted Indian Tribes identified as having potential interest in the proposed action.

By letter dated December 10, 2009, the NRC staff notified you that it will comply with its obligations under Section 106 of the National Historic Preservation Act of 1966, as amended, (NHPA) using the process set forth in 36 CFR 800.8(c) in lieu of the procedures set forth in 36 CFR 800.3 through 36 CFR 800.6. Pursuant to 36 CFR 800.8(c), the NRC staff is using the preparation of the DSEIS required by the National Environmental Policy Act of 1969, as amended, (NEPA), to comply with its obligations under Section 106 of the NHPA.

R. Allen   -2-

In the context of NEPA, under which the DSEIS was prepared, the NRC preliminary determination is that the impact of the two new proposed nuclear units on historical and archaeological resources remains moderate, as concluded in the ESP FEIS. In addition, SNC has entered into a Memorandum of Understanding with the Georgia State Historic Preservation Officer (SHPO). Under the provisions of the National Historic Preservation Act, the NRC preliminary determination is that, consistent with the determination in the ESP FEIS, the proposed project will affect, but not adversely affect, historic properties. Note that in Chapter 2 of the DSEIS you will find a discussion of the areas of potential effect, and impacts to historic properties from construction and operation are discussed in Chapters 4 and 5.

The NRC plans to hold a public meeting to go over the analysis and results in the DSEIS on October 7, 2010, at Augusta Technical College, Waynesboro Campus, 216 Highway 24 South, Waynesboro, Georgia 30830. The meeting will convene at 7:00 p.m. and will continue until 10 p.m., as necessary. In addition, the meeting will be preceded by an open house session from 6:00 p.m. to 7:00 p.m., during which members of the public may meet and talk with NRC staff members on an informal basis. You and your staff are invited to attend.

Pursuant to 10 CFR 51.92 and 36 CFR 800.2(c), the NRC wishes to ensure that Indian Tribes that might have an interest in any potential historic properties in the areas of potential effect are afforded the opportunity to identify their concerns, provide advice on the identification and evaluation of historic properties including those of traditional, religious, and cultural importance; and if necessary, participate in the resolution of any adverse effects to such properties.

In accordance with our December 10, 2009, letter, the NRC staff is forwarding the DSEIS for your review and comments. Pursuant to 36 CFR 800.8(c), we are requesting your comments on the DSEIS, specifically, on our preliminary conclusions regarding historic properties. Please provide any information or comments you may have on the DSEIS during the comment period, which ends on November 24, 2010. The NRC may consider additional comments after the comment period, to the extent practicable. Comments should be submitted either by mail to the Chief, Rules, Announcements, and Directives Branch, Division of Administrative Services, Office of Administration, Mailstop TWB-05-B01M, Washington, D.C. 20555-0001 or via e-mail to Vogtle.COLAEIS@nrc.gov. Your comments will be addressed in the final SEIS.

Appendix F

R. Allen   - 3 -

If you have any questions or require additional information, please contact Ms. Mallecia Sutton, NRC Environmental Project Manager at (301) 415-0673 or via e-mail to Mallecia.Sutton@nrc.gov.

                                              Sincerely,

                                              /RA/

                                          Gregory P. Hatchett, Chief
                                          Environmental Projects Branch 1
                                          Division of Site and Environmental Reviews
                                          Office of New Reactors

Docket Nos. 52-025
              52-026
Enclosure:
As stated

cc: See next page

Appendix F

September 2, 2010

Ms. Gingy (Virginia) Hail
NAGPRA Contact
Chickasaw Nation
P.O. Box 1548
Ada, OK 74883

SUBJECT: SECTION 106 CONSULTATION AND NOTIFICATION OF THE ISSUANCE AND REQUEST FOR COMMENTS ON THE DRAFT SUPPLEMENTAL ENVIRONMENTAL IMPACT STATEMENT FOR THE VOGTLE ELECTRIC GENERATING PLANT, UNITS 3 AND 4 COMBINED LICENSE APPLICATION

Dear Ms. Hail:

On behalf of the Nuclear Regulatory Commission (NRC) staff, I am forwarding a copy of the "Draft Supplemental Environmental Impact Statement for Combined Licenses (COLs) for Vogtle Electric Generating Plant Units 3 and 4," for your review and comments. The NRC is reviewing the application submitted by Southern Nuclear Operating Company, Inc. (SNC) and several co-applicants for two combined licenses (COLs) to construct and operate two new nuclear units at the Vogtle Electric Generating Plant site in Burke County, GA. As part of its review of the proposed action, the NRC staff has prepared the draft supplemental environmental impact statement (DSEIS) to include an analysis of relevant environmental issues, including potential impacts to historic properties. The DSEIS documents the NRC determination regarding the environmental impacts at the proposed site from the construction and operation of two new nuclear units.

This DSEIS is a supplement to the Final Environmental Impact Statement (FEIS) for the early site permit (ESP) issued on August 26, 2009, to SNC and the same co-applicants. An ESP is a Commission approval of a site suitable for construction and operation of one or more new nuclear units. Under Title 10 of the *Code of Federal Regulations* (10 CFR) CFR 51.50(c), a COL applicant referencing an ESP need not submit information or analyses regarding environmental issues that were resolved in the ESP EIS, except to the extent the COL applicant has identified any new and potentially significant information. Accordingly, in preparing the DSEIS, the NRC staff considered whether new and significant information has been identified, including with respect to potential impacts to historic properties. The NRC staff conducted an environmental audit at the site and reviewed historic and archaeological records. The NRC staff also contacted Indian Tribes identified as having potential interest in the proposed action.

By letter dated December 10, 2009, the NRC staff notified you that it will comply with its obligations under Section 106 of the National Historic Preservation Act of 1966, as amended, (NHPA) using the process set forth in 36 CFR 800.8(c) in lieu of the procedures set forth in 36 CFR 800.3 through 36 CFR 800.6. Pursuant to 36 CFR 800.8(c), the NRC staff is using the preparation of the DSEIS required by the National Environmental Policy Act of 1969, as amended, (NEPA), to comply with its obligations under Section 106 of the NHPA.

Appendix F

G. Hail   - 2 -

In the context of NEPA, under which the DSEIS was prepared, the NRC preliminary determination is that the impact of the two new proposed nuclear units on historical and archaeological resources remains moderate, as concluded in the ESP FEIS. In addition, SNC has entered into a Memorandum of Understanding with the Georgia State Historic Preservation Officer (SHPO). Under the provisions of the National Historic Preservation Act, the NRC preliminary determination is that, consistent with the determination in the ESP FEIS, the proposed project will affect, but not adversely affect, historic properties. Note that in Chapter 2 of the DSEIS you will find a discussion of the areas of potential effect, and impacts to historic properties from construction and operation are discussed in Chapters 4 and 5.

The NRC plans to hold a public meeting to go over the analysis and results in the DSEIS on October 7, 2010, at Augusta Technical College, Waynesboro Campus, 216 Highway 24 South, Waynesboro, Georgia 30830. The meeting will convene at 7:00 p.m. and will continue until 10 p.m., as necessary. In addition, the meeting will be preceded by an open house session from 6:00 p.m. to 7:00 p.m., during which members of the public may meet and talk with NRC staff members on an informal basis. You and your staff are invited to attend.

Pursuant to 10 CFR 51.92 and 36 CFR 800.2(c), the NRC wishes to ensure that Indian Tribes that might have an interest in any potential historic properties in the areas of potential effect are afforded the opportunity to identify their concerns, provide advice on the identification and evaluation of historic properties including those of traditional, religious, and cultural importance; and if necessary, participate in the resolution of any adverse effects to such properties.

In accordance with our December 10, 2009, letter, the NRC staff is forwarding the DSEIS for your review and comments. Pursuant to 36 CFR 800.8(c), we are requesting your comments on the DSEIS, specifically, on our preliminary conclusions regarding historic properties. Please provide any information or comments you may have on the DSEIS during the comment period, which ends on November 24, 2010. The NRC may consider additional comments after the comment period, to the extent practicable. Comments should be submitted either by mail to the Chief, Rules, Announcements, and Directives Branch, Division of Administrative Services, Office of Administration, Mailstop TWB-05-B01M, Washington, D.C. 20555-0001 or via e-mail to Vogtle.COLAEIS@nrc.gov. Your comments will be addressed in the final SEIS.

G. Hail   - 3 -

If you have any questions or require additional information, please contact Ms. Mallecia Sutton, NRC Environmental Project Manager at (301) 415-0673 or via e-mail to Mallecia.Sutton@nrc.gov.

                        Sincerely,

                        /RA/

                        Gregory P. Hatchett, Chief
                        Environmental Projects Branch 1
                        Division of Site and Environmental Reviews
                        Office of New Reactors

Docket Nos. 52-025
             52-026
Enclosure:
As stated

cc: See next page

Appendix F

September 2, 2010

Mr. Bill Anoatubby, Governor
Chickasaw Nation of Oklahoma
P.O. Box 1548
Ada, OK 74821-1548

SUBJECT: SECTION 106 CONSULTATION AND NOTIFICATION OF THE ISSUANCE AND REQUEST FOR COMMENTS ON THE DRAFT SUPPLEMENTAL ENVIRONMENTAL IMPACT STATEMENT FOR THE VOGTLE ELECTRIC GENERATING PLANT, UNITS 3 AND 4 COMBINED LICENSE APPLICATION

Dear Governor Anoatubby:

On behalf of the Nuclear Regulatory Commission (NRC) staff, I am forwarding a copy of the "Draft Supplemental Environmental Impact Statement for Combined Licenses (COLs) for Vogtle Electric Generating Plant Units 3 and 4," for your review and comments. The NRC is reviewing the application submitted by Southern Nuclear Operating Company, Inc. (SNC) and several co-applicants for two combined licenses (COLs) to construct and operate two new nuclear units at the Vogtle Electric Generating Plant site in Burke County, GA. As part of its review of the proposed action, the NRC staff has prepared the draft supplemental environmental impact statement (DSEIS) to include an analysis of relevant environmental issues, including potential impacts to historic properties. The DSEIS documents the NRC determination regarding the environmental impacts at the proposed site from the construction and operation of two new nuclear units.

This DSEIS is a supplement to the Final Environmental Impact Statement (FEIS) for the early site permit (ESP) issued on August 26, 2009, to SNC and the same co-applicants. An ESP is a Commission approval of a site suitable for construction and operation of one or more new nuclear units. Under Title 10 of the *Code of Federal Regulations* (10 CFR) CFR 51.50(c), a COL applicant referencing an ESP need not submit information or analyses regarding environmental issues that were resolved in the ESP EIS, except to the extent the COL applicant has identified any new and potentially significant information. Accordingly, in preparing the DSEIS, the NRC staff considered whether new and significant information has been identified, including with respect to potential impacts to historic properties. The NRC staff conducted an environmental audit at the site and reviewed historic and archaeological records. The NRC staff also contacted Indian Tribes identified as having potential interest in the proposed action.

By letter dated December 10, 2009, the NRC staff notified you that it will comply with its obligations under Section 106 of the National Historic Preservation Act of 1966, as amended, (NHPA) using the process set forth in 36 CFR 800.8(c) in lieu of the procedures set forth in 36 CFR 800.3 through 36 CFR 800.6. Pursuant to 36 CFR 800.8(c), the NRC staff is using the preparation of the DSEIS required by the National Environmental Policy Act of 1969, as amended, (NEPA), to comply with its obligations under Section 106 of the NHPA.

Governor Anoatubby   - 2 -

In the context of NEPA, under which the DSEIS was prepared, the NRC preliminary determination is that the impact of the two new proposed nuclear units on historical and archaeological resources remains moderate, as concluded in the ESP FEIS. In addition, SNC has entered into a Memorandum of Understanding with the Georgia State Historic Preservation Officer (SHPO). Under the provisions of the National Historic Preservation Act, the NRC preliminary determination is that, consistent with the determination in the ESP FEIS, the proposed project will affect, but not adversely affect, historic properties. Note that in Chapter 2 of the DSEIS you will find a discussion of the areas of potential effect, and impacts to historic properties from construction and operation are discussed in Chapters 4 and 5.

The NRC plans to hold a public meeting to go over the analysis and results in the DSEIS on October 7, 2010, at Augusta Technical College, Waynesboro Campus, 216 Highway 24 South, Waynesboro, Georgia 30830. The meeting will convene at 7:00 p.m. and will continue until 10 p.m., as necessary. In addition, the meeting will be preceded by an open house session from 6:00 p.m. to 7:00 p.m., during which members of the public may meet and talk with NRC staff members on an informal basis. You and your staff are invited to attend.

Pursuant to 10 CFR 51.92 and 36 CFR 800.2(c), the NRC wishes to ensure that Indian Tribes that might have an interest in any potential historic properties in the areas of potential effect are afforded the opportunity to identify their concerns, provide advice on the identification and evaluation of historic properties including those of traditional, religious, and cultural importance; and if necessary, participate in the resolution of any adverse effects to such properties.

In accordance with our December 10, 2009, letter, the NRC staff is forwarding the DSEIS for your review and comments. Pursuant to 36 CFR 800.8(c), we are requesting your comments on the DSEIS, specifically, on our preliminary conclusions regarding historic properties. Please provide any information or comments you may have on the DSEIS during the comment period, which ends on November 24, 2010. The NRC may consider additional comments after the comment period, to the extent practicable. Comments should be submitted either by mail to the Chief, Rules, Announcements, and Directives Branch, Division of Administrative Services, Office of Administration, Mailstop TWB-05-B01M, Washington, D.C. 20555-0001 or via e-mail to Vogtle.COLAEIS@nrc.gov. Your comments will be addressed in the final SEIS.

Appendix F

Governor Anoatubby   - 3 -

If you have any questions or require additional information, please contact Ms. Mallecia Sutton, NRC Environmental Project Manager at (301) 415-0673 or via e-mail to Mallecia.Sutton@nrc.gov.

              Sincerely,

              /RA/

              Gregory P. Hatchett, Chief
              Environmental Projects Branch 1
              Division of Site and Environmental Reviews
              Office of New Reactors

Docket Nos. 52-025
      52-026
Enclosure:
As stated

cc: See next page

Appendix F

September 2, 2010

Mr. Charles Thurmond, NAGPRA Contact
Georgia Tribe of Eastern Cherokee
P.O. Box 1324
Clayton, GA 30525

SUBJECT: SECTION 106 CONSULTATION AND NOTIFICATION OF THE ISSUANCE AND REQUEST FOR COMMENTS ON THE DRAFT SUPPLEMENTAL ENVIRONMENTAL IMPACT STATEMENT FOR THE VOGTLE ELECTRIC GENERATING PLANT, UNITS 3 AND 4 COMBINED LICENSE APPLICATION

Dear Mr. Thurmond:

On behalf of the Nuclear Regulatory Commission (NRC) staff, I am forwarding a copy of the "Draft Supplemental Environmental Impact Statement for Combined Licenses (COLs) for Vogtle Electric Generating Plant Units 3 and 4," for your review and comments. The NRC is reviewing the application submitted by Southern Nuclear Operating Company, Inc. (SNC) and several co-applicants for two combined licenses (COLs) to construct and operate two new nuclear units at the Vogtle Electric Generating Plant site in Burke County, GA. As part of its review of the proposed action, the NRC staff has prepared the draft supplemental environmental impact statement (DSEIS) to include an analysis of relevant environmental issues, including potential impacts to historic properties. The DSEIS documents the NRC determination regarding the environmental impacts at the proposed site from the construction and operation of two new nuclear units.

This DSEIS is a supplement to the Final Environmental Impact Statement (FEIS) for the early site permit (ESP) issued on August 26, 2009, to SNC and the same co-applicants. An ESP is a Commission approval of a site suitable for construction and operation of one or more new nuclear units. Under Title 10 of the *Code of Federal Regulations* (10 CFR) CFR 51.50(c), a COL applicant referencing an ESP need not submit information or analyses regarding environmental issues that were resolved in the ESP EIS, except to the extent the COL applicant has identified any new and potentially significant information. Accordingly, in preparing the DSEIS, the NRC staff considered whether new and significant information has been identified, including with respect to potential impacts to historic properties. The NRC staff conducted an environmental audit at the site and reviewed historic and archaeological records. The NRC staff also contacted Indian Tribes identified as having potential interest in the proposed action.

By letter dated December 10, 2009, the NRC staff notified you that it will comply with its obligations under Section 106 of the National Historic Preservation Act of 1966, as amended, (NHPA) using the process set forth in 36 CFR 800.8(c) in lieu of the procedures set forth in 36 CFR 800.3 through 36 CFR 800.6. Pursuant to 36 CFR 800.8(c), the NRC staff is using the preparation of the DSEIS required by the National Environmental Policy Act of 1969, as amended, (NEPA), to comply with its obligations under Section 106 of the NHPA.

Appendix F

C. Thurmond   - 2 -

In the context of NEPA, under which the DSEIS was prepared, the NRC preliminary determination is that the impact of the two new proposed nuclear units on historical and archaeological resources remains moderate, as concluded in the ESP FEIS. In addition, SNC has entered into a Memorandum of Understanding with the Georgia State Historic Preservation Officer (SHPO). Under the provisions of the National Historic Preservation Act, the NRC preliminary determination is that, consistent with the determination in the ESP FEIS, the proposed project will affect, but not adversely affect, historic properties. Note that in Chapter 2 of the DSEIS you will find a discussion of the areas of potential effect, and impacts to historic properties from construction and operation are discussed in Chapters 4 and 5.

The NRC plans to hold a public meeting to go over the analysis and results in the DSEIS on October 7, 2010, at Augusta Technical College, Waynesboro Campus, 216 Highway 24 South, Waynesboro, Georgia 30830. The meeting will convene at 7:00 p.m. and will continue until 10 p.m., as necessary. In addition, the meeting will be preceded by an open house session from 6:00 p.m. to 7:00 p.m., during which members of the public may meet and talk with NRC staff members on an informal basis. You and your staff are invited to attend.

Pursuant to 10 CFR 51.92 and 36 CFR 800.2(c), the NRC wishes to ensure that Indian Tribes that might have an interest in any potential historic properties in the areas of potential effect are afforded the opportunity to identify their concerns, provide advice on the identification and evaluation of historic properties including those of traditional, religious, and cultural importance; and if necessary, participate in the resolution of any adverse effects to such properties.

In accordance with our December 10, 2009, letter, the NRC staff is forwarding the DSEIS for your review and comments. Pursuant to 36 CFR 800.8(c), we are requesting your comments on the DSEIS, specifically, on our preliminary conclusions regarding historic properties. Please provide any information or comments you may have on the DSEIS during the comment period, which ends on November 24, 2010. The NRC may consider additional comments after the comment period, to the extent practicable. Comments should be submitted either by mail to the Chief, Rules, Announcements, and Directives Branch, Division of Administrative Services, Office of Administration, Mailstop TWB-05-B01M, Washington, D.C. 20555-0001 or via e-mail to Vogtle.COLAEIS@nrc.gov. Your comments will be addressed in the final SEIS.

C. Thurmond - 3 -

If you have any questions or require additional information, please contact Ms. Mallecia Sutton, NRC Environmental Project Manager at (301) 415-0673 or via e-mail to Mallecia.Sutton@nrc.gov.

                                  Sincerely,

                                  /RA/

                                Gregory P. Hatchett, Chief
                                Environmental Projects Branch 1
                                Division of Site and Environmental Reviews
                                Office of New Reactors

Docket Nos. 52-025
              52-026
Enclosure:
As stated

cc: See next page

Appendix F

September 2, 2010

Mr. Tarpie Yargee
Alabama-Quassarte Tribal Town
P.O. Box 187
Wetumka, OK 74883

SUBJECT: SECTION 106 CONSULTATION AND NOTIFICATION OF THE ISSUANCE AND REQUEST FOR COMMENTS ON THE DRAFT SUPPLEMENTAL ENVIRONMENTAL IMPACT STATEMENT FOR THE VOGTLE ELECTRIC GENERATING PLANT, UNITS 3 AND 4 COMBINED LICENSE APPLICATION

Dear Mr. Yargee:

On behalf of the Nuclear Regulatory Commission (NRC) staff, I am forwarding a copy of the "Draft Supplemental Environmental Impact Statement for Combined Licenses (COLs) for Vogtle Electric Generating Plant Units 3 and 4," for your review and comments. The NRC is reviewing the application submitted by Southern Nuclear Operating Company, Inc. (SNC) and several co-applicants for two combined licenses (COLs) to construct and operate two new nuclear units at the Vogtle Electric Generating Plant site in Burke County, GA. As part of its review of the proposed action, the NRC staff has prepared the draft supplemental environmental impact statement (DSEIS) to include an analysis of relevant environmental issues, including potential impacts to historic properties. The DSEIS documents the NRC determination regarding the environmental impacts at the proposed site from the construction and operation of two new nuclear units.

This DSEIS is a supplement to the Final Environmental Impact Statement (FEIS) for the early site permit (ESP) issued on August 26, 2009, to SNC and the same co-applicants. An ESP is a Commission approval of a site suitable for construction and operation of one or more new nuclear units. Under Title 10 of the *Code of Federal Regulations* (10 CFR) CFR 51.50(c), a COL applicant referencing an ESP need not submit information or analyses regarding environmental issues that were resolved in the ESP EIS, except to the extent the COL applicant has identified any new and potentially significant information. Accordingly, in preparing the DSEIS, the NRC staff considered whether new and significant information has been identified, including with respect to potential impacts to historic properties. The NRC staff conducted an environmental audit at the site and reviewed historic and archaeological records. The NRC staff also contacted Indian Tribes identified as having potential interest in the proposed action.

By letter dated December 10, 2009, the NRC staff notified you that it will comply with its obligations under Section 106 of the National Historic Preservation Act of 1966, as amended, (NHPA) using the process set forth in 36 CFR 800.8(c) in lieu of the procedures set forth in 36 CFR 800.3 through 36 CFR 800.6. Pursuant to 36 CFR 800.8(c), the NRC staff is using the preparation of the DSEIS required by the National Environmental Policy Act of 1969, as amended, (NEPA), to comply with its obligations under Section 106 of the NHPA.

T. Yargee   - 2 -

In the context of NEPA, under which the DSEIS was prepared, the NRC preliminary determination is that the impact of the two new proposed nuclear units on historical and archaeological resources remains moderate, as concluded in the ESP FEIS. In addition, SNC has entered into a Memorandum of Understanding with the Georgia State Historic Preservation Officer (SHPO). Under the provisions of the National Historic Preservation Act, the NRC preliminary determination is that, consistent with the determination in the ESP FEIS, the proposed project will affect, but not adversely affect, historic properties. Note that in Chapter 2 of the DSEIS you will find a discussion of the areas of potential effect, and impacts to historic properties from construction and operation are discussed in Chapters 4 and 5.

The NRC plans to hold a public meeting to go over the analysis and results in the DSEIS on October 7, 2010, at Augusta Technical College, Waynesboro Campus, 216 Highway 24 South, Waynesboro, Georgia 30830. The meeting will convene at 7:00 p.m. and will continue until 10 p.m., as necessary. In addition, the meeting will be preceded by an open house session from 6:00 p.m. to 7:00 p.m., during which members of the public may meet and talk with NRC staff members on an informal basis. You and your staff are invited to attend.

Pursuant to 10 CFR 51.92 and 36 CFR 800.2(c), the NRC wishes to ensure that Indian Tribes that might have an interest in any potential historic properties in the areas of potential effect are afforded the opportunity to identify their concerns, provide advice on the identification and evaluation of historic properties including those of traditional, religious, and cultural importance; and if necessary, participate in the resolution of any adverse effects to such properties.

In accordance with our December 10, 2009, letter, the NRC staff is forwarding the DSEIS for your review and comments. Pursuant to 36 CFR 800.8(c), we are requesting your comments on the DSEIS, specifically, on our preliminary conclusions regarding historic properties. Please provide any information or comments you may have on the DSEIS during the comment period, which ends on November 24, 2010. The NRC may consider additional comments after the comment period, to the extent practicable. Comments should be submitted either by mail to the Chief, Rules, Announcements, and Directives Branch, Division of Administrative Services, Office of Administration, Mailstop TWB-05-B01M, Washington, D.C. 20555-0001 or via e-mail to Vogtle.COLAEIS@nrc.gov. Your comments will be addressed in the final SEIS.

Appendix F

T. Yargee   - 3 -

If you have any questions or require additional information, please contact Ms. Mallecia Sutton, NRC Environmental Project Manager at (301) 415-0673 or via e-mail to Mallecia.Sutton@nrc.gov.

Sincerely,

/RA/

Gregory P. Hatchett, Chief
Environmental Projects Branch 1
Division of Site and Environmental Reviews
Office of New Reactors

Docket Nos. 52-025
            52-026
Enclosure:
As stated

cc: See next page

Appendix F

September 2, 2010

Mr. Pare Bowlegs
Seminole Nation of Oklahoma
P.O. Box 1498
Wewoka, OK 74884

SUBJECT: SECTION 106 CONSULTATION AND NOTIFICATION OF THE ISSUANCE AND REQUEST FOR COMMENTS ON THE DRAFT SUPPLEMENTAL ENVIRONMENTAL IMPACT STATEMENT FOR THE VOGTLE ELECTRIC GENERATING PLANT, UNITS 3 AND 4 COMBINED LICENSE APPLICATION

Dear Mr. Bowlegs:

On behalf of the Nuclear Regulatory Commission (NRC) staff, I am forwarding a copy of the "Draft Supplemental Environmental Impact Statement for Combined Licenses (COLs) for Vogtle Electric Generating Plant Units 3 and 4," for your review and comments. The NRC is reviewing the application submitted by Southern Nuclear Operating Company, Inc. (SNC) and several co-applicants for two combined licenses (COLs) to construct and operate two new nuclear units at the Vogtle Electric Generating Plant site in Burke County, GA. As part of its review of the proposed action, the NRC staff has prepared the draft supplemental environmental impact statement (DSEIS) to include an analysis of relevant environmental issues, including potential impacts to historic properties. The DSEIS documents the NRC determination regarding the environmental impacts at the proposed site from the construction and operation of two new nuclear units.

This DSEIS is a supplement to the Final Environmental Impact Statement (FEIS) for the early site permit (ESP) issued on August 26, 2009, to SNC and the same co-applicants. An ESP is a Commission approval of a site suitable for construction and operation of one or more new nuclear units. Under Title 10 of the *Code of Federal Regulations* (10 CFR) CFR 51.50(c), a COL applicant referencing an ESP need not submit information or analyses regarding environmental issues that were resolved in the ESP EIS, except to the extent the COL applicant has identified any new and potentially significant information. Accordingly, in preparing the DSEIS, the NRC staff considered whether new and significant information has been identified, including with respect to potential impacts to historic properties. The NRC staff conducted an environmental audit at the site and reviewed historic and archaeological records. The NRC staff also contacted Indian Tribes identified as having potential interest in the proposed action.

By letter dated December 10, 2009, the NRC staff notified you that it will comply with its obligations under Section 106 of the National Historic Preservation Act of 1966, as amended, (NHPA) using the process set forth in 36 CFR 800.8(c) in lieu of the procedures set forth in 36 CFR 800.3 through 36 CFR 800.6. Pursuant to 36 CFR 800.8(c), the NRC staff is using the preparation of the DSEIS required by the National Environmental Policy Act of 1969, as amended, (NEPA), to comply with its obligations under Section 106 of the NHPA.

Appendix F

P. Bowlegs   - 2 -

In the context of NEPA, under which the DSEIS was prepared, the NRC preliminary determination is that the impact of the two new proposed nuclear units on historical and archaeological resources remains moderate, as concluded in the ESP FEIS. In addition, SNC has entered into a Memorandum of Understanding with the Georgia State Historic Preservation Officer (SHPO). Under the provisions of the National Historic Preservation Act, the NRC preliminary determination is that, consistent with the determination in the ESP FEIS, the proposed project will affect, but not adversely affect, historic properties. Note that in Chapter 2 of the DSEIS you will find a discussion of the areas of potential effect, and impacts to historic properties from construction and operation are discussed in Chapters 4 and 5.

The NRC plans to hold a public meeting to go over the analysis and results in the DSEIS on October 7, 2010, at Augusta Technical College, Waynesboro Campus, 216 Highway 24 South, Waynesboro, Georgia 30830. The meeting will convene at 7:00 p.m. and will continue until 10 p.m., as necessary. In addition, the meeting will be preceded by an open house session from 6:00 p.m. to 7:00 p.m., during which members of the public may meet and talk with NRC staff members on an informal basis. You and your staff are invited to attend.

Pursuant to 10 CFR 51.92 and 36 CFR 800.2(c), the NRC wishes to ensure that Indian Tribes that might have an interest in any potential historic properties in the areas of potential effect are afforded the opportunity to identify their concerns, provide advice on the identification and evaluation of historic properties including those of traditional, religious, and cultural importance; and if necessary, participate in the resolution of any adverse effects to such properties.

In accordance with our December 10, 2009, letter, the NRC staff is forwarding the DSEIS for your review and comments. Pursuant to 36 CFR 800.8(c), we are requesting your comments on the DSEIS, specifically, on our preliminary conclusions regarding historic properties. Please provide any information or comments you may have on the DSEIS during the comment period, which ends on November 24, 2010. The NRC may consider additional comments after the comment period, to the extent practicable. Comments should be submitted either by mail to the Chief, Rules, Announcements, and Directives Branch, Division of Administrative Services, Office of Administration, Mailstop TWB-05-B01M, Washington, D.C. 20555-0001 or via e-mail to Vogtle.COLAEIS@nrc.gov. Your comments will be addressed in the final SEIS.

Appendix F

P. Bowlegs   - 3 -

If you have any questions or require additional information, please contact Ms. Mallecia Sutton, NRC Environmental Project Manager at (301) 415-0673 or via e-mail to Mallecia.Sutton@nrc.gov.

Sincerely,

/RA/

Gregory P. Hatchett, Chief
Environmental Projects Branch 1
Division of Site and Environmental Reviews
Office of New Reactors

Docket Nos. 52-025
            52-026

Enclosure:
As stated

cc: See next page

Appendix F

September 2, 2010

Mr. Michell Hicks, Principal Chief
Eastern Band of Cherokee Indians
P.O. Box 455
Qualla Boundary
Cherokee, NC 28719

SUBJECT: SECTION 106 CONSULTATION AND NOTIFICATION OF THE ISSUANCE AND REQUEST FOR COMMENTS ON THE DRAFT SUPPLEMENTAL ENVIRONMENTAL IMPACT STATEMENT FOR THE VOGTLE ELECTRIC GENERATING PLANT, UNITS 3 AND 4 COMBINED LICENSE APPLICATION

Dear Chief Hicks:

On behalf of the Nuclear Regulatory Commission (NRC) staff, I am forwarding a copy of the "Draft Supplemental Environmental Impact Statement for Combined Licenses (COLs) for Vogtle Electric Generating Plant Units 3 and 4," for your review and comments. The NRC is reviewing the application submitted by Southern Nuclear Operating Company, Inc. (SNC) and several co-applicants for two combined licenses (COLs) to construct and operate two new nuclear units at the Vogtle Electric Generating Plant site in Burke County, GA. As part of its review of the proposed action, the NRC staff has prepared the draft supplemental environmental impact statement (DSEIS) to include an analysis of relevant environmental issues, including potential impacts to historic properties. The DSEIS documents the NRC determination regarding the environmental impacts at the proposed site from the construction and operation of two new nuclear units.

This DSEIS is a supplement to the Final Environmental Impact Statement (FEIS) for the early site permit (ESP) issued on August 26, 2009, to SNC and the same co-applicants. An ESP is a Commission approval of a site suitable for construction and operation of one or more new nuclear units. Under Title 10 of the *Code of Federal Regulations* (10 CFR) CFR 51.50(c), a COL applicant referencing an ESP need not submit information or analyses regarding environmental issues that were resolved in the ESP EIS, except to the extent the COL applicant has identified any new and potentially significant information. Accordingly, in preparing the DSEIS, the NRC staff considered whether new and significant information has been identified, including with respect to potential impacts to historic properties. The NRC staff conducted an environmental audit at the site and reviewed historic and archaeological records. The NRC staff also contacted Indian Tribes identified as having potential interest in the proposed action.

By letter dated December 10, 2009, the NRC staff notified you that it will comply with its obligations under Section 106 of the National Historic Preservation Act of 1966, as amended, (NHPA) using the process set forth in 36 CFR 800.8(c) in lieu of the procedures set forth in 36 CFR 800.3 through 36 CFR 800.6. Pursuant to 36 CFR 800.8(c), the NRC staff is using the preparation of the DSEIS required by the National Environmental Policy Act of 1969, as amended, (NEPA), to comply with its obligations under Section 106 of the NHPA.

Appendix F

Chief Hicks   - 2 -

In the context of NEPA, under which the DSEIS was prepared, the NRC preliminary determination is that the impact of the two new proposed nuclear units on historical and archaeological resources remains moderate, as concluded in the ESP FEIS. In addition, SNC has entered into a Memorandum of Understanding with the Georgia State Historic Preservation Officer (SHPO). Under the provisions of the National Historic Preservation Act, the NRC preliminary determination is that, consistent with the determination in the ESP FEIS, the proposed project will affect, but not adversely affect, historic properties. Note that in Chapter 2 of the DSEIS you will find a discussion of the areas of potential effect, and impacts to historic properties from construction and operation are discussed in Chapters 4 and 5.

The NRC plans to hold a public meeting to go over the analysis and results in the DSEIS on October 7, 2010, at Augusta Technical College, Waynesboro Campus, 216 Highway 24 South, Waynesboro, Georgia 30830. The meeting will convene at 7:00 p.m. and will continue until 10 p.m., as necessary. In addition, the meeting will be preceded by an open house session from 6:00 p.m. to 7:00 p.m., during which members of the public may meet and talk with NRC staff members on an informal basis. You and your staff are invited to attend.

Pursuant to 10 CFR 51.92 and 36 CFR 800.2(c), the NRC wishes to ensure that Indian Tribes that might have an interest in any potential historic properties in the areas of potential effect are afforded the opportunity to identify their concerns, provide advice on the identification and evaluation of historic properties including those of traditional, religious, and cultural importance; and if necessary, participate in the resolution of any adverse effects to such properties.

In accordance with our December 10, 2009, letter, the NRC staff is forwarding the DSEIS for your review and comments. Pursuant to 36 CFR 800.8(c), we are requesting your comments on the DSEIS, specifically, on our preliminary conclusions regarding historic properties. Please provide any information or comments you may have on the DSEIS during the comment period, which ends on November 24, 2010. The NRC may consider additional comments after the comment period, to the extent practicable. Comments should be submitted either by mail to the Chief, Rules, Announcements, and Directives Branch, Division of Administrative Services, Office of Administration, Mailstop TWB-05-B01M, Washington, D.C. 20555-0001 or via e-mail to Vogtle.COLAEIS@nrc.gov. Your comments will be addressed in the final SEIS.

Appendix F

Chief Hicks   - 3 -

If you have any questions or require additional information, please contact Ms. Mallecia Sutton, NRC Environmental Project Manager at (301) 415-0673 or via e-mail to Mallecia.Sutton@nrc.gov.

          Sincerely,

          /RA/

          Gregory P. Hatchett, Chief
          Environmental Projects Branch 1
          Division of Site and Environmental Reviews
          Office of New Reactors

Docket Nos. 52-025
           52-026

Enclosure:
As stated

cc: See next page

Appendix F

September 2, 2010

Ms. Karen Kaniatobe
Director of the Cultural/Historical
 Preservation Department
Absentee-Shawnee Tribe of Oklahoma
2025 S. Gordon Cooper Drive
Shawnee, OK 74801

SUBJECT: SECTION 106 CONSULTATION AND NOTIFICATION OF THE ISSUANCE AND REQUEST FOR COMMENTS ON THE DRAFT SUPPLEMENTAL ENVIRONMENTAL IMPACT STATEMENT FOR THE VOGTLE ELECTRIC GENERATING PLANT, UNITS 3 AND 4 COMBINED LICENSE APPLICATION

Dear Ms. Kaniatobe:

On behalf of the Nuclear Regulatory Commission (NRC) staff, I am forwarding a copy of the "Draft Supplemental Environmental Impact Statement for Combined Licenses (COLs) for Vogtle Electric Generating Plant Units 3 and 4," for your review and comments. The NRC is reviewing the application submitted by Southern Nuclear Operating Company, Inc. (SNC) and several co-applicants for two combined licenses (COLs) to construct and operate two new nuclear units at the Vogtle Electric Generating Plant site in Burke County, GA. As part of its review of the proposed action, the NRC staff has prepared the draft supplemental environmental impact statement (DSEIS) to include an analysis of relevant environmental issues, including potential impacts to historic properties. The DSEIS documents the NRC determination regarding the environmental impacts at the proposed site from the construction and operation of two new nuclear units.

This DSEIS is a supplement to the Final Environmental Impact Statement (FEIS) for the early site permit (ESP) issued on August 26, 2009, to SNC and the same co-applicants. An ESP is a Commission approval of a site suitable for construction and operation of one or more new nuclear units. Under Title 10 of the *Code of Federal Regulations* (10 CFR) CFR 51.50(c), a COL applicant referencing an ESP need not submit information or analyses regarding environmental issues that were resolved in the ESP EIS, except to the extent the COL applicant has identified any new and potentially significant information. Accordingly, in preparing the DSEIS, the NRC staff considered whether new and significant information has been identified, including with respect to potential impacts to historic properties. The NRC staff conducted an environmental audit at the site and reviewed historic and archaeological records. The NRC staff also contacted Indian Tribes identified as having potential interest in the proposed action.

By letter dated December 10, 2009, the NRC staff notified you that it will comply with its obligations under Section 106 of the National Historic Preservation Act of 1966, as amended, (NHPA) using the process set forth in 36 CFR 800.8(c) in lieu of the procedures set forth in 36 CFR 800.3 through 36 CFR 800.6. Pursuant to 36 CFR 800.8(c), the NRC staff is using the preparation of the DSEIS required by the National Environmental Policy Act of 1969, as amended, (NEPA), to comply with its obligations under Section 106 of the NHPA.

Appendix F

K. Kaniatobe   - 2 -

In the context of NEPA, under which the DSEIS was prepared, the NRC preliminary determination is that the impact of the two new proposed nuclear units on historical and archaeological resources remains moderate, as concluded in the ESP FEIS. In addition, SNC has entered into a Memorandum of Understanding with the Georgia State Historic Preservation Officer (SHPO). Under the provisions of the National Historic Preservation Act, the NRC preliminary determination is that, consistent with the determination in the ESP FEIS, the proposed project will affect, but not adversely affect, historic properties. Note that in Chapter 2 of the DSEIS you will find a discussion of the areas of potential effect, and impacts to historic properties from construction and operation are discussed in Chapters 4 and 5.

The NRC plans to hold a public meeting to go over the analysis and results in the DSEIS on October 7, 2010, at Augusta Technical College, Waynesboro Campus, 216 Highway 24 South, Waynesboro, Georgia 30830. The meeting will convene at 7:00 p.m. and will continue until 10 p.m., as necessary. In addition, the meeting will be preceded by an open house session from 6:00 p.m. to 7:00 p.m., during which members of the public may meet and talk with NRC staff members on an informal basis. You and your staff are invited to attend.

Pursuant to 10 CFR 51.92 and 36 CFR 800.2(c), the NRC wishes to ensure that Indian Tribes that might have an interest in any potential historic properties in the areas of potential effect are afforded the opportunity to identify their concerns, provide advice on the identification and evaluation of historic properties including those of traditional, religious, and cultural importance; and if necessary, participate in the resolution of any adverse effects to such properties.

In accordance with our December 10, 2009, letter, the NRC staff is forwarding the DSEIS for your review and comments. Pursuant to 36 CFR 800.8(c), we are requesting your comments on the DSEIS, specifically, on our preliminary conclusions regarding historic properties. Please provide any information or comments you may have on the DSEIS during the comment period, which ends on November 24, 2010. The NRC may consider additional comments after the comment period, to the extent practicable. Comments should be submitted either by mail to the Chief, Rules, Announcements, and Directives Branch, Division of Administrative Services, Office of Administration Mailstop TWB-05-B01M, Washington, D.C. 20555-0001 or via e-mail to Vogtle.COLAEIS@nrc.gov. Your comments will be addressed in the final SEIS.

K. Kaniatobe   - 3 -

If you have any questions or require additional information, please contact Ms. Mallecia Sutton, NRC Environmental Project Manager at (301) 415-0673 or via e-mail to Mallecia.Sutton@nrc.gov.

                                         Sincerely,

                                         /RA/

                                         Gregory P. Hatchett, Chief
                                         Environmental Projects Branch 1
                                         Division of Site and Environmental Reviews
                                         Office of New Reactors

Docket Nos. 52-025
                 52-026

Enclosure:
As stated

cc: See next page

Appendix F

September 2, 2010

Ms. Debbie Thomas
Tribal Historic Preservation Officer
NAGPRA Coordinator
Alabama-Coushatta Tribe of Texas
571 State Park Road 56
Livingston, TX 77351

SUBJECT: SECTION 106 CONSULTATION AND NOTIFICATION OF THE ISSUANCE AND REQUEST FOR COMMENTS ON THE DRAFT SUPPLEMENTAL ENVIRONMENTAL IMPACT STATEMENT FOR THE VOGTLE ELECTRIC GENERATING PLANT, UNITS 3 AND 4 COMBINED LICENSE APPLICATION

Dear Ms. Thomas:

On behalf of the Nuclear Regulatory Commission (NRC) staff, I am forwarding a copy of the "Draft Supplemental Environmental Impact Statement for Combined Licenses (COLs) for Vogtle Electric Generating Plant Units 3 and 4," for your review and comments. The NRC is reviewing the application submitted by Southern Nuclear Operating Company, Inc. (SNC) and several co-applicants for two combined licenses (COLs) to construct and operate two new nuclear units at the Vogtle Electric Generating Plant site in Burke County, GA. As part of its review of the proposed action, the NRC staff has prepared the draft supplemental environmental impact statement (DSEIS) to include an analysis of relevant environmental issues, including potential impacts to historic properties. The DSEIS documents the NRC determination regarding the environmental impacts at the proposed site from the construction and operation of two new nuclear units.

This DSEIS is a supplement to the Final Environmental Impact Statement (FEIS) for the early site permit (ESP) issued on August 26, 2009, to SNC and the same co-applicants. An ESP is a Commission approval of a site suitable for construction and operation of one or more new nuclear units. Under Title 10 of the *Code of Federal Regulations* (10 CFR) CFR 51.50(c), a COL applicant referencing an ESP need not submit information or analyses regarding environmental issues that were resolved in the ESP EIS, except to the extent the COL applicant has identified any new and potentially significant information. Accordingly, in preparing the DSEIS, the NRC staff considered whether new and significant information has been identified, including with respect to potential impacts to historic properties. The NRC staff conducted an environmental audit at the site and reviewed historic and archaeological records. The NRC staff also contacted Indian Tribes identified as having potential interest in the proposed action.

By letter dated December 10, 2009, the NRC staff notified you that it will comply with its obligations under Section 106 of the National Historic Preservation Act of 1966, as amended, (NHPA) using the process set forth in 36 CFR 800.8(c) in lieu of the procedures set forth in 36 CFR 800.3 through 36 CFR 800.6. Pursuant to 36 CFR 800.8(c), the NRC staff is using the preparation of the DSEIS required by the National Environmental Policy Act of 1969, as amended, (NEPA), to comply with its obligations under Section 106 of the NHPA.

D. Thomas   - 2 -

In the context of NEPA, under which the DSEIS was prepared, the NRC preliminary determination is that the impact of the two new proposed nuclear units on historical and archaeological resources remains moderate, as concluded in the ESP FEIS. In addition, SNC has entered into a Memorandum of Understanding with the Georgia State Historic Preservation Officer (SHPO). Under the provisions of the National Historic Preservation Act, the NRC preliminary determination is that, consistent with the determination in the ESP FEIS, the proposed project will affect, but not adversely affect, historic properties. Note that in Chapter 2 of the DSEIS you will find a discussion of the areas of potential effect, and impacts to historic properties from construction and operation are discussed in Chapters 4 and 5.

The NRC plans to hold a public meeting to go over the analysis and results in the DSEIS on October 7, 2010, at Augusta Technical College, Waynesboro Campus, 216 Highway 24 South, Waynesboro, Georgia 30830. The meeting will convene at 7:00 p.m. and will continue until 10 p.m., as necessary. In addition, the meeting will be preceded by an open house session from 6:00 p.m. to 7:00 p.m., during which members of the public may meet and talk with NRC staff members on an informal basis. You and your staff are invited to attend.

Pursuant to 10 CFR 51.92 and 36 CFR 800.2(c), the NRC wishes to ensure that Indian Tribes that might have an interest in any potential historic properties in the areas of potential effect are afforded the opportunity to identify their concerns, provide advice on the identification and evaluation of historic properties including those of traditional, religious, and cultural importance; and if necessary, participate in the resolution of any adverse effects to such properties.

In accordance with our December 10, 2009, letter, the NRC staff is forwarding the DSEIS for your review and comments. Pursuant to 36 CFR 800.8(c), we are requesting your comments on the DSEIS, specifically, on our preliminary conclusions regarding historic properties. Please provide any information or comments you may have on the DSEIS during the comment period, which ends on November 24, 2010. The NRC may consider additional comments after the comment period, to the extent practicable. Comments should be submitted either by mail to the Chief, Rules, Announcements, and Directives Branch, Division of Administrative Services, Office of Administration Mailstop TWB-05-B01M, Washington, D.C. 20555-0001 or via e-mail to Vogtle.COLAEIS@nrc.gov. Your comments will be addressed in the final SEIS.

Appendix F

D. Thomas - 3 -

If you have any questions or require additional information, please contact Ms. Mallecia Sutton, NRC Environmental Project Manager at (301) 415-0673 or via e-mail to Mallecia.Sutton@nrc.gov.

                                        Sincerely,

                                        /RA/

                                      Gregory P. Hatchett, Chief
                                      Environmental Projects Branch 1
                                      Division of Site and Environmental Reviews
                                      Office of New Reactors

Docket Nos. 52-025
              52-026

Enclosure:
As stated

cc: See next page

Appendix F

September 2, 2010

Mrs. Joyce A. Bear, NAGPRA Contact
Muscogee (Creek) Nation of Oklahoma
P.O. Box 580
Okmulgee, OK   74447

SUBJECT: SECTION 106 CONSULTATION AND NOTIFICATION OF THE ISSUANCE AND REQUEST FOR COMMENTS ON THE DRAFT SUPPLEMENTAL ENVIRONMENTAL IMPACT STATEMENT FOR THE VOGTLE ELECTRIC GENERATING PLANT, UNITS 3 AND 4 COMBINED LICENSE APPLICATION

Dear Mrs. Bear:

On behalf of the Nuclear Regulatory Commission (NRC) staff, I am forwarding a copy of the "Draft Supplemental Environmental Impact Statement for Combined Licenses (COLs) for Vogtle Electric Generating Plant Units 3 and 4," for your review and comments. The NRC is reviewing the application submitted by Southern Nuclear Operating Company, Inc. (SNC) and several co-applicants for two combined licenses (COLs) to construct and operate two new nuclear units at the Vogtle Electric Generating Plant site in Burke County, GA. As part of its review of the proposed action, the NRC staff has prepared the draft supplemental environmental impact statement (DSEIS) to include an analysis of relevant environmental issues, including potential impacts to historic properties. The DSEIS documents the NRC determination regarding the environmental impacts at the proposed site from the construction and operation of two new nuclear units.

This DSEIS is a supplement to the Final Environmental Impact Statement (FEIS) for the early site permit (ESP) issued on August 26, 2009, to SNC and the same co-applicants. An ESP is a Commission approval of a site suitable for construction and operation of one or more new nuclear units. Under Title 10 of the *Code of Federal Regulations* (10 CFR) CFR 51.50(c), a COL applicant referencing an ESP need not submit information or analyses regarding environmental issues that were resolved in the ESP EIS, except to the extent the COL applicant has identified any new and potentially significant information. Accordingly, in preparing the DSEIS, the NRC staff considered whether new and significant information has been identified, including with respect to potential impacts to historic properties. The NRC staff conducted an environmental audit at the site and reviewed historic and archaeological records. The NRC staff also contacted Indian Tribes identified as having potential interest in the proposed action.

By letter dated December 10, 2009, the NRC staff notified you that it will comply with its obligations under Section 106 of the National Historic Preservation Act of 1966, as amended, (NHPA) using the process set forth in 36 CFR 800.8(c) in lieu of the procedures set forth in 36 CFR 800.3 through 36 CFR 800.6. Pursuant to 36 CFR 800.8(c), the NRC staff is using the preparation of the DSEIS required by the National Environmental Policy Act of 1969, as amended, (NEPA), to comply with its obligations under Section 106 of the NHPA.

Appendix F

J. Bear   - 2 -

In the context of NEPA, under which the DSEIS was prepared, the NRC preliminary determination is that the impact of the two new proposed nuclear units on historical and archaeological resources remains moderate, as concluded in the ESP FEIS. In addition, SNC has entered into a Memorandum of Understanding with the Georgia State Historic Preservation Officer (SHPO). Under the provisions of the National Historic Preservation Act, the NRC preliminary determination is that, consistent with the determination in the ESP FEIS, the proposed project will affect, but not adversely affect, historic properties. Note that in Chapter 2 of the DSEIS you will find a discussion of the areas of potential effect, and impacts to historic properties from construction and operation are discussed in Chapters 4 and 5.

The NRC plans to hold a public meeting to go over the analysis and results in the DSEIS on October 7, 2010, at Augusta Technical College, Waynesboro Campus, 216 Highway 24 South, Waynesboro, Georgia 30830. The meeting will convene at 7:00 p.m. and will continue until 10 p.m., as necessary. In addition, the meeting will be preceded by an open house session from 6:00 p.m. to 7:00 p.m., during which members of the public may meet and talk with NRC staff members on an informal basis. You and your staff are invited to attend.

Pursuant to 10 CFR 51.92 and 36 CFR 800.2(c), the NRC wishes to ensure that Indian Tribes that might have an interest in any potential historic properties in the areas of potential effect are afforded the opportunity to identify their concerns, provide advice on the identification and evaluation of historic properties including those of traditional, religious, and cultural importance; and if necessary, participate in the resolution of any adverse effects to such properties.

In accordance with our December 10, 2009, letter, the NRC staff is forwarding the DSEIS for your review and comments. Pursuant to 36 CFR 800.8(c), we are requesting your comments on the DSEIS, specifically, on our preliminary conclusions regarding historic properties. Please provide any information or comments you may have on the DSEIS during the comment period, which ends on November 24, 2010. The NRC may consider additional comments after the comment period, to the extent practicable. Comments should be submitted either by mail to the Chief, Rules, Announcements, and Directives Branch, Division of Administrative Services, Office of Administration, Mailstop TWB-05-B01M, Washington, D.C. 20555-0001 or via e-mail to Vogtle.COLAEIS@nrc.gov. Your comments will be addressed in the final SEIS.

Appendix F

J. Bear   - 3 -

If you have any questions or require additional information, please contact Ms. Mallecia Sutton, NRC Environmental Project Manager at (301) 415-0673 or via e-mail to Mallecia.Sutton@nrc.gov.

Sincerely,

/RA/

Gregory P. Hatchett, Chief
Environmental Projects Branch 1
Division of Site and Environmental Reviews
Office of New Reactors

Docket Nos. 52-025
             52-026
Enclosure:
As stated

cc: See next page

Appendix F

September 2, 2010

Mr. Chadwick Smith, Principal Chief
Cherokee Nation of Oklahoma
P.O. Box 948
Tahlequa, OK 74465

SUBJECT: SECTION 106 CONSULTATION AND NOTIFICATION OF THE ISSUANCE AND REQUEST FOR COMMENTS ON THE DRAFT SUPPLEMENTAL ENVIRONMENTAL IMPACT STATEMENT FOR THE VOGTLE ELECTRIC GENERATING PLANT, UNITS 3 AND 4 COMBINED LICENSE APPLICATION

Dear Chief Smith:

On behalf of the Nuclear Regulatory Commission (NRC) staff, I am forwarding a copy of the "Draft Supplemental Environmental Impact Statement for Combined Licenses (COLs) for Vogtle Electric Generating Plant Units 3 and 4," for your review and comments. The NRC is reviewing the application submitted by Southern Nuclear Operating Company, Inc. (SNC) and several co-applicants for two combined licenses (COLs) to construct and operate two new nuclear units at the Vogtle Electric Generating Plant site in Burke County, GA. As part of its review of the proposed action, the NRC staff has prepared the draft supplemental environmental impact statement (DSEIS) to include an analysis of relevant environmental issues, including potential impacts to historic properties. The DSEIS documents the NRC determination regarding the environmental impacts at the proposed site from the construction and operation of two new nuclear units.

This DSEIS is a supplement to the Final Environmental Impact Statement (FEIS) for the early site permit (ESP) issued on August 26, 2009, to SNC and the same co-applicants. An ESP is a Commission approval of a site suitable for construction and operation of one or more new nuclear units. Under Title 10 of the *Code of Federal Regulations* (10 CFR) CFR 51.50(c), a COL applicant referencing an ESP need not submit information or analyses regarding environmental issues that were resolved in the ESP EIS, except to the extent the COL applicant has identified any new and potentially significant information. Accordingly, in preparing the DSEIS, the NRC staff considered whether new and significant information has been identified, including with respect to potential impacts to historic properties. The NRC staff conducted an environmental audit at the site and reviewed historic and archaeological records. The NRC staff also contacted Indian Tribes identified as having potential interest in the proposed action.

By letter dated December 10, 2009, the NRC staff notified you that it will comply with its obligations under Section 106 of the National Historic Preservation Act of 1966, as amended, (NHPA) using the process set forth in 36 CFR 800.8(c) in lieu of the procedures set forth in 36 CFR 800.3 through 36 CFR 800.6. Pursuant to 36 CFR 800.8(c), the NRC staff is using the preparation of the DSEIS required by the National Environmental Policy Act of 1969, as amended, (NEPA), to comply with its obligations under Section 106 of the NHPA.

Appendix F

Chief Smith   - 2 -

In the context of NEPA, under which the DSEIS was prepared, the NRC preliminary determination is that the impact of the two new proposed nuclear units on historical and archaeological resources remains moderate, as concluded in the ESP FEIS. In addition, SNC has entered into a Memorandum of Understanding with the Georgia State Historic Preservation Officer (SHPO). Under the provisions of the National Historic Preservation Act, the NRC preliminary determination is that, consistent with the determination in the ESP FEIS, the proposed project will affect, but not adversely affect, historic properties. Note that in Chapter 2 of the DSEIS you will find a discussion of the areas of potential effect, and impacts to historic properties from construction and operation are discussed in Chapters 4 and 5.

The NRC plans to hold a public meeting to go over the analysis and results in the DSEIS on October 7, 2010, at Augusta Technical College, Waynesboro Campus, 216 Highway 24 South, Waynesboro, Georgia 30830. The meeting will convene at 7:00 p.m. and will continue until 10 p.m., as necessary. In addition, the meeting will be preceded by an open house session from 6:00 p.m. to 7:00 p.m., during which members of the public may meet and talk with NRC staff members on an informal basis. You and your staff are invited to attend.

Pursuant to 10 CFR 51.92 and 36 CFR 800.2(c), the NRC wishes to ensure that Indian Tribes that might have an interest in any potential historic properties in the areas of potential effect are afforded the opportunity to identify their concerns, provide advice on the identification and evaluation of historic properties including those of traditional, religious, and cultural importance; and if necessary, participate in the resolution of any adverse effects to such properties.

In accordance with our December 10, 2009, letter, the NRC staff is forwarding the DSEIS for your review and comments. Pursuant to 36 CFR 800.8(c), we are requesting your comments on the DSEIS, specifically, on our preliminary conclusions regarding historic properties. Please provide an information or comments you may have on the DSEIS during the comment period, which ends on November 24, 2010. The NRC may consider additional comments after the comment period, to the extent practicable. Comments should be submitted either by mail to the Chief, Rules, Announcements, and Directives Branch, Division of Administrative Services, Office of Administration Mailstop TWB-05-B01M, Washington, D.C. 20555-0001 or via e-mail to Vogtle.COLAEIS@nrc.gov. Your comments will be addressed in the final SEIS.

Appendix F

Chief Smith   - 3 -

If you have any questions or require additional information, please contact Ms. Mallecia Sutton, NRC Environmental Project Manager at (301) 415-0673 or via e-mail to Mallecia.Sutton@nrc.gov.

                                      Sincerely,

                                      /RA/

                                    Gregory P. Hatchett, Chief
                                    Environmental Projects Branch 1
                                    Division of Site and Environmental Reviews
                                    Office of New Reactors

Docket Nos. 52-025
              52-026

Enclosure:
As stated

cc: See next page

Appendix F

September 2, 2010

Mr. Willard Steele, Deputy THPO
Seminole Tribe of Florida
Ah-Tah-Thi-Ki Museum
HC 61, Box 21A
Clewiston, FL 33440

SUBJECT: SECTION 106 CONSULTATION AND NOTIFICATION OF THE ISSUANCE AND REQUEST FOR COMMENTS ON THE DRAFT SUPPLEMENTAL ENVIRONMENTAL IMPACT STATEMENT FOR THE VOGTLE ELECTRIC GENERATING PLANT, UNITS 3 AND 4 COMBINED LICENSE APPLICATION

Dear Steele:

On behalf of the Nuclear Regulatory Commission (NRC) staff, I am forwarding a copy of the "Draft Supplemental Environmental Impact Statement for Combined Licenses (COLs) for Vogtle Electric Generating Plant Units 3 and 4," for your review and comments. The NRC is reviewing the application submitted by Southern Nuclear Operating Company, Inc. (SNC) and several co-applicants for two combined licenses (COLs) to construct and operate two new nuclear units at the Vogtle Electric Generating Plant site in Burke County, GA. As part of its review of the proposed action, the NRC staff has prepared the draft supplemental environmental impact statement (DSEIS) to include an analysis of relevant environmental issues, including potential impacts to historic properties. The DSEIS documents the NRC determination regarding the environmental impacts at the proposed site from the construction and operation of two new nuclear units.

This DSEIS is a supplement to the Final Environmental Impact Statement (FEIS) for the early site permit (ESP) issued on August 26, 2009, to SNC and the same co-applicants. An ESP is a Commission approval of a site suitable for construction and operation of one or more new nuclear units. Under Title 10 of the *Code of Federal Regulations* (10 CFR) CFR 51.50(c), a COL applicant referencing an ESP need not submit information or analyses regarding environmental issues that were resolved in the ESP EIS, except to the extent the COL applicant has identified any new and potentially significant information. Accordingly, in preparing the DSEIS, the NRC staff considered whether new and significant information has been identified, including with respect to potential impacts to historic properties. The NRC staff conducted an environmental audit at the site and reviewed historic and archaeological records. The NRC staff also contacted Indian Tribes identified as having potential interest in the proposed action.

By letter dated December 10, 2009, the NRC staff notified you that it will comply with its obligations under Section 106 of the National Historic Preservation Act of 1966, as amended, (NHPA) using the process set forth in 36 CFR 800.8(c) in lieu of the procedures set forth in 36 CFR 800.3 through 36 CFR 800.6. Pursuant to 36 CFR 800.8(c), the NRC staff is using the preparation of the DSEIS required by the National Environmental Policy Act of 1969, as amended, (NEPA), to comply with its obligations under Section 106 of the NHPA.

Appendix F

W. Steele   - 2 -

In the context of NEPA, under which the DSEIS was prepared, the NRC preliminary determination is that the impact of the two new proposed nuclear units on historical and archaeological resources remains moderate, as concluded in the ESP FEIS. In addition, SNC has entered into a Memorandum of Understanding with the Georgia State Historic Preservation Officer (SHPO). Under the provisions of the National Historic Preservation Act, the NRC preliminary determination is that, consistent with the determination in the ESP FEIS, the proposed project will affect, but not adversely affect, historic properties. Note that in Chapter 2 of the DSEIS you will find a discussion of the areas of potential effect, and impacts to historic properties from construction and operation are discussed in Chapters 4 and 5.

The NRC plans to hold a public meeting to go over the analysis and results in the DSEIS on October 7, 2010, at Augusta Technical College, Waynesboro Campus, 216 Highway 24 South, Waynesboro, Georgia 30830. The meeting will convene at 7:00 p.m. and will continue until 10 p.m., as necessary. In addition, the meeting will be preceded by an open house session from 6:00 p.m. to 7:00 p.m., during which members of the public may meet and talk with NRC staff members on an informal basis. You and your staff are invited to attend.

Pursuant to 10 CFR 51.92 and 36 CFR 800.2(c), the NRC wishes to ensure that Indian Tribes that might have an interest in any potential historic properties in the areas of potential effect are afforded the opportunity to identify their concerns, provide advice on the identification and evaluation of historic properties including those of traditional, religious, and cultural importance; and if necessary, participate in the resolution of any adverse effects to such properties.

In accordance with our December 10, 2009, letter, the NRC staff is forwarding the DSEIS for your review and comments. Pursuant to 36 CFR 800.8(c), we are requesting your comments on the DSEIS, specifically, on our preliminary conclusions regarding historic properties. Please provide any information or comments you may have on the DSEIS during the comment period, which ends on November 24, 2010. The NRC may consider additional comments after the comment period, to the extent practicable. Comments should be submitted either by mail to the Chief, Rules, Announcements, and Directives Branch, Division of Administrative Services, Office of Administration, Mailstop TWB-05-B01M, Washington, D.C. 20555-0001 or via e-mail to Vogtle.COLAEIS@nrc.gov. Your comments will be addressed in the final SEIS.

Appendix F

W. Steele - 3 -

If you have any questions or require additional information, please contact Ms. Mallecia Sutton, NRC Environmental Project Manager at (301) 415-0673 or via e-mail to Mallecia.Sutton@nrc.gov.

Sincerely,

/RA/

Gregory P. Hatchett, Chief
Environmental Projects Branch 1
Division of Site and Environmental Reviews
Office of New Reactors

Docket Nos. 52-025
  52-026
Enclosure:
As stated

cc: See next page

Appendix F

September 02, 2010

Mr. Kenneth H. Carleton
THPO/Tribal Archaeologist
Mississippi Band of Choctaw Indians
P.O. Box 6257/ 101 Industrial Road
Choctaw, MS   39350

SUBJECT: SECTION 106 CONSULTATION AND NOTIFICATION OF THE ISSUANCE AND REQUEST FOR COMMENTS ON THE DRAFT SUPPLEMENTAL ENVIRONMENTAL IMPACT STATEMENT FOR THE VOGTLE ELECTRIC GENERATING PLANT, UNITS 3 AND 4 COMBINED LICENSE APPLICATION

Dear Mr. Carleton:

On behalf of the Nuclear Regulatory Commission (NRC) staff, I am forwarding a copy of the "Draft Supplemental Environmental Impact Statement for Combined Licenses (COLs) for Vogtle Electric Generating Plant Units 3 and 4," for your review and comments. The NRC is reviewing the application submitted by Southern Nuclear Operating Company, Inc. (SNC) and several co-applicants for two combined licenses (COLs) to construct and operate two new nuclear units at the Vogtle Electric Generating Plant site in Burke County, GA. As part of its review of the proposed action, the NRC staff has prepared the draft supplemental environmental impact statement (DSEIS) to include an analysis of relevant environmental issues, including potential impacts to historic properties. The DSEIS documents the NRC determination regarding the environmental impacts at the proposed site from the construction and operation of two new nuclear units.

This DSEIS is a supplement to the Final Environmental Impact Statement (FEIS) for the early site permit (ESP) issued on August 26, 2009, to SNC and the same co-applicants. An ESP is a Commission approval of a site suitable for construction and operation of one or more new nuclear units. Under Title 10 of the *Code of Federal Regulations* (10 CFR) CFR 51.50(c), a COL applicant referencing an ESP need not submit information or analyses regarding environmental issues that were resolved in the ESP EIS, except to the extent the COL applicant has identified any new and potentially significant information. Accordingly, in preparing the DSEIS, the NRC staff considered whether new and significant information has been identified, including with respect to potential impacts to historic properties. The NRC staff conducted an environmental audit at the site and reviewed historic and archaeological records. The NRC staff also contacted Indian Tribes identified as having potential interest in the proposed action.

By letter dated December 10, 2009, the NRC staff notified you that it will comply with its obligations under Section 106 of the National Historic Preservation Act of 1966, as amended, (NHPA) using the process set forth in 36 CFR 800.8(c) in lieu of the procedures set forth in 36 CFR 800.3 through 36 CFR 800.6. Pursuant to 36 CFR 800.8(c), the NRC staff is using the preparation of the DSEIS required by the National Environmental Policy Act of 1969, as amended, (NEPA), to comply with its obligations under Section 106 of the NHPA.

K. H. Carleton - 2 -

In the context of NEPA, under which the DSEIS was prepared, the NRC preliminary determination is that the impact of the two new proposed nuclear units on historical and archaeological resources remains moderate, as concluded in the ESP FEIS. In addition, SNC has entered into a Memorandum of Understanding with the Georgia State Historic Preservation Officer (SHPO). Under the provisions of the National Historic Preservation Act, the NRC preliminary determination is that, consistent with the determination in the ESP FEIS, the proposed project will affect, but not adversely affect, historic properties. Note that in Chapter 2 of the DSEIS you will find a discussion of the areas of potential effect, and impacts to historic properties from construction and operation are discussed in Chapters 4 and 5.

The NRC plans to hold a public meeting to go over the analysis and results in the DSEIS on October 7, 2010, at Augusta Technical College, Waynesboro Campus, 216 Highway 24 South, Waynesboro, Georgia 30830. The meeting will convene at 7:00 p.m. and will continue until 10 p.m., as necessary. In addition, the meeting will be preceded by an open house session from 6:00 p.m. to 7:00 p.m., during which members of the public may meet and talk with NRC staff members on an informal basis. You and your staff are invited to attend.

Pursuant to 10 CFR 51.92 and 36 CFR 800.2(c), the NRC wishes to ensure that Indian Tribes that might have an interest in any potential historic properties in the areas of potential effect are afforded the opportunity to identify their concerns, provide advice on the identification and evaluation of historic properties including those of traditional, religious, and cultural importance; and if necessary, participate in the resolution of any adverse effects to such properties.

In accordance with our December 10, 2009, letter, the NRC staff is forwarding the DSEIS for your review and comments. Pursuant to 36 CFR 800.8(c), we are requesting your comments on the DSEIS, specifically, on our preliminary conclusions regarding historic properties. Please provide any information or comments you may have on the DSEIS during the comment period, which ends on November 24, 2010. The NRC may consider additional comments after the comment period, to the extent practicable. Comments should be submitted either by mail to the Chief, Rules, Announcements, and Directives Branch, Division of Administrative Services, Office of Administration Mailstop TWB-05-B01M, Washington, D.C. 20555-0001 or via e-mail to Vogtle.COLAEIS@nrc.gov. Your comments will be addressed in the final SEIS.

Appendix F

K. H. Carleton   - 3 -

If you have any questions or require additional information, please contact Ms. Mallecia Sutton, NRC Environmental Project Manager at (301) 415-0673 or via e-mail to Mallecia.Sutton@nrc.gov.

                                        Sincerely,

                                        /RA/

                                        Gregory P. Hatchett, Chief
                                        Environmental Projects Branch 1
                                        Division of Site and Environmental Reviews
                                        Office of New Reactors

Docket Nos. 52-025
             52-026

Enclosure:
As stated

cc: See next page

Appendix F

September 2, 2010

Ms. Stephanie Rolin
NAGRA Contact
Poarch Band of Creek Indians
5811 Jack Springs Road
Atmore, AL 36502

SUBJECT: SECTION 106 CONSULTATION AND NOTIFICATION OF THE ISSUANCE AND REQUEST FOR COMMENTS ON THE DRAFT SUPPLEMENTAL ENVIRONMENTAL IMPACT STATEMENT FOR THE VOGTLE ELECTRIC GENERATING PLANT, UNITS 3 AND 4 COMBINED LICENSE APPLICATION

Dear Ms. Rolin:

On behalf of the Nuclear Regulatory Commission (NRC) staff, I am forwarding a copy of the "Draft Supplemental Environmental Impact Statement for Combined Licenses (COLs) for Vogtle Electric Generating Plant Units 3 and 4," for your review and comments. The NRC is reviewing the application submitted by Southern Nuclear Operating Company, Inc. (SNC) and several co-applicants for two combined licenses (COLs) to construct and operate two new nuclear units at the Vogtle Electric Generating Plant site in Burke County, GA. As part of its review of the proposed action, the NRC staff has prepared the draft supplemental environmental impact statement (DSEIS) to include an analysis of relevant environmental issues, including potential impacts to historic properties. The DSEIS documents the NRC determination regarding the environmental impacts at the proposed site from the construction and operation of two new nuclear units.

This DSEIS is a supplement to the Final Environmental Impact Statement (FEIS) for the early site permit (ESP) issued on August 26, 2009, to SNC and the same co-applicants. An ESP is a Commission approval of a site suitable for construction and operation of one or more new nuclear units. Under Title 10 of the *Code of Federal Regulations* (10 CFR) CFR 51.50(c), a COL applicant referencing an ESP need not submit information or analyses regarding environmental issues that were resolved in the ESP EIS, except to the extent the COL applicant has identified any new and potentially significant information. Accordingly, in preparing the DSEIS, the NRC staff considered whether new and significant information has been identified, including with respect to potential impacts to historic properties. The NRC staff conducted an environmental audit at the site and reviewed historic and archaeological records. The NRC staff also contacted Indian Tribes identified as having potential interest in the proposed action.

By letter dated December 10, 2009, the NRC staff notified you that it will comply with its obligations under Section 106 of the National Historic Preservation Act of 1966, as amended, (NHPA) using the process set forth in 36 CFR 800.8(c) in lieu of the procedures set forth in 36 CFR 800.3 through 36 CFR 800.6. Pursuant to 36 CFR 800.8(c), the NRC staff is using the preparation of the DSEIS required by the National Environmental Policy Act of 1969, as amended, (NEPA), to comply with its obligations under Section 106 of the NHPA.

Appendix F

S. Rolin - 2 -

In the context of NEPA, under which the DSEIS was prepared, the NRC preliminary determination is that the impact of the two new proposed nuclear units on historical and archaeological resources remains moderate, as concluded in the ESP FEIS. In addition, SNC has entered into a Memorandum of Understanding with the Georgia State Historic Preservation Officer (SHPO). Under the provisions of the National Historic Preservation Act, the NRC preliminary determination is that, consistent with the determination in the ESP FEIS, the proposed project will affect, but not adversely affect, historic properties. Note that in Chapter 2 of the DSEIS you will find a discussion of the areas of potential effect, and impacts to historic properties from construction and operation are discussed in Chapters 4 and 5.

The NRC plans to hold a public meeting to go over the analysis and results in the DSEIS on October 7, 2010, at Augusta Technical College, Waynesboro Campus, 216 Highway 24 South, Waynesboro, Georgia 30830. The meeting will convene at 7:00 p.m. and will continue until 10 p.m., as necessary. In addition, the meeting will be preceded by an open house session from 6:00 p.m. to 7:00 p.m., during which members of the public may meet and talk with NRC staff members on an informal basis. You and your staff are invited to attend.

Pursuant to 10 CFR 51.92 and 36 CFR 800.2(c), the NRC wishes to ensure that Indian Tribes that might have an interest in any potential historic properties in the areas of potential effect are afforded the opportunity to identify their concerns, provide advice on the identification and evaluation of historic properties including those of traditional, religious, and cultural importance; and if necessary, participate in the resolution of any adverse effects to such properties.

In accordance with our December 10, 2009, letter, the NRC staff is forwarding the DSEIS for your review and comments. Pursuant to 36 CFR 800.8(c), we are requesting your comments on the DSEIS, specifically, on our preliminary conclusions regarding historic properties. Please provide any information or comments you may have on the DSEIS during the comment period, which ends on November 24, 2010. The NRC may consider additional comments after the comment period, to the extent practicable. Comments should be submitted either by mail to the Chief, Rules, Announcements, and Directives Branch, Division of Administrative Services, Office of Administration Mailstop TWB-05-B01M, Washington, D.C. 20555-0001 or via e-mail to Vogtle.COLAEIS@nrc.gov. Your comments will be addressed in the final SEIS.

Appendix F

S. Rolin   - 3 -

If you have any questions or require additional information, please contact Ms. Mallecia Sutton, NRC Environmental Project Manager at (301) 415-0673 or via e-mail to Mallecia.Sutton@nrc.gov.

                                    Sincerely,

                                    /RA/

                                    Gregory P. Hatchett, Chief
                                    Environmental Projects Branch 1
                                    Division of Site and Environmental Reviews
                                    Office of New Reactors

Docket Nos. 52-025
            52-026

Enclosure:
As stated

cc: See next page

Appendix F

September 2, 2010

Carol Bernstein
Savannah District
U.S. Army Corps of Engineers
1000 West Oglethorpe Avenue
Savannah, GA 31401-3640

SUBJECT: NOTIFICATION OF THE ISSUANCE AND REQUEST FOR COMMENTS ON THE DRAFT SUPPLEMENTAL ENVIRONMENTAL IMPACT STATEMENT FOR THE VOGTLE ELECTRIC GENERATING PLANT, UNITS 3 AND 4 COMBINED LICENSE APPLICATION

Dear Ms. Bernstein:

The U.S. Nuclear Regulatory Commission (NRC) staff has completed NUREG -1947; "Draft Supplemental Environmental Impact Statement for Combined Licenses (COLs) for the Vogtle Electric Generating Plant Units 3 and 4." The NRC is reviewing the application submitted by Southern Nuclear Operating Company, Inc. (SNC) and several co-applicants for two (COLs) to construct and operate two new nuclear units at the Vogtle Electric Generating Plant site in Burke County, GA. As part of its review of the proposed action, the NRC staff has prepared the draft supplemental environmental impact statement (DSEIS) to include an analysis of relevant environmental issues. The DSEIS documents the NRC determination regarding the environmental impacts at the proposed site from the construction and operation of two new nuclear units. This notice advises the public that the DSEIS is available for public inspection at the NRC Public Documents Room or from the Publicly Available Records component of the NRC Agency-wide Documents Access and Management System (ADAMS). ADAMS is accessible from the NRC website at http://www.nrc.gov/reading-rm/adams.html (the Public Electronic Reading Room) and directly from the NRC website at www.nrc.gov. In addition, the Burke County Library, 130 Highway 24 South, Waynesboro, GA has agreed to make the DSEIS available for public inspection.

This DSEIS is a supplement to the Final EIS for the early site permit (ESP) issued on August 26, 2009, to SNC and the same co-applicants. An ESP is a Commission approval of a site suitable for construction and operation of one or more new nuclear units. Under Title 10 of the *Code of Federal Regulations* (10 CFR) CFR 51.50(c), a COL applicant referencing an ESP need not submit information or analyses regarding environmental issues that were resolved in the ESP EIS, except to the extent the COL applicant has identified any new and potentially significant information. Accordingly, in preparing the DSEIS, the NRC staff considered whether new and significant information has been identified.

Appendix F

C. Bernstein   - 2 -

The NRC plans to hold a public meeting on the DSEIS at the Augusta Technical College, Waynesboro Campus Auditorium, 216 Hwy 24 South, Waynesboro, GA 30830 on Thursday, October 7, 2010. The meeting will convene at 7:00 p.m. and will continue until 10:00 p.m., as necessary. For your information, the meeting will be transcribed and will include: (1) a presentation of the contents of the DSEIS and (2) the opportunity for interested government agencies, organizations, and individuals to provide comments on the DSEIS report. Additionally, the meeting will be preceded by an open house session from 6 p.m. to 7 p.m., during which members of the public may meet and talk with NRC staff members on an informal basis. You and your staff are invited to attend.

As discussed in Section 11.7 of the DSEIS, the staff's preliminary recommendation is that the COL should be issued. This preliminary recommendation is based on (1) the Environmental Report (ER) submitted by Southern Nuclear Operating Company, as revised, and responses to staff requests for additional information; (2) the staff's review conducted for the early site permit referenced by the COL application and the staff assessment documented in the ESP environmental impact statement; (3) consultation with Federal, State, and Tribal agencies; (4) the staff's own independent review of potential new and significant information available since preparation and publication of the ESP EIS; and (5) the assessments summarized in the DSEIS, including the potential mitigation measures identified.

Please provide any information or comments you may have on the DSEIS during the comment period, which ends on November 24, 2010. The NRC may consider additional comments after the comment period, to the extent practicable. Comments should be submitted either by mail to the Chief, Rules, Announcements, and Directives Branch, Division of Administrative Services, Office of Administration, Mailstop TWB-05-B01M, Washington, D.C. 20555-0001 or via e-mail to Vogtle.COLAEIS@nrc.gov. Your comments will be addressed in the final SEIS.

Appendix F

C. Bernstein   - 3 -

A separate notice of filing of the DSEIS will be placed in the *Federal Register* through the U.S. Environmental Protection Agency (EPA). If you have any questions regarding this matter, please contact Ms. Mallecia Sutton, NRC Environmental Project Manager at 301-415-0673 or via e-mail to Mallecia.Sutton@nrc.gov.

                                    Sincerely,

                                    /RA/

                                    Gregory P. Hatchett, Chief
                                    Environmental Projects Branch 1
                                    Division of Site and Environmental Reviews
                                    Office of New Reactors

Docket Nos.: 52-025
              52-026

Enclosure:
As stated

cc w/encl: See next page

Appendix F

September 3, 2010

Mr. David Bernhart
National Marine Fisheries Service
Southeast Regional Office
263 13th Avenue South
St. Petersburg, FL 33701

SUBJECT: NOTIFICATION OF THE ISSUANCE AND REQUEST FOR COMMENTS ON THE DRAFT SUPPLEMENTAL ENVIRONMENTAL IMPACT STATEMENT FOR VOGTLE ELECTRIC GENERATING PLANT, UNITS 3 AND 4 COMBINED LICENSE APPLICATION

Dear Mr. Bernhart:

On behalf of the U.S. Nuclear Regulatory Commission (NRC) staff, I am forwarding a copy of NUREG -1947, "Draft Supplemental Environmental Impact Statement for Combined Licenses (COLs) for the Vogtle Electric Generating Plant Units 3 and 4." The NRC is reviewing the application submitted by Southern Nuclear Operating Company, Inc. (SNC) and several co-applicants for two COLs to construct and operate two new nuclear units at the Vogtle Electric Generating Plant site in Burke County, GA. As part of its review of the proposed action, the NRC staff has prepared the draft supplemental environmental impact statement (DSEIS) to include an analysis of relevant environmental issues. The DSEIS documents the NRC determination regarding the environmental impacts at the proposed site from the construction and operation of two new nuclear units.

The DSEIS is available for public inspection at the NRC Public Documents Room or from the Publicly Available Records component of the NRC Agency-wide Documents Access and Management System (ADAMS). ADAMS is accessible from the NRC website at http://www.nrc.gov/reading-rm/adams.html (the Public Electronic Reading Room) and directly from the NRC website at www.nrc.gov. In addition, the Burke County Library, 130 Highway 24 South, Waynesboro, GA has agreed to make the DSEIS available for public inspection.

This DSEIS is a supplement to the Final EIS for the early site permit (ESP) issued on August 26, 2009, to SNC and the same co-applicants. An ESP is a Commission approval of a site suitable for construction and operation of one or more new nuclear units. Under Title 10 of the *Code of Federal Regulations* (10 CFR) CFR 51.50(c), a COL applicant referencing an ESP need not submit information or analyses regarding environmental issues that were resolved in the ESP EIS, except to the extent the COL applicant has identified any new and potentially significant information. Accordingly, in preparing the DSEIS, the NRC staff considered whether new and significant information has been identified.

Appendix F

D. Bernhart - 2 -

During the ESP environmental review, the NRC consulted with the Southeast Regional Office and, by letter dated August 11, 2008 (Enclosure 1), received concurrence on a biological assessment evaluating the impacts of construction and operation of two new reactors at the VEGP site on the shortnose sturgeon. The draft SEIS's analysis of impacts to the shortnose sturgeon did not change from the characterization in the ESP FEIS (NUREG-1872) and remains small with no additional mitigation warranted. The Staff has concluded that the COL action involves similar impacts to the same Federally listed species in the same geographic area as analyzed in the ESP, that no new species have been listed or proposed and no new critical habitat designated or proposed for the action area, and that with respect to potential impacts to the shortnose sturgeon, no relevant information has changed regarding the project since the earlier BA was submitted. Therefore, pursuant to 50 C.F.R. § 402.12(g), the Staff hereby proposes to incorporate that biological assessment by reference. Enclosed is a copy of the draft SEIS, NUREG-1947, along with a CD containing the environmental impact statement for the ESP, NUREG-1872, to aid your review.

The NRC plans to hold a public meeting on the DSEIS at the Augusta Technical College, Waynesboro Campus Auditorium, 216 Hwy 24 South, Waynesboro, GA 30830 on Thursday, October 7, 2010. The meeting will convene at 7:00 p.m. and will continue until 10:00 p.m., as necessary. For your information, the meeting will be transcribed and will include: (1) a presentation of the contents of the DSEIS and (2) the opportunity for interested government agencies, organizations, and individuals to provide comments on the DSEIS report. Additionally, the meeting will be preceded by an open house session from 6 p.m. to 7 p.m., during which members of the public may meet and talk with NRC staff members on an informal basis. You and your staff are invited to attend.

To ensure compliance with Section 7 of the Endangered Species Act of 1973 (ESA) and fulfill consultation requirements as required by the Fish and Wildlife Coordination Act (FWCA), please provide any information and comments you consider appropriate under the provisions of the ESA or FWCA during the comment period, which ends on November 24, 2010. With respect to the incorporation by reference of the ESP biological assessment as discussed above, if no response from the Southeast Regional Office is received during the comment period, the NRC will consider the consultation closed. Comments should be submitted either by mail to the Chief, Rules, Announcements, and Directives Branch, Division of Administrative Services, Office of Administration, Mailstop TWB-05-B01M, Washington, D.C. 20555-0001 or via e-mail to Vogtle.COLAEIS@nrc.gov. Your comments will be addressed in the final SEIS.

Appendix F

D. Bernhart   - 3 -

A separate notice of filing of the DSEIS will be placed in the *Federal Register* through the U.S. Environmental Protection Agency (EPA). If you have any questions regarding this matter, please contact Ms. Mallecia Sutton, NRC Environmental Project Manager at 301-415-0673 or via e-mail to Mallecia.Sutton@nrc.gov.

Sincerely,

/RA/

Gregory P. Hatchett, Chief
Environmental Projects Branch 1
Division of Site and Environmental Reviews
Office of New Reactors

Docket Nos.: 52-025
              52-026

Enclosure:
As stated

cc w/encl: See next page

Appendix F

September 3, 2010

Mr. Robert D. Perry
Special Projects Manager
Office of Environmental Programs
South Carolina Department of
 Natural Resources
1000 Assembly Street, Room 310A
P.O. Box 167
Columbia, SC 29202

SUBJECT: NOTIFICATION OF THE ISSUANCE AND REQUEST FOR COMMENTS ON THE DRAFT SUPPLEMENTAL ENVIRONMENTAL IMPACT STATEMENT FOR THE VOGTLE ELECTRIC GENERATING PLANT, UNITS 3 AND 4 COMBINED LICENSE APPLICATION REVIEW

On behalf of the Nuclear Regulatory Commission (NRC) staff, I am forwarding a copy of NUREG-1947; Draft Supplemental Environmental Impact Statement for Combined Licenses (COLs) for Vogtle Electric Generating Plant Units 3 and 4 for your review and comments. The NRC is reviewing the application submitted by Southern Nuclear Operating Company, Inc. (SNC) and several co-applicants for two COLs to construct and operate two new nuclear units at the Vogtle Electric Generating Plant (VEGP) site in Burke County, GA. As part of its review of the proposed action, the NRC staff has prepared the DSEIS to include an analysis of relevant environmental issues.

The NRC staff completed the DSEIS and the associated *Federal Register* Notice of Availability. The notice advises the public that the DSEIS is available for public inspection at the NRC Public Documents Room or from the Publicly Available Records component of the NRC Agency-wide Documents Access and Management System (ADAMS). ADAMS is accessible from the NRC Website at http://www.nrc.gov/reading-rm/adams.html, which provides access through the NRC Electronic Reading Room link. The accession number in ADAMS for the DSEIS is ML102370278. The DSEIS can also be found at the NRC VEGP COL-specific webpage at http://www.nrc.gov/reactors/new-reactors/col/vogtle.html. In addition, the Burke County Library located at 130 Hwy 24 South, Waynesboro, GA 30830 has agreed to maintain a copy of the DSEIS and make it available for public inspection.

This DSEIS is a supplement to the Final EIS for the early site permit (ESP) issued on August 26, 2009, to SNC and the same co-applicants. An ESP is a Commission approval of a site suitable for construction and operation of one or more new nuclear units. Under Title 10 of the Code of Federal Regulations (10 CFR) CFR 51.50(c), a COL applicant referencing an ESP need not submit information or analyses regarding environmental issues that were resolved in the ESP EIS, except to the extent the COL applicant has identified any new and potentially significant information. Accordingly, in preparing the DSEIS, the NRC staff considered whether new and significant information has been identified.

Appendix F

R. Perry - 2 -

The NRC plans to hold a public meeting to present the analysis and results of the DSEIS on October 7, 2010, at the Augusta Technical College, Waynesboro Campus, 216 Hwy 24 South, Waynesboro, GA 30830. The meeting will convene at 7:00 p.m., and will continue until 10:00 p.m., as necessary. For your information, the meeting will be transcribed and will include a presentation of the contents of the DSEIS and the opportunity for interested government agencies, organizations, and individuals to provide comments on the draft report. Additionally, the meeting will be preceded by an open house session from 6:00 p.m. to 7:00 p.m. during which members of the public may meet and talk with NRC staff members on an informal basis. You and your staff are invited to attend.

As discussed in Section 11.7 of the DSEIS, the staff's preliminary recommendation is that the COLs and requested Limited Work Authorization (LWA) should be issued. This preliminary recommendation is based on (1) the Environmental Report (ER) submitted by Southern Nuclear Operating Company, as revised; and responses to staff requests for additional information; (2) the staff's review conducted for the early site permit referenced by the COL application and the staff assessment documented in the ESP environmental impact statement (EIS); (3) consultation with Federal, State, Tribal and local agencies; (4) the staff's own independent review of potential new and significant information available since preparation and publication of the ESP EIS, and; (5) the assessments summarized in the DSEIS, including the potential mitigation measures identified. Finally, the staff concludes that the requested LWA construction activities defined at 10 CFR 50.10(a) and described in the site redress plan would not result in any significant adverse environmental impacts that cannot be redressed.

Please provide any information or comments on the DSEIS that you consider appropriate during the comment period, which ends on November 24, 2010. Please include in these comments any information you consider appropriate consistent with the provisions of the Fish and Wildlife Coordination Act. The NRC may consider additional comments after the comment period ends to the extent practicable. Comments should be submitted either by mail to the Chief, Rules, Announcements, and Directives Branch, Division of Administrative Services, Office of Administration, Mailstop TWB-05-B01M, Washington DC 20555-0001 or by e-mail to Vogtle.COLAEIS@nrc.gov.

Appendix F

R. Perry - 3 -

A separate notice of filing of the DSEIS will be placed in the *Federal Register* through the U.S. Environmental Protection Agency. If you have any questions or require additional information, please contact Ms. Mallecia Sutton, NRC Environmental Project Manager at (301) 415-0673 or via e-mail to Mallecia.Sutton@nrc.gov.

Sincerely,

/RA/

Gregory P. Hatchett, Chief
Environmental Projects Branch 1
Division of Site and Environmental Reviews
Office of New Reactors

Docket Nos.: 52-025
    52-026

Enclosures:
As stated

cc: See next page

Appendix F

September 3, 2010

Ms. Sandra Tucker
Field Supervisor
Georgia Ecological Services
U.S. Fish and Wildlife Service
105 West Park Drive
Athens, GA. 30607

SUBJECT: NOTIFICATION OF THE ISSUANCE AND REQUEST FOR COMMENTS ON THE DRAFT SUPPLEMENTAL ENVIRONMENTAL IMPACT STATEMENT FOR THE VOGTLE ELECTRIC GENERATING PLANT, UNITS 3 AND 4 COMBINED LICENSES APPLICATION

Dear Ms. Tucker:

On behalf of the Nuclear Regulatory Commission (NRC) staff, I am forwarding a copy of the "Draft Supplemental Environmental Impact Statement for Combined Licenses (COLs) for the Vogtle Electric Generating Plant, Units 3 and 4," for your review and comments. The NRC is reviewing the application submitted by Southern Nuclear Operating Company, Inc. (SNC) and several co-applicants for two COLs to construct and operate two new nuclear units at the VEGP site in Burke County, GA. As part of its review of the proposed action, the NRC staff has prepared the DSEIS to include an analysis of relevant environmental issues.

This notice advises the public that the draft report is available for public inspection at the NRC Public Document Room or from the Publicly Available Records component of the NRC Agency-wide Documents Access and Management System (ADAMS). ADAMS is accessible from the NRC Website at http://www.nrc.gov/reading-rm/adams.html, which provides access through the NRC Electronic Reading Room link. The accession number in ADAMS for the DSEIS is ML102370278. The DSEIS can also be found at the NRC Vogtle Electric Generating Plant COL-specific webpage at http://www.nrc.gov/reactors/new-reactors/col/vogtle.html. The Burke County Library located at 130 Hwy 24 South, Waynesboro, GA 30830 has agreed to maintain a copy of the DSEIS and make it available for public inspection. A separate notice of filing of the DEIS will be placed in the *Federal Register* through the U.S. Environmental Protection Agency.

This DSEIS is a supplement to the Final EIS for the early site permit (ESP) issued on August 26, 2009, to SNC and the same co-applicants. An ESP is a Commission approval of a site suitable for construction and operation of one or more new nuclear units. Under Title 10 of the Code of Federal Regulations (10 CFR) CFR 51.50(c), a COL applicant referencing an ESP need not submit information or analyses regarding environmental issues that were resolved in the ESP EIS, except to the extent the COL applicant has identified any new and potentially significant information. Accordingly, in preparing the DSEIS, the NRC staff considered whether new and significant information has been identified.

Appendix F

S. Tucker   - 2 -

The notice also informs the public that the NRC plans to hold a public meeting to present the analysis and results of the DSEIS on October 7, 2010, at the Augusta Technical College, Waynesboro Campus, 216 Hwy 24 South, Waynesboro, GA 30830. The meeting will convene at 7:00 p.m., and will continue until 10:00 p.m., as necessary. For your information, the meeting will be transcribed and will include a presentation of the contents of the DSEIS and the opportunity for interested government agencies, organizations, and individuals to provide comments on the draft report. Additionally, the meeting will be preceded by an open house session from 6:00 p.m. to 7:00 p.m. during which members of the public may meet and talk with NRC staff members on an informal basis. You and your staff are invited to attend.

During the ESP environmental review, the NRC consulted with your office and, by letter dated September 19, 2008 (Enclosure 1), received concurrence on a biological assessment evaluating the impacts of site preparation and preliminary construction at the VEGP site on potentially occurring Federally listed threatened or endangered species. The draft SEIS's analysis of impacts to potentially occurring Federally listed threatened or endangered species did not change from the characterization in the ESP FEIS (NUREG-1872). The Staff is preparing a biological assessment documenting potential impacts on potentially occurring Federally listed threatened or endangered species as a result of operation of the proposed new units and construction and operation of the proposed transmission line right-of-way associated with the development of the VEGP site, and will be providing that assessment for your consideration.

To ensure compliance with Section 7 of the Endangered Species Act of 1973 (ESA) and fulfill consultation requirements as required by the Fish and Wildlife Coordination Act (FWCA), please provide any information and comments you consider appropriate under the provisions of the ESA or FWCA during the comment period, which ends on November 24, 2010. The NRC may consider additional comments after the comment period ends to the extent practicable. Comments should be submitted either by mail to the Chief, Rules, Announcements, and Directives Branch, Division of Administrative Services, Office of Administration, Mailstop TWB-05-B01M, Washington DC 20555-0001 or by e-mail to Vogtle.COLAEIS@nrc.gov.

Appendix F

S. Tucker   - 3 -

If you have any questions regarding this matter, please contact Ms. Mallecia Sutton, NRC Environmental Project Manager at 301-415-0673 or by e-mail to Mallecia.Sutton@nrc.gov.

Sincerely,

/RA/

Gregory P. Hatchett, Chief
Environmental Projects Branch 1
Division of Site and Environmental Reviews
Office of New Reactors

Docket Nos.: 52-025
              52-026

Enclosures:
As stated

cc: See next page

Appendix F

VogtleEISCEmails

From:	Bryant J. Celestine [celestine.bryant@actribe.org]
Sent:	Wednesday, October 06, 2010 9:01 AM
To:	VogtleCOLAEIS Resource
Subject:	Draft SEIS

On behalf of Mikko Oscola Clayton Sylestine and the Alabama-Coushatta Tribe, our appreciation is expressed on your efforts to consult us regarding the draft Supplemental Environmental Impact Statement for the Vogtle Electric Generating Plant, Units 3 and 4 Combined License Application in Burke County.

Our Tribe maintains ancestral associations within the state of Georgia despite the absence of written documentation to completely identify Tribal activities, villages, trails, or burial sites. However, it is our objective to ensure significances of Native American ancestry, especially of Alabama-Coushatta Tribal origin, are administered with the utmost considerations.

Upon review of your September 2, 2010 submission, we reiterate our January 7, 2010 electronic message to decline the opportunity to participate in this consultation. Burke County currently exists beyond our scope of interest for the state of Georgia. No known impacts to religious, cultural, or historical assets of the Alabama-Coushatta Tribe of Texas will occur in conjunction with this proposal. No further consultation with our Tribe regarding this project is anticipated at this time.

Should you require further assistance, please do not hesitate to contact us.

Sincerely,

Bryant J. Celestine
Historic Preservation Officer
Alabama-Coushatta Tribe of Texas
571 State Park Rd 56
Livingston, TX 77351
936 - 563 - 1181
celestine.bryant@actribe.org

Appendix F

REGION 4
SAM NUNN
ATLANTA FEDERAL CENTER
61 FORSYTH STREET
ATLANTA GEORGIA 30303-8960
November 15, 2010

Chief, Rulemaking and Directives Branch
Office of Administration
Mail Stop: TWB-05-B01M
U.S. Nuclear Regulatory Commission
Washington, DC 20555-0001

RE: EPA Review and Comments
Draft Supplemental Environmental Impact Statement (DSEIS) for the
Combined Licenses (COLs) for Vogtle Electric Generating Plant Units 3 and 4
Construction and Operation, Application for Combined Licenses (COLs), NUREG-1947
CEQ No. 20100351

Dear Sir:

The U.S. Environmental Protection Agency (EPA) has reviewed the Draft Supplemental Environmental Impact Statement (DSEIS) for the Combined Licenses (COLs) for Vogtle Electric Generating Plant Units 3 and 4, pursuant to Section 102(2)(C) of the National Environmental Policy Act (NEPA), and Section 309 of the Clean Air Act. The purpose of this letter is to inform you of the results of our review, and our detailed comments are enclosed.

Southern Nuclear Operating Company, Inc. (Southern) and four co-applicants applied for combined construction permits and operating licenses (combined licenses or COLs) for Vogtle Electric Generating Plant (VEGP) Units 3 and 4. The proposed action is NRC issuance of COLs for two new nuclear power reactor units (Units 3 and 4) at the VEGP site near Waynesboro, Georgia.

EPA previously reviewed and submitted written comments regarding the Draft and Final Environmental Impact Statements (EISs) for the Early Site Permit (ESP) for the new units, and for the Joint Public Notice for the U.S. Army Corps of Engineers (USACE) Permit. Since these documents stated that there were no transmission line impacts, our comments at that time pertained to the plant site only. The USACE permit action on an Individual Permit application pursuant to Section 404 of the Clean Water Act, and Section 401 water quality certification for the Plant VEGP expansion were finalized in September 2010. The current DSEIS provides updated information and focuses on the proposed issuance of the COLs to authorize construction and operation of the new units and ancillary facilities.

The NRC issued an Early Site Permit (ESP) on August 26, 2009, approving the VEGP site as suitable for the construction of Units 3 and 4. NRC issuance of a Limited Work Authorization

Appendix F

(LWA) enabled specific pre-construction activities at the site to begin. The NRC is currently reviewing the Westinghouse AP1000 pressurized reactor design in a design certification process.

Radioactive waste storage and disposal are ongoing concerns with existing and proposed nuclear power plants. The NRC approved final revisions to the Waste Confidence findings and regulation (10 CFR Part 51.23) in September 2010. This update expresses confidence that commercial high-level radioactive waste and spent fuel generated by any reactor "...*can be stored safely and without significant environmental impacts for at least 30 years beyond the licensed life for operation (which may include the term of a revised or renewed license) of that reactor.*" This refers to storage in a spent fuel basin or at either onsite or offsite independent spent fuel storage installations.

Since appropriate storage of spent fuel assemblies and other radioactive wastes is necessary to prevent environmental impacts, the FSEIS should provide a thorough consideration of impacts resulting from such storage. Given the uncertainty regarding ultimate disposal at a repository, on-site storage may continue for many years.

Southern indicated that there would be an operations-related three percent increase in the thermal discharge flow in the DSEIS. The NRC determined that the thermal plume would remain small compared to the width of the Savannah River at this location, and that it would not impede fish passage in the river. The Final Supplemental Environmental Impact Statement (FSEIS) should include a graph of the plume showing the temperature profile, and a discussion of how the increase will (or will not) cause a violation of Georgia's water quality standard for temperature at the point of discharge.

In addition, the design and location of the proposed new cooling water intake structure has changed. The NRC determined that this new location would not alter conclusions presented in the previous ESP FEIS. Continuing measures to limit bioentrainment and other impacts to aquatic species from surface water withdrawals and discharges should be referenced in the FSEIS, and should continue to be addressed as the project progresses, in compliance with the NPDES Permit.

The FSEIS should include further information regarding plans to reduce Greenhouse Gases (GHGs) and other air emissions during construction of the facility. Specifically, energy efficiency and renewable energy should be a consideration in the construction and operation of facility buildings, equipment, and vehicles. We also recommend that the FSEIS explicitly reference the draft guidance from CEQ related to evaluating GHGs in Federal actions, describe the elements of the draft guidance, and to the relevant extent, provide the assessments suggested by the guidance. Based on your analysis using the CEQ NEPA Guidance, further data collection may be necessary in the future.

Based on EPA's review of the DSEIS, the document received a rating of EC-2, meaning that the EPA review identified environmental concerns. (A summary of EPA's rating definitions is enclosed.) In particular, EPA recommends that the FSEIS include updated information about radioactive waste storage and disposal, impacts of macro-right-of-way transmission lines, a consideration of GHGs using CEQ's draft guidance for GHGs, and a discussion of opportunities to reduce GHG and other air emissions during construction and operation of the facility. In

2

Appendix F

addition, the FSEIS should include a status update regarding the Westinghouse AP1000 certification review.

Thank you for your continuing coordination with us. We look forward to reviewing the FSEIS. If you have any questions or need additional information, please contact Ramona McConney of my staff at (404) 562-9615.

Sincerely,

Heinz J. Mueller, Chief
NEPA Program Office
Office of Policy and Management

Enclosures: EPA Review and Comments
Summary of Rating Definitions and Follow Up Action

Appendix F

EPA Review and Comments Regarding
Draft Supplemental Environmental Impact Statement (DSEIS) for the
Combined Licenses (COLs) for Vogtle Electric Generating Plant Units 3 and 4
Construction and Operation, Application for Combined Licenses (COLs), NUREG-1947
CEQ No. 20100351

General

This DSEIS provides updated information (subsequent to the ESP FEIS) regarding preconstruction activities and environmental data, and focuses on the proposed issuance of COLs for the two new reactor units and ancillary facilities.

In the DSEIS, the NRC concludes that there are no new and significant data or changes to conclusions since the ESP FEIS regarding the following: land-use impacts, meteorology and air quality impacts, water quality impacts, terrestrial and aquatic ecosystems, socioeconomic impacts, historic and cultural resource impacts, environmental justice, nonradiological health impacts, radiological impacts of normal operations, environmental impacts of postulated accidents.

Alternatives

Alternatives in the DSEIS include the no-action alternative, energy source alternatives and system design alternatives. The NRC's evaluation of alternative sites is documented in the EIS for the ESP, which EPA previously reviewed and submitted comments.

Radioactive wastes

Appropriate on-site storage of spent fuel assemblies and other radioactive waste is necessary to prevent environmental impacts. Given the uncertainty regarding ultimate disposal at a repository, on-site storage may continue for a longer term than currently expected.

Yucca Mountain was formerly considered a possible final repository for spent nuclear fuel, but this plan was withdrawn by the U.S. Department of Energy by the motion of March 3, 2010. The abandonment of the plan to create a Yucca Mountain permanent geologic repository has been recently countered by NRC's Atomic Safety and Licensing Board. If another repository in the contiguous United States (other than Yucca Mountain) is ever selected, the environmental impact estimates from the transportation of spent reactor fuel to the repository should be calculated as required under 42 USC 4321 Fuel Cycle, Transportation, and Decommissioning.

In the Waste Confidence Rule (10 CFR 51.23), the Commission generically determined that the spent fuel generated by any reactor can be safely stored on-site for at least 30 years beyond the licensed operating life of the reactor. The NRC approved final revisions to the Waste Confidence findings and regulation in September 2010, extending the storage period until *"...30 years beyond the licensed life for operation (which may include the term of a revised or renewed license) of that reactor"* in its spent fuel basin or at either onsite or offsite independent spent fuel storage installations.

4

Appendix F

The FSEIS should clarify the impact of this revision on the proposed project, as this new determination finds that spent nuclear fuel can be stored safely and securely without significant environmental impacts for at least 60 years after operation at any nuclear power plant. EPA recommends that the FSEIS cite any new analyses for longer-term storage regarding scientific knowledge relating to spent fuel storage and disposal. The FSEIS should also mention any developments with the Presidential Blue Ribbon Commission on alternatives for dealing with high-level radioactive waste, if there are such updates before FSEIS publication.

We understand that shipping casks have not yet been designed for the spent fuel from advanced reactor designs such as the Westinghouse AP1000. Information in the Early Site Permit Environmental Report Sections and Supporting Documentation (INEEL 2003) indicated that advanced light water reactor (LWR) fuel designs would not be significantly different from existing LWR designs; therefore, current shipping cask designs were used for the analysis of Westinghouse AP1000 reactor spent fuel shipments. EPA recommends that when shipping casks are designed for the spent fuel for the Westinghouse AP1000, the analysis should be repeated.

EPA understands that concerns have been raised by the NRC that certain structural components of the revised AP1000 shield building may not be suitable to withstand design loads. The shield building is designed to protect the reactor's primary containment from severe weather and other events, as well as serving as a radiation barrier and also supporting an emergency cooling water tank. It is EPA's understanding that the NRC is currently reviewing the remainder of the next-generation reactor's design certification amendment application, and that Westinghouse is expected to make design modifications and conduct safety testing to ensure the shield building design can meet its safety functions.

The FSEIS should address the status of the Westinghouse AP1000 certification review and related issues, particularly the analysis of the structural integrity of the AP1000. We understand that the Safety Evaluation Report will address these issues in even more detail, and that the certification review may be completed as soon as December 2010. EPA understands that Revision 15 of the AP1000 design is codified in 10 CFR Part 52, Appendix D. EPA concurs with NRC's plan to conduct an additional environmental review if changes result in the final design being significantly different from the design considered in the DEIS.

Transmission lines

We note that the NRC considers transmission lines to be "preconstruction" activities (discussed in the EIS for the ESP), and that preconstruction activities are considered in the context of cumulative impacts. EPA is concerned about the impacts of transmission lines and supporting infrastructure for the project and, in accordance with NEPA, considers these activities as part of the project, and not a separate action.

The DSEIS (pages 3-7 and 3-8) discusses the construction of a new transmission line through a "macro-right-of-way." This term should be defined in the text, with details given regarding the proposed extent and impacts of this new transmission line. The FSEIS should also clarify whether there are plans to issue a Limited Work Authorization (LWA) for these lines pursuant to the NRC's LWA process.

5

Appendix F

Wetlands and Streams

Jurisdictional determinations for all site wetlands are complete, with the exception of the required metes and bounds survey. A joint application package was submitted for all permits under the jurisdiction of the USACE (Section 404, Section 10, and Dredge and Fill) on January 7, 2010.

EPA reviewed the impacts to wetlands and streams in response to the USACE's public notice for the Clean Water Act Section 404 permit application, and transmitted a comment letter in accordance with Section 404 coordination procedures. We note that the Dredge and Fill discharge permit was for the transmission line corridor.

NPDES Permitting

Southern indicated that there would be an operations-related three percent increase in the thermal discharge flow. The NRC determined that the thermal plume would remain small compared to the width of the Savannah River at this location, and that it would not impede fish passage in the river (Section 5.4.2). In addition, the design and location of the proposed new cooling water intake structure has changed. The NRC determined that this new location would not alter conclusions in the previous ESP FEIS. Pursuant to our review, the following areas need clarification:

- *Temperature:* The discussion of the 3% increase in the thermal discharge should include a graph of the plume showing the temperature profile, and a discussion of how the increase will (or will not) cause a violation of Georgia's water quality standard for temperature at the point of discharge.

- *Cooling Water Intake:* For clarity, the FSEIS should restate the requirements for the cooling water intake structure.

Greenhouse Gases (GHGs)

We appreciate your discussion of climate change and GHGs in the DSEIS. The DSEIS states that the majority of the potential carbon dioxide (CO_2) emissions of the proposed nuclear power plant would be the life cycle contributions associated with the uranium fuel cycle (Section 7.2). The DSEIS notes that such emissions primarily result from the operation of fossil-fueled power plants that provide the electricity needed to manufacture the nuclear fuel.

CEQ Draft Guidance on GHG Analysis within NEPA: On February 18, 2010, the Council on Environmental Quality (CEQ) proposed four steps to modernize and reinvigorate NEPA. In particular, the CEQ issued draft guidance for public comment on, among other issues, when and how Federal agencies must consider greenhouse gas emissions and climate change in their proposed actions.
(Reference: http://www.whitehouse.gov/administration/eop/ceq/initiatives/nepa)

The draft guidance explains how Federal agencies should analyze the environmental impacts of greenhouse gas emissions and climate change when they describe the environmental impacts of a

6

Appendix F

proposed action under NEPA. It provides practical tools for agency reporting, including a presumptive threshold of 25,000 metric tons of carbon dioxide equivalent (CO_2e) emissions from the proposed action to trigger a quantitative analysis, and instructs Federal agencies regarding how to assess the effects of climate change on the proposed action and their design. The draft guidance does not apply to land and resource management actions and does not propose to regulate greenhouse gases.

While this guidance is not yet final (and thus, not required), we recommend that the FSEIS explicitly reference the draft guidance, describe the elements of the draft guidance, and to the relevant extent, provide the assessments suggested by the guidance. (Note that the discussion in Section 7.2 and referencing the Sovacool paper (see footnote 1 below) regarding the derivation of 447,000 metric tons/year of CO_2 emissions from a 1000 MW nuclear power plant is difficult to follow. For example, we could not find the "1 percent to 5 percent" citation noted as being in the Sovacool paper. It would be helpful to show a detailed derivation of the amount of direct and indirect CO_2-equivalent emissions expected specifically from this project.)

EPA also recommends a discussion of best management practices (BMPs) to reduce GHGs and other air emissions during construction and operation of the facility. Specifically, clean energy options such as energy efficiency and renewable energy should be a consideration in the use of construction and maintenance equipment and vehicles. For example, equipment and vehicles that use conventional petroleum (e.g., diesel) should incorporate clean diesel technologies and fuels to reduce emissions of GHGs and other pollutants, and should adhere to anti-idling policies to the extent possible. Alternate fuel vehicles (e.g., natural gas, electric) are also possibilities.

(1) Sovacool, BK. Valuing the Greenhouse Gas Emissions for Nuclear Power: A Critical Survey. Energy Policy 36 (2008) 2940 - 2953.

Diesel Exhaust

In addition to the EPA's concerns regarding climate change effects and GHG emissions, the National Institute for Occupational Safety and Health (NIOSH) has determined that diesel exhaust is a potential human carcinogen, based on a combination of chemical, genotoxicity, and carcinogenicity data. In addition, acute exposures to diesel exhaust have been linked to health problems such as eye and nose irritation, headaches, nausea, and asthma.

Although every construction site is unique, common actions can reduce exposure to diesel exhaust. EPA recommends that the following actions be considered for construction equipment:

- Using low-sulphur diesel fuel (less than 0.05% sulphur).
- Retrofit engines with an exhaust filtration device to capture DPM before it enters the workplace.
- Position the exhaust pipe so that diesel fumes are directed away from the operator and nearby workers, thereby reducing the fume concentration to which personnel are exposed.
- A catalytic converter reduces carbon monoxide, aldehydes, and hydrocarbons in diesel fumes. These devices must be used with low sulphur fuels.
- Ventilate wherever diesel equipment operates indoors. Roof vents, open doors and windows, roof fans, or other mechanical systems help move fresh air through work areas.

7

Appendix F

As buildings under construction are gradually enclosed, remember that fumes from diesel equipment operating indoors can build up to dangerous levels without adequate ventilation.
- Attach a hose to the tailpipe of a diesel vehicle running indoors and exhaust the fumes outside, where they cannot reenter the workplace. Inspect hoses regularly for defects and damage.
- Use enclosed, climate-controlled cabs pressurized and equipped with high efficiency particulate air (HEPA) filters to reduce operators' exposure to diesel fumes. Pressurization ensures that air moves from inside to outside. HEPA filters ensure that any air coming in is filtered first.
- Regular maintenance of diesel engines is essential to keep exhaust emissions low. Follow the manufacturer's recommended maintenance schedule and procedures. Smoke color can signal the need for maintenance. For example, blue/black smoke indicates that an engine requires servicing or tuning.
- Work practices and training can help reduce exposure. For example, measures such as turning off engines when vehicles are stopped for more than a few minutes; training diesel-equipment operators to perform routine inspection and maintenance of filtration devices.
- When purchasing a new vehicle, ensure that it is equipped with the most advanced emission control systems available.
- With older vehicles, use electric starting aids such as block heaters to warm the engine, avoid difficulty starting, and thereby reduce diesel emissions.
- Respirators are only an interim measure to control exposure to diesel emissions. In most cases an N95 respirator is adequate. Respirators are for interim use only, until primary controls such as ventilation can be implemented. Workers must be trained and fit-tested before they wear respirators. Personnel familiar with the selection, care, and use of respirators must perform the fit testing. Respirators must bear a National Institute of Occupational Safety and Health (NIOSH) approval number. Never use paper masks or surgical masks without NIOSH approval numbers.

Endangered and Threatened Species

The DSEIS states that a biological assessment documenting potential impact on the federally listed threatened or endangered terrestrial special as a result of operation of the proposed new units and proposed transmission line is in development. The FSEIS should provided updated information on this assessment.

Historic Preservation

We appreciate the thorough discussion of cultural and historic resources in the DSEIS. Pursuant to the location of a historic cemetery on the VEGP site, Southern entered into a Memorandum of Understanding (SHPO) with the Georgia State Historic Preservation Office (SHPO). We also note SCE&G's cultural resources awareness training and inadvertent discovery procedure training for staff working at the site. The FSEIS should include an update of coordination activities with the SHPO.

8

Appendix F

SUMMARY OF RATING DEFINITIONS AND FOLLOW UP ACTION[*]

Environmental Impact of the Action

LO-Lack of Objections
The EPA review has not identified any potential environmental impacts requiring substantive changes to the proposal. The review may have disclosed opportunities for application of mitigation measures that could be accomplished with no more than minor changes to the proposal.

EC-Environmental Concerns
The EPA review has identified environmental impacts that should be avoided in order to fully protect the environment. Corrective measures may require changes to the preferred alternative or application of mitigation measures that can reduce the environmental impacts. EPA would like to work with the lead agency to reduce these impacts.

EO-Environmental Objections
The EPA review has identified significant environmental impacts that must be avoided in order to provide adequate protection for the environment. Corrective measures may require substantial changes to the preferred alternative or consideration of some other project alternative (including the no action alternative or a new alternative). EPA intends to work with the lead agency to reduce these impacts.

EU-Environmentally Unsatisfactory
The EPA review has identified adverse environmental impacts that are of sufficient magnitude that they are unsatisfactory from the standpoint of public health or welfare or environmental quality. EPA intends to work with the lead agency to reduce these impacts. If the potential unsatisfactory impacts are not corrected at the Draft EIS sate, this proposal will be recommended for referral to the CEQ.

Adequacy of the Impact Statement

Category 1-Adequate
The EPA believes the draft EIS adequately sets forth the environmental impact(s) of the preferred alterative and those of the alternatives reasonably available to the project or action. No further analysis or data collecting is necessary, but the reviewer may suggest the addition of clarifying language or information.

Category 2-Insufficient Information
The draft EIS does not contain sufficient information for the EPA to fully assess the environmental impacts that should be avoided in order to fully protect the environment, or the EPA reviewer has identified new reasonably available alternatives that are within the spectrum of alternatives analyzed in the draft EIS, which could reduce the environmental impacts of the action. The identified additional information, data, analyses, or discussion should be included in the Draft EIS.

Category 3-Inadequate
EPA does not believe that the draft EIS adequately assesses potentially significant environmental impacts of the action, or the EPA reviewer has identified new, reasonably available alternatives that are outside of the spectrum of alternatives analyzed in the draft EIS, which should be analyzed in order to reduce the potentially significant environmental impacts. EPA believes that the identified additional information, data analyses, or discussions are of such a magnitude that they should have full public review at a draft stage. EPA does not believe that the draft EIS is adequate for the purposes of the NEPA and/or Section 309 review, and thus should be formally revised and made available for public comment in a supplemental or revised draft EIS. On the basis of the potential significant impacts involved, this proposal could be a candidate for referral to the CEQ.

[*] From EPA Manual 1640 Policy and Procedures for the Review of the Federal Actions Impacting the Environment

Appendix F

United States Department of the Interior

OFFICE OF THE SECRETARY
Office of Environmental Policy and Compliance
Richard B. Russell Federal Building
75 Spring Street, S.W.
Atlanta, Georgia 30303

ER10/0767

November 29, 2010

Chief, Rules, Announcements, and Directives Branch
Office of Administration
Mail Stop: TWB-05-B01M
U.S. Nuclear Regulatory Commission
Washington, DC 20555-0001

Re: Comments for the Draft Environmental Impact Statement (DEIS) for Vogtle Nuclear Plant Units 3 and 4, Application for Combined Licenses (COLs), NUREG-1947, Burke County, Georgia

The Department of the Interior (Department) has reviewed the Draft Environmental Impact Statement (DEIS) of the Nuclear Regulatory Commission (NRC) for the proposed addition of two nuclear reactors (Units 3 and 4) at the Vogtle Electric Generating Plant (VEGP). The license applicant is Southern Nuclear Operating Company, Inc. (Southern), on behalf of itself and four co-applicants (two private and two municipal utilities). The project involves building two pressurized water nuclear reactors and associated facilities adjacent to the existing VEGP Units 1 and 2. The VEGP site is located in Burke County, Georgia, approximately 26 mi southeast of Augusta, Georgia. The reactors would draw cooling water from the Savannah River. Constructing the new reactors and associated on-site facilities would disturb about 556 acres at the VEGP site. The exact route of new transmission lines associated with the new reactors is not yet determined, but would extend from the VEGP west into Jefferson County, and then north into Warren and McDuffie Counties. Our comments follow.

Threatened and Endangered Species

By letter dated September 19, 2008, we concurred with the findings of NRC's Biological Assessment for the effects of early site preparation and preliminary construction activities at the VEGP site. The list of species protected under the Endangered Species Act (ESA) that occur in the project area has not changed since September 2008, and includes the wood stork, red-cockaded woodpecker, indigo snake, and Canby's dropwort. The DEIS indicates that the NRC is preparing a second Biological Assessment for construction and operations effects. As transmission line corridors and other pertinent construction details are more precisely defined,

Appendix F

please coordinate directly with the US Fish and Wild Life Service's Coastal Georgia Sub-office supervisor, Strant Colwell, at (912) 832-8739, to conclude the ESA consultation process for the project.

The Department had been concerned about the possible impacts of dredging the channel for barge delivery of reactors, containment vessels, and other large equipment; however, the DEIS notes (page 7-6) that Southern will instead deliver large components and materials by rail, and will not construct a barge slip or seek dredging of the Savannah River navigation channel. This change in the project plans eliminates our concerns related to ESA-protected aquatic species, such as the robust redhorse.

Avian Protection Plan

The DEIS notes that bird collisions with tall structures and transmission lines are among the impacts of building and operating the proposed project (pages 4-6 and 5-3), but does not describe mitigation measures for these impacts. The Department recommends that the NRC and Southern coordinate with us and the Georgia Department of Natural Resources Wildlife Division in the development of an Avian Protection Plan (APP). The Migratory Bird Treaty Act (MBTA) prohibits take of migratory birds except when specifically authorized by the Department of the Interior. The regulations implementing the MBTA (50 CFR Part 21) do not provide for permits authorizing take of migratory birds that may be killed or injured by activities that are otherwise lawful, such as by the construction and operation of power transmission lines. The Bald and Golden Eagle Protection Act provides for very limited issuance of permits that authorize take of eagles when such take is associated with otherwise lawful activities, is unavoidable despite implementation of advanced conservation practices, and is compatible with the goal of stable or increasing eagle breeding populations. The overall goal of the APP would be to minimize avian mortality associated with the proposed facilities.

The Department appreciates the opportunity to comment on this project. If you have questions or concerns about our comments, I can be reached on (404) 331-4524 or via email at gregory_hogue@ios.doi.gov.

Sincerely yours,

Gregory Hogue
Regional Environmental Officer

cc: Jerry Ziewitz – FWS
    Brenda Johnson – USGS
    David Vela - NPS
    OEPC – WASH

Appendix F

February 24, 2011

Ms. Sandra Tucker
Field Supervisor
Georgia Ecological Services
U.S. Fish and Wildlife Service
105 West Park Drive, Suite D
Athens, GA 30606

SUBJECT: BIOLOGICAL ASSESSMENT FOR THREATENED AND ENDANGERED SPECIES AND DESIGNATED CRITICAL HABITAT FOR THE VOGTLE ELECTRIC GENERATING PLANT, UNITS 3 AND 4 COMBINED LICENSES APPLICATION

Dear Ms. Tucker:

The U.S Nuclear Regulatory Commission (NRC) has prepared the enclosed Biological Assessment (BA) associated with Southern Nuclear Operating Company, Inc. (Southern) and its four co-applicants request for combined licenses (COLs) for Vogtle Electric Generating Plant (VEGP) Units 3 and 4. The assessment examines the potential impacts of construction and operation of the facility on threatened or endangered species. The purpose of this letter is to request the U.S Fish and Wildlife Service's (FWS) concurrence with the NRC staff's determination in the assessment that threatened and endangered species are not likely to be adversely affected by the proposed action.

The proposed action is NRC issuance of COLs for two new nuclear power reactor units at the VEGP Site near Waynesboro, GA. The BA evaluates the effects of the proposed action on four Federally listed threatened or endangered species identified in your October 20, 2010, letter. The Federally listed species are: (1) one plant: Canby's dropwort (*Oxypolis canbyi*), (2) two birds: the wood stork (*Mycteria americana*) and red-cockaded woodpecker (*Picoides borealis*) and (3) one reptile: eastern indigo snake (*Drymarchon couperi*). In developing the BA, the NRC staff performed research, reviewed information provided by the applicant, and relied on information provided by FWS (i.e., current listings of species provided by the FWS Field Office, Brunswick, GA) in reaching its conclusion.

The FWS previously reviewed the NRC staff's BA developed in connection with Southern's VEGP, Units 3 and 4 Early Site Permit (ESP) request. The VEGP ESP Site is located adjacent to the existing VEGP, Units 1 and 2. The proposed Federal action at that time was issuance of a permit for a site suitable for constructing and operating additional nuclear power facilities and to conduct site preparation and limited construction activities under provisions of Title 10, Part 52 of the *Code of Federal Regulations*. Because issuance of COLs would authorize both construction and operation of the proposed new units, the enclosed assessment addresses the potential impact to threatened and endangered species, including impacts associated with construction and operation of offsite transmission lines.

Appendix F

S. Tucker - 2 -

The Federally listed species considered in the BA for the ESP included (1) three plants: smooth coneflower (*Echinacea laevigata*), Canby's dropwort (*Oxypolis canbyi*), and relict trillium (*Trillium reliquum*), (2) two birds: the wood stork (*Mycteria americana*) and red-cockaded woodpecker (*Picoides borealis*), (3) one reptile: American alligator (*Alligator mississippiensis*), and (4) one amphibian: flatwoods salamander (*Ambystoma cingulatum*). The USFWS reviewed the BA associated with the ESP and in a letter dated September 19, 2008, concluded that "…. that the species under the jurisdiction of the Service have been adequately addressed for limited site-preparation activities at the Vogtle site." The ESP and limited work authorization was subsequently approved by the NRC on August 26, 2009.

If you have any questions regarding this BA or the staff's request, please contact Ms. Mallecia Sutton, NRC Environmental Project Manager via telephone at 301-415-0673 or via e-mail to Mallecia.Sutton@nrc.gov.

Sincerely,

/RA/

Gregory Hatchett, Chief
Environmental Projects Branch 1
Division of Site and Environmental Reviews
Office of New Reactors

Docket Nos.: 52-025
52-026

Enclosure:
As stated

cc w/o encl: See next page

Appendix F

S. Tucker   - 2 -

The Federally listed species considered in the BA for the ESP included (1) three plants: smooth coneflower (*Echinacea laevigata*), Canby's dropwort (*Oxypolis canbyi*), and relict trillium (*Trillium reliquum*), (2) two birds: the wood stork (*Mycteria americana*) and red-cockaded woodpecker (*Picoides borealis*), (3) one reptile: American alligator (*Alligator mississippiensis*), and (4) one amphibian: flatwoods salamander (*Ambystoma cingulatum*). The USFWS reviewed the BA associated with the ESP and in a letter dated September 19, 2008, concluded that "…. that the species under the jurisdiction of the Service have been adequately addressed for limited site-preparation activities at the Vogtle Site." The ESP and limited work authorization was subsequently approved by the NRC on August 26, 2009.

If you have any questions regarding this BA or the staff's request, please contact Ms. Mallecia Sutton, NRC Environmental Project Manager via telephone at 301-415-0673 or via e-mail to Mallecia.Sutton@nrc.gov.

                                    Sincerely,

                                    /RA/

                                    Gregory Hatchett, Chief
                                    Environmental Projects Branch 1
                                    Division of Site and Environmental Reviews
                                    Office of New Reactors

Docket Nos.: 52-025
             52-026

Enclosure:
As stated

cc w/o encl: See next page

Distribution:
Public   P Moulding   N Kuntzelman   Sackschewsky (PNNL)   OPA
N Chokshi   T Chandler, R1   G Hatchett   M Sutton   G Hawkins
K Leigh(PNNL) S Flanders   MCain, SRI   S Coffin (NWE1)   K Clark, R2
RidsNroDser   RidsNroDnrl

ADAMS Accession No: ML103410229 [Ml103410233-pkg]   NRO-002

Office	NRO/DSER/PM	DSER/LA/RAP1	DSER/RENV	OGC	DSER/BC
Name	MSutton	GHawkins	NKuntzelman	PMoulding(NLO subject to edits)	GHatchett
Date	12/7/2010	12/8/2010	12/8/2010	1/26/2011	2/24/2011

OFFICIAL RECORD COPY

Appendix F

Southern Nuclear - Vogtle Mailing List

Mr. M. Stanford Blanton
Esquire
Balch and Bingham, LLP
P.O. Box 306
Birmingham, AL 35201

Ms. Michele Boyd
Legislative Director
Energy Program
Public Citizens Critical Mass Energy
 and Environmental Program
215 Pennsylvania Avenue, SE
Washington, DC 20003

Mr. Marvin Fertel
 Senior Vice President
 and Chief Nuclear Officer
Nuclear Energy Institute
1776 I Street, NW
Suite 400
Washington, DC 20006-3708

Lucious Abram
County Commissioner
Office of the County Commissioner
Burke County Commission
PO Box 1626
Waynesboro, GA 30830

O. C. Harper, IV
Vice President - Resources Planning and
Nuclear Development
Georgia Power Company
241 Ralph McGill Boulevard
Atlanta, GA 30308

Mr. Steven M. Jackson
Senior Engineer - Power Supply
Municipal Electric Authority of Georgia
1470 Riveredge Parkway, NW
Atlanta, GA 30328-4684

Mr. Louis B. Long
Vice President Technical Support
Southern Nuclear Operating Company, Inc.
P.O. Box 1295
Birmingham, AL 35201-1295

Director
Consumer's Utility
Counsel Division
Governor's Office of Consumer Affairs
2 Martin Luther King, Jr. Drive
Plaza Level East, Suite 356
Atlanta, GA 30334-4600

Mr. Arthur H. Domby, Esquire
Troutman Sanders
Nations Bank Plaza
 600 Peachtree Street, NE
Suite 200
Atlanta, GA 30308-2216

Mr. Jeffrey T. Gasser
Executive Vice President
Southern Nuclear Operating Company, Inc.
P.O. Box 1295
Birmingham, AL 35201-1295

Laurence Bergen
Oglethorpe Power Corp.
2100 E Exchange Pl,
PO Box 1349
Tucker, GA 30085-1349

Mr. Charles R. Pierce
Vogtle Deployment Licensing Manager
Southern Nuclear Operating Co., Inc.
PO Box 1295
Birmingham, AL 35201-1295

Resident Inspector
Vogtle Plant
8805 River Road
Waynesboro, GA 30830

Appendix F

Southern Nuclear - Vogtle Mailing List

Resident Manager
Mr. Reece McAlister
Executive Secretary
Georgia Public Service Commission
Atlanta, GA 30334

Mr. Thomas O. McCallum
Site Development Project Engineer
Southern Nuclear Operating Co., Inc.
PO Box 1295
Birmingham, AL 35201-1295

Mr. Joseph (Buzz) Miller
Executive Vice President
Southern Nuclear Operating Company, Inc.
P.O. Box 1295
Birmingham, AL 35201-1295

Mr. Thomas Moorer
Environmental Project Manager
Southern Nuclear Operating Co., Inc.
PO Box 1295
Birmingham, AL 35201-1295

David Bernhart
Assistant Regional Administrator
 for Protect Resources
National Marine Fisheries Service
263 13th Avenue South
St. Petersburg, Florida 33701

Sam Booher
Concerned Citizen
4387 Roswell Drive
Augusta, GA 30907

Claude Howard
394 Nathaniel Howard Rd
Waynesboro, GA 30830

Glenn Carroll
Nuclear Watch South
PO Box 8574
Atlanta, GA 31106

Oglethorpe Power Corporation
Alvin W. Vogtle Nuclear Plant
7821 River Road
Waynesboro, GA 30830

Mr. Jerry Smith
Commissioner
 District 8
Augusta-Richmond County Commission
1332 Brown Road
Hephzibah, GA 30815

Mr. Robert E. Sweeney
IBEX ESI
4641 Montgomery Avenue
Suite 350
Bethesda, MD 20814

Bentina C. Terry
Southern Nuclear Operating Company, Inc.
PO Box 1295, BIN B-022
Birmingham, AL 35201-1295

Courtney Hanson
250 Georgia Avenue
Atlanta, GA 30312

Annie Laura Stephens
146 Nathanield Howard Rd
Waynesboro, GA 30830

Lucious Abrams
Burke County Commissioner
2032 Bough Red Hill
Keysville, GA 30816

Richard H. Byne
537 Jones Avenue
Waynesboro, GA 30830

Appendix F

Southern Nuclear - Vogtle Mailing List

Tommy Mitchell
Burke County Schools
352 Southside Dr
Waynesboro, GA 30830

George DeLoach
Mayor of Waynesboro
201 Oak Lane
Waynesboro, GA 30830

Robin Baxley
Best Office Solutions
142 S. Liberty St
Waynesboro, GA 30830

Appendix F

Email
APH@NEI.org (Adrian Heymer)
awc@nei.org (Anne W. Cottingham)
robin@bestofficesolutions.net (Robin Y. Baxley)
BrinkmCB@westinghouse.com (Charles Brinkman)
deloachjane@hotmail.com (George DeLoach)
chris.maslak@ge.com (Chris Maslak)
crpierce@southernco.com (C.R. Pierce)
cwaltman@roe.com (C. Waltman)
david.hinds@ge.com (David Hinds)
david.lewis@pillsburylaw.com (David Lewis)
dlochbaum@UCSUSA.org (David Lochbaum)
erg-xl@cox.net (Eddie R. Grant)
frankq@hursttech.com (Frank Quinn)
greshaja@westinghouse.com (James Gresham)
james.beard@gene.ge.com (James Beard)
jgutierrez@morganlewis.com (Jay M. Gutierrez)
jim.riccio@wdc.greenpeace.org (James Riccio)
jim@ncwarn.org (Jim Warren)
JJNesrsta@cpsenergy.com (James J. Nesrsta)
Joseph_Hegner@dom.com (Joseph Hegner)
KSutton@morganlewis.com (Kathryn M. Sutton)
kwaugh@impact-net.org (Kenneth O. Waugh)
lynchs@gao.gov (Sarah Lynch - Meeting Notices Only)
maria.webb@pillsburylaw.com (Maria Webb)
mark.beaumont@wsms.com (Mark Beaumont)
matias.travieso-diaz@pillsburylaw.com (Matias Travieso-Diaz)
mcaston@southernco.com (Moanica Caston)
media@nei.org (Scott Peterson)
mike_moran@fpl.com (Mike Moran)
tmitche@eburke.k12.ga.us (Tommy Mitchell)
nirsnet@nirs.org (Michael Mariotte)
eogleyoliver@gmail.com (Emma Ogley-Oliver)
patriciaL.campbell@ge.com (Patricia L. Campbell)
paul.gaukler@pillsburylaw.com (Paul Gaukler)
bobby@waud.org (Bobbie Paul)
Paul@beyondnuclear.org (Paul Gunter)
phinnen@entergy.com (Paul Hinnenkamp)
pshastings@duke-energy.com (Peter Hastings)
RJB@NEI.org (Russell Bell)
RKTemple@cpsenergy.com (R.K. Temple)
Crivard (?)
roberta.swain@ge.com (Roberta Swain)
burkechamber@roelco.net (Ashley Roberts)
sandra.sloan@areva.com (Sandra Sloan)
sfrantz@morganlewis.com (Stephen P. Frantz)
sbooher@aol.com (Sam Booher)
steven.hucik@ge.com (Steven Hucik)
tomccall@southernco.com (Tom McCallum)
diannevalentin@gmail.com (Dianne Valentin)
pvince20@gmail.com (Patricia Vincent)
waraksre@westinghouse.com (Rosemarie E. Warak)

Appendix F

Biological Assessment

U.S. Fish and Wildlife Service

Vogtle Electric Generating Plant
Combined Licenses Application

U.S. Nuclear Regulatory Commission Combined Licenses Application
Docket Nos. 52-025; 52-026

Burke County, Georgia

February 2011

U.S. Nuclear Regulatory Commission
Rockville, Maryland

Appendix F

Contents

1.0	Introduction 1
2.0	VEGP Site Description 5
	2.1 Wildlife Habitat 6
3.0	Proposed Federal Actions 8
4.0	Potential Environmental Impacts 8
	4.1 Construction Impacts 8
	4.2 Operational Impacts 10
5.0	Evaluation of Impacts on Threatened or Endangered Species 15
	5.1 Red-Cockaded Woodpecker – Endangered 16
	5.2 Wood Stork – Endangered 18
	5.3 Canby's Dropwort – Endangered 20
	5.4 Eastern Indigo Snake – Threatened 22
6.0	Cumulative Effects 23
	6.1 VEGP Site 23
	6.2 Transmission Line ROW 25
	6.3 Summary 26
7.0	Conclusions 26
8.0	References 28

Figures

Figure 1. Proposed VEGP Site Footprint 35
Figure 2. Representative Delineated Corridor 36

Tables

Table 1. Federally Listed Species Potentially Occurring on and in the Vicinity of the VEGP Site and the Proposed Transmission Line Right-of-Way 5

Table 2. Federally Listed Species Potentially Affected by Operation of the Proposed Units 3 and 4 at the VEGP Site and Construction and Operation of the Proposed Transmission Line Right of Way 27

ii

Appendix F

Abbreviations/Acronyms

ac	acre(s)
AP1000	Advanced Passive 1000
APP	Avian Protection Program
BA	biological assessment
CCAA	Candidate Conservation Agreement with Assurances
CFR	Code of Federal Regulations
cm	centimeter(s)
COL	combined license
CWS	circulating water system
dBA	decibel(s) (acoustic)
DOE	U.S. Department of Energy
EA	environmental assessment
Eco-Sciences	Eco-Sciences of Georgia
EMFs	electromagnetic fields
EPP	environmental protection plan
EPRI	Electric Power Research Institute
ESA	Endangered Species Act
ESP	early site permit
FONSI	Finding of No Significant Impact
FR	Federal Register
ft	foot/feet
FWS	U.S. Fish and Wildlife Service
GDNR	Georgia Department of Natural Resources
GEIS	generic environmental impact statement
GPC	Georgia Power Company
GTC	Georgia Transmission Corporation
ha	hectare(s)
in.	inch(es)
kg/ha/mo	kilograms per hectare per month
km	kilometer(s)
kV	kilovolt(s)
lbs/ac/mo	pounds per acre per month
LWA	Limited Work Authorization
m	meter(s)

Appendix F

mi	mile(s)
MW(t)	megawatts thermal
NEPA	National Environmental Policy Act of 1969, as amended
NRC	U.S. Nuclear Regulatory Commission
NRCS	Natural Resources Conservation Service
Plant Wilson	Allen B. Wilson Combustion Turbine Plant
RDC	Representative Delineated Corridor
ROW	right(s)-of-way
SCDNR	South Carolina Department of Natural Resources
SCE&G	South Carolina Electric and Gas
SEIS	supplemental environmental impact statement
SERPPAS	Southeast Regional Partnership for Planning and Sustainability
SPL	sound pressure level
Southern	Southern Nuclear Operating Company, Inc.
TDS	total dissolved solids
TRC	Third Rock Consultants, LLC
USACE	U.S. Army Corps of Engineers
VEGP	Vogtle Electric Generating Plant
Westinghouse	Westinghouse Electric Company, LLC

Appendix F

1.0 Introduction

The U.S. Nuclear Regulatory Commission (NRC) is reviewing an application from Southern Nuclear Operating Company, Inc. (Southern), acting on behalf of itself and several co-applicants (i.e., Georgia Power Company [GPC], Oglethorpe Power Corporation, Municipal Electric Authority of Georgia, and the City of Dalton, Georgia) for combined licenses (COLs) to construct and operate two Westinghouse Electric Company, LLC (Westinghouse) Advanced Passive 1000 (AP1000) pressurized water reactors (Units 3 and 4) on the site of the Vogtle Electric Generating Plant (VEGP) in Burke County, Georgia. The VEGP Site and existing facilities are owned and operated by GPC, Oglethorpe Power Corporation, Municipal Electric Authority of Georgia, and the City of Dalton, Georgia. Southern is the licensee and operator of the existing VEGP Units 1 and 2, and has been authorized by the VEGP co-owners to apply for COLs to construct and operate two additional units (Units 3 and 4) at the VEGP Site.

On August 26, 2009, the NRC approved issuance of an early site permit (ESP) and a limited work authorization (LWA) for two additional nuclear units at the VEGP Site (NRC 2009) to Southern and the same four co-applicants. This approval was supported by information contained in NUREG-1872, *Final Environmental Impact Statement for an Early Site Permit (ESP) at the Vogtle Electric Generating Plant Site, Volumes* 1 and 2 and errata (NRC 2008a). The ESP resolved many safety and environmental issues and allowed Southern to "bank" the VEGP ESP Site for up to 20 years. The LWA authorized Southern to conduct certain limited construction activities at the site in accordance with Title 10 of the Code of Federal Regulations (CFR), Sections 50.10 and 52.24(c). As permitted by NRC regulations, the COL application references the VEGP ESP.

Southern's COL application addressed the impacts of constructing and operating two new nuclear units at the existing VEGP Site in Burke County, Georgia. The VEGP Site is approximately 42 km (26 mi) south of Augusta, Georgia. The proposed COL site is completely within the confines of the existing VEGP Site, with the new units to be constructed and operated adjacent to the existing Units 1 and 2 (Figure 1). In October 2009, as part of the COL application, Southern requested a second LWA that would authorize installation of reinforcing steel, sumps, drain lines, and other embedded items along with placement of concrete for the nuclear island foundation base slab.

Independent of the COL application and LWA request, Southern and GPC intend to construct and operate a new 500-kV transmission line to serve the proposed Units 3 and 4. The two new units would use some combination of the new and existing transmission lines. The exact route of the new transmission line has not been determined, but the new transmission line right-of-

Appendix F

way (ROW) would be routed northwest from the VEGP Site, passing west of Fort Gordon, a U.S.

Army facility west of Augusta, Georgia, and then north to the Thomson substation. The Thomson substation is located about 32 km (20 mi) west of Augusta, Georgia. The transmission line ROW would be approximately 46 m (150 ft) wide and approximately 97 km (60 mi) long (NRC 2008a). The new transmission line would require approximately 390 towers (NRC 2008a). Each tower would require foundation excavations. Transmission line siting in Georgia is regulated under Title 22 of the Georgia Code. Construction and operation of the potential transmission line is not authorized by the NRC and approval of that activity is thus not part of the NRC's determination on the COL application. However, that activity is considered in the environmental review in assessing potential impacts of the major Federal action of issuing the requested COLs. Using the Electric Power Research Institute-Georgia Transmission Corporation (EPRI-GTC) Transmission Line Siting Methodology (EPRI-GTC 2006), Southern and GPC (GPC 2007) identified a set of potential transmission routes within what they termed the Representative Delineated Corridor (RDC), as depicted in Figure 2. The RDC was used as the basis for environmental impact analysis. Although the precise route for the planned new transmission line has not yet been determined, it will be within the RDC.

As permitted by NRC regulations in 10 CFR Part 52, which contains NRC's reactor licensing regulations, the COL application references the VEGP ESP. In accordance with the applicable provisions of 10 CFR Part 51, which are the NRC regulations implementing the National Environmental Policy Act of 1969 (NEPA), NRC is required to prepare a supplemental environmental impact statement (SEIS) as part of its review of a COL application referencing an ESP. As required by 10 CFR 51.26, the NRC published the draft SEIS for public comment in the *Federal Register* (FR) on September 3, 2010.

During April, May, and June, 2010, Southern submitted requests for three ESP license amendments associated with the previously authorized LWA construction activities. These amendment requests sought authorization to use Category 1 and Category 2 backfill materials from additional onsite sources, including three new borrow areas, and to change the classification of engineered backfill over the side slopes of the excavations for Units 3 and 4 (Southern 2010a, b, c, d). NRC prepared environmental assessments (EA) and Findings of No Significant Impact (FONSI) for each license amendment request (NRC 2010a, b, c). These ESP license amendments were issued in May 2010 (NRC 2010d), June 2010 (NRC 2010e), and July 2010 (NRC 2010f). The ESP license amendments requesting authorization to use backfill materials from three new borrow areas resulted in changes to the construction footprint on the VEGP Site. The change in the site preparation footprint for additional borrow areas resulted in an additional 108 ha (267 ac) that was cleared and excavated for backfill material.

Appendix F

The SEIS, together with the ESP EIS (NRC 2008a), the ESP hearing proceedings, and the ESP license amendment EAs, provides the NRC staff's evaluation of the environmental effects of constructing and operating two new AP1000 reactors at the VEGP Site.

During the review of the ESP application, as part of the NRC's responsibilities under Section 7 of the Endangered Species Act (ESA), the NRC staff prepared a biological assessment (BA) documenting potential impacts on the Federally listed threatened or endangered species as a result of the site preparation (including construction of the onsite portion of the new 500-kV transmission line) and construction of Units 3 and 4 on the VEGP Site. The BA was submitted to U.S. Fish and Wildlife Service (FWS) on January 25, 2008 (NRC 2008b), and FWS concurred with the findings on September 19, 2008 (FWS 2008).

The NRC staff has concluded that, with respect to site preparation activities and construction of Units 3 and 4 on the VEGP Site (including construction of the onsite portion of the proposed transmission line), the COL action involves similar impacts to the same Federally listed species in the same geographic area as analyzed in the ESP; that no new species have been listed or proposed and no new critical habitat designated or proposed for the action area; and that, with respect to potential impacts to listed species from the activities previously analyzed, no relevant information has changed regarding the project since the earlier BA was submitted. Therefore, pursuant to 50 CFR 402.12(g), the ESA of 1973, as amended, the NRC staff proposes to incorporate the earlier BA by reference. Furthermore, NRC has prepared this BA to document potential impacts on Federally listed threatened or endangered terrestrial species resulting from operation of Units 3 and 4, including potential impacts anticipated from construction and operation of the proposed transmission line ROW. Operation of the transmission lines includes maintenance activities, such as herbicide applications, tree removal, and mowing.

In a letter dated January 7, 2010, NRC requested that the FWS Field Office in Brunswick, Georgia, provide information regarding Federally listed species and critical habitat that may have changed since the 2008 consultation (NRC 2010g). On February 12, 2010, FWS provided a response letter indicating listed species under FWS had been adequately addressed for limited site-preparation activities on the VEGP Site (FWS 2010a). On October 20, 2010, FWS provided an updated list of Federally listed threatened or endangered species that can be expected to occur in the project area (FWS 2010b). In addition to the federally listed species, FWS provided information on the bald eagle (*Haliaeetus leucocephalus*) and the gopher tortoise (*Gopherus polyphemus*) in the response letter.

The bald eagle was Federally delisted under the ESA in August 2007. In May 2007, National Bald Eagle Management Guidelines were published to assist in understanding protections afforded to and prohibitions related to the bald eagle under the Bald Eagle Act (FWS 2010b). There are bald eagle nests in Jefferson and McDuffie Counties in Georgia, and one known location of an active nest in McDuffie County in the vicinity of the proposed new transmission line (FWS 2010b). GPC stated that it would ensure the new transmission line ROW would not

3

Appendix F

come within 180 m (600 ft) of this known bald eagle nesting site (GPC 2007). Eagle nests on transmission/distribution structures or other electrical equipment have not been documented in Georgia (GPC 2006): nevertheless, one of GPC's procedures in its Avian Protection Program (APP) includes contacting the FWS to advise the agency of the situation and to obtain additional instructions or permits, if an eagle's nest is encountered on a transmission/distribution structure (GPC 2006). Potential impacts to the bald eagle related to construction and operation of proposed Units 3 and 4, including impacts from construction and operation of the proposed transmission line, are discussed in the ESP EIS (NRC 2008a).

The gopher tortoise is a Georgia state threatened species and is currently under review by the FWS to be listed as threatened (FWS 2010b). There are no known populations of the gopher tortoise on the VEGP Site or within the proposed transmission corridor (GDNR 2009; FWS 2010b). Southern submitted a draft Candidate Conservation Agreement with Assurances (CCAA) for the gopher tortoise at the VEGP Site. This CCAA is currently under review by FWS (SERPPAS 2010). The draft CCAA does not include the offsite portions of the proposed transmission line. In the October 20, 2010 letter to NRC, FWS recommended that tortoise surveys be included in surveys that are conducted where sandhills habitat exists. FWS stated that there are several areas within the proposed transmission line corridor that have sandhills habitat that may contain gopher tortoises (FWS 2010b). Potential impacts to the gopher tortoise related to construction and operation of the proposed Units 3 and 4, including impacts from construction and operation of the proposed transmission line, will be included in the final COL SEIS.

Pursuant to Section 7(c) of the ESA of 1973, as amended, NRC has prepared this BA, which examines the potential impacts of facility operation related to the proposed Units 3 and 4 at the VEGP Site on threatened or endangered species, including potential impacts from transmission line construction and operation activities. This BA evaluates the effects of the proposed action on four Federally listed threatened or endangered species identified by FWS in its October 20, 2010, letter that may occur on or in the vicinity of the VEGP Site and/or in habitats crossed by the proposed transmission line (Table 1). The consultation is between NRC and FWS.

4

Appendix F

Table 1. Federally Listed Species Potentially Occurring on and in the Vicinity of the VEGP Site and the Proposed Transmission Line Right-of-Way

Scientific Name	Common Name	Federal Status[a]
Vascular Plant		
Oxypolis canbyi	Canby's dropwort	E
Birds		
Mycteria americana	wood stork	E
Picoides borealis	red-cockaded woodpecker	E
Reptile		
Drymarchon couperi	Eastern Indigo Snake	T

a. Federal status rankings determined by the FWS under the Endangered Species Act:
E = Endangered, T = Threatened.
Source: FWS 2010b

2.0 VEGP Site Description

The VEGP Site is located on the Savannah River shoreline approximately 24 km (15 mi) east-northeast of Waynesboro, Georgia, and 42 km (26 mi) southeast of Augusta, Georgia. The existing site consists of two Westinghouse pressurized water reactors, a turbine building, a switchyard, intake and discharge structures, and support buildings. Two generating units (Units 1 and 2) are currently operating at the site (Figure 1). The Allen B. Wilson Combustion Turbine Plant (Plant Wilson), a six-unit, oil-fueled combustion turbine facility built in 1974 and owned by GPC, and ancillary structures and systems related to Units 1 and 2 also are located onsite. The existing Units 1 and 2 and Plant Wilson would not be affected by this action.

The footprint for Units 3 and 4 is in a previously disturbed area adjacent to the existing VEGP Units 1 and 2 (Figure 1). The existing Units 1 and 2 and the proposed Units 3 and 4 would share certain support structures such as office buildings and water, wastewater, and waste-handling facilities; however, the new intake and discharge facilities for Units 3 and 4 would be separate from the intake and discharge facilities for Units 1 and 2. Each proposed Westinghouse AP1000 reactor would have a rated thermal power level of 3400 megawatts thermal MW(t) (NRC 2008a). For the circulating water cooling system for Units 3 and 4, Southern proposed natural-draft cooling towers, and for the service water system, mechanical-draft cooling towers.

The VEGP Site is approximately 1282.5 ha (3169 ac) in size and is located in the sandhills of the Upper Coastal Plain Region, approximately 48 km (30 mi) southeast of the Fall Line (Eco-Sciences 2007; NRC 2008a). The site has 12 soil types and several major habitat types, including ponds, pine plantations, native upland pines, and the bottomland hardwoods that are

5

Appendix F

found along stream drainages onsite and adjacent to the Savannah River (NRCS 2003; TRC 2006).

Directly across the Savannah River from the VEGP Site is the Savannah River Site, a U.S. Department of Energy (DOE) facility with restricted access (NRC 2008a). River swamp, bottomland hardwood, and upland pine-hardwood communities occur on the Savannah River Site within 10 km (6 mi) of the VEGP Site (NRC 2008a). The Savannah River Swamp comprises about 3800 ha (9400 ac) and borders the Savannah River on the southwestern edge of the Savannah River Site, adjacent to the VEGP Site (Wike et al. 2006).

2.1 Wildlife Habitat

The VEGP Site is characterized by low, gently rolling sandy hills. Scrub oaks, including turkey (*Quercus laevis*), post (*Q. stellata*), and willow oak (*Q. phellos*), and longleaf pine (*Pinus palustris*) occur in the upland wooded areas that were not previously cultivated. Red oak (*Q. rubra*), water oak (*Q. nigra*), and maple (*Acer* sp.) dominate the lowland hardwood areas. Bald cypress (*Taxodium distichum*) and water tupelo (*Nyssa aquatica*) characterize the Savannah River floodplain.

The longleaf pine-scrub oak community is found on ridge tops as well as south and west slopes in undisturbed upland areas on the VEGP Site. Common canopy species in this habitat include longleaf pine, turkey oak, and bluejack oak (*Q. incana*). The north and east slopes in the undisturbed uplands support the more mesic oak-hickory community. The canopy in this community is mainly composed of white oak (*Q. alba*), white ash (*Fraxinus americana*), mockernut hickory (*Carya alba*), and flowering dogwood (*Cornus florida*). A few turkey oaks and a scattering of shortleaf pine (*P. echinata*) are also present (TRC 2006). A steep bluff separates the dry upland forest from the intermittently flooded bottomland along the Savannah River. Common canopy species include oak, mockernut hickory, tuliptree (*Liriodendron tulipifera*), sweetgum (*Liquidambar styraciflua*), American elm (*Ulmus americana*), basswood (*Tilia americana*), and Florida maple (*A. barbatum*). The planted pine plantations on the VEGP Site are of various ages and differ in the stocking rates. The plantations vary from a nearly closed canopy with very little understory, to areas that resemble old fields with only scattered pine. Loblolly (*P. taeda*) and longleaf pines are the primary overstory species (TRC 2006). Pine plantations are managed through prescribed burning every 3 to 5 years, timber thinning after 20 years, and aesthetic cuts after thinning. Burning is limited to 25 to 30 percent of the upland and planted pine acreage each year (NRC 2008a).

The wetlands associated with the VEGP Site include those near the Savannah River, as well as those near ponds and streams located onsite. Principal water bodies onsite include Mallard Pond and two streams in the southern portion of the VEGP Site (Figure 1). Southern contracted with Eco-Sciences of Georgia (Eco-Sciences) to survey the VEGP Site in December 2006 to determine where jurisdictional waters of the United States occur. Approximately 69 ha (170 ac)

6

Appendix F

of potential jurisdictional wetlands were identified on the site during the Eco-Sciences survey (NRC 2008a). These include 48 wetlands, 6 perennial streams, 13 intermittent streams, and 3 ephemeral streams.

The proposed transmission line ROW is within the Piedmont and Coastal Plain Physiographic Regions of Georgia. The Piedmont is characterized by rolling hills and irregular plains. The soils are finely textured and can be highly erodible. The Coastal Plain is composed of mostly flat areas with some rolling hills with well-drained soils (GPC 2007). Using the Electric Power Research Institute-Georgia Transmission Corporation (EPRI-GTC) Transmission Line Siting Methodology (EPRI-GTC 2006), Southern and GPC identified a set of potential transmission routes within the RDC (Figure 2) (GPC 2007) that was used as the basis for environmental impact analysis. The RDC ranges from approximately 1.6 km (1 mi) to a little of 5 km (3 mi) in width and is approximately 80 km (50 mi) long. The actual routing of the 45m (150 ft) wide, up to about 97 km (60 mi) long transmission ROW would be within the RDC. The siting model takes into consideration important features, including residential and other developed areas, mining activities, wetlands and sensitive land uses, cultural resources, and endangered and other species of special interest. GPC conducted an aerial field verification of the RDC, and identified a narrowing of the modeled corridor to avoid wetlands and stream crossings and reduce the overall length and land area that potentially would be affected. The RDC depicts areas in which a transmission line should minimize adverse impact on people, places, and cultural resources; protect water resources, plants, and animals; maximize co-location of the new line; and balance these considerations to reduce the overall impact of the transmission line (GPC 2007).

In siting the new transmission line ROW, GPC would consult with the Georgia State Historic Preservation Officer, FWS, the Georgia Department of Natural Resources (GDNR), and the U.S. Army Corps of Engineers (Southern 2008). If wetlands are disturbed, construction would be conducted in accordance with necessary State and Federal permits to protect wetland areas (Southern 2008).

There are no U.S. Forest Service Wilderness Areas, Wild/Scenic Rivers, Wildlife Refuges, State Parks, or National Parks within the RDC (GPC 2007). The Savannah River and Brier Creek, a tributary of the Savannah River, are the primary waterways located in the RDC. The general wildlife habitats within the RDC include forested land, planted pine stands, open land, and open water. The exact habitat types within the new 500-kV transmission line ROW are not known at this time, but it is assumed they comprise similar habitats to those on the VEGP Site. GPC has estimated the total acreage for a 46-m (150-ft)-wide hypothetical representative ROW within the RDC to be 416 ha (1029 ac) (Southern 2007).

Appendix F

3.0 Proposed Federal Actions

The proposed Federal action is issuance of COLs, under the provisions of 10 CFR Part 52, for two AP1000 reactors at the VEGP Site, and an LWA for requested construction activities. The ESP EIS (NRC 2008a) disclosed the staff's analysis of the environmental impacts that could result from the construction and operation of these two new units. The draft COL SEIS (NRC 2010i) evaluated whether any new and potentially significant information has been identified that would alter the staff's conclusions regarding issues resolved in the ESP proceeding. In the draft ESP EIS and the COL SEIS, the NRC staff evaluated the impacts of construction and operation of two AP1000 units, with a total combined thermal power rating of 6800 MW(t). The proposed units would use a closed-cycle cooling system and require a single natural draft cooling tower for each unit.

4.0 Potential Environmental Impacts

This section provides information on the terrestrial impacts related to operation of the proposed Units 3 and 4 at the VEGP Site, including potential impacts from construction and operation of the proposed transmission line ROW. Construction and operation activities associated with the issuance of the COLs and LWA, including cumulative impacts, that could affect the Federally protected terrestrial species based on habitat affinities and life-history characteristics and the nature and spatial and temporal considerations of the activity are listed below:

- Construction
    - Transmission line ROW clearing and grading
    - Installation of new or upgraded transmission lines and towers

- Operation
    - Vegetation control in the transmission line ROW
    - Transmission line repairs or upgrades
    - Avian collisions with structures
    - Cooling tower operation.

4.1 Construction Impacts

The exact extent and types of wildlife habitats within the proposed new transmission line ROW are not known. Currently, Southern and GPC are evaluating the actual ROW alternatives for the transmission line within the RDC. The proposed transmission line ROW would be routed northwest from the VEGP Site, passing through Jefferson, McDuffie and Warren Counties. The ROW would pass west of Fort Gordon, and then continue north to the Thomson substation,

Appendix F

which is approximately 32 km (20 mi) west of Augusta, Georgia. It is anticipated that the transmission line would be about 46 m (150 ft) wide and 97 km (60 mi) long and would cover approximately 416 ha (1029 ac) (Southern 2007). A hypothetical transmission line ROW that represents what the GPC believes is a feasible route within the RDC was identified as part of a 2007 study (GPC 2007). Based on the GPC analysis, habitats within the ROW could include approximately 60 ha (148 ac) of forested habitat, 37 ha (91.5 ac) of forested wetlands, 133 ha (329 ac) of planted pine, 2.6 ha (6.4 ac) of open water, and 64 ha (158 ac) of open land (GPC 2007). Other land-use categories identified as potentially being impacted, such as mine/quarry, utility, transportation, and row crops, provide little value as wildlife habitat. Construction activities would avoid wetlands to the extent practicable. In the event that wetlands are encountered, construction would be conducted in accordance with the necessary permits obtained to protect wetland areas (GPC 2007).

A wide variety of wildlife common to Georgia is expected to occur within the transmission line ROW. The greatest extent of wildlife diversity is expected to occur within areas that support an interspersion of native upland, wetland, and aquatic habitats, and less diversity is expected in disturbed or developed lands. Lower-quality wildlife habitat is represented by areas cleared for utilities, roads, agricultural and residential development; and disturbed habitats such as pastureland, and open land.

Potential impacts on Federally listed threatened and endangered species from construction on the proposed transmission line ROW would include loss of habitat (temporary and permanent), presence of humans, heavy-equipment operation, traffic, noise, and avian collisions. The use of heavy equipment would likely displace or destroy wildlife that inhabit the areas that will be developed. Larger and more mobile animals would likely flee the area, while less mobile animals such as reptiles, amphibians, and small mammals would be at greater risk of death. Although the surrounding forest and wetland habitat would be available for displaced animals, the movement of wildlife into surrounding areas would increase competition for available space and could result in increased predation and decreased fecundity for certain species. These conditions could lead to a temporary localized reduction in population size for particular species. When construction activities are completed, species that can adapt to disturbed or developed areas may readily re-colonize portions of the site where suitable habitat remains, is replanted, or restored.

Forests or forested wetlands within the corridors would be converted to and maintained in an herbaceous or scrub-shrub condition. Species dependent on forest habitats or those that are sensitive to forest fragmentation could decline or be displaced, such as the red-cockaded woodpecker (*Picoides borealis*). Wildlife also would be affected by equipment noise and traffic, and birds could be injured if they collide with new transmission towers and conductors or the equipment used to install these components. However, increased noise levels associated with installation of the transmission lines would be of short duration and likely intermittent. Thus, the

9

Appendix F

impact on wildlife from noise is expected to be temporary and minor. Similarly, the potential for traffic-related wildlife mortality also is expected to be low because relatively small crews would spend only a limited time in each area as construction progresses over large geographic areas.

GPC would site the transmission line in accordance with Georgia Code Title 22, Section 22-3-161. GPC's procedures for implementing this code include consultation with FWS as well as an evaluation of impacts to special habitats (including wetlands) and threatened and endangered species. In addition, GPC would comply with all applicable laws, regulations, and permit requirements, and would use good engineering and construction practices (Southern 2008). GPC has developed an APP that includes guidelines for siting new transmission lines. When siting new transmission lines, substations, or other GPC facilities, available information on migratory and resident bird populations will be taken into account to ensure that the lines or facilities will have as little adverse impact as practicable on these bird species (GPC 2006).

In areas where agencies are concerned about the safety of protected birds, consideration of appropriate siting and placement will reduce the likelihood of collisions. When possible, areas with known bird concentrations will be avoided, and such vegetation or topographic characteristics that would naturally lead to shielding the birds from collision will be used. If this is not possible, installing visibility devices also may reduce the risk of collision. Examples of these devices are marker balls or other line visibility devices placed in varying configurations, depending on the line or locations. The effectiveness of these devices has been validated by Federal and state agencies in conjunction with Edison Electric Institute (GPC 2006).

When designing power transmission lines in high–bird-use areas or on Federal Lands, GPC construction standards for transmission, distribution, and substation equipment and facilities will reflect the most appropriate and practicable "raptor-safe" stands for new construction consistent with available information. The objective is to provide 1.5 m (60 in.) between energized conductors and grounded hardware, or to insulate energized hardware if such spacing is not possible. The design standards are consistent with raptor-safe specifications recommended by Federal wildlife agencies (GPC 2006).

4.2 Operational Impacts

Potential impacts on terrestrial habitats and Federally listed species related to the operation of the proposed Units 3 and 4 may result from cooling-system operation and operation of the transmission system. The proposed cooling system for Units 3 and 4 is a closed-cycle system employing natural draft cooling towers. The heat would be transferred to the atmosphere in the form of water vapor and drift. Vapor plumes and drift may affect wildlife habitat. In addition, bird collisions and noise-related impacts are possible with natural draft cooling towers.

Electric transmission systems potentially can affect terrestrial habitat and Federally listed species through ROW maintenance, bird collisions with transmission lines, and electromagnetic

10

Appendix F

fields (EMFs). Southern estimates that one additional 500-kV transmission line would be necessary to distribute the additional power generated by Units 3 and 4 (Southern 2008). Maintenance activities on the new transmission line ROW would be the responsibility of GPC (Southern 2008). Each of these topics is discussed in the following paragraphs.

4.2.1 Impacts on Vegetation

Impacts on Federally listed species may result from cooling tower drift, icing, fogging, or increased humidity. Through the process of evaporation, the total dissolved solids (TDS) concentration in the circulating water system (CWS) increases. A small percentage of the water in the CWS is released into the atmosphere as fine droplets containing elevated levels of TDS that can be deposited on nearby vegetation. Operation of the CWS would be based on four-cycles of concentration, which means the TDS in the make-up water would be concentrated approximately four times before being released.

Depending on the make-up source water body, the TDS concentration in the drift can contain high levels of salts that, under certain conditions and for certain species, can be damaging. Vegetation stress can be caused from drift with high levels of deposited TDS, either directly by deposition onto foliage or indirectly from the accumulation in the soils. The maximum estimated cumulative deposition rate is less than 10.0 kg/ha/mo (9 lbs/ac/mo) at 490 m (1600 ft) north of the cooling towers (NRC 2008a). The location of the maximum deposition rate is in the vicinity of the proposed switchyard for Units 3 and 4, which is more than 1.6 km (1 mi) from the northern site boundary. General guidelines for predicting effects of drift deposition on plants suggest that many species have thresholds for visible leaf damage in the range of 10 to 20 kg/ha/mo (9 to 18 lbs/ac/mo) on leaves during the growing season (NRC 1996). The maximum deposition for the proposed Units 3 and 4 is below the level that could cause visible leaf damage in many common species.

Southern expects the longest vapor plume associated with the new towers would be 10 km (6 mi), but would only occur 3.9 percent of the time (NRC 2008a). The longest plume length would occur in the winter months and the shortest in the summer months. Ground-level fogging and icing do not occur currently at the cooling towers for the existing Units 1 and 2 and are not expected to occur at the new cooling towers associated with the proposed Units 3 and 4.

4.2.2 Bird Collisions with Cooling Towers

The natural draft cooling towers associated with the proposed Units 3 and 4 would be 180 m (600 ft) high (Southern 2008). The VEGP Site is located adjacent to the Savannah River, and although migratory birds pass through the vicinity of the VEGP Site, it is not located on a major American flyway. No formal bird collision surveys have been conducted at the VEGP Site. However, the Environmental Protection Plan (EPP) for VEGP Units 1 and 2 stipulates that any excessive bird-impact events be reported to NRC within 24 hours (Southern 1989). No

11

Appendix F

excessive bird-impact events have been reported onsite. The conclusion presented in the *Generic Environmental Impact Statement (GEIS) for License Renewal of Nuclear Plants* is that bird collisions with natural draft cooling towers are of small significance at all operating nuclear plants, including those with multiple cooling towers (NRC 1996).

4.2.3 Noise

The effects of noise on most wildlife species are not well understood partly because noise disturbance cannot be generalized across species or genera, and there may be response differences among individuals or groups of individuals of the same species (Larkin 1996; AMEC Americas Limited 2005). An animal's response to noise can depend on a variety of factors including the noise level, frequency distribution, duration, background noise, time of year, animal activity, age, and sex (AMEC Americas Limited 2005). The potential effects of noise on wildlife include acute or chronic physiological damage to the auditory system; increased energy expenditure; physical injury incurred during panic responses; and interference with normal activities, such as feeding; and impaired communications among individuals and groups (AMEC Americas Limited 2005). The impacts of these effects might include habitat loss through avoidance, reduced reproductive success, and mortality. Long-term noise thresholds have not been established for wildlife; evidence for habituation is limited; long-term effects are generally unknown; and how observed behavioral and physiological response might be manifested ecologically and demographically are poorly understood (AMEC Americas Limited 2005).

The noise levels from natural-draft cooling tower operation and diesel generators are estimated to be approximately 55 decibels (dBA) SPL (sound pressure level) at 300 m (1000 ft) (NRC 2008a). Researchers have found that dBA measurements contain frequencies that are out of the hearing bandwidth of birds and some mammals and are not inclusive of the total hearing range for other animals. Consequently, the dBA weighting system does not accurately characterize sound exposure or hearing response for wildlife (Dooling 2002; AMEC Americas Limited 2005). Natural-draft cooling towers emit broadband noise that is spectrally very similar to environmental (wind) noise. In the case of relatively flat spectra, the spectrum level of cooling tower and diesel generator noise, given the estimated dBA SPL, would be approximately 15 dB SPL. Cooling tower noise does not change appreciably with time (i.e., it is at steady state), and the estimated noise level at 300 m (984 ft) is well below the 80 to 85-dBA SPL threshold at which birds and small mammals are startled or frightened (Golden et al. 1980). Using the startle criterion reported by Golden et al. (1980), the noise level expected to be generated by cooling tower and diesel generator operations would only approach startle levels in the immediate vicinity (within 5 m [16.4 ft]) for noise with approximately 60 dBA SPL at 300 m [984 ft]) of the tower or generator. In addition, birds and other animals show habituation to acoustic deterrents (complex sounds designed with spectral components to be within the hearing band of the target animal). Thus, noise generated by natural draft cooling towers would be unlikely to disturb

Appendix F

transient wildlife beyond the VEGP Site perimeter fence, which is over 300 m (984 ft) from the towers. Seasonal or long-term resident wildlife could be expected to habituate to cooling tower and generator noise.

Impacts to species as a result of their response to noise (i.e., ranging from startle to avoidance) within the distance of the VEGP perimeter fence, if any, would be negligible because of the large expanses of open habitat available into which mobile wildlife species could move if disturbed. In addition, the new towers would be near the existing VEGP Unit 1 and 2 facilities, where wildlife have likely acclimated to typical operating facility noise levels. Consequently, the potential for startle and avoidance responses by wildlife posed by the incremental noise resulting from the operation of the two new natural-draft cooling towers for the proposed Units 3 and 4 and other facilities at the VEGP Site would be minimal.

4.2.4 Transmission Line Right-of-Way Management (Cutting and Herbicide Application)

Southern stated that the same vegetation management practices currently employed by GPC for the existing Units 1 and 2 transmission line ROWs (such as hand-cutting on an as-needed basis) would be applied to the proposed new 500-kV transmission line ROW (Southern 2008).

GPC performs aerial inspections of transmission line ROWs five times each year to support routine maintenance activities. These surveys are normally conducted using a helicopter. The noise may startle and temporarily displace wildlife. However, these impacts are of short durations and occur in very localized areas. Woody growth is cleared from transmission line ROWs on a 5-year maintenance cycle. This cycle may vary based on public concerns, local ordinances, line maintenance, or environmental considerations. Vegetation management includes use of herbicides, hand tools, and light equipment. Hand cutting or herbicides are used in areas that cannot be mowed either because it is impractical or because of environmental concerns. Herbicide use is conducted in accordance with manufacturer specifications and by licensed applicators. Any spills of fuel and/or lubricants that occur as a result of equipment use in the transmission line ROWs are immediately cleaned up and reported. GPC cooperates with GDNR to manage sites considered environmentally sensitive within the transmission line ROWs (Southern 2008). GPC has developed recommendations for maintenance practices for the protection of pitcher plants, caves, nests, rookeries, and habitat such as rock outcrops that occur within GPC transmission line ROWs (Southern 2007).

GPC also has developed an APP that includes recommendations on procedures for GPC personnel to follow if a Federally Endangered Species nest is encountered within the transmission line ROW. The GPC Environmental Field Service office will provide GPC staff with FWS-compliant guidelines and/or recommendations for management of these nests (GPC 2006).

13

Appendix F

Avian mortalities resulting from collisions with conductors, guy wires, and overhead ground (static) wires have not been specifically documented on GPC system components, but are known to occur on other utilities' systems and communication systems. GPC has installed spiral vibration dampers to increase visibility on some of the transmission lines, especially along the coastal areas where the wood stork is known to nest and forage (GPC 2006). Section 4.1 of the EPP for the existing Units 1 and 2 stipulates that any excessive bird-impact events be reported to NRC within 24 hours (Southern 1989). Transmission line and ROW maintenance personnel have not reported bird deaths attributed to collisions or contact with Units 1 and 2 transmission lines (Southern 2008).

EPRI (1993) notes that factors appearing to influence the rate of avian impacts with structures are diverse and related to bird behavior, the structure attributes, and weather. Structure height, location, configuration, and lighting also appear to play a role in avian mortality. Weather such as low cloud ceilings, advancing fronts, and fog also contribute to this phenomenon. Larger birds such as waterfowl are more prone to collide with transmission lines, especially when they cross wetland areas used by large concentrations of birds (EPRI 1993).

EPRI (1993) documents electrocution of large birds, particularly eagles, as a source of mortality that could be significant to listed species. However, electrocutions do not normally occur on lines whose voltages are greater than 69 kV because the distance between lines is too great to be spanned by birds (EPRI 1993). The voltage of the proposed new transmission line is greater than 69 kV; therefore, bald eagles and other large bird populations should not be noticeably affected by transmission-line electrocutions. GPC has implemented an APP to monitor and address the impacts of transmission lines on birds. Any impact events would be coordinated with GPC's Environmental Field Services and, if necessary, coordination also would involve FWS (GPC 2006).

4.2.5   Impact of EMFs on Flora and Fauna

Electromagnetic fields (EMFs) are unlike other agents that have an adverse impact (e.g., toxic chemicals and ionizing radiation) in that dramatic acute effects cannot be demonstrated and long-term effects, if they exist, are subtle (NRC 1996). As discussed in the GEIS (NRC 1996), a careful review of biological and physical studies of EMFs did not reveal consistent evidence linking harmful effects with field exposures. Thus, the conclusion presented in the GEIS (NRC 1996) was that the impacts of EMFs on terrestrial flora and fauna were of small significance at operating nuclear power plants, including transmission systems with variable numbers of transmission lines. Since 1997, over a dozen studies have been published that looked at cancer in animals that were exposed to EMFs for all or most of their lives (Moulder 2003). These studies have found no evidence that EMFs cause any specific types of cancer in rats or mice (Moulder 2003).

14

Appendix F

5.0 Evaluation of Impacts on Threatened or Endangered Species

This section describes Federally listed threatened or endangered terrestrial species and designated and proposed critical habitat that may occur on or in the vicinity of the VEGP Site and/or in habitats that would be crossed by the proposed transmission line ROW (Table 1). This list is composed of the Federally listed species identified in the October 20, 2010, FWS letter to NRC (FWS 2010b).

Surveys for species of interest, including those Federally listed species classified as threatened or endangered, proposed for listing, or candidate species were performed in spring, summer, and fall 2005 at the VEGP Site by Third Rock Consultants, LLC (TRC). The surveys were conducted on 675 ha (1669 ac) of the 1283 ha (3169 ac) that comprise the VEGP Site (TRC 2006). The American alligator (*Alligator mississippiensis*) was the only Federally listed species observed on the VEGP Site during the 2005 surveys. One adult alligator was observed in Mallard Pond during the summer survey (TRC 2006). It is Federally listed as threatened because it is similar in appearance to the endangered American crocodile (*Crocodylus acutus*). It is not included in this assessment based on input from FWS in its October 20, 2010 letter to NRC (FWS 2010b). Furthermore, based on the contents of the October 2010 letter, three other species that were addressed in the ESP BA (the smooth coneflower, relict trillium, and flatwoods salamander) were not further considered in this assessment because they were not identified as occurring in the project area or the proposed transmission line ROW.

The RDC is based on the EPRI-GTC siting model, developed in Georgia, to identify a reasonable corridor for locating the proposed 500 kV transmission line. The siting model takes into consideration important features, including wetlands and sensitive land uses and endangered and other species of special interest. The RDC represents a narrowing of the modeled corridor to avoid wetlands and stream crossings and reduce the overall length and land area potentially affected (GPC 2007). GPC would site the transmission line in accordance with Georgia Code Title 22, Section 22-3-161, and has developed an APP that includes provisions for siting new transmission lines (GPC 2006). GPC's procedures for implementing this code include consultation with FWS as well as an evaluation of impacts to special habitats (including wetlands) and threatened and endangered species (Southern 2008). At this time, on-the-ground surveys for Federally listed species have not been conducted in the RDC.

Four Federally listed terrestrial plant and animal species may occur on or in the vicinity of the VEGP Site and/or in the vicinity of the RDC (FWS 2010b). These four species – the red cockaded woodpecker (*Picoides borealis*), the wood stork (*Mycteria americana*), Canby's dropwort (*Oxypolis canbyi*), and the Eastern indigo snake (*Drymarchon couperi*) – are discussed below. No designated or proposed critical habitat for terrestrial species occurs on or in the general area of the site or the RDC.

Appendix F

5.1 Red-Cockaded Woodpecker – Endangered

The red-cockaded woodpecker (*Picoides borealis*), was listed by the FWS as endangered in 1970 (35 FR 16047). The red-cockaded woodpecker's historic range extended from north Florida to New Jersey and Maryland, as far west as Texas and Oklahoma, and inland to Missouri, Kentucky, and Tennessee. This species has been extirpated in New Jersey, Maryland, Tennessee, Missouri, and Kentucky (FWS 2007a), and currently, it is estimated that about 6000 family groups of red-cockaded woodpeckers, or 15,000 birds, remain from Florida north to Virginia and west to southeast Oklahoma and eastern Texas. Critical habitat has not been established for red-cockaded woodpeckers (FWS 2007b). In 1998, there were 665 family groups of red-cockaded woodpeckers in Georgia (GDNR 1999).

The red-cockaded woodpecker is endemic to open, mature, and old growth pine ecosystems in the southeastern United States. Red-cockaded woodpeckers require open pine woodlands and savannahs with large old pines for nesting and roosting habitat for family groups (clusters). Large old pines are required as cavity trees because the cavities are excavated completely within inactive heartwood and the higher incidence of heartwood decay in older trees greatly facilitates excavation. Cavity trees must be in open stands with little or no hardwood midstory and few or no overstory hardwoods. Suitable foraging habitat consists of mature pines with an open canopy, low densities of small pines, little or no hardwood or pine midstory, few or no overstory hardwoods, and abundant native bunchgrass and forb groundcovers (FWS 2003).

Red-cockaded woodpeckers are a cooperatively breeding species, living in family groups that typically consist of a breeding pair with or without one or two male helpers. In red-cockaded woodpeckers (and other cooperative breeders), a large pool of helpers is available to replace breeders when they die. Helpers do not disperse very far and typically occupy vacancies on their natal territory or a neighboring one (FWS 2003). A typical territory for an active group ranges from approximately 51 to 80 ha (125 to 200 ac), but can be as large as 240 ha (600 ac). The size of the particular territory is related to both habitat quality and population density (FWS 2007a). Dispersal is primarily undertaken by young birds; mate loss and an apparent avoidance of inbreeding sometimes cause adults to disperse, and adults may also occasionally move to neighboring territories for unknown reasons (Walters et al. 1988). In a North Carolina study, females dispersed a maximum of 31.4 km (19.5 mi) and males a maximum of 21.1 km (13.1 mi) (Walters et al. 1988).

In June 2007, Southern enrolled approximately 380 ha (940 ac) of the VEGP Site in the GDNR Safe-Harbor Program for red-cockaded woodpeckers (Southern 2010c, e). Safe-Harbor Agreements are arrangements that encourage voluntary management for red-cockaded woodpeckers while protecting the participating landowners and their rights for development in the event these woodpeckers become established on the private property. Landowners entering into safe-harbor agreements must establish a baseline number of individuals that would be maintained in the event that they are observed. Currently, Southern has no baseline

16

Appendix F

responsibilities under the red-cockaded woodpecker safe-harbor agreement because there are no active clusters or nest trees onsite, and there are no red-cockaded woodpecker clusters on neighboring lands within foraging distance (Southern 2010c, e; NRC 2010h).

Surveys at the VEGP Site conducted in February 2006 found no occurrence of red-cockaded woodpeckers onsite (NRC 2008a). There are no recorded occurrences of the red-cockaded woodpecker in Burke County, Georgia (GDNR 2007, GDNR 2009), and no active colonies exist within 16 km (10 mi) of the VEGP Site in South Carolina (SCDNR 2007; SCDNR 2009; Wike et al. 2006). There are no known occurrences of the red-cockaded woodpecker in the proposed RDC (GDNR 2007; GDNR 2009). However, red-cockaded woodpeckers are listed as having the potential to occur in the project area (FWS 2010b). The red-cockaded woodpecker has been recorded on Fort Gordon (Mitchell 1999), which is located in Richmond County adjacent to the RDC. In 1998, there were two active groups on Fort Gordon representing less than 1 percent of the total number of groups in Georgia. At this time, surveys for red-cockaded woodpeckers have not been conducted in the RDC, and it is not known if suitable nesting or foraging habitats exist in the vicinity of the proposed 500-kV transmission line ROW.

Red-cockaded woodpeckers are found mainly in large stands of old longleaf pine, and this type of habitat would not be disturbed during operation of Units 3 and 4. Based on the distance to the closest known active colony, and the fact that red-cockaded woodpeckers have not been recorded on the VEGP Site or in the general vicinity of the site, it is unlikely that red-cockaded woodpeckers would be affected during operational activities onsite.

Clearing activities (e.g., tree removal, noise, increased habitat fragmentation, etc.) in the transmission line ROW have the potential to affect the red-cockaded woodpecker and its habitat. Because the final transmission line ROW would be narrow (46-m [150-ft] wide), the actual extent of clearing would be limited, thereby minimizing the potential for impact on redcockaded woodpeckers. However, increased habitat fragmentation and/or removal of cavity trees could negatively impact the red-cockaded woodpecker. GPC would site the transmission line ROW in accordance with Georgia Code Title 22, Section 22-3-161. GPC's procedures for implementing this code include consultation with FWS. GPC also has developed an APP that includes guidelines for siting new transmission lines. Available information on resident bird populations will be taken into account to ensure that the lines will have as little adverse impact as practicable on bird populations (GPC 2006).

Potential operational impacts associated with the transmission line ROW maintenance include mowing close enough to an active colony to disturb the nesting effort and removing trees during side clearing or building access roads. GPC has implemented procedures that recommend identification of all active colony areas within 3.2 km (2 mi) of a transmission line ROW and to identify active "hot-spots" within 229 m (750 ft) of a ROW. GPC recommends maintenance activities around "hot-spots" be conducted during non-breeding periods (Southern 2007). Avian mortalities resulting from collisions with conductors, guy wires, and overhead ground (static)

17

Appendix F

wires have not been specifically documented on the GPC system components. However, electrocution of birds is unlikely on lines with voltages greater than 69 kV because the distance between lines is too great to be spanned by birds (EPRI 1993). Therefore, it is unlikely that operational impacts would adversely affect the red-cockaded woodpecker.

In summary, based on the distance to the closest known active colony, and the fact that red-cockaded woodpeckers have not been recorded on the VEGP Site, it is unlikely that red-cockaded woodpeckers are foraging on the VEGP Site, and there is no evidence of nesting onsite. It is unlikely that red-cockaded woodpeckers would be encountered during operational activities onsite with the exception of possible transient individuals. There are no known occurrences of red-cockaded woodpeckers within the RDC; however, on-the-ground surveys have not been conducted at this time. If nest trees are removed during clearing for the proposed transmission line, red-cockaded woodpeckers could be affected. However, as previously noted, there are no known nest locations within the RDC. GPC has procedures to protect red-cockaded woodpeckers encountered during maintenance activities, and electrocution of birds is unlikely. Therefore, operation of the transmission system is not likely to adversely affect the red-cockaded woodpecker.

Based on the available information, the NRC staff has determined that operation of the proposed Units 3 and 4 and construction and operation of the proposed transmission system may affect, but are not likely to adversely affect, the red-cockaded woodpecker.

5.2 Wood Stork – Endangered

Breeding populations of the wood stork (*Mycteria americana*), which are Federally listed as endangered, currently occur or have recently occurred only in Florida, Georgia, South Carolina, and North Carolina (FWS 2007c). From 1975 to 1984, Georgia averaged three colonies and had an average total of 210 nesting pairs. Beginning in 1992, surveys in Georgia were expanded, and 1091 breeding pairs were documented at nine colonies. In 2005, 1817 breeding pairs were documented at 19 colonies. In 2006, there were 1928 breeding pairs at 21 colonies. Wood storks have nested at 43 different locations in the Georgia coastal plain, and the number of colonies averaged 14 during the years from 1997 to 2007 (FWS 2007c). No critical habitat has been designated for this species (FWS 2007d).

The wood stork is a highly colonial species, usually nesting and feeding in flocks. Its habitat includes freshwater and brackish wetlands, and it normally nests in bald cypress or red mangrove (*Rhizophora mangle*) swamps. At freshwater sites, nests are often constructed in bald cypress and swamp tupelo (*Nyssa biflora*). Wood storks in Georgia and South Carolina lay eggs from March to late May, with fledging occurring in July and August (FWS 1997).

Wood storks have a unique feeding technique (tacto-location) and typically require higher prey concentrations than other birds. They tend to rely on depressions in marshes or swamps where

Appendix F

prey can become concentrated during low-water periods (FWS 1997). A study from a wood stork colony in east-central Georgia found the diet was mostly composed of fish, including sunfishes (*Lepomis* spp.), bowfin (*Amia calva*), redfin pickerel (*Esox americanus americanus*), and lake chubsuckers (*Erimyzon* spp.) (FWS 1997).

Although forage areas may be 60 to 70 km (37 to 43 mi) from the colony, 85 percent are within 19 km (12 mi) (Coulter and Bryan 1993). Wood storks in east-central Georgia forage in a wide variety of wetland habitats, including hardwood and cypress swamps, ponds, marshes, drainage ditches, and flooded logging roads. Typical wood stork foraging sites have reduced quantities of both submerged and emergent macrophytes. The water in the foraging areas is either still or very slowly moving, and the depth is normally between 5 and 41 cm (2 and 16 in.). It has been suggested storks may have difficultly feeding in water with a depth more than 50 cm (20 in.) (Coulter and Bryan 1993).

Differences among seasons, rainfall, and surface-water patterns often cause storks to change where and when certain habitats are used for nesting, feeding, or roosting. These hydrological changes may cause storks to shift the timing or intensity of feeding at a local wetland, or cause entire regional populations of birds to make large geographic shifts between one year and the next. Successful colonies are those that are in regions where birds have options to feed under a variety of rainfall and surface-water conditions. Maintaining a wide range of feeding site options requires that many different types of wetlands, both large and small, and relatively long and short annual hydro-periods be available for foraging (FWS 1997).

Wood storks have the potential to occur in the project area (FWS 2010b). However, no wood storks were identified in the VEGP threatened and endangered species surveys completed in 2005, and there are no known records of wood storks occurring on the VEGP Site or within the RDC (NRC 2008a; TRC 2006; GDNR 2007; GDNR 2009). The closest known wood stork colonies to the VEGP Site are located in Jenkins and Screvin Counties, Georgia, which are south of the project area. The Birdsville colony is located at Big Dukes Pond, a 570-ha (1400-ac) cypress swamp, which is 12.6 km (7.8 mi) northwest of Millen in Jenkins County, Georgia. The VEGP Site is approximately 45 km (28 mi) from the Birdsville colony. The Chew Mill Pond colony in Jenkins County is approximately 6 km (3.7 mi) southwest of the Birdsville colony. Chew Mill Pond has a history of being a wood stork foraging site and a wading bird rookery. Researchers consider it to be an overflow or satellite colony of the Birdsville colony (Wike et al. 2006). The Jacobsons Landing colony in Screven County is approximately 43 km (27 mi) southeast of the VEGP Site. In 1996, it contained an estimated 40 wood stork nests. The distance from the VEGP Site to these colonies is within the maximum radius that wood storks travel during daily feeding flights (i.e., 60 to 70 km [37 to 43 mi]) (Coulter and Bryan 1993). Foraging wood storks have been recorded throughout Burke County, Georgia (Coulter and Bryan 1993; Wike et al. 2006), and in the Savannah River Swamp on DOE's Savannah River Site in South Carolina, which is adjacent to the VEGP Site (Wike et al. 2006).

19

Appendix F

Wood storks were reported in the vicinity of the Savannah River Site before the site was established in 1952, and before the discovery of the Birdsville colony. Storks have been followed from the Birdsville colony to the Savannah River Site. However, data from the aerial wood stork surveys of the Savannah River Swamp and the studies at the Birdsville colony suggest that the Savannah River Swamp probably is not used extensively during the breeding or pre-fledging phases of the Birdsville colony. Most of the observations of storks on the Savannah River Site occur during the late-nestling or the post-fledging period, which occurs between June and September. Some of the birds observed foraging in the Savannah River Swamp may be storks from farther south, either non-breeders or birds that already have finished breeding for the year (Wike et al. 2006).

Foraging habitats for wood storks exist on the VEGP Site and in the RDC, and wood storks have been seen within 3.2 km (2 mi) of the site in the Savannah River Swamp and on Fort Gordon, which is adjacent to a portion of the RDC. In the October 20, 2010, letter from FWS to NRC, FWS noted that there are no documented occurrences of wood stork rookeries in the project area; however, FWS stated that foraging wood storks may occur in the project streams and wetlands, and their locations should be noted (FWS 2010b). Foraging from June to September on the VEGP Site and on the RDC appears possible in wetland areas along stream drainages, ponds, drainage ditches. However, there are no records of wood stork colonies in the RDC or on the VEGP Site or within 32 km (20 mi) of the site and the proposed transmission line. This species does not likely nest in the RDC or on the VEGP Site. The wood stork is highly mobile and impacts associated with foraging during operation on the VEGP Site and construction and operation activities within the proposed transmission line ROW would be negligible.

GPC maintenance recommendations include identifying all active nesting wood stork colony rookeries that are within 1.6 km (1 mi) of a transmission line ROW. In areas within 230 m (750 ft) of an active rookery, GPC recommends mowing during the non-nesting season (Southern 2007). Therefore, activities related to the maintenance of the transmission line ROW are not expected to adversely affect the wood stork.

Based on the available information, the NRC staff has determined that operation of the proposed Units 3 and 4 and construction and operation of the proposed transmission system may affect, but are not likely to adversely affect, the wood stork.

5.3 Canby's Dropwort – Endangered

Canby's dropwort (*Oxypolis canbyi*) was listed as endangered by the FWS in 1986 (51 FR 6690). This species is native to the Coastal Plain from Delaware (historical only), Maryland, North Carolina, South Carolina, and Georgia. Historically, this plant was found in Burke, Dooly, Lee, and Sumter Counties in Georgia. There is no critical habitat designated for this species (FWS 1990).

Appendix F

Canby's dropwort has been found in a variety of habitats, including ponds dominated by pond cypress (*Taxodium ascendens*), grass-sedge-dominated Carolina bays, wet-pine savannahs, shallow-pineland ponds, and cypress-pine swamps or sloughs. The largest and most vigorous populations occur in open bays or ponds, which are wet throughout most of the year and have little or no canopy cover. Sites occupied by this species generally have infrequent and shallow inundations (5 to 30 cm [2 to 12 in.]). The species water requirements are narrow, with too little or too much water being detrimental (FWS 1990). Suitable habitat is normally on a sandy loam or loam soil underlain by a clay layer, which along with the slight gradient of the areas results in the retention of water.

Canby's dropwort has the potential to occur in the project area (FWS 2010b). However, Canby's dropwort was not found on the VEGP Site during the 2005 threatened and endangered species surveys, and there are no historical records of it occurring onsite (NRC 2008a, TRC 2006). There are two historical records of occurrence in Burke County around Waynesboro, Georgia (51 FR 6690), and these populations are currently thought to be extirpated (FWS 1990). There are no recorded occurrences within 16 km (10 mi) of the VEGP Site (GDNR 2007, GDNR 2009). Known soil types that support populations of Canby's dropwort are Rembert loam, Portsmouth loam, McColl loam, Grady loam, Coxville fine sandy loam, and Rains sandy loam. These soil types are similar in that they have a medium-to-high organic matter content, a high water table, and are deep, poorly drained, and acidic (FWS 1990). None of these soil types occur on the VEGP Site. Soil types found on the site include soils in the Chastain-Tawcaw association; Lucy, Osier, and Bibb soils; the Tawcaw-Shellbluff association; and Fuquay, Bonifay, and Troup series soils (NRCS 2003). It is unlikely that the VEGP Site contains suitable habitat for Canby's dropwort. Because of the lack of suitable habitat, it is unlikely there would be adverse impacts during operational activities at the VEGP Site.

There are no known occurrences of Canby's dropwort within the RDC. The nearest known occurrence is about 5.6 km (3.5 mi) from the RDC in Burke County (GDNR 2007). Soils known to support Canby's dropwort occur in the RDC (USGS 2001). These soils are associated with pond or wetland areas. GPC has committed to avoiding wetlands to the extent practicable during construction. In the event that wetlands are encountered, construction would be conducted in accordance with the necessary permits to protect wetland areas (GPC 2007). Therefore, it is unlikely that Canby's dropwort will be adversely affected during construction and operation activities along the transmission line ROW. GPC has implemented transmission line ROW maintenance procedures that include hand cutting in areas, such as wetlands, that have special environmental concerns (Southern 2008). In the October 20, 2010, letter from FWS to NRC, FWS noted that there are no documented occurrences of Canby's dropwort in the direct project area; however, FWS recommends that Canby's dropwort should be surveyed for, if habitat is encountered (FWS 2010b).

21

Appendix F

Based on the available information, the NRC staff has determined that operation of the proposed Units 3 and 4 and construction and operation of the proposed transmission system may affect, but are not likely to adversely affect, Canby's dropwort.

5.4 Eastern Indigo Snake – Threatened

The eastern indigo snake (*Drymarchon couperi*) was Federally listed as threatened by FWS in 1978 (FWS 1978). Historically, the eastern indigo snake occurred through Florida and in the coastal plain of Georgia, Alabama, and Mississippi (FWS 2006). Most, if not all, of the remaining viable populations of the eastern indigo snake occur in Georgia and Florida. Diemer and Speak (1983) conducted a 2-year study to survey the distribution of the eastern indigo snake and to characterize and delineate its habitat in Georgia. Results from this study indicated that the stronghold for the species was in a contiguous block of approximately 41 southeastern and south-central Georgia counties. The status and distribution in Georgia was recently reviewed by Stevenson (2006). He determined that populations of eastern indigo snakes still remain widespread in Georgia with recent records from 25 of the original 41 counties identified in the study by Deimer and Speak (1983). There are no historic or recent records for the upper Coastal Plain or Fall Line sandhill region of Georgia, including Burke, McDuffie, Jefferson, and Warren Counties (FWS 2006; Deimer and Speake 1983; Stevenson 2006). In its October 20, 2010, letter to NRC, FWS noted that there are no documented occurrences of the indigo snake in the area; however, FWS recommends that any pedestrian surveys of sandhill habitats, especially those with gopher tortoise burrows, should include cursory indigo snake surveys (FWS 2010b).

The eastern indigo snake occupies a broad range of habitats, including pine flatwoods, scrubby flatwoods, high pine, dry prairie, edges of freshwater marshes, agricultural fields, and human altered habitats (FWS 1982). In the northern parts of its range, including southeastern Georgia, eastern indigo snakes are tied to the use of gopher tortoise burrows and longleaf pine habitat (FWS 2006). The gopher tortoise burrows are used by the eastern indigo snakes not only to protect against cold in the winter and heat in the summer, but also for foraging, nesting, mating, and shelter prior to shedding (FWS 2006). Habitat use often varies seasonally between upland and wetland areas in Georgia (FWS 2006). Movement between habitat types may relate to the needs for thermal refugia, differences in habitat use by the juveniles and adults, or seasonal differences in availability of food resources. For these reasons, it is particularly vulnerable to habitat fragmentation (FWS 2006).

The eastern indigo snake is not documented in Burke County or any of the counties crossed by the proposed transmission line ROW. Suitable habitat may occur in the RDC, and gopher tortoise burrows are in the vicinity. However, the project area is outside the historic and current range of the eastern indigo snake.

22

Appendix F

Based on the available information, the NRC staff has determined that operation of the proposed Units 3 and 4 and construction and operation of the proposed transmission system may affect, but are not likely to adversely affect, the eastern indigo snake.

6.0 Cumulative Effects

Construction and operation of two new nuclear units at the VEGP Site were evaluated to determine the magnitude of their contribution to regional cumulative adverse impacts on terrestrial ecological resources. An assessment of potential impacts caused by plant construction was made for important terrestrial species (animal and plant) and habitats (as defined in the publication *Standard Review Plans for Environmental Reviews for Nuclear Power Plants* [NRC 2000]) by evaluating the impact of construction in light of other past, present, and future actions in the region. An assessment of potential impacts caused by plant operation was made for resource attributes normally affected by cooling tower operation, transmission line operation, and ROW maintenance. For this analysis, the geographic region encompassing past, present, and foreseeable future actions is the area immediately surrounding the VEGP Site, including adjoining sections of the Savannah River bottomland. GPC completed a transmission line study in 2007 to identify potential ROWs for the proposed 500-kV transmission line (GPC 2007). For the analysis of cumulative impacts related to the addition of the transmission line and its ROW, the geographic region encompassing past, present, and foreseeable future actions is the original study area identified by the GPC (GPC 2007).

6.1 VEGP Site

Approximately 353 ha (873 ac) of land would be disturbed by construction of the proposed Units 3 and 4 (NRC 2010i), including hardwood forest, planted pine plantations, open fields, and previously disturbed industrial areas. An estimated 3.7 ha (9.23 ac) of wetlands habitat on the site would be disturbed (USACE 2010). Most of the wetlands acreage involved would be in the Savannah River floodplain. The amount of wetland acreage that would be disturbed represents about 5 percent of the total 69 ha (170 ac) of wetlands currently present onsite. There are no Federally listed threatened or endangered species that would be adversely affected during construction of the proposed Units 3 and 4 (NRC 2008b; FWS 2008).

The area around the VEGP Site is rural and primarily forested and farmland. The habitats that would be disturbed at VEGP are not considered to be critical for the survival of any species, including those that are Federally protected. In addition, the percent of wetlands that would be disturbed represents only a small portion of the available wetlands in the vicinity of the site. Therefore, the staff concludes that the impact of development of the VEGP Site on the cumulative habitat loss and important species in the region associated with construction impacts would be negligible.

There are five fossil-fueled power generating stations within 145 km (90 mi) of the VEGP Site: the South Carolina Electric and Gas (SCE&G) Urquhart station, 34 km (21 mi) from the VEGP

23

Appendix F

Site; the SCE&G D area powerhouse station, 32 km (20 mi) from the VEGP Site; the GPC Plant McIntosh, 134 km (83 mi) from the VEGP Site; the GPC Port Wentworth, 124 km (77 mi) from the VEGP Site; and Plant Wilson, located on the VEGP Site. Fossil-fueled power plants release a variety of emissions to the air, including carbon dioxide, mercury, nitrous oxides, and sulfur dioxide. Nitrous oxides and sulfur dioxides can combine with water to form acid rain, which can lead to erosion and changes in soil pH levels. Mercury can deposit on soils and surface water, which may then be taken up by terrestrial plant and animal species, and poses the risk of bioaccumulation in the soil. For these reasons, these fossil-fueled power plants are likely to have current and future impacts to the environment on the VEGP Site and surrounding area (NRC 2008a).

There are three non-power generating plants that are on the Savannah River within the geographic area: the International Paper Corporation, the Savannah Industrial and Domestic Water plant, and the Beaufort-Jasper Water and Sewer authority wastewater treatment plant chemical discharges and the resulting bioaccumulation from these plants have the potential to have impacts on the surrounding area, including vegetation, wildlife, and wetlands (NRC 2008a).

DOE's Savannah River Site could impact terrestrial habitats, including habitats used by Federally listed threatened or endangered species. The Savannah River Site facility includes non-operational nuclear reactors, a currently operational coal-fired generating plant, and a proposed facility to convert weapons-grade plutonium into nuclear reactor fuel. The Savannah River Site, when originally constructed, added runoff from additional roads and impervious surfaces, increased development on wetlands and riparian zones, and decreased forest habitat. Current operations at the Savannah River Site, through chemical discharges and water withdrawal, could also have a cumulative impact on the geographic area. Future actions, such as additional construction and maintenance of buildings and facilities could affect the VEGP Site and the surrounding area (NRC 2008a).

Because the proposed Units 3 and 4 are nuclear plants, there would be little additional impact to the nearby environment from airborne releases typical of fossil fuel or other industrial facilities. Therefore, even when combined with emissions from the facilities described above, the operation of Units 3 and 4 would not result in unacceptable deposition rates of airborne pollutants. Furthermore, terrestrial habitat loss or alteration for the proposed action would be confined primarily to the VEGP Site. This loss or alteration of habitat, even in combination with chemical discharges and habitat modification associated with the other facilities in the region as discussed above, would not destabilize terrestrial resources, including Federally listed threatened or endangered species.

No other past, present, or future actions in the region were identified that could significantly affect Federally listed threatened or endangered species and critical habitat in ways similar to those associated with the proposed Units 3 and 4 site cooling tower operation (cooling tower

24

Appendix F

noise, drift from cooling towers, and bird collisions with cooling towers). The impacts associated with cooling tower operation were considered to be negligible for the VEGP Site; the cumulative adverse impact of these types of activities in the region also would be considered to be minor. Consequently, the NRC staff concludes that contributions of VEGP Site cooling tower operation to cumulative impacts on Federally listed threatened or endangered species and critical habitat in the region would be minimal.

6.2 Transmission Line ROW

The exact extent and type of wildlife habitat within the proposed new transmission line ROW is not known at this time because Southern and the GPC are evaluating ROW alternatives within the RDC. It is anticipated that the transmission line would cross Burke, Jefferson, McDuffie, and Warren Counties and would be 45 m (150 ft) wide and 97 km (60 mi) long (NRC 2008a). There are no U.S. Forest Service Wilderness Areas, Wild/Scenic Rivers or Wildlife Refuges, or State or National Parks within the RDC (GPC 2007). If possible, wetland areas would be avoided in the routing (GPC 2007).

A hypothetical transmission line ROW that represents what the GPC believes is a feasible route within the RDC was identified as part of a 2007 study (GPC 2007). Based on the GPC analysis, habitats within the ROW could include approximately 60 ha (148 ac) of forested habitat, 37 ha (91.5 ac) of forested wetlands, 133 ha (329 ac) of planted pine, 2.6 ha (6.4 ac) of open water, and 64 ha (158 ac) of open land (GPC 2007). Other land-use categories identified as potentially being impacted, such as mine/quarry, utility, transportation, and row crops, provide little value as wildlife habitat. In the region surrounding the proposed transmission line ROW, there are approximately 18,085 ha (44,688 ac) of forest, 16,956 ha (41,898 ac) of forested wetlands, 1354 ha (3346 ac) of open water, and 17,262 ha (42,656 ac) of open land (GPC 2007). Assuming the actual routing would be similar to the hypothetical route, the number of acres of forested habitat, forested wetlands, open water, open land, and planted pine forest that would be affected represent a very small portion of the available habitat. If the actual route would be similar to the hypothetical route, impacts on wildlife habitat in the region would be negligible. However, if the actual route differs from the hypothetical route, wildlife habitat impacts could either be greater or smaller.

There are no known occurrences of Federally listed threatened and endangered species within the RDC. However, suitable habitat for the red-cockaded woodpecker (*Picoides borealis*), wood stork (*Mycteria americana*), Canby's dropwort (*Oxypolis canbyi*), and the eastern indigo snake (*Drymarchon couperi*) could exist within the RDC. The GPC would site the transmission line in accordance with Georgia Code Title 22, Section 22-3-161. Part of the GPC procedures for implementing this regulation include consultation with FWS and GDNR and an evaluation of impacts to special habitats and threatened and endangered species. In addition, the GPC has guidelines for transmission line maintenance practices for nests and rookeries in Georgia (Southern 2007), has developed an APP that provides guidance for minimizing impacts to bird

25

Appendix F

species when siting new transmission lines (GPC 2006), would use good engineering and construction practices, and would comply with all applicable laws, regulations, and permit requirements (Southern 2008). Based on this review, cumulative impacts on important species and habitat loss in the region associated with construction of the transmission line ROW would be negligible.

No other past, present, or future actions in the region were identified that could significantly affect Federally listed threatened or endangered species and critical habitat in ways similar to those associated with transmission line operation and ROW maintenance (i.e., bird collisions with transmission lines, flora and fauna affected by EMFs and ROW maintenance, and floodplains and wetlands affected by ROW maintenance). Therefore, because these impacts were considered negligible for the VEGP Site transmission line operation and ROW maintenance, the cumulative adverse impacts of these types of activities in the region also would be minor. Consequently, the staff concludes that the contribution of transmission line operation and the maintenance of transmission line ROWs to cumulative impacts on wildlife and wildlife habitat in the region would be minimal.

6.3 Summary

The cumulative terrestrial resource impacts of the proposed action, including to Federally listed threatened or endangered species, may be detectable, but they are expected to be minor and not destabilizing to the resource. Therefore, the NRC staff concludes that cumulative impacts to terrestrial resources resulting from construction and operation of the proposed Units 3 and 4, including consideration of impacts from transmission line ROW construction and operation, would be minor.

7.0 Conclusions

The potential impacts to the protected species listed in Table 1 from operating the proposed Units 3 and 4 at the VEGP Site, considered cumulatively with the potential impacts of construction and operation of the offsite transmission line, are shown in Table 2. The known distributions and records of these species, in combination with the potential ecological impacts of the proposed action on the species, their habitat, and their prey, have been considered in making the impact determinations in this BA.

26

Appendix F

Table 2. Federally Listed Species Potentially Affected by Operation of the Proposed Units 3 and 4 at the VEGP Site and Construction and Operation of the Proposed Transmission Line Right of Way

Scientific Name	Common Name	Federal Status	Determination
Birds			
Mycteria americana	wood stork	E	May affect, not likely to adversely affect
Picoides borealis	red-cockaded woodpecker	E	May affect, not likely to adversely affect
Reptile			
Drymarchon couperi	Eastern Indigo Snake	T	May affect, not likely to adversely affect
Vascular Plant			
Oxypolis canbyi	Canby's dropwort	E	May affect, not likely to adversely affect

27

Appendix F

8.0 References

10 CFR Part 50. Code of Federal Regulations, Title 10, *Energy*, Part 50, "Domestic Licensing of Production and Utilization Facilities."

10 CFR Part 51. Code of Federal Regulations, Title 10, *Energy*, Part 51, "Environmental Protection Regulations for Domestic Licensing and Related Regulatory Functions."

10 CFR Part 52. Code of Federal Regulations, Title 10, *Energy*, Part 52, "Permits, Licenses, Certifications, and Licenses Approvals for Nuclear Power Plants."

35 FR 16047. October 13, 1970. "Conservation of Endangered Species and Other Fish or Wildlife." *Federal Register*.

51 FR 6690. February 25, 1986. "Endangered and Threatened Wildlife and Plants: Determination of *Oxypolis Canbyi* (Canby's dropwort) to be an Endangered Species." *Federal Register*.

AMEC Americas Limited. 2005. *Mackenzie Gas Project: Effects of Noise on Wildlife*. Prepared for Imperial Oil Resources Ventures Limited. Accessed June 11, 2008 at http://www.ngps.nt.ca/upload/proponent/imperial%20oil%20resources%20ventures%20limited/birdfield_wildlife/documents/noise_wildlife_report_filed.pdf.

Coulter M.C. and A.L. Bryan. 1993. "Foraging Ecology of Wood Storks (*Mycteria americana*) in East-Central Georgia I. Characteristics of Foraging Sites." *Colonial Waterbirds* 16(1):59-70.

Diemer J.E. and D.W. Speake. 1983. "The Distribution of the Eastern Indigo Snake, *Drymarchon corais couperi*, in Georgia." Journal of Herpetology 17(3): 256-264.

Dooling R. 2002. *Avian Hearing and the Avoidance of Wind Turbines*. NREL/TP-500-30844, National Renewwable Energy Laboratory, Golden, Colorado. Accessed July 8, 2008 at www.nrel.gov/docs/fy02osti/30844.pdf.

Eco-Sciences of Georgia (Eco-Sciences). 2007. Jurisdictional Waters Report, Vogtle Electric Generating Plant. Found in *Southern Nuclear Operating Company, Vogtle Early Site Permit Application, Response to Requests for Additional Information on the Environmental Report*. Letter report from Southern Nuclear Operating Company (Birmingham, Alabama) to the U.S. Nuclear Regulatory Commission (Washington, D.C.). Accession number ML0760460323 Electric Power Research Institute (EPRI). 1993. Proceedings: Avian Interactions with Utility

Appendix F

Structure, International Workshop, September 13-16, 1992, Miami, Florida. EPRI TR-103268, Palo Alto, California.

Electric Power Research Institute & Georgia Transmission Corp. 2006. EPRI-GTC Overhead Electric Transmission Line Siting Methodology,"., February 2006

Georgia Department of Natural Resources (GDNR). 1999. *A Conservation Plan for Red-Cockaded Woodpeckers (Picoides borealis) on Private Lands in Georgia.* Georgia Department of Natural Resources, Wildlife Resources Division, Nongame/Natural Heritage Section, Nongame-Endangered Wildlife Program, Forsyth, Georgia. Available at http://www.jonesctr.org/conservation/monitoring_mapping/Gahcp11a.pdf. Accessed on August 7, 2007.

Georgia Department of Natural Resources (GDNR). 2007. E-mail from Greg Krakow (Georgia Department of Natural Resources) to Amanda Stegen (Pacific Northwest National Laboratory, Terrestrial Ecology Scientist), regarding "Digital Rare Element Data for Pacific [Northwest] National Laboratory." Accession No. ML070851855.

Georgia Department of Natural Resources (GDNR). 2009. E-mail and letter from Katrina Morris (Georgia Department of Natural Resources) to Mallecia Sutton (Nuclear Regulatory Commission), regarding Natural Heritage Database occurrences within the VEGP Boundary and the Transmission Line Macrocorridor. Accession No. ML100490042.

Georgia Power Company (GPC). 2006. *Avian Protection Program for Georgia Power Company, Rev 1.* March 14, 2006. Accession No. ML063000228.

Georgia Power Company (GPC). 2007. *Corridor Study - Thomson Vogtle 500-kV Transmission Project.* Atlanta, Georgia. Accession No. ML070460368.

Golden J., R.P. Ouellette, S. Saari, and P.N. Cheremisinoff. 1980. *Environmental Impact Data Book.* Ann Arbor Science Publishers, Inc., Ann Arbor, Michigan.

Larkin R.P. 1996. *Effects of Military Noise on Wildlife: A Literature Review.* Technical Report 96/21, U.S. Army Corps of Engineers Research Laboratory, Champaign, Illinois. Accessed at http://nhsbig.inhs.uiuc.edu/bioacoustics/noise_and_wildlife.txt.

Mitchell W.A. 1999. *Species Profile: Wood stork (Mycteria americana) on Military Installations in the Southeastern United States.* Technical Report SERDP-99-2, U.S. Army Engineer Research and Development Center, Vicksburg, Mississippi.

29

Appendix F

Moulder J.E. 2003. *Electromagnetic Fields and Human Health: Power Lines and Cancer FAQs.* Accessed February 4, 2008, at http://www.faqs.org/faqs/medicine/powerlines-cancer-faq/.

Natural Resources Conservation Service (NRCS). 2003. *Soil Survey Geographic (SSURGO) Database for Burke County, GA.* USDA Natural Resources Conservation Service, National Cartography and Geospatial Center, Fort Collins, Colorado. Available at http://SoilDataMart.nrcs.usda.gov.

South Carolina Department of Natural Resources (SCDNR). 2007. E-mail from Julie Holling (SCDNR) to Amanda Stegen (Pacific Northwest National Laboratory, Terrestrial Ecology Scientist) Federal Threatened and Endangered Species in the Vicinity of Vogtle Electric Generating Plant," April 18, 2007. Accession No. ML071230462.

South Carolina Department of Natural Resources (SCDNR). 2009. E-mail from Julie Holling (SCDNR) to Mallecia Sutton (Nuclear Regulatory Commission) Threatened and Endangered Species in the Vicinity of Vogtle Electric Generating Plant, December 15, 2009. Accession No. ML093491132.

Southeast Regional Partnership for Planning and Sustainability (SERPPAS). 2010. Candidate Conservation Agreement for the Gopher Tortoise First Annual Report: October 1, 2008 – September 30, 2009. Published on February 2010. Accessed at www.serppas.org/.../GTCCA%20First%20Annual%20Report%202008-2009.pdf on November 15, 2010.

Southern Nuclear Operating Company, Inc. (Southern). 1989. *Environmental Protection Plan, Appendix B to the Facility Operating License No. NPF-68 and Facility Operating License No. NPF-81, Vogtle Electric Generating Plant, Units 1 and 2.* Southern Nuclear, Docket Nos. 50-424 and 50-425. March 31, 1989. Southern Company, Birmingham, Alabama. Accession No. ML012350369.

Southern Nuclear Operating Company, Inc. (Southern). 2007. E-mail from Southern Nuclear Operating Company to U.S. Nuclear Regulatory Commission regarding Southern Nuclear Operating Company, Vogtle Early Site Permit Application, Response to Follow-up Requests for Additional Information on the Environmental Report (AR-07-0924). Attachment: Maintenance Recommendations for Caves, Nests, and Rookeries (ML071710160). Letter report from Southern Nuclear Operating Company (Birmingham, Alabama) to the U.S. Nuclear Regulatory Commission (Washington, D.C.). May 14, 2007. Southern Company, Birmingham, Alabama. Accession No. ML071510102.

Appendix F

Southern Nuclear Operating Company, Inc. (Southern). 2008. Southern Nuclear Operating Company, Vogtle Early Site Permit Application, Revision 4. Southern Company, Birmingham, Alabama. Accession No. ML081020073.

Southern Nuclear Operating Company (Southern). 2010a. Southern Nuclear Operating Company, Early Site Permit Site Safety Analysis Report Change Request, Vogtle Electric Generating Plant Units 3 and 4, Use of Category 1 and 2 Backfill Material for Additional Onsite Areas, on an Exigent Basis for Units 3 and 4. Letter ND-10-0795 dated April 20, 2010. Southern Company, Birmingham, Alabama. Accession No. ML101120089.

Southern Nuclear Operating Company (Southern). 2010b. Southern Nuclear Operating Company, Vogtle Electric Generating Plant Units 3 and 4, Early Site Permit Site Safety Analysis Report Amendment Request, Revised Site Safety Analysis Report Markup for Onsite Sources of Backfill. Letter ND-10-0960 dated May 13, 2010. Southern Company, Birmingham, Alabama. Accession No. ML101340649.

Southern Nuclear Operating Company (Southern). 2010c. Southern Nuclear Operating Company, Vogtle Electric Generating Plant Units 3 and 4, Early Site Permit Site Safety Analysis Report Amendment Request, Revised Site Safety Analysis Report Markup for Onsite Sources of Backfill, Part 2. Letter ND-10-1005 dated May 24, 2010. Southern Company` Birmingham, Alabama. Accession No. ML101470212.

Southern Nuclear Operating Company (Southern). 2010d. Southern Nuclear Operating Company, Vogtle Electric Generating Plant Units 3 and 4, Site Safety Analysis Report License Amendment Request, Revise Backfill Geometry. Letter ND-10-0964 dated May 24, 2010. Southern Company, Birmingham, Alabama. Accession No. ML101470213.

Southern Nuclear Operating Company (Southern). 2010e. Southern Nuclear Operating Company, Vogtle Electric Generating Plant Units 3 and 4, Combined License Application, Post New and Significant Audit Supporting Information. Letter ND-10-0923 dated May 10, 2010. Birmingham, Alabama. Accession No. ML101320256.

Stevenson, D.J. 2006. *Distribution and Status of the Eastern Indigo Snake (Drymarchon couperi) in Georgia: 2006*. Unpublished report to the Georgia Department of Natural Resources Nongame and Endangered Wildlife Program, Forsyth, Georgia.

Third Rock Consultants, LLC (TRC). 2006. *Threatened and Endangered Species Survey Final Report. Vogtle Electric Generating Plant and Associated Transmission Corridors*. Lexington, Kentucky.

31

Appendix F

U.S. Army Corps of Engineers (USACE). 2010. Joint Public Notice Savanna District/State of Georgia RE: Application for a Department of the Army Permit. February 2, 2010. Accessed on November 3 at www.sac.usace.army.mil/assets/pdf/.../PNs20100212/200701837%20JPN.pdf

U.S. Fish and Wildlife Service (FWS). 1978. "Endangered and Threatened Wildlife and Plants. Listing of the Eastern Indigo Snake as a Threatened Species." Federal Register 43:4026.

U.S. Fish and Wildlife Service (FWS). 1982. *Eastern Indigo Snake Recovery Plan*. Atlanta, Georgia. 23 pp.

U.S. Fish and Wildlife Service (FWS). 1990. *Canby's Dropwort Recovery Plan*. Atlanta, Georgia.

U.S. Fish and Wildlife Service (FWS). 1997. *Revised Recovery Plan for the U.S. Breeding Population of the Wood Stork*. Atlanta, Georgia.

U.S. Fish and Wildlife Service (FWS). 1999. Federally-Listed Species in South Carolina Counties. Accessed at http://www.fws.gov/southeast/es/SCarolina.htm on April 25, 2007.

U.S. Fish and Wildlife Service (FWS). 2003. *Recovery Plan for the Red-Cockaded Woodpecker (Picoides borealis)*. Second Revision. Atlanta, Georgia. 296 pp.

U.S. Fish and Wildlife Service (FWS). 2006. *Eastern Indigo Snake (Drymarchon couperi): 5-Year Summary and Evaluation*. Southeast Region, Mississippi Ecological Services Field Office, Jackson, Mississippi.

U.S. Fish and Wildlife Service (FWS). 2007a. Red-Cockaded Woodpecker. Accessed at http://www.fws.gov/ncsandhills/rcw.htm on April 4, 2007.

U.S. Fish and Wildlife Service (FWS). 2007b. Species Profile: Red-Cockaded Woodpecker (*Picoides borealis*). Accessed at http://ecos.fws.gov/speciesProfile/SpeciesReport.do?spcode=B04F on April 17, 2007.

U.S. Fish and Wildlife Service (FWS). 2007c. Wood Stork (*Mycteria americana*) 5-Year Review: Summary and Evaluation. Southeast Region, Jacksonville Ecological Services Field Office. Jacksonville, Florida.

U.S. Fish and Wildlife Service (FWS). 2007d. Species Profile: Wood Stork (*Mycteria americana*). Accessed at http://ecos.fws.gov/speciesProfile/SpeciesReport.do?spcode=B06O on April 26, 2007.

Appendix F

U.S. Fish and Wildlife Service (FWS). 2008. Letter from U.S. Fish and Wildlife Service, Athens, Georgia, to U.S. Nuclear Regulatory Commission, Washington, D.C. Re: USFWS Log# 08-FA-0473. September 19, 2008. Accession No. ML082760694.

U.S. Fish and Wildlife Service (FWS). 2010a. Letter from U.S. Fish and Wildlife Service, Athens, Georgia, to U.S. Nuclear Regulatory Commission, Washington, D.C. Re: USFWS Log# 2009-1387. February 12, 2010. Accession No. ML100500426.

U.S. Fish and Wildlife Service (FWS). 2010b. Letter from U.S. Fish and Wildlife Service, Townsend, Georgia, to U.S. Nuclear Regulatory Commission, Washington, D.C. Re: USFWS Log# 2010-1254. October 20, 2010. Accession No. ML103010076.

U.S. Geological Survey (USGS). 2001. *National Land Cover Database Zone 55 Land Cover Layer*. Accessed February 11, 2010, at http://www.mrlc.gov/nlcd.php. Accession No. ML100601081.

U.S. Nuclear Regulatory Commission (NRC). 1996. *Generic Environmental Impact Statement for License Renewal of Nuclear Plants*. NUREG-1437, Volumes 1 and 2, U.S. Nuclear Regulatory Commission, Washington, D.C.

U.S. Nuclear Regulatory Commission (NRC). 2000. *Standard Review Plans for Environmental Reviews for Nuclear Power Plants*. NUREG-1555, Vol. 1, Washington, D.C. Includes 2007 revisions.

U.S. Nuclear Regulatory Commission (NRC). 2008a. *Final Environmental Impact Statement for an Early Site Permit (ESP) at the Vogtle Electric Generating Plant Site*. NUREG-1872, Vols. 1 and 2, and Errata, Washington, D.C.

U.S. Nuclear Regulatory Commission (NRC). 2008b. Letter from NRC to FWS dated January 25, 2008. Washington, D.C. Accession No. ML080100512.

U.S. Nuclear Regulatory Commission (NRC). 2009. *Vogtle Electric Generating Plant Early Site Permit No. ESP-004*. Washington, D.C. Accession No. ML092290157.

U.S. Nuclear Regulatory Commission (NRC). 2010a. *Vogtle Electric Generating Plant ESP Site Early Site Permit and Limited Work Authorization Environmental Assessment and Finding of No Significant Impact*. Docket No. 52-011. Accession No. ML101380114.

U.S. Nuclear Regulatory Commission (NRC). 2010b. *Vogtle Electric Generating Plant ESP Site Early Site Permit and Limited Work Authorization Environmental Assessment and Finding of No Significant Impact*. Docket No. 52-011. Accession No. ML101670592.

Appendix F

U.S. Nuclear Regulatory Commission (NRC). 2010c. *Vogtle Electric Generating Plant ESP Site Early Site Permit and Limited Work Authorization Environmental Assessment and Finding of No Significant Impact.* Docket No. 52-011. Accession No. ML101660076.

U.S. Nuclear Regulatory Commission (NRC). 2010d. May 21, 2010, letter from C. Patel, Senior Project Manager, NRC, to J.A. Miller, Executive Vice President, SNC, Subject: Vogtle Electric Generating Plant ESP Site - Issuance of Exigent Amendment Regarding Request for Changes to the Site Safety Analysis Report. Washington, D.C. Accession No. ML101400509.

U.S. Nuclear Regulatory Commission (NRC). 2010e. June 25, 2010, Letter from T. Spicher, Project Manager, NRC, to J.A. Miller, Executive Vice President, SNC, Subject: Vogtle Electric Generating Plant ESP Site - Issuance of Amendment Regarding Request for Changes to the Site Safety Analysis Report Regarding Onsite Sources of Backfill. Washington, D.C. Accession No. ML101760370.

U.S. Nuclear Regulatory Commission (NRC). 2010f. July 9, 2010, Letter from T. Spicher, Project Manager, NRC, to J.A. Miller, Executive Vice President, SNC, Subject: Vogtle Electric Generating Plant ESP Site, Subject: Issuance of Amendment RE: Request for changes to the classification of backfill over the side slopes of Units 3 and 4 excavations. Washington, D.C. Accession No. ML101870522.

U.S. Nuclear Regulatory Commission (NRC). 2010g. Letter from NRC to FWS dated January 7, 2010. Washington, D.C. Accession No. ML092600684.

U.S. Nuclear Regulatory Commission (NRC). 2010h. Memorandum regarding the Site Audit Summary Concerning Environmental Impacts Associated with Acquisition of Additional Backfill Material for the Vogtle Electric Generating Plant Site Combined License Application Review. Washington, D.C. Accession No. ML101550095.

U.S. Nuclear Regulatory Commission (NRC). 2010i. *Draft Environmental Impact Statement for Combined Licenses (COLs) at the Vogtle Electric Generating Plant Site. Units 3 and 4. Draft Report for Comment.* NUREG-1947, Washington, D.C.

Walters J.R., P.D. Doerr, and J.H. Carter, III. 1988. "The Cooperative Breeding System of the Red-Cockaded Woodpecker." *Ethology* 78:275-305.

Wike L.D., F.D. Martin, E.A. Nelson, N.V. Halverson, J.J. Mayer, M.H. Paller, R.S. Riley, M.G. Serrato, and W.L. Specht. 2006. *SRS Ecology: Environmental Information Document.* WSRC-TR-2005-00201, Savannah River Site, Aiken, South Carolina.

34

Appendix F

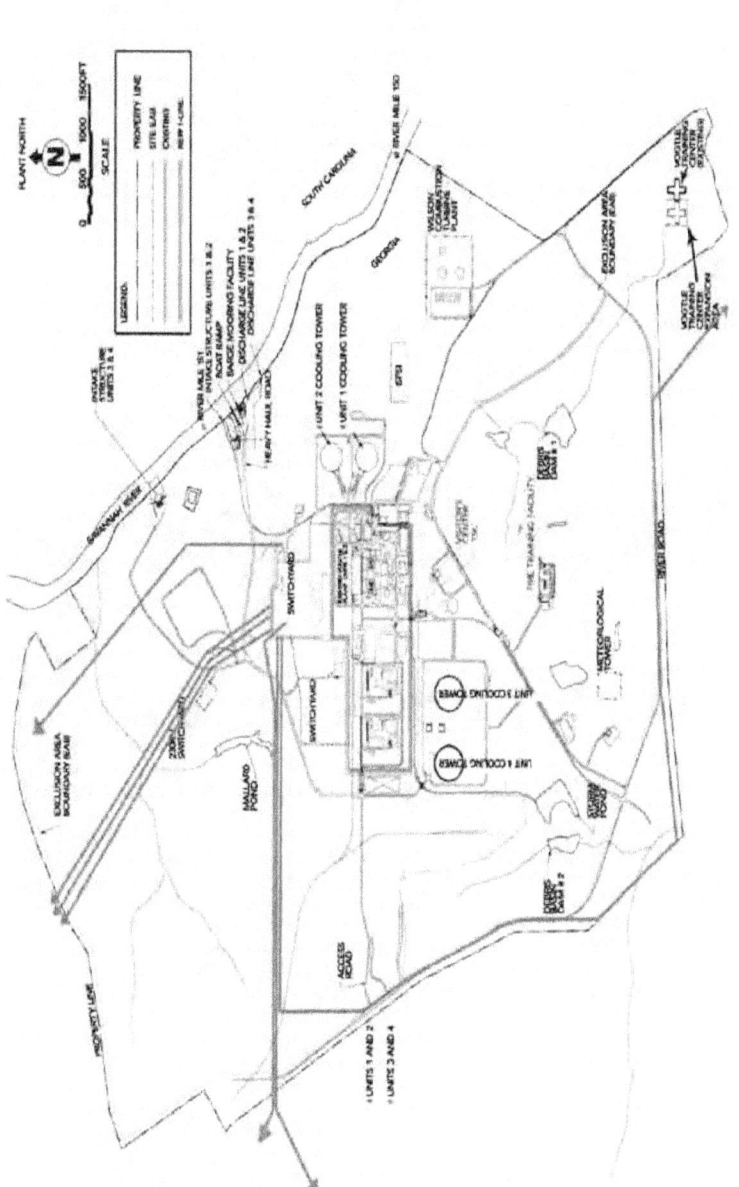

Figure 1. Proposed VEGP Site Footprint

Appendix F

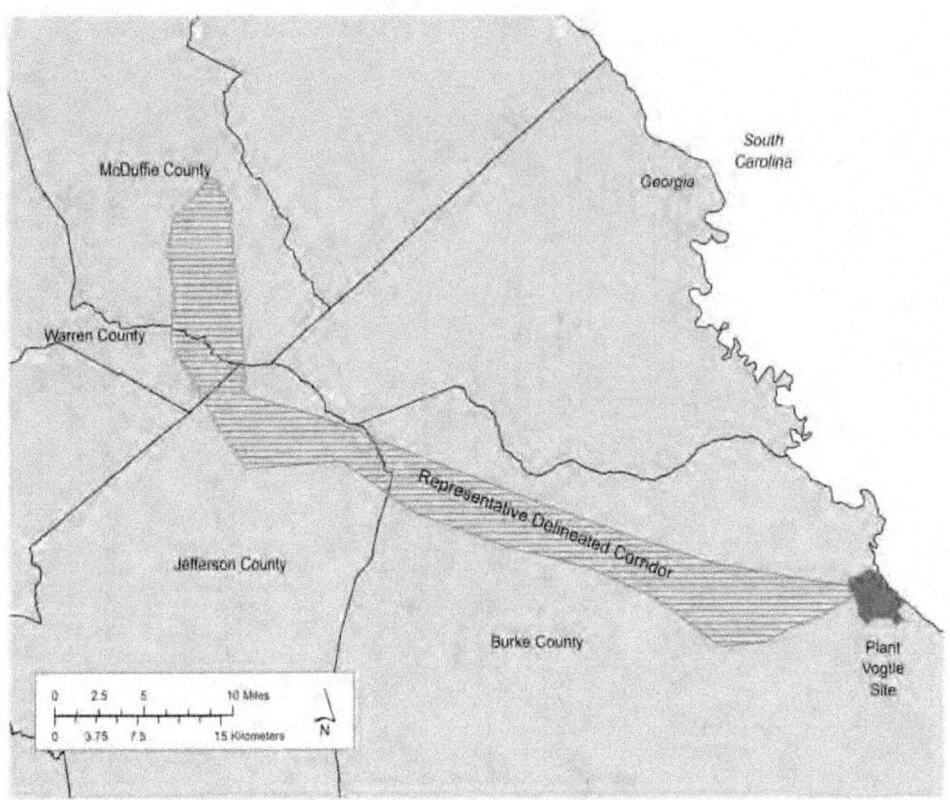

Figure 2. Representative Delineated Corridor

Appendix F

March 2, 2011

Mr. David Bernhart
Assistant Regional Administrator
 for Protected Resources
National Marine Fisheries Service
Southeast Regional Office
263 13th Avenue South
St. Petersburg, FL 33701

SUBJECT: CONFERENCE CONSULTATION FOR THE ATLANTIC STURGEON
FOR THE VOGTLE ELECTRIC GENERATING PLANT, UNITS 3 AND 4
COMBINED LICENSES APPLICATION

Dear Mr. Bernhart:

The U.S. Nuclear Regulatory Commission (NRC) is reviewing an application, submitted on March 31, 2008, from Southern Nuclear Operating Company, Inc (Southern) and its four co-applicants for combined licenses (COLs) to construct and operate two Westinghouse AP1000 pressurized water reactors at the Vogtle Electric Generating Plant (VEGP) site in Burke County, GA. The COL application referenced an early site permit (ESP) for the VEGP site that was issued to Southern and its co-applicants in 2009. As part of the ESP process, the NRC staff developed a draft and final environmental impact statement.

As part of the NRC's responsibilities under Section 7 of the Endangered Species Act (ESA), the NRC staff prepared a biological assessment (BA) in connection with the VEGP ESP review documenting potential impacts on the shortnose sturgeon (*Acipenser brevirostrum*) as a result of preconstruction site-development activities of the two new units at the VEGP site. That BA, which was submitted to your office on January 25, 2008, concluded that the proposed action is not likely to adversely affect the shortnose sturgeon. The National Marine Fisheries Service (NMFS) concurred with that determination in a letter dated August 11, 2008. In a letter dated September 3, 2010, the NRC confirmed with your office that the ESP-stage consultation encompassed the proposed actions included in the COL application.

The shortnose sturgeon was the only applicable listed or proposed species under the purview of the NMFS during the NRC staff's ESP-stage consultation. On October 6, 2010, NMFS, published in the Federal Register (75 FR 61904), a proposed rule for listing the Carolina and South Atlantic distinct population segments of the Atlantic sturgeon (*Acipenser oxyrinchus oxyrinchus*) as endangered under the ESA. To address this development, the NRC has prepared the enclosed document which describes the potential effects of the construction and operation of two new nuclear units at the VEGP site on the Atlantic sturgeon and serves as our conference consultation under Title 50 of the Code of Federal Regulations (CFR) Part 402, subpart B, Section 402.10 (50 CFR 402). This document is limited to consultation on the Atlantic sturgeon and does not affect the prior NRC or NMFS assessment regarding the shortnose sturgeon. The NRC is requesting NMFS concurrence with the NRC staff's determination that the proposed action is unlikely to adversely affect the Atlantic sturgeon.

Appendix F

D. Bernhart   - 2 -

If you have any questions regarding this consultation letter or the staff's request, please contact Ms. Mallecia Sutton, NRC Environmental Project Manager via telephone at 301-415-0673 or via e-mail to Mallecia.Sutton@nrc.gov.

                              Sincerely,

                              /RA/

                              Gregory Hatchett, Chief
                              Environmental Projects Branch 1
                              Division of Site and Environmental Reviews
                              Office of New Reactors

Docket Nos.: 52-025
                52-026

Enclosure:
As stated

cc w/o encl: See next page

Appendix F

D. Bernhart   - 2 -

If you have any questions regarding this consultation letter or the staff's request, please contact Ms. Mallecia Sutton, NRC Environmental Project Manager via telephone at 301-415-0673 or via e-mail to Mallecia.Sutton@nrc.gov.

Sincerely,

/RA/

Gregory Hatchett, Chief
Environmental Projects Branch 1
Division of Site and Environmental Reviews
Office of New Reactors

Docket Nos.: 52-025
              52-026

Enclosure:
As stated

cc w/o encl: See next page

Distribution:
Public
P Moulding
 N Kuntzelman
Sackschewsky (PNNL)   OPA
N Chokshi
T Chandler, R1
G Hatchett
M Sutton
G Hawkins
K Leigh(PNNL)
S Flanders
MCain, SRI
S Coffin (NWE1)
K Clark, R2
RidsNroDser
RidsNroDnrl

ADAMS Accession No:ML110460152 or ML110460317-pkg				NRO-002
Office	NRO/DSER/PM	DSER/LA/RAP1	DSER/RENV	OGC	DSER/BC
Name	MSutton	GHawkins	NKuntzelman	PMoulding/NLO subject to edits	GHatchett
Date	02/15/11	03/01/11	03/02/11	02/25/11	03/02/11

OFFICIAL RECORD COPY

Appendix F

Southern Nuclear - Vogtle Mailing List

Mr. M. Stanford Blanton
Esquire
Balch and Bingham, LLP
P.O. Box 306
Birmingham, AL 35201

Ms. Michele Boyd
Legislative Director
Energy Program
Public Citizens Critical Mass Energy
 and Environmental Program
215 Pennsylvania Avenue, SE
Washington, DC 20003

Mr. Marvin Fertel
 Senior Vice President
  and Chief Nuclear Officer
Nuclear Energy Institute
1776 I Street, NW
Suite 400
Washington, DC 20006-3708

Lucious Abram
County Commissioner
Office of the County Commissioner
Burke County Commission
PO Box 1626
Waynesboro, GA 30830

O. C. Harper, IV
Vice President - Resources Planning and
Nuclear Development
Georgia Power Company
241 Ralph McGill Boulevard
Atlanta, GA 30308

Mr. Steven M. Jackson
Senior Engineer - Power Supply
Municipal Electric Authority of Georgia
1470 Riveredge Parkway, NW
Atlanta, GA 30328-4684

Mr. Louis B. Long
Vice President Technical Support
Southern Nuclear Operating Company, Inc.
P.O. Box 1295
Birmingham, AL 35201-1295

Director
Consumer's Utility
Counsel Division
Governor's Office of Consumer Affairs
2 Martin Luther King, Jr. Drive
Plaza Level East, Suite 356
Atlanta, GA 30334-4600

Mr. Arthur H. Domby, Esquire
Troutman Sanders
Nations Bank Plaza
 600 Peachtree Street, NE
Suite 200
Atlanta, GA 30308-2216

Mr. Jeffrey T. Gasser
Executive Vice President
Southern Nuclear Operating Company, Inc.
P.O. Box 1295
Birmingham, AL 35201-1295

Laurence Bergen
Oglethorpe Power Corp.
2100 E Exchange Pl,
PO Box 1349
Tucker, GA 30085-1349

Mr. Charles R. Pierce
Vogtle Deployment Licensing Manager
Southern Nuclear Operating Co., Inc.
PO Box 1295
Birmingham, AL 35201-1295

Resident Inspector
Vogtle Plant
8805 River Road
Waynesboro, GA 30830

Appendix F

Southern Nuclear - Vogtle Mailing List

Resident Manager
Mr. Reece McAlister
Executive Secretary
Georgia Public Service Commission
Atlanta, GA 30334

Mr. Thomas O. McCallum
Site Development Project Engineer
Southern Nuclear Operating Co., Inc.
PO Box 1295
Birmingham, AL 35201-1295

Mr. Joseph (Buzz) Miller
Executive Vice President
Southern Nuclear Operating Company, Inc.
P.O. Box 1295
Birmingham, AL 35201-1295

Mr. Thomas Moorer
Environmental Project Manager
Southern Nuclear Operating Co., Inc.
PO Box 1295
Birmingham, AL 35201-1295

David Bernhart
Assistant Regional Administrator
 for Protect Resources
National Marine Fisheries Service
263 13th Avenue South
St. Petersburg, Florida 33701

Sam Booher
Concerned Citizen
4387 Roswell Drive
Augusta, GA 30907

Claude Howard
394 Nathaniel Howard Rd
Waynesboro, GA 30830

Glenn Carroll
Nuclear Watch South
PO Box 8574
Atlanta, GA 31106

Oglethorpe Power Corporation
Alvin W. Vogtle Nuclear Plant
7821 River Road
Waynesboro, GA 30830

Mr. Jerry Smith
Commissioner
 District 8
Augusta-Richmond County Commission
1332 Brown Road
Hephzibah, GA 30815

Mr. Robert E. Sweeney
IBEX ESI
4641 Montgomery Avenue
Suite 350
Bethesda, MD 20814

Bentina C. Terry
Southern Nuclear Operating Company,Inc.
PO Box 1295, BIN B-022
Birmingham, AL 35201-1295

Courtney Hanson
250 Georgia Avenue
Atlanta, GA 30312

Annie Laura Stephens
146 Nathanield Howard Rd
Waynesboro, GA 30830

Lucious Abrams
Burke County Commissioner
2032 Bough Red Hill
Keysville, GA 30816

Richard H. Byne
537 Jones Avenue
Waynesboro, GA 30830

Appendix F

Southern Nuclear - Vogtle Mailing List

Tommy Mitchell
Burke County Schools
352 Southside Dr
Waynesboro, GA 30830

George DeLoach
Mayor of Waynesboro
201 Oak Lane
Waynesboro, GA 30830

Robin Baxley
Best Office Solutions
142 S. Liberty St
Waynesboro, GA 30830

Appendix F

Email
APH@NEI.org (Adrian Heymer)
awc@nei.org (Anne W. Cottingham)
robin@bestofficesolutions.net (Robin Y. Baxley)
BrinkmCB@westinghouse.com (Charles Brinkman)
deloachjane@hotmail.com (George DeLoach)
chris.maslak@ge.com (Chris Maslak)
crpierce@southernco.com (C.R. Pierce)
cwaltman@roe.com (C. Waltman)
david.hinds@ge.com (David Hinds)
david.lewis@pillsburylaw.com (David Lewis)
dlochbaum@UCSUSA.org (David Lochbaum)
erg-xl@cox.net (Eddie R. Grant)
frankq@hursttech.com (Frank Quinn)
greshaja@westinghouse.com (James Gresham)
james.beard@gene.ge.com (James Beard)
jgutierrez@morganlewis.com (Jay M. Gutierrez)
jim.riccio@wdc.greenpeace.org (James Riccio)
jim@ncwarn.org (Jim Warren)
JJNesrsta@cpsenergy.com (James J. Nesrsta)
Joseph_Hegner@dom.com (Joseph Hegner)
KSutton@morganlewis.com (Kathryn M. Sutton)
kwaugh@impact-net.org (Kenneth O. Waugh)
lynchs@gao.gov (Sarah Lynch - Meeting Notices Only)
maria.webb@pillsburylaw.com (Maria Webb)
mark.beaumont@wsms.com (Mark Beaumont)
matias.travieso-diaz@pillsburylaw.com (Matias Travieso-Diaz)
mcaston@southernco.com (Moanica Caston)
media@nei.org (Scott Peterson)
mike_moran@fpl.com (Mike Moran)
tmitche@eburke.k12.ga.us (Tommy Mitchell)
nirsnet@nirs.org (Michael Mariotte)
eogleyoliver@gmail.com (Emma Ogley-Oliver)
patriciaL.campbell@ge.com (Patricia L. Campbell)
paul.gaukler@pillsburylaw.com (Paul Gaukler)
bobby@waud.org (Bobbie Paul)
Paul@beyondnuclear.org (Paul Gunter)
phinnen@entergy.com (Paul Hinnenkamp)
pshastings@duke-energy.com (Peter Hastings)
RJB@NEI.org (Russell Bell)
RKTemple@cpsenergy.com (R.K. Temple)
Crivard (?)
roberta.swain@ge.com (Roberta Swain)
burkechamber@roelco.net (Ashley Roberts)
sandra.sloan@areva.com (Sandra Sloan)
sfrantz@morganlewis.com (Stephen P. Frantz)
sbooher@aol.com (Sam Booher)
steven.hucik@ge.com (Steven Hucik)
tomccall@southernco.com (Tom McCallum)
diannevalentin@gmail.com (Dianne Valentin)
pvince20@gmail.com (Patricia Vincent)
waraksre@westinghouse.com (Rosemarie E. Warak)

Appendix F

Analysis Regarding Potential Impacts on Atlantic Sturgeon (*Acipenser oxyrinchus oxyrinchus*)

Background

The U.S. Nuclear Regulatory Commission (NRC) is reviewing an application from Southern Nuclear Operating Company, Inc. (Southern), acting on behalf of itself and co-applicants (Georgia Power Company [GPC], Oglethorpe Power Corporation, Municipal Electric Authority of Georgia, and the City of Dalton, Georgia). The application is for combined licenses (COLs) to construct and operate two Westinghouse Electric Company, LLC (Westinghouse) Advanced Passive 1000 (AP1000) pressurized water reactors (i.e., Units 3 and 4) on the site of the Vogtle Electric Generating Plant (VEGP) in Burke County, Georgia. The COL application (Southern 2009) referenced an early site permit (ESP) for the VEGP site that was issued to Southern and the same co-applicants in 2009 (NRC 2009a). As part of the ESP process the NRC staff developed a draft and final environmental impact statement (EIS) (NRC 2007 and 2008a).

As part of the NRC's responsibilities under Section 7 of the Endangered Species Act (ESA), the NRC staff prepared a biological assessment (BA) in connection with the VEGP ESP review. The BA, which documented potential impacts on the shortnose sturgeon (*Acipenser brevirostrum*) as a result of preconstruction site-development activities of two new units at the VEGP site, was submitted to the National Marine Fisheries Service (NMFS) on January 25, 2008, (NRC 2008b). In the BA, the staff concluded that the overall impact of preconstruction-related activities (including constructing the intake and discharge systems and modifying the barge slip) would be temporary and unlikely to adversely impact shortnose sturgeon in the Savannah River. In its draft and final EIS (NRC 2007, 2008a) supporting the review of the ESP application, the NRC staff also analyzed the impacts of operation of two new nuclear units at the VEGP site and concluded that operation is unlikely to adversely impact shortnose sturgeon.

NMFS reviewed the BA and the September 2007 draft ESP EIS (NRC 2007) and, in a letter dated August 11, 2008, (NMFS 2008), concluded that "... effects on the species caused by exclusion from and temporary loss of spawning habitat due to construction activities are expected to be insignificant..." NMFS's basis for this conclusion was that, "... neither the water depths, substrate bottom type, time of year for construction [i.e., outside of the spawning season], nor the shape of the river at this location are conducive to shortnose sturgeon spawning. Shortnose sturgeon generally do not inhabit this section of the Savannah River at this time of year [i.e., outside of the spawning season]; sturgeon are generally found upstream from the site during the proposed construction months and no spawning studies have observed them in the river adjacent to the Vogtle Site." Further, based on its review of the draft ESP EIS, NMFS indicated that, "... the potential effect from thermal discharge will be insignificant as it is expected that fish and other organisms would avoid the elevated temperatures, as they can move through this part of the river unencumbered by any structures or physical features that would retain them in the plume; this also reduces the likelihood of cold shock when moving outside of the plume." NMFS concluded that, "... the risk of sturgeon impingement within the intake structures will be discountable due to the very small chance of sturgeon being trapped." Finally, NMFS concluded "... potential effects from chemical effluents will be insignificant." In summary, after considering impacts of both construction and operation of two new units at the VEGP site, NMFS concluded that the proposed action is not likely to adversely affect shortnose sturgeon.

The shortnose sturgeon was the only applicable listed or proposed species under the purview of the NMFS during the NRC staff's ESP-stage consultation. On October 6, 2010, NMFS

Appendix F

published in the Federal Register (75 FR 61904) a proposed rule for listing the Carolina and South Atlantic distinct population segments of the Atlantic sturgeon (*Acipenser oxyrinchus oxyrinchus*) as endangered under the ESA. To address this development, this document describes the potential effects of the construction and operation of two new nuclear units at the VEGP site on the Atlantic sturgeon, and serves as our conference consultation under Title 50 of the Code of Federal Regulations (CFR) Part 402, subpart B, Section 402.10 (50 CFR 402). This document is limited to consultation on the Atlantic sturgeon and does not affect the prior NRC or NMFS assessment regarding the shortnose sturgeon. In a letter dated September 3, 2010 (NRC 2010a), NRC notified NMFS of the issuance and request for comments for the Vogtle draft supplemental EIS (SEIS) for the COL application. The letter further stated that no relevant information had changed regarding the project since the earlier BA was submitted. The NRC staff has incorporated by reference the ESP-stage consultation with respect to the shortnose sturgeon, pursuant to 50 CFR 402.12(g). However, because of the similarities between the Atlantic sturgeon and the shortnose sturgeon, material supporting the previous consultation is referenced or included here as appropriate.

Description of the Action

NRC is reviewing an application, submitted on March 31, 2008, from Southern and the aforementioned co-applicants for COLs to construct and operate two Westinghouse AP1000 pressurized water reactors at the VEGP site in Burke County, Georgia. The VEGP site and existing facilities are owned and operated by GPC, Oglethorpe Power Corporation, Municipal Electric Authority of Georgia, and the City of Dalton, Georgia. Southern is the licensee and operator of the existing VEGP, Units 1 and 2 and has been authorized by the VEGP co-owners to apply for COLs for the new Units 3 and 4.

On August 26, 2009, NRC approved issuance to Southern and co-applicants of an ESP and a Limited Work Authorization (LWA) for two additional nuclear units at the VEGP site (NRC 2009a). This approval was supported by information contained in NUREG-1872, Final Environmental Impact Statement for an Early Site Permit (ESP) at the Vogtle Electric Generating Plant Site (ESP EIS) (NRC 2008a) and errata. The ESP EIS considered the environmental issues and impacts of constructing and operating two new nuclear units at the VEGP site. Issuance of the ESP allowed Southern to "bank" the VEGP ESP site for up to 20 years. The LWA authorized Southern to conduct certain limited construction activities at the site in accordance with 10 CFR 50.10 and 52.24(c). As permitted by NRC regulations, Southern's COL application references the ESP.

Southern has performed, or plans to initiate, the following site-preparation activities for the two new Units 3 and 4 at the VEGP site which were considered in the BA prepared for the shortnose sturgeon and in the ESP EIS:

- Prepare the site for construction of the facilities (including such activities as clearing, grading, constructing temporary access roads, and preparing borrow areas),

- Install temporary construction support facilities (including items such as warehouses, shop facilities, utilities, concrete mixing plants, docking and unloading facilities, and construction-support buildings),

- Excavate for facility structures,

Appendix F

- Construct service facilities (including items such as roadways, paving, railroad spurs, fencing, exterior utility and lighting systems, transmission lines, and sanitary sewage treatment facilities), and

- Construct structures, systems, and components that do not prevent or mitigate the consequences of postulated accidents that could cause undue risk to the health and safety of the public. These structures, systems, and components include, but are not limited to the following:
  - Cooling towers
  - Intake and discharge structures
  - Circulating water lines
  - Fire protection equipment
  - Switchyard and onsite interconnections.

The ESP BA concerning the shortnose sturgeon also described modification of a barge slip (NRC 2008b). Since then, Southern has decided not to modify the barge slip because large components will be delivered by rail (Southern 2010a) thus precluding the need to modify the barge slip.

Under 10 CFR Part 52, which contains NRC's reactor licensing regulations and in accordance with the applicable provisions of 10 CFR Part 51, which are the NRC regulations implementing the National Environmental Policy Act of 1969 (NEPA), the NRC is required to prepare a SEIS (NRC 2010b) as part of its review of a COL application referencing an ESP. As required by 10 CFR 51.26, the NRC published a notice of availability of the draft SEIS for public comment in the *Federal Register* (FR) on September 3, 2010, (75 FR 54145). The SEIS, together with the ESP EIS (NRC 2008a), the ESP hearing proceedings, and specifically the NRC staff's prefiled testimony (NRC 2009b), and environmental assessments for three ESP license amendments concerning onsite backfill activities authorized by the LWA, (NRC 2010c, NRC 2010d, NRC 2010e) provide the NRC staff's evaluation of the environmental effects of constructing and operating two AP1000 reactors at the VEGP site.

VEGP Site Description

The VEGP site is located in Burke County, Georgia, adjacent to the Savannah River between river kilometers (RKM) 241 and 244 (river miles [RM] 150 and 152). The site is approximately 24 km (15 mi) east-northeast of Waynesboro, Georgia and 42 km (26 mi) southeast of Augusta, Georgia (see Figure 1). The proposed COL site is completely within the confines of the existing VEGP site with the new units to be constructed and operated adjacent to the existing Units 1 and 2 (Figure 2). A more detailed site description was provided in the ESP BA (NRC 2008b).

Appendix F

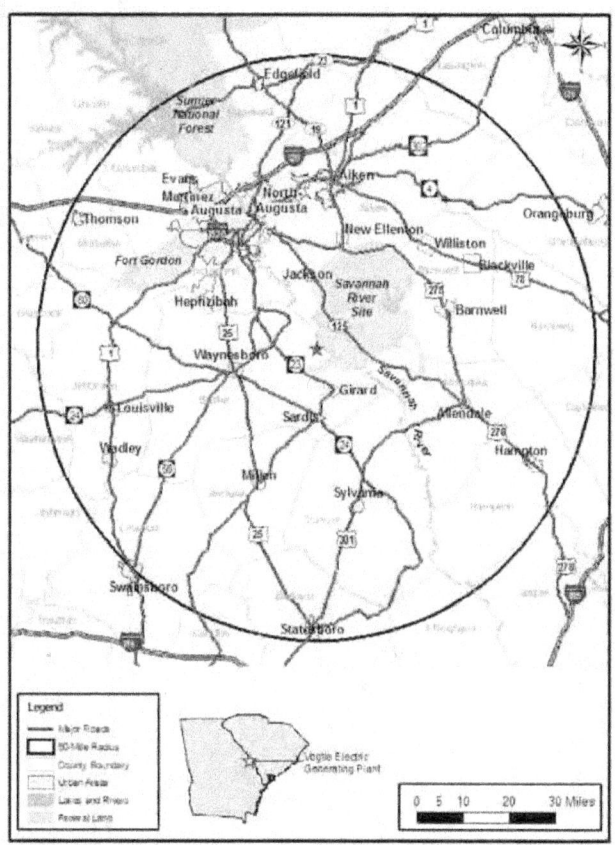

Figure 1. VEGP Site and the Vicinity within an 80-km (50-mi) Radius (Southern 2007)

Appendix F

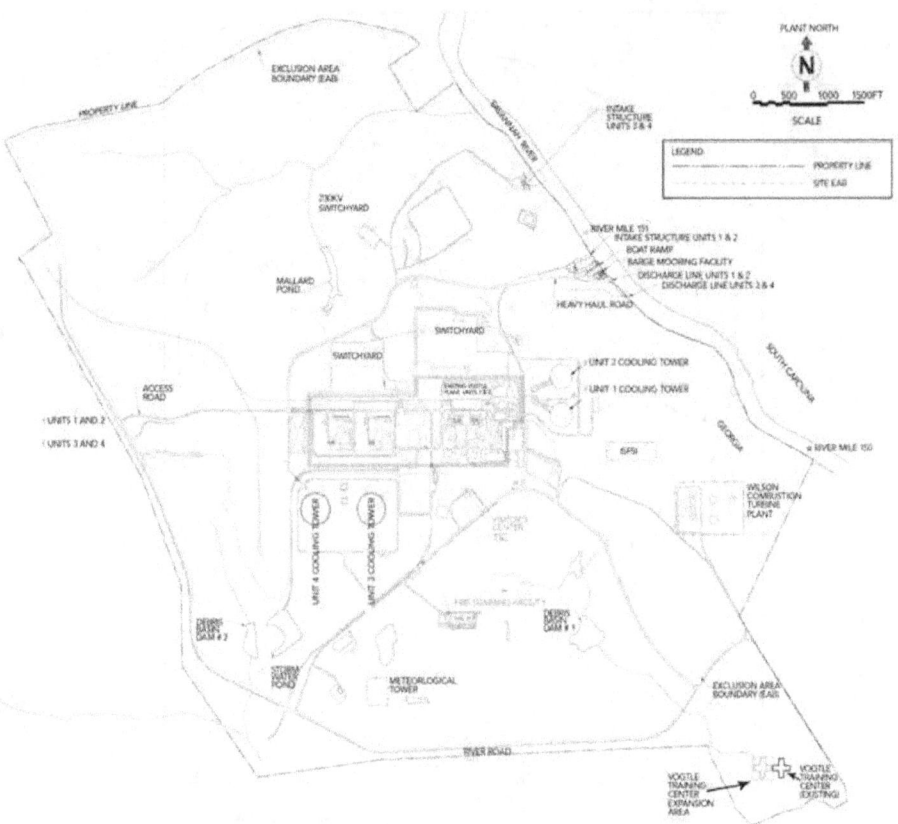

Figure 2. VEGP Site Footprint with the Existing and Proposed Nuclear Units (Southern 2010b)

Potential Environmental Impacts from Preconstruction Site-Preparation Activities

The activities that could potentially affect the habitat for the Atlantic sturgeon during construction of the intake and discharge structures are the same as those described in the ESP BA (NRC 2008b), with the exception of the construction of a barge slip, dredging from the barge slip to the Savannah River Navigation Channel, and maintenance dredging of the Savannah River Navigation Channel, which are no longer planned to occur (Southern 2010a).

On September 29, 2010, the Department of the Army issued an individual Section 10/404 permit (Permit Number SAS-2007-01837) to Southern authorizing impacts to 9.23 acres of jurisdictional wetland, 734 linear feet of stream (only the Georgia side of the Savannah River, equivalent of 1.42 acres of open water), and 0.07 acre of ephemeral stream in the southeast corner of the site near the debris basins (USACE 2010a). Southern also received a Section 401 Water Quality Certification from the Georgia Department of Natural Resources (GDNR) dated June 1, 2010, (USACE 2010a).

5

Appendix F

The design and location of the cooling water intake structure for proposed Units 3 and 4 has changed since the original BA was sent to NMFS in January 2008. The cooling water intake structure has been repositioned upstream approximately 46 m (150 ft), which places it approximately 650 m (2130 ft) upstream of the existing intakes for Units 1 and 2 and approximately 427 m (1400 ft) downstream of the outlet to the unnamed tributary of Mallard Pond. Southern also described a change in the dimensions of the intake structure (Southern 2010b); this change will lower the intake structure floor from elevation 38.1 m to 32.0 m (125 to 105 ft). In addition, there will be a slight bend (i.e., approximately 30 degrees) about halfway down the canal to orient the mouth of the intake canal perpendicular to the river. Figure 3 illustrates the revised intake structure and the wetlands in its vicinity. The design changes (Southern 2010b) do not substantially modify the width of the intake canal or the length of the canal extending beyond the existing river bank. The new location and design modifications did not alter the basis for the NRC staff's analysis of construction impacts in the COL SEIS.

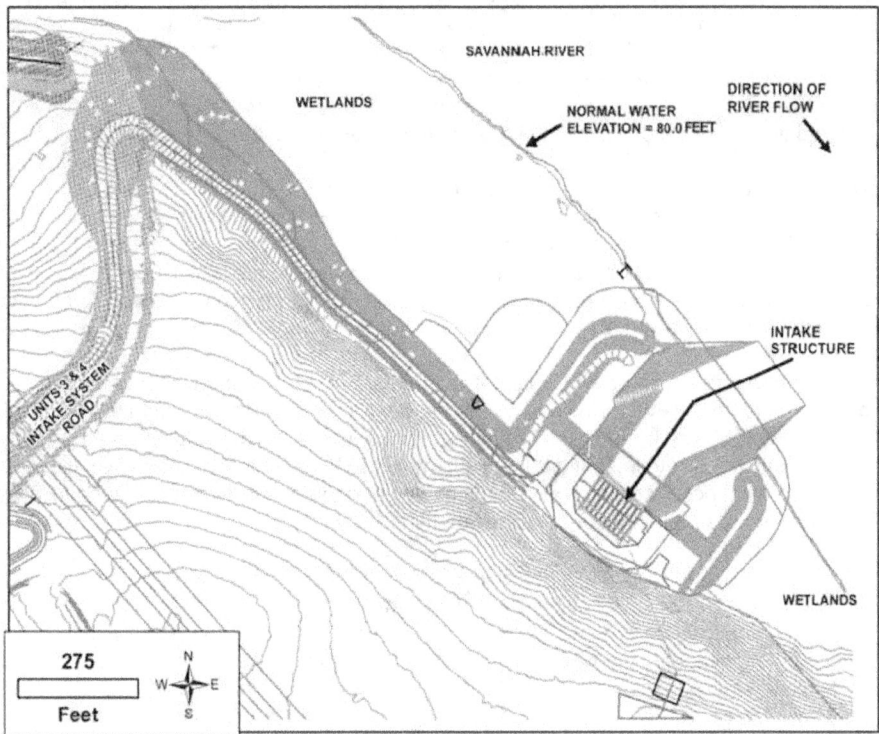

Figure 3. Revised Intake Structure and Surrounding Wetlands (Southern 2010b)

Appendix F

As discussed in the ESP BA (NRC 2008b), the proposed discharge structure will be placed near the southwest bank of the Savannah River, extending about 15 m (50 ft) into the river (Southern 2007). Details related to the design and placement of the discharge structure did not change.

Potential Environmental Impacts of Operational Activities

The potential impacts to the Atlantic sturgeon from the operation of the proposed Units 3 and 4 would include the loss of habitat from the consumption of water from the Savannah River, the entrainment of fish eggs or larvae, impingement against intake screens, the discharge of heated effluents, the discharge of chemicals, and the physical impact of bottom scouring from the discharge into the Savannah River.

Although the design and location of the cooling water intake structure has changed, the orientation of the mouth of the intake canal in relation to the river (perpendicular) has not changed. There is a slight bend in the intake canal (approximately 30 degrees) as shown in Figure 3; however, the orientation of the mouth of the intake canal relative to the river will not change. The new location of the intake canal is in habitat similar to that in the previous location (i.e., on a straight portion of the river and in the same floodplain.) No changes were made to the water withdrawal rates, through-screen velocities, traveling screen mesh size, or the hydraulic zone of influence, which are the main factors that would impact entrainment or impingement rates of aquatic biota during operation of the cooling water intake structure (Southern 2010b).

The staff evaluated the potential for fish, including the Atlantic sturgeon to be affected by the withdrawal of water from the Savannah River in the ESP EIS (NRC 2008a). The combined normal withdrawal rate of 2.35 m^3/s (83 cfs) for both VEGP Units 3 and 4 represents 0.9 percent of the average river discharge measured at the Augusta gauge. This is significantly less than the U.S. Environmental Protection Agency (EPA) national performance requirement of 5 percent for a cooling water intake structure located in a freshwater river or stream.

The staff also considered in the ESP EIS, the percentage of water withdrawn during normal operations for the proposed Units 3 and 4 from the Savannah River at Drought Level 3 river flow levels (108 m^3/s [3800 cfs]). At normal withdrawal rates, Units 3 and 4 would withdraw 2.2 percent of the river flow at the Drought Level 3 flow rates (NRC 2008a). Historically, these drought levels have occurred for short periods of time and this withdrawal rate is a small fraction of the water in the Savannah River at this location in the river.

As part of the evaluation process for the ESP EIS and the COL SEIS, the NRC staff considered several factors related to the operation of the discharge structure: (1) the physical and thermal characteristics of the plume in relation to the receiving water body, (2) the potential for cold shock, and (3) impacts from the discharge of chemicals from operation of the two proposed units. Regarding the physical and thermal characteristics of the plume in relation to the receiving water body, at the location of the discharge outfall and at a Drought Level 3 flow rate, the Savannah River is approximately 95-m (312-ft) wide (NRC 2008a). In its COL Environmental Report (ER), Southern (2009) indicated that there would be a 3 percent increase in the discharge flow beyond what was assessed in the ESP EIS. Using the same conservative assumptions employed in the ESP EIS analysis, this change would result in only a small increase in the size of the 2.8°C (5°F)-above-ambient isotherm, from 4.6 m (15 ft) to 5.2 m (17 ft) in width and from 29.6 m (97 ft) to 33.6 m (110 ft) in length (NRC 2010b). Because the estimated extent of the thermal plume remains small in relation to the width of the Savannah River at the VEGP site, the staff concluded the thermal plume still would not impede fish passage up and down the river. The staff concluded that consistent with the reasoning

7

Appendix F

identified by the ESP EIS analysis, fish and other organisms likely would avoid the elevated temperatures and would be able to move through this part of the river unencumbered by any structures or physical features that would retain them in the plume. In addition, the staff determined that the thermal plume would not create a barrier to the upstream or downstream movement of migratory fish (NRC 2010b).

Operation of the proposed Units 3 and 4 could potentially result in cold shock, which occurs when aquatic organisms that have become acclimated to warm water such as fish in a power plant's discharge canal are exposed suddenly to a lower temperature. The staff concluded that cold shock would be less likely to occur at the VEGP site because multiple units would be operating, thus lowering the possibility of simultaneous shutdown of all the units. In addition, the volume of the discharge plume would be very small in comparison with the river flow (NRC 2008a).

Regarding the discharge of chemicals from operation of the two proposed units, the cooling water will be treated with biocides and chemicals to control scaling, corrosion, and solids deposition. Operation of the cooling towers would be based on four cycles of concentration, which means that the total dissolved solids in the make-up water would be concentrated four times before being discharged. Thus, the levels of solids and organics in the cooling tower blowdown would be approximately four times higher than ambient or upstream concentrations. Cooling water chemical treatment for the proposed Units 3 and 4 would be similar to that used for the existing units. The final plant discharge from the proposed Units 3 and 4 would be composed of circulating service water blowdown and other site wastewater streams, including sanitary waste, miscellaneous low-volume waste, and treated liquid radwaste. Blowdown from the cooling towers would be discharged to a common blowdown sump to provide retention time for settling of solids or treatment, if required to remove biocide residuals before the water is discharged to the Savannah River. Calculations performed by Southern and confirmed by the staff give an estimated in-river dilution factor of 60 to 120 during periods of average Savannah River discharge, depending on the time of the year and the river flow rate (NRC 2008a).

The use of chemicals in the existing VEGP Units 1 and 2 is regulated by the GDNR, as set forth in a National Pollutant Discharge Elimination System (NPDES) permit. The chemical concentrations at the outfall for the existing units meet the NPDES limits. The chemical concentrations from Units 3 and 4 are anticipated to be the same as those for Units 1 and 2. No impacts to the aquatic ecology of the Savannah River have been observed from the operation of Units 1 and 2 and no impacts are anticipated from operation of Units 3 and 4. Southern would be required to obtain a NPDES permit from GDNR prior to operation of Units 3 and 4. To protect the aquatic environment, the NPDES permit will specify discharge limits for the various water-treatment chemicals. The NRC staff has determined that impacts to the aquatic environment from chemical discharges to the Savannah River during operation would be minimal (NRC 2008a).

Appendix F

Life History of Atlantic Sturgeon

Based on information published by Marcy et al. (2005), the staff identified the Atlantic sturgeon as being present in the Middle Savannah River Basin. The Atlantic sturgeon is a member of the family Acipenseridae, which is a long-lived group of ancient anadromous and freshwater fishes. Historically, the Atlantic sturgeon was present in 38 rivers in the United States, ranging from St. Croix, Maine, to the Saint Johns River in Florida. Historical spawning populations were confirmed in 35 of the rivers. Currently, Atlantic sturgeon populations are present in 35 rivers and spawning occurs in at least 20 rivers, including the Savannah River (ASSRT 2007)

Although the life history of the Atlantic sturgeon has been studied intensely since the 1970s, important aspects of the life history are still unknown. Generally, the Atlantic sturgeon is anadromous and spends the majority of its life in marine waters, but it reproduces in a freshwater habitat. Spawning is believed to occur in flowing water between the salt wedge and the fall line of large rivers. Like the shortnose sturgeon, spawning adults generally migrate upriver during the spring (February to March) in southern rivers. A fall-spawning migration also may occur in some southern rivers (ASSRT 2007). This appears to have first been reported by Smith (1985) indicating the occurrence of a fall run of fish that are in spawning condition in the south. Smith et al. (1984) note that the fall-run fish are typically smaller than those caught in the spring. Collins et al. (2000) provided additional evidence of a fall spawning period in the Ashepoo, Combahee, and Edisto river basins in South Carolina. This finding was based on movements of two male fish that spent the summer in the lower Edisto River and then moved upriver to RKM 190 during October 1998. In addition, a female Atlantic sturgeon that had recently spawned was captured near RKM 56 of the Edisto River during the fall during this study; however, no spawning sites were confirmed.

Atlantic sturgeon eggs are highly adhesive and are deposited on the bottom substrate, usually on hard surfaces. Hatching occurs within approximately 94 to 140 hours after egg deposition at temperatures of 20°C and 18°C (68°F and 64.4°F), respectively. Embryos (age 1 to 8 days old) tend to seek cover and stay near the bottom after hatching (Kynard and Horgan 2002). When the yolk-sac larval stage is complete (after 8 to 12 days), the larvae move downstream over a 6- to 12-day period to rearing grounds. Larvae are demersal and stay near the bottom of the water column (ASSRT 2007). During the first half of their migration, movement is limited to the night and during the day, they use the bottom (e.g., a gravel matrix) as refugia. As the larvae develop further, migration occurs during both the day and the night (Kynard and Horgan 2002). Juvenile sturgeon eventually arrive in estuarine waters, where they remain for months or years. Sub-adults may move to coastal waters and may make long migrations (ASSRT 2007).

Status of Atlantic Sturgeon in the Savannah River

Atlantic sturgeon have been found in the Savannah River, with records documenting 70 individuals having been captured since 1999 (ASSRT 2007). It appears that they are spawning in the river, although specific spawning locations have not been identified. In 1997, a single running ripe male was found at the base of the dam near Augusta in the late summer (ASSRT 2007) pointing to a potential fall migration in the Savannah also.

Ichthyoplankton studies conducted during a four-year period (1982-1985) near the Savannah River Site which is across the river from the VEGP site resulted in a total of 43 sturgeon larvae being collected. The larvae were taken from the river between RM 120 and 176. Differentiating shortnose sturgeon larvae from Atlantic sturgeon larvae is difficult because of the similarity in appearance; however, a total of 31 of the 43 sturgeon larvae were identified as Atlantic

Appendix F

sturgeon. Of the 31 larvae, four were identified as being collected from near the top of the water column. The remainder were from near the bottom. The Atlantic sturgeon larvae were collected during April. Sampling was conducted from February through July, so a fall spawning season would not have been noticed (Paller et al. 1986). In addition, Collins et al. (2000) documented an early larval *Acipenser* sp., tentatively identified as an Atlantic sturgeon located at RKM 42 (RM 26) in the Savannah River.

Cumulative Impacts

On November 15, 2010, the U.S. Army Corps of Engineers published a draft General Re-Evaluation Report (GRR) (USACE 2010b) and a Tier II EIS (USACE 2010c) related to determining the feasibility of improvements to the Federal navigation project at Savannah Harbor. The GRR and EIS assess mitigation plans for alternative channel depths from -42 to -48 ft mean lower low water. The Savannah Harbor expansion project has the potential to result in the loss of several hundred acres of habitat for fish that use the estuary. Many mitigation measures are being considered in connection with this project, including building a fish-way round the New Savannah Bluff Lock and Dam at Augusta, Georgia, which would open up an additional 32 km (20 mi) of habitat upstream of the dam (USACE 2010c). As explained previously, construction of the proposed units at the VEGP site would temporarily affect less than 0.6 ha (1.5 ac) of sturgeon migratory habitat. Water withdrawal rates during operation would be less than 1 percent of Savannah River flow during average flow conditions and the small zone of influence would have a negligible impact on pelagic spawning (NRC 2008a). Furthermore, the proposed activities associated with the VEGP expansion would not impede the mitigation measures being considered for the Savannah River expansion project. Accordingly, construction and operation of the proposed VEGP units would not have an adverse cumulative impact on important fish species when considered together with the Savannah Harbor expansion project.

Evaluation of Potential Impacts from Preconstruction Site-Preparation Activities

The construction activities previously described are expected to have minimal impacts on the aquatic ecology of the Savannah River. The extent of benthic habitat altered during construction of the intake canal would be small because most of the major construction activities would occur in the floodplain. Likewise, there would be limited disturbance of the benthic habitat during construction of the discharge structure. Disruption of silt and debris and its subsequent movement downstream during construction is expected to be minor because siltation curtains and cofferdams will be used, as discussed in the ESP BA. Noise impacts from pile-driving activities would be transient. Fish, including Atlantic sturgeon that may be inhabiting the river in the vicinity of the construction activities, would likely leave temporarily or avoid the Georgia side of the river. This temporary habitat loss would be a very small percentage of the total aquatic habitat in this area of the Savannah River.

The NRC staff has concluded that, because of the limited scope of the activities and the best management practices employed by Southern, site preparation activities addressed in this analysis would be temporary and would be unlikely to adversely affect Atlantic sturgeon.

Evaluation of Potential Impacts from Operational Activities

The operational impacts previously described are expected to have minimal impact on the aquatic ecology of the Savannah River. The anticipated volume of water to be withdrawn from

10

Appendix F

the river by the closed-cycle cooling system is a small fraction (1.2 percent) of the water in the river.

The anticipated approach velocities (about 3 cm/sec [0.1 ft/sec]) in the proposed intake canal and a designed through-screen intake velocity of less than 15 cm/sec (0.5 ft/sec) are low enough that healthy Atlantic sturgeon would be able to avoid impingement. Further, the staff is not aware of any documented case of healthy Atlantic sturgeon being impinged at any nuclear power station along the Atlantic coast including stations that employ once-through cooling systems. Sturgeon that migrate both upstream and downstream in the Savannah River are accustomed to flow rates higher than 15 cm/sec (0.5 ft/sec). An impingement study undertaken from March 10, 2008 through February 26, 2009 at VEGP Units 1 and 2 which are similar in design to the proposed Units 3 and 4, resulted in a total of 168 organisms being impinged (GPC 2009). Extrapolation of the results for a full year (365 days) of cooling-water withdrawal provided an estimate of 2580 impinged organisms with a biomass of 15 kg (33.1 lbs). No sturgeon were impinged.

An entrainment study undertaken by Southern from March 10, 2008 through July 29, 2008, resulted in entrainment of a total of 910 fish eggs and larvae from 23 taxa, representing 13 taxonomic families (GPC 2008). No sturgeon eggs or larvae were collected in either the source water or the entrainment samples.

According to the Atlantic Sturgeon Status Review Team, it is believed that the inherent behavior of larval sturgeon to maintain an active migration and to seek deep water plays a role in helping them to avoid intake structures (ASSRT 2007). Thus, they would not be susceptible to entrainment or impingement.

The size of the modeled thermal plume is small in comparison to the width of the Savannah River at the VEGP site; therefore, the plume created by operations at VEGP would not create a barrier to the upstream or downstream migration of fish species, including the Atlantic sturgeon, in the Savannah River.

Chemical discharges at the outfall for the existing Units 1 and 2 meet the limits specified in the NPDES permit and the discharge from the proposed Units 3 and 4 will be similar. No impacts to the aquatic ecology of the Savannah River have been observed from the operation of Units 1 and 2, and no impact from chemical discharges from Units 3 and 4 would be expected for Atlantic sturgeon.

Conclusion

Based on its review of the proposed action and the biology of the Atlantic sturgeon, the staff concludes that the overall impact of the VEGP Units 3 and 4 construction- and operation-related activities would be unlikely to adversely affect Atlantic sturgeon in the Savannah River.

References

10 CFR Part 50. Code of Federal Regulations, Title 10, *Energy*, Part 50, "Domestic Licensing of Production and Utilization Facilities."

10 CFR Part 51. Code of Federal Regulations, Title 10, *Energy*, Part 51, "Environmental Protection Regulations for Domestic Licensing and Related Regulatory Functions."

11

Appendix F

10 CFR Part 52. Code of Federal Regulations, Title 10, *Energy*, Part 52, "Early Site Permits, Standard Design Certifications, and Combined License for Nuclear Power Plants."

50 CFR Part 402. Code of Federal Regulations, Title 50, *Wildlife and Fisheries*, Part 402, "Interagency cooperation –Endangered Species Act of 1973, as amended ."

75 FR 54145 "Environmental Impact Statements; Notice of Availability, EIS No. 20100351, Draft EIS, NRC, GA, Vogtle Electric Generating Plant Units 3 and 4, Construction and Operation, Application for Combined Licenses (COLs)." NUREG 1947. Vol. 75, No. 171. September 3, 2010.

75 FR 61904. "Endangered and Threatened Wildlife and Plants; Proposed Listings for Two Distinct Population Segments of Atlantic Sturgeon (*Acipenser oxyrinchus oxyrinchus*) in the Southeast." National Oceanic and Atmospheric Administration. October 6, 2010.

Atlantic Sturgeon Status Review Team (ASSRT). 2007. *Status Review of Atlantic Sturgeon (Acipenser oxyrinchus oxyrinchus)*. Report to National Marine Fisheries Service, Northeast Regional Office. February 23, 2007. Available at www.nero.noaa.gov/prot_res/CandidateSpeciesProgram/AtlSturgeonStatusReviewReport.pdf.

Collins MR, TIJ Smith, WC Post, and O Pashuk. 2000. "Habitat Utilization and Biological Characteristics of Adult Atlantic Sturgeon in Two South Carolina Rivers." *Transactions of the American Fisheries Society* 129:982-988.

Endangered Species Act (ESA) of 1973. 16 USC 1531, et seq.

Georgia Power Company (GPC). 2008. *Entrainment Assessment at the Plant Vogtle Electric Generating Plant, Waynesboro, Georgia.* Atlanta, Georgia.

Georgia Power Company (GPC). 2009. *Final Report of Fish Impingement at the Plant Vogtle Electric Generating Plant, Waynesboro, Georgia.* Atlanta, Georgia.

Kynard B and M Horgan. 2002. Ontogenetic Behavior and Migration of Atlantic Sturgeon, *Acipenser oxyrinchus oxyrinchus*, and Shortnose Sturgeon, *A. brevirostrum*, with Notes on Social Behavior. Environmental Biology of Fishes 63:137-150.

Marcy, Jr. B.C., D.E. Fletcher, F.D. Martin, M. Paller, and M.J.M. Reichert. 2005. *Fishes of the Middle Savannah River Basin*. The University of Georgia Press, Athens, Georgia

National Environmental Policy Act of 1969, as amended (NEPA). 42 USC 4321, et seq.

National Marine Fisheries Service (NMFS). 2008. Letter from Roy E. Crabtree, Ph.D., Regional Administrator to William Burton, NRC, dated August 11, 2008, "A Biological Assessment for the Shortnose Sturgeon for the Vogtle Electric Generating Plant Early Site Permit Application." Accession No. ML082480450.

Paller MH, BM Saul, and DV Osteen. 1986. *Distribution and Abundance of Ichthyoplankton in the Mid-Reaches of the Savannah River and Selected Tributaries.* Primary Report Number DIST.-86-798, Environmental and Chemical Sciences, Inc., Aiken, South Carolina.

Appendix F

Smith TIJ. 1985. The Fishery, Biology, and Management of Atlantic Sturgeon, *Acipenser oxyrhynchus*, in North America. *Environmental Biology of Fishes* 14(1):61-72.

Smith, TIJ, DE Marchette and GF Ulrich. 1984. The Atlantic Sturgeon Fishery in South Carolina. *North American Journal of Fisheries Management* 4:164-176, 1984.

Southern Nuclear Operating Company, Inc. (Southern). 2007. *Southern Nuclear Operating Company, Vogtle Early Site Permit Application: Environmental Report, Rev. 2.* Southern Company, Birmingham, Alabama.

Southern Nuclear Operating Company, Inc. (Southern). 2009. *Vogtle Electric Generating Plant, Units 3 and 4, COL Application: Part 3. Environmental Report.* Revision 1, August 23, 2009, Southern Company, Birmingham, Alabama. Accession No. ML092740400.

Southern Nuclear Operating Company, Inc. (Southern). 2010a. Large Component Transportation Method Decision. February 19, 2010. Southern Company, Birmingham, Alabama. Accession No. ML100550033.

Southern Nuclear Operating Company, Inc. (Southern). 2010b. Response to Request for Additional Information Letter on Environmental Issues, January 8, 2010. Southern Company, Birmingham, Alabama. Accession No. ML100120479.

U.S. Army Corps of Engineers. 2010a. Department of the Army Permit to expand the existing Vogtle Electric Generating Plant, Permit Number SAS-2007-01837, issued September 29, 2010. U.S. Army Corps of Engineers, Savannah District, Regulatory Branch, Savannah, GA.

U.S. Army Corps of Engineers. 2010b. *Draft General Re-Evaluation Report for Savannah Harbor Expansion Project. Chatham County, Georgia and Jasper County, South Carolina.* Dated 15 November 2010. Accessed at January 6, 2010 at http://www.sas.usace.army.mil/shexpan/SHEPreport.html on January 6, 2010.

U.S. Army Corps of Engineers Project. 2010c. *Draft Tier II Environmental Impact Statement for the Savannah Harbor Expansion Project. Chatham County, Georgia and Jasper County, South Carolina.* Accessed January 6, 2010 at http://www.sas.usace.army.mil/shexpan/SHEPTierII.html on January 6, 2010.

U.S. Nuclear Regulatory Commission (NRC). 2007. *Draft Environmental Impact Statement for an Early Site Permit (ESP) at the Vogtle Electric Generating Plant Site.* NUREG-1872, Washington, D.C.

U.S. Nuclear Regulatory Commission (NRC). 2008a. *Final Environmental Impact Statement for an Early Site Permit (ESP) at the Vogtle Electric Generating Plant Site.* NUREG-1872, Washington, D.C.

U.S. Nuclear Regulatory Commission (NRC). 2008b. Letter from NRC to NMFS dated January 25, 2008. Subject: Biological Assessment for the Shortnose Sturgeon for the Vogtle Electric Generating Plant Early Site Permit Application. Accession Number ML080070538 (cover letter) and ML080100588 (attachment).

Appendix F

U.S. Nuclear Regulatory Commission (NRC). 2009a. Vogtle Electric Generating Plant Early Site Permit No. ESP-004. Washington, D.C. Accession No. ML092290157.

U.S. Nuclear Regulatory Commission (NRC). 2009b. NRC Staff Prefiled Testimony (Revised) – Dr. Michael T. Masnik, Anne R. Kuntzleman, Rebekah H. Krieg, Dr. Christopher B. Cook, and Lance W. Vail Concerning Environmental Contention EC 1.2. February 26, 2009.

U.S. Nuclear Regulatory Commission. 2010a. Letter from NRC to NMFS dated September 3, 2010. Subject: Notification of the Issuance and the Request for Comments on the Draft Supplemental Environmental Impact Statement for Vogtle Generating Plant, Units 3 and 4 Combined License Application. Accession Number ML102320162.

U.S. Nuclear Regulatory Commission (NRC). 2010b. *Draft Supplemental Environmental Impact Statement for Combined Licenses (COLs) for Vogtle Electric Generating Plant Units 3 and 4.* NUREG-1947, Washington, D.C.

U.S. Nuclear Regulatory Commission (NRC). 2010c. *Vogtle Electric Generating Plant ESP Site Early Site Permit and Limited Work Authorization Environmental Assessment and Finding of No Significant Impact.* Docket No. 52-011, Washington, D.C. Accession No. ML101380114.

U.S. Nuclear Regulatory Commission (NRC). 2010d. *Vogtle Electric Generating Plant ESP Site Early Site Permit and Limited Work Authorization Environmental Assessment and Finding of No Significant Impact.* Docket No. 52-011, Washington, D.C. Accession No. ML101670592.

U.S. Nuclear Regulatory Commission (NRC). 2010e. *Vogtle Electric Generating Plant ESP Site Early Site Permit and Limited Work Authorization Environmental Assessment and Finding of No Significant Impact.* Docket No. 52-011. Accession No. ML101660076.

Appendix F

F.1 Reference

U.S. Nuclear Regulatory Commission (NRC). 2008. *Final Environmental Impact Statement for an Early Site Permit (ESP) at the Vogtle Electric Generating Plant Site.* NUREG-1872, Vol. 2, Washington, D.C.

Appendix G

Supporting Documentation for Radiological Dose Assessment

Appendix G

Supporting Documentation for Radiological Dose Assessment

Appendix G of the Vogtle Electric Generating Plant early site permit (ESP) environmental impact statement (EIS) (NRC 2008) provided information regarding the methodology and input data for dose estimates to the public from liquid effluents, from gaseous effluents, cumulative dose estimates, and dose estimates to biota from liquid and gaseous effluents. Southern Nuclear Operating Company, Inc. (Southern) indicated in the Environmental Report (ER) included in its combined operating license (COL) application that there is no new and significant information regarding construction, operation, and cumulative radiological impacts (Southern 2009). During its review of the COL application, the NRC staff independently verified that there is no new and significant information related to radiological impacts (see Sections 4.9, 5.9, and 7.8) by reviewing Southern's ER, auditing Southern's process for identifying new and significant information, examining other information available at the site audit, and considering applicable regulations and reference documents. While the ESP EIS is based on information from Revision 15 of the AP1000 Design Control Document (DCD) (Westinghouse 2005), this SEIS is based on information from Revision 17 of the DCD (Westinghouse 2008). No significant changes in radiation doses result from using the information from Revision 17 of the DCD rather than information provided in Revision 15. Based on this review, the staff determined that the information presented in Appendix G of the ESP EIS remains valid.

G.1 References

Southern Nuclear Operating Company (Southern). 2009. *Vogtle Electric Generating Plant, Units 3 and 4, COL Application, Part 3 Environmental Report*. Revision 1, September 23, 2009. Accession No. ML092740400.

U.S. Nuclear Regulatory Commission (NRC). 2008a. *Final Environmental Impact Statement for an Early Site Permit (ESP) at the Vogtle Electric Generating Plant Site*. NUREG-1872, Vols. 1, 2, and Errata, Washington, D.C. Accession Nos. ML082240145; ML082240165, ML082260203; ML082550040.

Westinghouse Electric Company, LLC (Westinghouse). 2005. *AP1000 Design Control Document*. AP1000 Document. APP-GW-GL-700, Revision 15, Westinghouse Electric Company, Pittsburgh, Pennsylvania. Package Accession No. ML053480403.

Appendix G

Westinghouse Electric Company LLC (Westinghouse). 2008. *AP1000 Design Control Document*. APP-GW-GL-700, Revision 17, Pittsburgh, Pennsylvania. Accession No. ML083230868.

Appendix H

Authorizations and Certifications

Appendix

Instructions and Questionnaire

Appendix H

Authorizations and Certifications

This appendix contains a list of the authorizations, permits, and certifications potentially required by Federal, State, regional, local and affected Native American Tribal agencies related to the site preparation, construction, and operation of the proposed Units 3 and 4 at the Vogtle Electric Generating Plant site. Tables 1.5-1 through 1.5-5 of the Environmental Report submitted by Southern Nuclear Operating Company, Inc. on September 23, 2009 (Southern 2009) to the U.S. Nuclear Regulatory Commission, as amended by information provided in Southern's response to a request for additional information (2010a) Southern's comments on the draft supplemental environmental impact statement (2010c), are reproduced in this appendix as Table H-1, Table H-2, Table H-4, Table H-5, and Table H-6. Table H-3 is reproduced from Table 1.4-1 in the Environmental Report for the Limited Work Authorization Request (Southern 2010b). Table H-1 also contains additional information, not provided by Southern, concerning Endangered Species Act consultations with the U.S. Fish and Wildlife Service and National Marine Fisheries Service. Tables H-2 and H-5 contain information concerning permits from the U.S. Army Corps of Engineers (USACE 2010).

Appendix H

Table H-1. Authorizations Required for Early Site Permit

Agency	Authority	Requirement	License/ Permit No.	Expiration Date	Activity Covered	Status
U.S. Fish and Wildlife Service (USFWS)	Endangered Species Act	Consultation regarding potential to adversely impact protected species (non-marine species)	NA	NA	Concurrence with no adverse impact or consultation on appropriate mitigation measures.	On Oct 12, 2006, the NRC wrote the USFWS describing the project and asking for a list of protected species and habitats at the proposed site and alternative sites, and for any information under the jurisdiction of the USFWS that the agency considered pertinent to the project. In a letter dated January 25, 2008, the NRC submitted a biological assessment to USFWS documenting potential impacts on threatened or endangered species as a result of the limited site-preparation activities at the VEGP site (In a letter dated September 19, 2008, the USFWS concurred with the NRC findings that limited site-preparation activities would not likely adversely affect threatened or endangered species at the VEGP site. By letter dated February 24, 2011, the NRC submitted a second biological assessment to USFWS concerning potential impacts of operation of Units 3 and 4 at the VEGP site and of construction and operation of the proposed new

Appendix H

Table H-1. (contd)

Agency	Authority	Requirement	License/ Permit No.	Expiration Date	Activity Covered	Status
National Marine Fisheries Service (NMFS)	Endangered Species Act	Consultation regarding potential to adversely impact protected species (marine species)	NA	NA	Concurrence with no adverse impact or consultation on appropriate mitigation measures.	transmission line. On Oct 12, 2006, the NRC wrote the NMFS describing the project and asking for a list of protected species and habitats at the proposed site and alternative sites, and for any information under the jurisdiction of the NMFS that the agency considered pertinent to the project. NMFS responded on Oct 24, 2006 with a list of federally protected species under the jurisdiction of NMFS in Georgia and Alabama. In a letter dated January 25, 2008, The NRC submitted a biological assessment to NMFS documenting potential impacts on shortnose sturgeon as a result of the limited site-preparation activities at the VEGP site. In a letter dated August 11, 2008, NMFS responded to the NRC biological assessment prepared for the ESP and concurred that the project is not likely to adversely affect the protected species under their jurisdiction.

Appendix H

Table H-1. (contd)

Agency	Authority	Requirement	License/Permit No.	Expiration Date	Activity Covered	Status
South Carolina Department of Archives and History	National Historic Preservation Act (36 CFR 800)	Consultation regarding potential to adversely affect historic resources	NA	NA	Confirm site construction or operation would not affect protected historic resources.	SNC has initiated preliminary discussions with permitting agency regarding permits and compliance actions relative to this issue.
Alabama Historical Commission	National Historic Preservation Act (36 CFR 800)	Consultation regarding potential to adversely affect historic resources	NA	NA	Confirm site construction or operation would not affect protected historic resources.	On Oct 12, 2006, the NRC wrote the Alabama Historical Commission describing the project and inviting the SHPO to consult with the NRC regarding the proposed project. The SHPO responded without comment on Oct 20, 2006.
Georgia Department of Natural Resources (GDNR)	National Historic Preservation Act (36 CFR 800)	Consultation regarding potential to adversely affect historic resources	NA	NA	Confirm site construction or operation would not affect protected historic resources.	On Oct 12, 2006, the NRC wrote the Georgia SHPO describing the project and inviting the SHPO to consult with the NRC regarding the proposed project. The Georgia SHPO responded on Dec 27, 2007 and provided their assessment of the eligibility of sites at VEGP and suggested measures to protect eligible sites during construction and after.

Appendix H

Table H-1. (contd)

Agency	Authority	Requirement	License/ Permit No.	Expiration Date	Activity Covered	Status
						SNC submitted a Memorandum of Understanding (MOU) to the Georgia SHPO on January 4, 2010 for review and approval. The MOU is for the installation of the river water intake piping and associated duct bank and to preserve the balance of archaeological site 9BK416 for future investigations as directed by the Georgia SHPO.
GDNR	Federal Clean Water Act (33 U.S.C. 1251 et seq.) (CWA)	Section 401 Certification	NA	NA	Compliance with water quality standards.	GDNR issued a 401 Certification by letter to SNC, dated June 1, 2010 (Southern 2010c).
Native American Nations: Cherokee Nation of Oklahoma Chickasaw Nation Chickasaw Nation of Oklahoma Georgia Tribe of Eastern Cherokee Alabama-Quassarte Tribal Town Seminole Nation of Oklahoma	Environmental Protection Regulations for Domestic Licensing and Related Regulatory Functions (10 CFR 51) require Protection of Historic Properties (36 CFR 800)	Consultation regarding protection of traditional Native American religious or cultural resources	NA	NA	Confirm that traditional Native American religious or cultural resources are protected	On Oct 12, 2006 and Oct 16, 2006 the NRC wrote the listed Native American groups describing the project and inviting them to consult with the NRC regarding the proposed project.

The Miccosukee Tribe responded on Oct 16, 2006 that it limited itself to matters within the State of Florida.

The United Keetoowah Band of Cherokee Indians in Oklahoma responded on Oct 22, 2006 that it had no objections to the referenced |

Appendix H

Table H-1. (contd)

Agency	Authority	Requirement	License/Permit No.	Expiration Date	Activity Covered	Status
Eastern Band of Cherokee Indians						project.
United Keetoowah Band of Cherokee Indians						The Seminole Nation of Oklahoma responded on Oct 13, 2006 that it was not interested in the project.
Poarch Band of Creek Indians						
Coushatta Tribe of Louisiana						
Absentee-Shawnee Tribe of Oklahoma						
Muscogee (Creek) Nation of Oklahoma						
Alabama-Coushatta Tribe of Texas						
Catawba Indian Tribe						
Seminole Tribe of Florida						
Mississippi Band of Choctaw Indians						

Table H-1. (contd)

Agency	Authority	Requirement	License/Permit No.	Expiration Date	Activity Covered	Status
Kialegee Tribal Town						
Miccosukee Tribe of Indians of Florida						
Thlopthlocco Tribal Town						
Muscogee (Creek) Nation						

Appendix H

Table H-2. Authorizations Required for Site Preparation Activities that Do Not Require a Limited Work Authorization

Agency	Authority	Requirement	License/Permit No.	Expiration Date	Activity Covered	Status
U.S. Army Corps of Engineers (USACE)	CWA	Section 404 Permit	SAS-2007-01837	9/30/2015	Disturbance or crossing wetland areas or navigable waters. For site and rail corridor upgrade.[a]	SNC has completed jurisdictional determinations for all site wetlands with the exception of the required metes and bounds survey. SNC has submitted a joint application package for all permits under the jurisdiction of the USACE (Section 404, Section 10, and Dredge and Fill) on January 7, 2010. The USACE issued to SNC an individual Department of the Army permit on September 30, 2010 (USACE 2010).
USACE	33 CFR 323	Dredge and Fill Discharge Permit	SAS-2007-01837	9/30/2015	Construction/modification of intake/discharge to Savannah River. For site and rail corridor upgrade.[a]	SNC has initiated preliminary discussions with permitting agency regarding permits and compliance actions relative to this issue. SNC has submitted a joint application package for all permits under the jurisdiction of the USACE (Section 404, Section 10, and Dredge and Fill) on January 7, 2010. The USACE issued to SNC an individual Department of the Army permit on September 30, 2010 (USACE 2010).

Appendix H

Table H-2. (contd)

Agency	Authority	Requirement	License/Permit No.	Expiration Date	Activity Covered	Status
USACE	Rivers and Harbors Act	Section 10 Permit	SAS-2007-01837	9/30/2015	Barge slip modification impacts to navigable waters of the U.S.	SNC has initiated preliminary discussions with permitting agency regarding permits and compliance actions relative to this issue. SNC has submitted a joint application package for all permits under the jurisdiction of the USACE (Section 404, Section 10, and Dredge and Fill) on January 7, 2010. In February 2010, SNC deleted the barge slip modification.
U.S. Department of Transportation (USDOT)	49 CFR 107, Subpart G	Certificate of Registration	051409 551 044R	06/30/2010	Transportation of hazardous materials.	USDOT has provided the certificate.
USFWS	Migratory Bird Treaty Act, 50 CFR 21	Federal permit			Adverse impacts on protected species and/or their nests. For site and rail corridor upgrade.[a]	SNC has initiated preliminary discussions with permitting agency regarding permits and compliance actions relative to this issue.
Federal Aviation Administration (FAA)	49 USC 1501 14 CFR 77	Construction Notice			Notice of erection of structures (>200 feet high) potentially impacting air navigation.	SNC has initiated preliminary discussions with permitting agency regarding permits and compliance actions relative to this issue.

Appendix H

Table H-2. (contd)

Agency	Authority	Requirement	License/Permit No.	Expiration Date	Activity Covered	Status
Georgia Public Service Commission (GPSC)	GA Public Utilities Act (O.C.G.A. Section 46-3-1 et seq.), GA Rules and Regulations 515-3-4-.07	Certificate of Public Convenience and Necessity			Present and future public convenience and necessity require the operation of such equipment or facility.	SNC received GPSC certification of the project on March 17, 2009.
GDNR	GA Endangered Wildlife Act (O.C.G.A. Section 27-3-130 et seq.), GA Rules and Regulations 391-4-10	Depredation Permit			Adverse impacts on state designated protected species and/or their habitat. For site and rail corridor.[a]	SNC has initiated preliminary discussions with permitting agency regarding permits and compliance actions relative to this issue.
GDNR	Federal Clean Air Act (CAA), GA Air Quality Act (O.C.G.A. Section 12-9-1 et seq.), GA Rules and Regulations 391-3-1	Part 70 Air Quality Construction Permit	1629-033-0039-S-01-0		Construction air emission sources.	Shaw was issued its SIP Air Quality permit on June 18, 2009. SNC PSD permit application currently under review by GA EPD.
GDNR	CWA, GA Water Quality Control Act	Revision of existing National Pollutant Discharge Elimination System Permit			Regulates limits of pollutants in liquid discharge to surface water	SNC has initiated preliminary discussions with permitting agency regarding permits and compliance actions relative to this issue.
GDNR	CWA, GA Water Quality Control Act (O.C.G.A. 12-5-20), GA Rules and Regulations 391-3-6	General Permit Registration for Storm Water Discharges Associated with Construction Activity for Common Development Projects.	GAR 100003	07/31/2013	Discharge storm water from site during construction	SNC does not expect to have to file for coverage under GAR 100003. No Erosion, Sedimentation and Pollution Control plans have been developed for submittal under GAR 100003.

Appendix H

Table H-2. (contd)

Agency	Authority	Requirement	License/ Permit No.	Expiration Date	Activity Covered	Status
GDNR	CWA, GA Water Quality Control Act (O.C.G.A. 12-5-20), GA Rules and Regulations 391-3-6	General Permit Registration for Storm Water Discharges Associated with Construction Activity for Infrastructure Construction Projects	GAR 100002	07/31/2013	Discharge storm water from linear construction sites (e.g., roadways and rail corridor)	SNC has developed Erosion, Sedimentation, and Pollution Control Plans and submitted Notices of Intent to GA EPD for coverage under GAR 100002.
GDNR	GA Safe Drinking Water Act (O.C.G.A. 12-5-170 et seq.), GA Rules and Regulations 391-3-5	Permit to operate a public water system			Operate a public, nontransient, non-community water system.	Based on discussions with GDNR, the potable water system for VEGP Units 3 & 4 will be a new system. SNC submitted the potable water permit application and construction design details on June 16, 2009. GDNR approved the construction design on July 14, 2009, which allows SNC to initiate construction of the potable water system and drill two wells for potable water use.
GDNR	GA Safe Drinking Water Act (O.C.G.A. 12-5-170 et seq.), GA Rules and Regulations 391-3-5	Permit to operate a public water system				
GDNR	GA Groundwater Use Act (O.C.G.A. 12-5-90 et seq.), GA Rules and Regulations 391-3-2-.03	Modification of Existing Permit to Use Groundwater	017-0003	08/06/2012	Consumptive use of 100,000 gallons per day or more of groundwater.	Received.

Appendix H

Table H-2. (contd)

Agency	Authority	Requirement	License/ Permit No.	Expiration Date	Activity Covered	Status
GDNR	GA Groundwater Use Act (O.C.G.A. 12-5-90 et seq.), GA Rules and Regulations 391-3-2-.09	Permit to Withdraw Groundwater	017-0006	03/13/2012	Dewater for foundation if needed for more than 60 days.	Received.
GDNR	GA Groundwater Use Act (O.C.G.A. 12-5-90 et seq.), GA Rules and Regulations 391-3-2-.14	Certification of Abandoned Wells			Abandoned wells have been filled, plugged and sealed.	SNC provided a notification to EPD regarding Well SW-5, one of two wells to be taken out of service, on February 18, 2009. The remaining well, MU-2a, is scheduled to be removed from service in 2012.
GDNR	GA Erosion and Sedimentation Act (O.C.G.A. Section 12-7-1 et seq), GA Rules and Regulations 391-3-7	Land Disturbing Activity Permit	GAR 100001	07/31/2013	Permission to conduct land disturbing activities of one acre or larger, or within 200 feet of the bank of any state waters. For site and rail corridor upgrace.[a]	SNC has developed Erosion, Sedimentation, and Pollution Control Plans and submitted Notices of Intent to GA EPD for coverage under GAR 100001.
GDNR	GA Comprehensive Solid Waste Management Act (O.C.G.A. 12-8-20 et seq.), GA Rules and Regulations 391-3-4-.06	Permit by Rule - Inert Landfill Permit			On-site disposal of solid waste consisting of earth and earth-like products, concrete, cured asphalt, rock, bricks, and land clearing debris.	SNC has initiated preliminary discussions with permitting agency regarding permits and compliance actions relative to this issue.
GDNR	GA Comprehensive Solid Waste Management Act (O.C.G.A.	Private Industry Landfill Permit			Onsite disposal of solid waste consisting of construction and demolition debris.	SNC has initiated preliminary discussions with permitting agency regarding permits and compliance actions relative to this issue.

Appendix H

Table H-2. (contd)

Agency	Authority	Requirement	License/ Permit No.	Expiration Date	Activity Covered	Status
GDNR	12-8-20 et seq.), GA Rules and Regulations 391-3-4	Solid Waste Handling Permit			Disposal of industrial solid wastes. Transportation of putrescible waste for disposal in a permitted landfill.	SNC has initiated preliminary discussions with permitting agency regarding permits and compliance actions relative to this issue.
GDNR	GA Comprehensive Solid Waste Management Act (O.C.G.A. 12-8-20 et seq.), GA Rules and Regulations 391-3-4					
GDNR	Federal Clean Air Act (FCAA), GA Air Quality Act (O.C.G.A. Section 12-9-1 et seq.), GA Rules and Regulations 391-3-1	Revision of existing Title V Operating Permit			Operation of air emission sources.	SNC submitted a request for modification to this permit along with the PSD/NSR permit application submitted on May 26, 2009.
Burke County Building Office	Burke County Code of Ordinances, Article VII, Sec. 26-331	Land Disturbing Activity Permit			All land disturbing activities within the boundaries of Burke County.	As a utility regulated by the GA PSC, SNC is exempt from having to submit a Land Disturbing Activity request to a Local Issuing Authority (Burke County). Instead, a Land Disturbing Activity request is submitted directly to the GA EPD through GAR 100001 and GAR 100002.

Appendix H

Table H-2. (contd)

Agency	Authority	Requirement	License/ Permit No.	Expiration Date	Activity Covered	Status
Burke County Building Office	Burke County Code of Ordinances, Article VII, Sec. 26-336	Building Permit			Construction, alteration, repair, or demolition of any building or structure within the boundaries of Burke County.	SNC has initiated preliminary discussions with permitting agency regarding permits and compliance actions relative to this issue.

(a) The VEGP rail spur was recently upgraded, and SNC will verify that additional upgrades are not needed. For completeness, this table assumes upgrades to the rail corridor will be made.

Appendix H

Table H-3. Permits and Authorizations Required for Limited Work Authorization Activities

Agency	Authority	Requirement	License/Permit No.	Expiration Date	Activity Covered	Status
GDNR	Georgia Groundwater Use Act (O.C.G.A. 12-5-90 et seq.); Georgia Rules and Regulation 391-3-2-09	Permit to withdraw groundwater	017-0006	03/13/2012	Dewater for foundation if needed for no more than 60 days.	Received
GDNR	Federal Clean Air Act; Georgia Air Quality Act (O.C.G.A. 12-9-1 et seq.); Georgia Rules and Regulation 391-3-1	Part 70 Air Quality Construction Permit	1629-033-0039-S-01-1		Construction of air emission sources.	Received
GDNR	Georgia Erosion and Sedimentation Act (O.C.G.A. 12-7-1 et seq.); Georgia Rules and Regulation 391-3-7	Land-Disturbing Activity Permit	GAR 100001	07/31/2013	Permission to conduct land disturbing activities of one acre or larger, or within 200 feet of the bank of any state waters. For site (and rail corridor) upgrades.	SNC has developed Erosion, Sedimentation, and Pollution Control Plans and submitted Notices of Intent to the Georgia Environmental Protection Division for coverage under GAR 100001
GDNR	Federal Clean Water Act (CWA); Georgia Water Quality Control Act (O.C.G.A. 12-5-31 et seq.); Georgia Rules and Regulation 391-3-6	Permit to discharge process waste water	GA0039276 (pending EPD issuance)	5 years from date of issuance	Ready-mix concrete batch plant process wastewater discharges	EPD has issued a draft permit for public comment. Issuance of final permit expected in March 2010.
GDNR	CWA, Georgia Water Quality Control Act (O.C.G.A. 12-5-31 et seq.); Georgia Rules and Regulation 391-3-6	Industrial Storm Water Permit	GAR 000000	07/31/2011	Permit to discharge storm water associated with industrial activity.	SNC is preparing to submit to EPD a Storm Water Pollution Prevention Plan and Notice of Intent for coverage under GAR 000000.

Appendix H

Table H-4. Authorizations Required for Redress Activities

Agency	Authority	Requirement	License/Permit No.	Expiration Date	Activity Covered	Status
USACE	CWA	Section 404 Permit			Disturbance or crossing wetland areas or navigable waters.	If redress activities were required SNC would seek the necessary permits.
USDOT	49 FR 107, Subpart G	Certificate of Registration			Transportation of hazardous materials.	If redress activities were required SNC would seek the necessary permits.
GDNR	CWA	Section 401 Certification			Compliance with water quality standards.	If redress activities were required SNC would seek the necessary permits.
GDNR	CWA, GA Water Quality Control Act (O.C.G.A. 12-5-20), GA Rules and Regulations 391-3-6	General Permit Registration for Storm Water Discharges Associated with Construction Activity for Common Development Projects			Discharge storm water from site during construction.	If redress activities were required SNC would seek the necessary permits.
GDNR	CWA, GA Water Quality Control Act (O.C.G.A. 12-5-20), GA Rules and Regulations 391-3-6	General Permit Registration for Storm Water Discharges Associated with Construction Activity for Infrastructure Construction Projects			Discharge storm water linear construction sites (e.g., roadways, transmission lines) during construction.	If redress activities were required SNC would seek the necessary permits.
GDNR	GA Erosion and Sedimentation Act (O.C.G.A. Section 12-7-1 et seq.), GA Rules and Regulations 391-3-7	Land Disturbing Activity Permit			Permission to conduct land disturbing activities of one acre or larger, or within 200 feet of the bank of any state waters. For site and rail corridor.	If redress activities were required SNC would seek the necessary permits.

Appendix H

Table H-4. (contd)

Agency	Authority	Requirement	License/Permit No.	Expiration Date	Activity Covered	Status
GDNR	CAA, GA Air Quality Act (O.C.G.A. Section 12-9-1 et seq.), GA Rules and Regulations 391-3-1	Part 70 Air Quality Construction Permit			Construction air emission sources.	If redress activities were required SNC would seek the necessary permits.
GDNR	GA Safe Drinking Water Act (O.C.G.A. 12-5-170 et seq.), GA Rules and Regulations 391-3-5	Notice of Termination (NOT) - Permit to operate a Public Water System			Operate a public, non-transient, non-community water system.	If redress activities were required SNC would seek the necessary permits.
GDNR	GA Safe Drinking Water Act (O.C.G.A. 12-5-170 et seq.), GA Rules and Regulations 391-3-5	NOT - Permit to operate a Public Water System			Operate a public, transient, non-community water system.	If redress activities were required SNC would seek the necessary permits.
GDNR	GA Groundwater Use Act (O.C.G.A. 12-5-90 et seq.), GA Rules and Regulations 391-3-2-.03	NOT - Permit to Use Groundwater			Consumptive use of 100,000 gallons per day or more of groundwater.	If redress activities were required SNC would seek the necessary permits.

Appendix H

Table H-4. (contd)

Agency	Authority	Requirement	License/Permit No.	Expiration Date	Activity Covered	Status
GDNR	GA Groundwater Use Act (O.C.G.A. 12-5-90 et seq.), GA Rules and Regulations 391-3-2-.09	Permit to Withdraw Groundwater			Dewater for foundation if needed for more than 60 days.	If redress activities were required SNC would seek the necessary permits.
GDNR	GA Groundwater Use Act (O.C.G.A. 12-5-90 et seq.), GA Rules and Regulations 391-3-2-.14	Certification of Abandoned Wells			Abandoned wells have been filled, plugged and sealed.	If redress activities were required SNC would seek the necessary permits.
GDNR	GA Comprehensive Solid Waste Management Act (O.C.G.A. 12-8-20 et seq.), GA Rules and Regulations 391-3-4-.06	Permit by Rule - Inert Landfill Permit			Onsite disposal of solid waste consisting of earth and earth-like products, concrete, cured asphalt, rock, bricks, and land clearing debris	If redress activities were required SNC would seek the necessary permits.
GDNR	GA Comprehensive Solid Waste Management Act (O.C.G.A. 12-8-20 et seq.), GA Rules and Regulations 391-3-4	Private Industry Landfill Permit			Onsite disposal of solid waste consisting of construction and demolition debris.	If redress activities were required SNC would seek the necessary permits.

Appendix H

Table H-4. (contd)

Agency	Authority	Requirement	License/ Permit No.	Expiration Date	Activity Covered	Status
GDNR	GA Comprehensive Solid Waste Management Act (O.C.G.A. 12-8-20 et seq.), GA Rules and Regulations 391-3-4	Solid Waste Handling Permit			Disposal of industrial solid wastes. Transportation of putrescible waste for disposal in a permitted landfill.	If redress activities were required SNC would seek the necessary permits.
Burke County Building Office	Burke County Code of Ordinances, Article VII, Sec. 26-331	Land Disturbing Activity Permit			All land disturbing activities within the boundaries of Burke County	If redress activities were required SNC would seek the necessary permits.
Burke County Building Office	Burke County Code of Ordinances, Article VII, Sec. 26-336	Building Permit			Construction, alteration, repair, or demolition of any building or structure within the boundaries of Burke County.	If redress activities were required SNC would seek the necessary permits.

Appendix H

Table H-5. Authorizations Required for Construction[a]

Agency	Authority	Requirement	License/Permit No.	Expiration Date	Activity Covered	Status
NRC	10 CFR 52, Subpart C	Combined Operating License			Safety-related construction for a nuclear power facility.	NRC issued LWA on August 26, 2009 as part of ESP-004.
	or	or				
	10 CFR 50.10(e)(3)	Limited Work Authorization	LWA is part of permit ESP-004	09/26/2029		
FAA	49 USC 1501 14 CFR 77	Construction Notice			Notice of erection or structures (>200 feet high) potentially impacting air navigation.	SNC has initiated preliminary discussions with permitting agency regarding permits and compliance actions relative to this issue.
USACE	CWA	Section 404 Permit	SAS-2007-01837	09/30/2015	Disturbance or crossing wetland areas or navigable waters. For transmission line corridor.	SNC has completed jurisdictional determinations for all site wetlands with the exception of the required metes and bounds survey. SNC submitted a joint application package for all permits under the jurisdiction of the USACE (Section 404, Section 10, and Dredge and Fill) on January 7, 2010. The USACE issued to SNC an individual Department of the Army permit on September 30, 2010 (USACE 2010).

Appendix H

Table H-5. (contd)

Agency	Authority	Requirement	License/Permit No.	Expiration Date	Activity Covered	Status
USACE	33 CFR 323	Dredge and Fill Discharge Permit	SAS-2007-01837	09/30/2015	Construction/modification of intake/discharge to Savannah River. For transmission line corridor.	SNC has initiated preliminary discussions with permitting agency regarding permits and compliance actions relative to this issue. SNC submitted a joint application package for all permits under the jurisdiction of the USACE (Section 404, Section 10, and Dredge and Fill) on January 7, 2010. The USACE issued to SNC an individual Department of the Army permit on September 30, 2010. (USACE 2010)
USFWS	Migratory Bird Treaty Act, 50 CFR 21	Federal Depredation Permit			Adverse impacts on protected species and/or their nests. For site transmission line corridor.	SNC has initiated preliminary discussions with permitting agency regarding permits and compliance actions relative to this issue.
GDNR	GA Endangered Wildlife Act (O.C.G.A. Section 27-3-130 et seq.), GA Rules and Regulations 391-4-10	Depredation permit			Designated protected species and/or their habitat. For transmission line corridor.	SNC has initiated preliminary discussions with permitting agency regarding permits and compliance actions relative to this issue.
GDNR	CAA, GA Air Quality Act (O.C.G.A. Section 12-9-1 et seq.), GA Rules and	Part 70 Air Quality Construction Permit	1629-033-0039-S-01-0		Construction air emission sources.	Shaw was issued its SIP Air Quality permit on June 18, 2009. Southern PSD permit application currently under review by GA EPD.

March 2011 H-21 NUREG-1947

Appendix H

Table H-5. (contd)

Agency	Authority	Requirement	License/Permit No.	Expiration Date	Activity Covered	Status
GDNR	CWA, GA Water Quality Control Act (O.C.G.A. 12-5-20), GA Rules and Regulations 391-3-6	General Permit Registration for Storm Water Discharges Associated with Construction Activity for Infrastructure Construction Projects	GAR 100002	07/31/2013	Discharge storm water from linear construction sites (e.g., roadways, transmission lines) during construction.	SNC has developed Erosion, Sedimentation, and Pollution Control Plans and submitted Notices of Intent to GA EPD for coverage under GAR 100002.
GDNR	GA Comprehensive Solid Waste Management Act (O.C.G.A. 12-8-20 et seq.), GA Rules and Regulations 391-3-4	Solid Waste Handling Permit			Disposal of industrial solid wastes. Transportation of putrescible waste for disposal in a permitted landfill.	SNC has initiated preliminary discussions with permitting agency regarding permits and compliance actions relative to this issue.
GDNR	GA Erosion and Sedimentation Act (O.C.G.A. Section 12-7-1 et seq.), GA Rules and Regulations 391-3-7	Land Disturbing Activity Permit	GAR 100001	07/31/2013	Permission to conduct land disturbing activities of one acre or larger, or within 200 feet of the bank of any state waters. For transmission line corridor.	SNC has developed Erosion, Sedimentation, and Pollution Control Plans and submitted Notices of Intent to GA EPD for coverage under GAR 100001.
GDNR	CWA, GA Water Quality Control Act (O.C.G.A. 12-5-20), GA Rules and Regulations 391-3-6	General Permit Registration for Storm Water Discharges Associated with Construction Activity for Common Development Projects	GAR 100003		Discharge storm water from site during construction.	SNC currently does not expect to have to file for coverage under GAR 100003. No Erosion, Sedimentation and Pollution Control plans have been developed for submittal under GAR 100003.

Appendix H

Table H-5. (contd)

Agency	Authority	Requirement	License/Permit No.	Expiration Date	Activity Covered	Status
Georgia Department of Transportation (GDOT)	23 CFR 1.23	Permit	—	—	Utility right-of-way easement.	SNC has initiated preliminary discussions with permitting agency regarding permits and compliance actions relative to this issue.
Burke County Building Office	Burke County Code of Ordinances, Article VII, Sec. 26-331	Land Disturbing Activity Permit	—	—	All land disturbing activities within the boundaries of Burke County.	As a utility regulated by the GA PSC, SNC is exempt from having to submit a Land Disturbing Activity request to a Local Issuing Authority (Burke County). Instead, a Land Disturbing Activity request is submitted directly to the GA EPD through GAR 100001 and GAR 100002.
Various county offices responsible for land disturbing activities	Jefferson, Warren, and McDuffie County Ordinances	Land Disturbing Activity Permit	—	—	Land disturbing activities within county boundaries for transmission line corridor.	As a utility regulated by the GA PSC, SNC is exempt from having to submit a Land Disturbing Activity request to a Local Issuing Authority (Jefferson, Warren and McDuffie Counties). Instead, a Land Disturbing Activity request is submitted directly to the GA EPD through GAR 100001 and GAR 100002.

(a) Assumes that SNC obtained the authorizations that Table 1.5-2 identifies.

Appendix H

Table H-6. Authorizations Required for Operation[a]

Agency	Authority	Requirement	License/Permit No.	Expiration Date	Activity Covered	Status
GDNR	CWA, GA Water Quality Control Act	Revision of existing National Pollutant Discharge Elimination System Permit			Regulates limits of pollutants in liquid discharge to surface water.	SNC has initiated preliminary discussions with permitting agency regarding permits and compliance actions relative to this issue.
GDNR	Federal Clean Air Act (CAA), GA Air Quality Act (O.C.G.A. Section 12-9-1 et seq.), GA Rules and Regulations 391-3-1	Revision of existing Title V Operating Permit			Operation of air emission sources.	SNC submitted a request for modification to this permit along with the PSD/NSR permit application submitted on May 26, 2009.
GDNR	GA Groundwater Use Act (O.C.G.A. 12-5-90 et seq.), GA Rules and Regulations 391-3-2-.03	Revision of existing Permit to Use Groundwater	017-0003	08/06/2010	Consumptive use of 100,000 gallons per day or more of groundwater.	Received.
GDNR	GA Water Quality Control Act (O.C.G.A. 12-5-31 et seq.), GA Rules and Regulations 391-3-6	Revision of existing Surface Water Withdrawal Permit to Withdraw, Divert or Impound Surface Water			Withdraw water from the Savannah River for cooling makeup and in-plant use.	SNC has initiated preliminary discussions with permitting agency regarding permits and compliance actions relative to this issue.

Appendix H

Table H-6. (contd)

Agency	Authority	Requirement	License/Permit No.	Expiration Date	Activity Covered	Status
State of Tennessee Department of Environment and Conservation Division of Radiological Health	Tennessee Department of Environment and Conservation Rule 1200-2-10.32	Revision of existing Tennessee Radioactive Waste License-for-Delivery			Transportation of radioactive waste into the State of Tennessee.	SNC has initiated preliminary discussions with permitting agency regarding permits and compliance actions relative to this issue.
State of Utah Department of Environmental Quality Division of Radiation Control	R313-26 of the Utah Radiation Control Rules	Revision of existing General Site Access Permit			Transportation of radioactive materials into the State of Utah.	SNC has initiated preliminary discussions with permitting agency regarding permits and compliance actions relative to this issue.
GPSC	GA Radiation Control Act (O.C.G.A. 31-13-1 et seq.), GA Rules and Regulations 391-3-17-.06	Revision of existing General Permit – Transportation of Radioactive Materials			Transportation of radioactive materials in the State of Georgia.	SNC has initiated preliminary discussions with permitting agency regarding permits and compliance actions relative to this issue.

(a) Assumes that SNC obtained the authorizations that Tables 1.5-2 and 1.5-4 identify.

Appendix H

H.1 References

Southern Nuclear Operating Company, Inc. (Southern). 2009. *Vogtle Electric Generating Plant, Units 3 and 4, COL Application: Part 3. Environmental Report.* Revision 1, September 23, 2009. Southern Company, Birmingham, Alabama. Accession No. ML092740400.

Southern Nuclear Operating Company, Inc. (Southern). 2010a. Response to Request for Additional Information Letter on Environmental Issues, January 8, 2010. Southern Company, Birmingham, Alabama. Accession No. ML100120479.

Southern Nuclear Operating Company (Southern). 2010b. Environmental Report in Support Revision 1 to Part 6 , Limited Work Authorization Request, of the Vogtle Electric Generating Plant Units 3 and 4 Combined License Application. Southern Company, Birmingham, Alabama. Accession No. ML100470600.

Southern Nuclear Operating Company (Southern). 2010c. Comments on Draft Supplemental Environmental Impact Statement, November 23,1010. Southern Company, Birmingham, Alabama. Accession No. ML103300035.

U.S. Army Corps of Engineers (USACE). 2010. Letter from Carol Bernstein (USACE Chief, Coastal Branch) to Thomas Moorer (Southern), "Subject:: Signed Department of the Army Permit for the expansion of the existing Vogtle Electric Generating Plant." SAS-2007-01837.

Appendix I

Vogtle Electric Generating Plant Site Characteristics, AP1000 Design Parameters and Site Interface Values

Appendix I

Vogtle Electric Generating Plant Site Characteristics, AP1000 Design Parameters and Site Interface Values

Appendix I of the Vogtle Electric Generating Plant (VEGP) early site permit (ESP) environmental impact statement (EIS) provides the site characteristics, AP1000 design parameters, and site interface values (NRC 2008). Table 3.0-1 of Southern Nuclear Operating Company, Inc.'s Environmental Report (ER), Revision 1, dated September 23, 2009 (Southern 2009), reproduced on the following pages as Table I-1, shows that most of the site characteristics, design parameters, and site interface values considered in this combined license (COL) application fall within those described in the ESP. These characteristics and parameters were used by the U.S. Nuclear Regulatory Commission staff in its independent evaluation of the new and significant information related to the environmental impacts of the proposed new units.

Appendix I

Table I-1. VEGP Site Characteristics, AP1000 Design Parameters, and Site Interface Values

Part I Site Characteristic Item	ESP Single Unit [Two Unit] Value		Description and Reference	COL Single Unit [Two Unit] Value		Comments
Airborne Effluent Release Point						
Minimum Distance to Exclusion Area Boundary (EAB)	½ mi (~800 m)		The lateral distance from the release point (power block area) to the modeled EAB for dose analysis.	½ mi (~800 m)		Unchanged from ESP.
Atmospheric Dispersion (χ/Q) (Accident)	The atmospheric dispersion coefficients used to estimate dose consequences of accident airborne releases.					
	Time (hour)	Site (χ/Q)	Atmospheric dispersion coefficients used to estimate dose consequences of accident airborne releases. (From Table 5-13 of the ESP EIS)	Time (hour)	Site (χ/Q)	Unchanged from ESP.
EAB (χ/Q)	0 - 2 EAB	7.38E-5 sec/m3		0 - 2 EAB	7.38E-5 sec/m3	
Low Population Zone (LPZ) (χ/Q)	0 - 8 LPZ 8 - 24 LPZ 24 - 96 LPZ 96 - 720 LPZ	1.40E-5 sec/m3 1.22E-5 sec/m3 9.15E-6 sec/m3 6.04E-6 sec/m3		0 - 8 LPZ 8 - 24 LPZ 24 - 96 LPZ 96 - 720 LPZ	1.40E-5 sec/m3 1.22E-5 sec/m3 9.15E-6 sec/m3 6.04E-6 sec/m3	
Gaseous Effluents Dispersion, Deposition (Annual Average)						
Atmospheric Dispersion (χ/Q)	See Table 3.0-2		The atmospheric dispersion coefficients used to estimate dose consequences of normal airborne releases.	χ/Q values as described in ESP		Unchanged from ESP.
Population Density						
Population density over the lifetime of the new units until 2090	Population density meets the guidance of RS-002, Attachment 3			Population density meets the guidance of RS-002, Attachment 3		Unchanged from ESP.

Appendix I

Table I-1. (contd)

Part I Site Characteristic Item	ESP Single Unit [Two Unit] Value	ESP Description and Reference	COL Single Unit [Two Unit] Value	Comments
Population density over the lifetime of the new units until 2090	Population density meets the guidance of RS-002, Attachment 3	Population density meets the guidance of RS-002, Attachment 3	Unchanged from ESP.	
EAB	Refer to Figure 2-1 in the EIS	The exclusion area boundary generally follows the plant property line.	Refer to Figure 3.2-1 in the ER	Unchanged from ESP.
LPZ	A 2-mile-radius from the midpoint between the containment buildings of Units 1 and 2	The LPZ is a circle with a radius of 2 miles, centered on the midpoint between Unit 1 and Unit 2 containment buildings	A 2-mile-radius from the midpoint between the containment buildings of Units 1 and 2	Unchanged from ESP.
Height	234 ft 0 in	The height from finished grade to the top of the tallest power block structure, excluding cooling towers	229 ft 0 in	(DCD Rev 17, Table 3.3-5) (Westinghouse 2008) The height affects aesthetic impacts and the potential for bird collisions. Because this height is lower than that analyzed in the ESP application, the impacts are bounded by those impacts

Appendix I

Table I-1. (contd)

Part II Site Characteristic Item	ESP Single Unit [Two Unit] Value	Description and Reference	COL Single Unit [Two Unit] Value	Comments
Facility Characteristics				
Foundation Embedment	39 ft 6 in to bottom of basemat from plant grade	The depth from finished grade to the bottom of the basemat for the most deeply embedded power block structure.	39 ft 6 in to bottom of basemat from plant grade	Unchanged from ESP.
Max Inlet Temp Condenser / Heat Exchanger	91°F	The maximum acceptable design circulating water temperature at the inlet to the condenser or cooling water system heat exchangers.	91°F	Unchanged from ESP.
Condenser / Heat Exchanger Duty	7.55E9 BTU/hr [1.51E10 BTU/hr]	Design value for the waste heat rejected to the circulating water system across the condensers. Selected value includes part of the service water system heat duty (from turbine equipment heat exchanger).	7.63E9 BTU/hr [1.53E10 BTU/hr]	The COL value was provided in Southern (2007) and was considered in the ESP analysis

Appendix I

Table I-1. (contd)

Part II Site Characteristic Item	ESP Single Unit [Two Unit] Value	Description and Reference	COL Single Unit [Two Unit] Value	Comments
Cooling Tower Temperature Range	25.2°F	The temperature difference between the hot water entering the tower and the cold water leaving the tower.	25.2°F	Unchanged from ESP.
Cooling Tower Cooling Water Flow Rate	600,000 gpm [1,200,000 gpm]	The total nominal cooling water flow rate through the condenser/heat exchangers.	631,000 gpm [1,262,000 gpm]	The COL value was provided in Southern (2007) and was considered in the ESP analysis.
Auxiliary Heat Sink				
Component Cooling Water (CCW) Heat Exchanger Duty	8.3E7 BTU/hr normal 2.96E8 BTU/hr shutdown [1.66E8 BTU/hr normal 5.92E8 BTU/hr shutdown]	The heat transferred from the CCW heat exchangers to the service water system for rejection to the environment.	8.3E7 BTU/hr normal 2.96E8 BTU/hr shutdown [1.66E8 BTU/hr normal 5.92E8 BTU/hr shutdown]	Unchanged from ESP.
Service Water System (SWS) Cooling Tower Cooling Water Flow Rate	9,000 gpm normal 18,000 gpm shutdown [18,000 gpm normal 36,000 gpm shutdown]	The total nominal cooling water flow rate through the SWS.	9,000 gpm normal 18,000 gpm shutdown [18,000 gpm normal 36,000 gpm shutdown]	Unchanged from ESP.
Plant Characteristics				
Rated Thermal Power (RTP)	3,400 MWt	The thermal power generated by the core.	3,400 MWt	Unchanged from ESP.

Appendix I

Table I-1. (contd)

Part II Site Characteristic Item	ESP Single Unit [Two Unit] Value	Description and Reference	COL Single Unit [Two Unit] Value	Comments
Rated Nuclear Steam Supply System (NSSS) Thermal Output	3,415 MWt [6,830 MWt]	The thermal power generated by the core plus heat from the reactor coolant pumps.	3,415 MWt [6,830 MWt]	Unchanged from ESP.
Average Fuel Enrichment	2.35 wt % to 4.45 wt % 4.51 wt %	Concentration of U-235 in fuel - initial load. Average concentration, in weight percent, of U-235 in reloads	2.35 wt % to 4.45 wt % 4.51 wt %	Unchanged from ESP.
Fuel Burn-up	60,000 MWd/MTU (design max) 48,700 MWd/MTU (expected)	Value derived by multiplying the reactor thermal power by time of irradiation divided by fuel mass (expressed in megawatt - days per metric ton of uranium fuel).	60,000 MWd/MTU (design max) 48,700 MWd/MTU (expected)	Unchanged from ESP.
Normal Releases				
Liquid Source Term	See Table G-1 of the EIS 0.26 curies total nuclides except tritium [0.52 curies]	The annual activity, by isotope, contained in routine liquid effluent streams.	0.26 curies total nuclides except tritium [0.52 curies]	Unchanged from ESP.

Appendix I

Table I-1. (contd)

Part II Site Characteristic Item	ESP Single Unit [Two Unit] Value	Description and Reference	COL Single Unit [Two Unit] Value	Comments
Tritium (liquid)	1010 curies [2020 curies]	The annual activity of tritium contained in routine liquid effluent streams	1010 curies [2020 curies]	Unchanged from ESP.
Gaseous Source Term	See Table G-4 of the EIS 11,000 curies total nuclides except tritium [22,000 total curies]	The annual activity, by isotope, contained in routine plant airborne effluent streams.	11,000 curies total nuclides except tritium [22,000 total curies]	Unchanged from ESP.
Tritium (gaseous)	350 curies [700 curies]	The annual activity of tritium contained in routine plant airborne effluent streams.	350 curies [700 curies]	Unchanged from ESP.
Solid Waste Activity	1764 curies [3528 curies]	The annual activity contained in solid radioactive wastes generated during routine plant operations.	1764 curies [3528 curies]	Unchanged from ESP.
Dry Active ("Solid") Waste Volume	4994 ft3 [9988 ft3]	The expected volume of solid radioactive wastes generated during routine plant operations.	4994 ft3 [9988 ft3]	Unchanged from ESP.

Appendix I

Table I-1. (contd)

Part III Site Characteristic Item	ESP Single Unit [Two Unit] Value	Description and Reference	COL Single Unit [Two Unit] Value	Comments
Accident Releases				
Elevation (Post Accident)	Groundlevel at edge of power block circle	The elevation above finished grade of the release point for accident sequence release analyses	Groundlevel at edge of power block circle	Unchanged from ESP
Gaseous Source Term (Post-Accident)	See ESP Application ER Table 7.1-11	The activity, by isotope, contained in post-accident airborne effluents.	See DCD, Rev 17, Table 15A-5 (Westinghouse 2008).	Doses resulting from design basis accidents (DBAs) are presented and discussed in ER Table 5.10-1 and SEIS Table 5-1.
Normal Plant Heat Sink (condenser and turbine auxiliary cooling)				
Cooling water system (CWS) Cooling Tower Acreage	38 acres [69.3 acres]	The land required for CWS natural draft cooling towers, including support facilities such as equipment sheds, basins, or canals,	38 acres [69.3 acres]	Unchanged from ESP
CWS Cooling Tower Approach Temperature	11°F	The difference between the cold water temperature leaving the tower and the ambient wet bulb temperature.	11°F	Unchanged from ESP

NUREG-1947   I-8   March 2011

Appendix I

Table I-1. (contd)

Part III Site Characteristic Item	ESP		COL	
	Single Unit [Two Unit] Value	Description and Reference	Single Unit [Two Unit] Value	Comments
CWS Cooling Tower Blowdown Temperature	91°F	The design maximum expected blowdown temperature at the point of discharge to the receiving water body.	91°F	Unchanged from ESP
CWS Cooling Tower Evaporation Rate	13,950 gpm (14,440 gpm) [27,900 gpm (28,880 gpm)]	The expected (and maximum) rate at which water is lost by evaporation from the cooling water systems.	14,550 gpm (15,280 gpm) [29,100 gpm (30,560 gpm)]	The COL value was provided in Southern (2007) and was considered in the ESP analysis.
CWS Cooling Tower Drift Rate	12 gpm [24 gpm]	The maximum rate at which water is lost by drift from the cooling water systems.	12.5 gpm [25 gpm]	The COL value was provided in Southern (2007) and was considered in the ESP analysis.
CWS Cooling Tower Height	600 ft	The vertical height above finished grade of the natural draft cooling tower.	600 ft	Unchanged from ESP.

Appendix I

Table I-1. (contd)

Part III Site Characteristic Item	ESP Single Unit [Two Unit] Value	Description and Reference	COL Single Unit [Two Unit] Value	Comments
CWS Cooling Tower Make-up Flow Rate	18,612 gpm (28,892 gpm) [37,224 gpm (57,784 gpm)]	The expected (and maximum) design rate of removal of water from the Savannah River to replace water losses from circulating water systems. The make-up flow rate is a calculated value based on the sum of the evaporation rate plus the blowdown flow rate plus drift.	19,412 gpm (30,572 gpm) [38,825 gpm (61,145)]	The COL value was provided in Southern (2007) and was considered in the ESP analysis.
CWS Cooling Tower Offsite Noise Levels	<30 to <40 dBa	The maximum expected sound level at the site boundary.	<30 to <40 dBa	Unchanged from ESP.
CWS Cooling Tower Heat Rejection Rate (Blowdown)	4650 gpm (expected), 14,440 gpm (max) @91°F [9300 gpm (expected), 28,880 gpm (max)] @91°F	The expected heat rejection rate to a receiving water body, expressed as flow rate in gallons per minute at a temperature in degrees Fahrenheit.	4850 gpm (expected) 15,280 gpm (max) @91°F [9700 gpm (expected) 30,560 gpm (max)] @ 91°F	The NRC staff analysis of the revised discharge rates is discussed in Section 5.3 of the SEIS.

NUREG-1947   I-10   March 2011

Appendix I

Table I-1. (contd)

Part III Site Characteristic Item	ESP		Description and Reference	COL	
	Single Unit [Two Unit] Value			Single Unit [Two Unit] Value	Comments
CWS Cooling Tower Maximum Consumption of Raw Water	14,452 gpm [28,904 gpm]		The expected maximum short-term consumptive use of water by the circulating water systems (evaporation and drift losses).	15,292 gpm [30,585 gpm]	The COL value was provided in Southern (2007) and was considered in the ESP analysis.
CWS Cooling Tower Expected Consumption of Raw Water	13,692 gpm [27,924 gpm]		The expected normal operating consumptive use of water by the circulating water systems (evaporation and drift losses).	14,562 gpm [29,125 gpm]	The COL value was provided in Southern (2007) and was considered in the ESP analysis.
SWS Cooling Tower Makeup Rate	269 gpm (1177 gpm) [537 gpm (2353 gpm)]		The expected (maximum) rate of removal of water from wells to replace water losses from auxiliary heat sink.	269 gpm (800 gpm) [537 gpm (1600 gpm)]	The COL value was provided in Southern (2007) and was considered in the ESP analysis.

Appendix I

Table I-1. (contd)

Part III Site Characteristic Item	ESP		COL	
	Single Unit [Two Unit] Value	Description and Reference	Single Unit [Two Unit] Value	Comments
Airborne Effluent Release Point				
Normal Dose Consequences to the Maximally Exposed Individual	Total body: 1.12 mrem [2.24 mrem]	The estimated annual design radiological dose consequences due to gaseous releases from normal operation of the plant (Table 3.0-1 of ESP Application ER Rev 4) is not correct. See Section 5.4.2.2.	Total body: 1.12 mrem [2.24 mrem]	Unchanged from ESP.
Post-Accident Dose Consequences	See Tables 5-14 in the ESP EIS.	The estimated design radiological dose consequences due to gaseous releases from postulated accidents.	See ER Table 5.10-1, SEIS Table 5-1.	Design-basis accidents were recalculated using updated information from DCD, Rev 17 (Westinghouse 2008). All dose consequences remained the same or decreased except those for a loss-of-coolant accident, which increased by 2.86 percent, but remains below the regulatory criterion of 25 rem.

NUREG-1947   I-12   March 2011

Appendix I

Table I-1. (contd)

Part III Site Characteristic Item	ESP Single Unit [Two Unit] Value	Description and Reference	COL Single Unit [Two Unit] Value	Comments
Normal Dose Consequences	10 CFR 50, App I, 10 CFR 20 40 CFR 190	The estimated design radiological dose consequences due to liquid effluent releases from normal operation of the plant.	10 CFR 50, App I, 10 CFR 20 40 CFR 190	Unchanged from ESP.
Plant Characteristics				
Total Acreage	310 acres for 2 units	The land area required to provide space for all plant facilities, including power block, switchyard, spent fuel storage, and administrative facilities.	376 acres for 2 units	Acreage increased by 66 acres. Acreages for many of the permanent facilities increased or decreased by a few acres between ESP and COL. The new acreage estimate includes the fire training facility and the simulator building, which were not included in previous estimates, and together account for 44 of the additional 66 acres. The NRC staff evaluation of this change is provided in Section 4.1 of the SEIS.

March 2011   I-13   NUREG-1947

Appendix I

Table I-1. (contd)

Part III Site Characteristic Item	ESP		COL	
	Single Unit [Two Unit] Value	Description and Reference	Single Unit [Two Unit] Value	Comments
Groundwater Consumptive Use	376 gpm (1570 gpm) [752 gpm (3140 gpm)]	The expected (maximum) rate of withdrawal of groundwater to serve the new units. (Table 3.0-1 in the ESP Application listed the expected gpm for 2 units as 762, which was a typographical error.)	376 gpm (1398.5 gpm) [752 gpm (2797 gpm)]	The COL value was provided in Southern (2007) and was considered in the ESP analysis.
Operation	345 [600]	The number of people required to operate and maintain the plant	400 [800]	The COL value was provided in Southern (2007) and was considered in the ESP analysis.
Refueling / Major Maintenance	1000	The additional number of temporary staff required to conduct refueling and major maintenance activities	1000	Unchanged from ESP.

Appendix I

Table I-1. (contd)

Part III Site Characteristic Item	ESP		COL	
	Single Unit [Two Unit] Value	Description and Reference	Single Unit [Two Unit] Value	Comments
Construction	1576 people monthly average [3152 people monthly average]	The monthly average estimated construction workforce staffing for two AP1000 units being constructed simultaneously. This assumes a site preparation schedule of 18 months, 48 months from first concrete to fuel load, with 6 months from fuel load to commercial operation and 12 months between commercial operation of each unit. This assumes 20.5 job hours per net kilowatt installed, giving credit for offsite modular construction. The peak number of construction workforce personnel could reach the 4400 range.	[3500], excluding SNC and NRC employees	The COL value was provided in Southern (2007) and was considered in the ESP analysis.

Appendix I

I.1 References

Southern Nuclear Operating Company, Inc. (Southern). 2007. Southern Nuclear Operating Company, Vogtle Early Site Permit Application, Comments on Draft Environmental Impact Statement. Letter from Southern Nuclear Operating Company (Birmingham, Alabama) to the U.S. Nuclear Regulatory Commission (Washington, DC), December 26, 2007. Southern Company, Birmingham, Alabama. Accession No. ML073620401.

Southern Nuclear Operating Company, Inc. (Southern). 2009. *Vogtle Electric Generating Plant, Units 3 and 4, COL Application, Part 3 Environmental Report.* Revision 1, September 23, 2009, Southern Company, Birmingham, Alabama. Accession No. ML092740400.

U.S. Nuclear Regulatory Commission (NRC). 2008. *Final Environmental Impact Statement for an Early Site Permit (ESP) at the Vogtle Electric Generating Plant Site.* NUREG-1872, Vols. 1 and 2, Washington, D.C.

Westinghouse Electric Company, LLC (Westinghouse). 2008. AP1000 Design Control Document. AP1000 Document. APP-GW-GL-700, Revision 17, Westinghouse Electric Company, Pittsburgh, Pennsylvania. Accession No. ML083230167.

Appendix J

Statements Made in the Environmental Report Considered in the U.S. Nuclear Regulatory Commission Staff's Environmental Review

Appendix J

Statements Made in the Environmental Report Considered in the U.S. Nuclear Regulatory Commission Staff's Environmental Review

Appendix J of the Vogtle Electric Generating Plant early site permit (ESP) environmental impact statement (EIS) (NRC 2008) outlined representations and assumptions in Southern Nuclear Operating Company, Inc.'s ESP environmental report that the U.S. Nuclear Regulatory Commission (NRC) staff relied upon to reach its conclusions in the ESP EIS. Appendix J of the ESP EIS was created primarily as a tool to help reviewers of a future construction permit or combined license (COL). The NRC staff relied on these representations and assumptions in assessing the environmental impacts associated with construction and operation of the proposed Units 3 and 4.

Southern submitted a COL application referencing an ESP in March 2008 (Southern 2008). The staff of the Southern Nuclear Operating Company, Inc. and the NRC considered Appendix J of the ESP EIS (NRC 2008) in their review of new and significant information. New and significant information considered in the staff's review of the COL application is addressed in the appropriate section of this supplemental EIS.

J.1 Reference

Southern Nuclear Operating Company (Southern). 2008. *Vogtle Electric Generating Plant, Units 3 and 4, COL Application.* Revision 0, March 28, 2008, Southern Company, Birmingham, Alabama.

U.S. Nuclear Regulatory Commission (NRC). 2008. *Final Environmental Impact Statement for an Early Site Permit (ESP) at the Vogtle Electric Generating Plant Site. Appendixes.* NUREG-1872, Vol. 2, Washington, D.C.

NRC FORM 335 (12-2010) NRCMD 3.7	U.S. NUCLEAR REGULATORY COMMISSION	1. REPORT NUMBER (Assigned by NRC, Add Vol., Supp., Rev., and Addendum Numbers, if any.)
BIBLIOGRAPHIC DATA SHEET *(See instructions on the reverse)*		NUREG-1947

2. TITLE AND SUBTITLE	3. DATE REPORT PUBLISHED	
Final Supplemental Environmental Impact Statement for Combined Licenses (COLs) for Vogtle Electric Generating Plant Units 3 and 4	MONTH	YEAR
	March	2011
	4. FIN OR GRANT NUMBER	

5. AUTHOR(S)	6. TYPE OF REPORT
See Appendix a of Report	Technical
	7. PERIOD COVERED *(Inclusive Dates)*

8. PERFORMING ORGANIZATION - NAME AND ADDRESS *(If NRC, provide Division, Office or Region, U.S. Nuclear Regulatory Commission, and mailing address; if contractor, provide name and mailing address.)*

Office of New Reactors, Division of Site and Environemental Reviews
U.S. Nuclear Regulatory Commission
Washington, DC 20555-0001

9. SPONSORING ORGANIZATION - NAME AND ADDRESS *(If NRC, type "Same as above"; if contractor, provide NRC Division, Office or Region, U.S. Nuclear Regulatory Commission, and mailing address.)*

same as above

10. SUPPLEMENTARY NOTES
Docket Nos. 52-025 and 52-026

11. ABSTRACT *(200 words or less)*

This final supplemental environmental impact statement (FSEIS) documents the U.S. Nuclear Regulatory Commission (NRC) staff's analysis and conclusion regarding the environmental impacts of constructing and operating two new nuclear units (Units 3 and 4) at the Vogtle Electric Generating Plant (VEGP) site near Waynesboro, Georgia, and the mitigation measures available for reducing and avoiding adverse environmental impacts.

The NRC staff recommendation to the Commission related to the environmental aspects of the proposed action is that the combined licenses (COLs) and limited work authorization (LWA) should be issued. This recommendation is based on (1) the applicant's environmental report (ER) and responses to staff requests for additional information; (2) the staff's review conducted for the referenced early site permit (ESP) application and the assessment documented in the ESP EIS; (3) consultation with Federal, State, and Tribal agencies,;(4) the staff's own independent review of potential new and significant information available since preparation and publication of the ESP EIS; and (5) the assessments summarized in the SEIS, including the potential mitigation measures identified and consideration of public comments received on the draft SEIS. Finally, the staff concludes that the requested LWA construction activities defined at 10 CFR 50.10(a) described in the site redress plan would not result in any significant adverse environmental impacts that cannot be redressed.

12. KEY WORDS/DESCRIPTORS *(List words or phrases that will assist researchers in locating the report.)*	13. AVAILABILITY STATEMENT
Vogtle Electric Generating Plants Unit 3 and 4 National Environmental Policy Act NEPA Supplemental Environmental Impact Statement Final SEIS Early Site Permit ESP COL Combined License New Reactors Limited Work Authorization	unlimited
	14. SECURITY CLASSIFICATION
	(This Page) unclassified
	(This Report) unclassified
	15. NUMBER OF PAGES
	16. PRICE

NRC FORM 335 (12-2010)